Geochemical Sediments and Landscapes

RGS-IBG Book Series

Geochemical Sediments and Landscapes

Edited by

David J. Nash and Sue J. McLaren

Blackwell
Publishing

BLACKWELL PUBLISHING
350 Main Street, Malden, MA 02148-5020, USA
9600 Garsington Road, Oxford OX4 2DQ, UK
550 Swanston Street, Carlton, Victoria 3053, Australia

The right of David J. Nash and Sue J. McLaren to be identified as the Authors of the Editorial Material in this Work has been asserted in accordance with the UK Copyright, Designs, and Patents Act 1988.

Designations used by companies to distinguish their products are often claimed as trademarks. All brand names and product names used in this book are trade names, service marks, trademarks, or registered trademarks of their respective owners. The publisher is not associated with any product or vendor mentioned in this book.

This publication is designed to provide accurate and authoritative information in regard to the subject matter covered. It is sold on the understanding that the publisher is not engaged in rendering professional services. If professional advice or other expert assistance is required, the services of a competent professional should be sought.

First published 2007 by Blackwell Publishing Ltd

2   2008

Library of Congress Cataloging-in-Publication Data

Geochemical sediments and landscapes / edited by David J. Nash, Sue J. McLaren.
    p. cm. – (RGS-IBG book series)
  Includes bibliographical references and index.
  ISBN 978-1-4051-2519-2 (hardcover : acid-free paper) 1. Sediments (Geology) – Analysis. 2. Geomorphology. I. Nash, David J. II. McLaren, Sue J.

  QE471.2.G462 2007
  551.3–dc22

                                                                        2007018801

A catalogue record for this title is available from the British Library.

Set in 10/12 pt Plantin
by SNP Best-set Typesetter Ltd., Hong Kong

The publisher's policy is to use permanent paper from mills that operate a sustainable forestry policy, and which has been manufactured from pulp processed using acid-free and elementary chlorine-free practices. Furthermore, the publisher ensures that the text paper and cover board used have met acceptable environmental accreditation standards.

For further information on
Blackwell Publishing, visit our website:
www.blackwellpublishing.com

Contents

Figures

Tables

Contributors

Dr Andrea Borsato – Museo Tridentino de Scienze Naturali, via Calepina 14, 38100 Trento, Italy. Email: borsato@mtsn.tn.it

Professor Allan R. Chivas – GeoQuEST Research Centre, School of Earth and Environmental Sciences, University of Wollongong, NSW 2522, Australia. Email: toschi@uow.edu.au

Professor Ronald I. Dorn – School of Geographical Sciences, Arizona State University, P.O. Box 870104, Tempe, Arizona 85287-0104, USA. Email: ronald.dorn@asu.edu

Professor Ian Fairchild – School of Geography, Earth and Environmental Sciences, University of Birmingham, Edgbaston, Birmingham B15 2TT, UK. Email: i.j.fairchild@bham.ac.uk

Dr Silvia Frisia – School of Environmental and Life Sciences, University of Newcastle, Callaghan, NSW 2308, Australia. Email: silvia.frisia@ newcastle.edu.au

Professor Eberhard Gischler – Institut für Geowissenschaften, Universität Frankfurt am Main, Senckenberganlage 32–34, Postfach 11 19 32, D-60054 Frankfurt am Main, Germany. Email: gischler@em. unifrankfurt.de

Professor Andrew S. Goudie – School of Geography, Centre for the Environment, University of Oxford, South Parks Road, Oxford, OX1 3QY, UK. Email: andrew.goudie@stx.ox.ac.uk

Dr Elaine Heslop – School of Geography, Centre for the Environment, University of Oxford, South Parks Road, Oxford, OX1 3QY, UK.

Dr John McAlister – School of Geography, Queens University, Belfast BT7 1NN, UK. Email: j.mcalister@qub.ac.uk

Dr Sue J. McLaren – Department of Geography, University of Leicester, University Road, Leicester LE1 7RH, UK. Email: sjm11@leicester.ac.uk

Dr David J. Nash – School of Environment and Technology, University of Brighton, Lewes Road, Brighton BN2 4GJ, UK. Email: d.j.nash@bton.ac.uk

Dr Allan Pentecost – Department of Life Sciences, Kings College London, Franklin-Wilkins Building, 150 Stamford St, London SE1 9NN, UK. Email: allan.pentecost@kcl.ac.uk

Professor Bernie J. Smith – School of Geography, Queens University, Belfast BT7 1NN, UK. Email: b.smith@qub.ac.uk

Dr Anna Tooth – Groundwater and Contaminated Land, The Environment Agency, Guildbourne House, Chatsworth Road, Worthing, West Sussex BN11 1LD, UK. Email: anna.tooth@environment-agency.gov.uk

Dr J. Stewart Ullyott – School of Environment and Technology, University of Brighton, Lewes Road, Brighton BN2 4GJ, UK. Email: j.s.ullyott@bton.ac.uk

Professor Eric P. Verrecchia – Institut de Géologie, Université de Neuchâtel, Rue Emile-Argand 11, CP 2, CH-2007 Neuchâtel, Switzerland. Email: eric.verrecchia@unine.ch

Dr Heather A. Viles – School of Geography, Centre for the Environment, University of Oxford, South Parks Road, Oxford, OX1 3QY, UK. Email: heather.viles@ouce.ox.ac.uk

Dr Mike Widdowson – Department of Earth Sciences, The Open University, Walton Hall, Milton Keynes MK7 6AA, UK. Email: m.widdowson@open.ac.uk

Professor V. Paul Wright – School of Earth, Ocean and Planetary Sciences, Cardiff University, Main Building, Park Place, Cardiff CF10 3YE, UK. Email: wrightvp@cardiff.ac.uk

Series Editors' Preface

The RGS-IBG Book Series only publishes work of the highest international standing. Its emphasis is on distinctive new developments in human and physical geography, although it is also open to contributions from cognate disciplines whose interests overlap with those of geographers. The Series places strong emphasis on theoretically-informed and empirically-strong texts. Reflecting the vibrant and diverse theoretical and empirical agendas that characterize the contemporary discipline, contributions are expected to inform, challenge and stimulate the reader. Overall, the RGS-IBG Book Series seeks to promote scholarly publications that leave an intellectual mark and change the way readers think about particular issues, methods or theories.

For details on how to submit a proposal please visit:
www.blackwellpublishing.com/pdf/rgsibg.pdf

Kevin Ward
University of Manchester, UK

Joanna Bullard
Loughborough University, UK

RGS-IBG Book Series Editors

Acknowledgements

In addition to the editors, who reviewed all the individual chapters, numerous external referees, selected for their expertise in specific geochemical sediments, provided constructive and conscientious reviews of the manuscript. These included: **Ana Alonso-Zarza**, Department of Petrology and Geochemistry, Universidad Complutense, Madrid, Spain; **Mark Bateman**, Department of Geography, University of Sheffield, UK; **Joanna Bullard**, Department of Geography, Loughborough University, UK; **Ian Candy**, Department of Geography, Royal Holloway, University of London, UK; **Frank Eckardt**, Department of Environmental and Geographical Science, University of Cape Town, South Africa; **Frank McDermott**, School of Geological Sciences, University College Dublin, Eire; **Martyn Pedley**, Department of Geography, University of Hull, UK; **Heather Viles**, School of Geography, Centre for the Environment, University of Oxford, UK; **John Webb**, Department of Earth Sciences, La Trobe University, Melbourne, Australia; and **Brian Whalley**, School of Geography, Archaeology and Palaeoecology, Queen's University Belfast, UK.

The majority of the photographs, line diagrams and tables within this volume are the authors' own. The following organisations and publishers are thanked for their permission to reproduce figures (which, in some instances, may have been redrawn or slightly modified): **Association des géologues du bassin de Paris**, for permission to reproduce Figure 4.9B (from Thiry, M. & Bertrand-Ayrault, M., 1988, 'Les grès de Fontainebleau: Genèse par écoulement de nappes phréatiques lors de l'entaille des vallées durant le Plio-Quaternaire et phénomènes connexes', *Bulletin d'Information des géologues du Bassin de Paris* 25, 25–40. © Association des géologues du bassin de Paris). **Cooperative Research Centre for Landscape Environments and Mineral Exploration**, for permission to reproduce Figure 3.9 (from Anand, R.R., 2005, 'Weathering history, landscape evolution and implications for exploration', In: Anand, R.R. & de Broekert, P. (Eds) (2005) *Regolith Landscape Evolution Across Australia*, pp. 2–40. © Cooperative Research Centre for Landscape Environments and Mineral Exploration). **Elsevier**, for permission to reproduce Figure

4.10C (from Summerfield, M.A., 1982, 'Distribution, nature and genesis of silcrete in arid and semi-arid southern Africa', *Catena Supplement* 1, 37–65. © Elsevier). **Quaternary Research Association**, for permission to reproduce Figure 7.2B (from Smart, P.L. & Francis, P.D., 1990, *Quaternary Dating Methods – A User's Guide.* © Quaternary Research Association). **E. Schweizerbart'sche Science Publishers**, for permission to reproduce Figure 3.11 (from Borger, H., 2000, *Mikromorphologie und Paläoenvironment: Die Mineralverwitterung als Zeugnis der cretazisch-tertiären Umwelt in Süddeutschland.* © E. Schweizerbart Science Publishers). **SEPM (Society for Sedimentary Geology)**, for permission to reproduce Figures 7.7A and 7.10E (from Genty, D. & Quinif, Y., 1996, 'Annually laminated sequences in the internal structure of some Belgian stalagmites – importance for paleoclimatology', *Journal of Sedimentary Research* **66**, 275–288. © Society for Sedimentary Geology). **Springer Science and Business Media**, for permission to reproduce Figures 10.4 and 10.8 (from Eugster, H.P. & Hardie, L.A., 1978, 'Saline lakes'. In: Lerman, A. (ed) *Lakes: Chemistry, Geology, Physics*, pp. 237–293. © Springer, New York). **UNESCO**, for permission to reproduce Figure 10.7 (from Valyashko, M.G., 1972, Playa lakes – a necessary stage in the development of a salt-bearing basin. In: Richter-Bernberg, G. (ed.) *Geology of Saline Deposits*, pp. 41–51. © UNESCO, Paris). **United States Geological Survey**, for permission to reproduce Figure 10.6 (from Eakin, T.E., Price, D. & Harrill, J.R., 1976, *Summary appraisals of the nation's ground-water resources – Great Basin region.* USGS Professional Paper 813-G. © United States Geological Survey). **John Wiley and Sons Ltd**, for permission to reproduce Figure 3.13 (from Thomas, M.F., 1994, *Geomorphology in the Tropics. A Study of Weathering and Denudation in Low Latitudes.* © John Wiley and Sons Ltd), Figure 4.10A (from Shaw, P.A. & Nash, D.J., 1998, Dual mechanisms for the formation of fluvial silcretes in the distal reaches of the Okavango Delta Fan, Botswana. *Earth Surface Processes and Landforms* **23**, 705–714 © John Wiley and Sons Ltd) and Figure 4.10B (modified from Ollier, C.D. & Pain, C.F., 1996, *Regolith, Soils and Landforms* © John Wiley and Sons Ltd).

Finally, our thanks go to the British Geomorphological Research Group (now **British Society for Geomorphology**) for supporting the working group from which this collection arose, and to **Jacqueline Scott, Angela Cohen** and **Rebecca du Plessis** at Blackwell Publishing for their patience and assistance during the long, painful gestation period leading to the publication of *Geochemical Sediments and Landscapes*.

David J. Nash
Sue J. McLaren
Brighton and Leicester, August 2007

Chapter One

Introduction: Geochemical Sediments in Landscapes

David J. Nash and Sue J. McLaren

1.1 Scope of This Volume

Geochemical sediments of various types are an often overlooked but extremely important component of global terrestrial environments. Where present, chemical sediments and residual deposits may control slope development and landscape evolution, increase the preservation potential of otherwise fragile sediments, provide important archives of environmental change, act as relative or absolute dating tools and, in some cases, be of considerable economic importance. Chemical sedimentation may occur in almost any terrestrial environment, providing there is a suitable dissolved mineral source, a mechanism to transfer the mineral in solution to a site of accumulation and some means of triggering precipitation. However, given the increased importance of chemical weathering in the tropics and sub-tropics, they tend to be most widespread in low-latitude regions (Goudie, 1973).

Despite their global significance, terrestrial geochemical sediments have not been considered collectively for over 20 years. Indeed, the last book to review the full suite of chemical sediments and residual deposits was Goudie and Pye's seminal volume *Chemical Sediments and Geomorphology* (Goudie and Pye, 1983a). Since then, selected geochemical sediments have been discussed in volumes such as Wright and Tucker (*Calcretes*; 1991), Martini and Chesworth (*Weathering, Soils and Palaeosols*; 1992), Ollier and Pain (*Regolith, Soils and Landforms*; 1996), Thiry and Simon-Coinçon (*Palaeoweathering, Palaeosurfaces and Related Continental Deposits*; 1999), Dorn (*Rock Coatings*; 1998), Taylor and Eggleton (*Regolith Geology and Geomorphology*; 2001), and Chen and Roach (*Calcrete: Characteristics, Distribution and Use in Mineral Exploration*; 2005). However, many of these texts tend to discuss geochemical sediments within either a geological or pedological framework, often with little attempt to position them in their geomorphological context. As will be seen in section 1.3 and many of the chapters in this volume, understanding the influence of landscape setting upon geochemical sedimentation is of paramount importance if the resulting chemical sediments and residua are to be correctly interpreted. The need for a follow-up volume to Goudie and Pye (1983a) became very apparent during meetings of the British Geomorphological Research Group (BGRG) fixed-term working group on *Terrestrial Geochemical Sediments and Geomorphology*, convened by the editors and Andrew Goudie, which ran between 2001 and 2004. Indeed, the majority of the authors within this collection were members of the working group, and all royalties from this book will go to the BGRG (now the British Society for Geomorphology).

The individual chapters within *Geochemical Sediments and Landscapes* focus largely on the relationships between geomorphology and geochemical sedimentation. Given the emphasis on landscape, the range of precipitates and residual deposits considered are mainly those which form in terrestrial settings. An exception is the chapter on beachrock and intertidal precipitates (Gischler, Chapter 11), which develop at the terrestrial–marine interface but, where present, have a significant impact upon coastal geomorphology and sedimentology. The definition of geochemical sediments used in the volume is a deliberately broad one, reflecting the wide range of environments under which chemical sedimentation can occur. As Goudie and Pye (1983b) suggest, geochemical sediments are conventionally defined as sedimentary deposits originating through inorganic chemical processes. This distinguishes them from clastic, volcaniclastic, biochemical and organic sediments. However, this definition is not especially useful, since the majority of the geochemical sediments reviewed here comprise a mixture of detrital clastic particles which are bound together by various intergranular chemical precipitates. Certainly, there are some very 'pure' chemical precipitates, such as speleothems (see Fairchild et al., Chapter 7) and some lacustrine deposits (Verrecchia, Chapter 9), but these are the exception rather than the rule. The conventional definition also places greatest emphasis on the role of physico-chemical processes in geochemical sedimentation. However, as will be seen from many chapters in this collection, biogeochemical processes are increasingly recognised as being of vital importance for the formation of a wide range of supposedly 'chemical' precipitates. Indeed, biological agencies may be directly implicated in the formation of many chemical sediments, and play a key role in the weathering and release of solutes for a wide range of other precipitates.

1.2 Organisation

Geochemical Sediments and Landscapes is organised into 14 chapters. These are arranged so that the main duricrusts (calcrete, laterite and silcrete) are discussed first (Chapters 2–4), followed by a consideration of deposits precipitated in various aeolian, slope, spring, fluvial, lake, cave and near-coastal environments (Chapters 5–12). The volume concludes with an overview of the range of techniques available for analysing geochemical sediments (McAlister and Smith, Chapter 13) and a general summary which includes a consideration of directions for future research (McLaren and Nash, Chapter 14).

The specific content of individual chapters, inevitably, reflects the primary research interests of the contributing authors. However, all contributors were requested, where appropriate, to include information about the nature and general characteristics, distribution, field occurrence, landscape relations, macro- and micromorphology, chemistry, mineralogy, mechanisms of formation or accumulation, and palaeoenvironmental significance of their respective geochemical sediment. Individual deposits are treated as discrete entities in their specific chapters. However, in recognition of the fact that individual chemical sediments may grade laterally or vertically into geochemically allied materials, for example along pH (e.g. calcrete and silcrete) or other environmental gradients (e.g. beachrock and coastal aeolianite), authors were also asked to highlight any significant relationships to other terrestrial geochemical sediments. Despite its title, the chapters within *Geochemical Sediments and Landscapes* do not include lengthy discussions of the physics of geochemical sedimentation; authors were instead asked to cite suitable references so that interested readers can access such materials.

1.3 Significance of Geochemical Sediments in Landscapes

The geochemical precipitates and residual deposits discussed within this volume are significant from a range of geomorphological, palaeoenvironmental and economic perspectives. From a geomorphological standpoint, the more indurated and resistant chemical sediments such as calcrete (Wright, Chapter 2), ferricrete (Widdowson, Chapter 3) and silcrete (Nash and Ullyott, Chapter 4) exert a major influence upon the topographic evolution of many parts of the world. This influence is most noticeable in tropical and sub-tropical areas because such duricrusts are most widespread in these regions (Goudie, 1973). Geochemical crusts that have developed over palaeosurfaces may be preserved as horizontal to sub-horizontal caprocks on plateaux and mesas (Goudie, 1984). Along the southern coast of South Africa, for example, silcrete and ferricrete accumulation within deeply weathered bedrock has led to the preservation of remnants of the post-Gondwana 'African Surface' (Summerfield, 1982, 1983a; Marker et al., 2002). In contrast, where geochemical sediment formation took place preferentially in a topographic low, usually as a product of groundwater-related cementation, relief and drainage inversion may occur if surrounding uncemented and less resistant materials are removed by erosion (Pain and Ollier, 1995). In Australia, silcretes developed within palaeochannels may now crop out in inverted relief (e.g. Barnes and Pitt, 1976; Alley et al., 1999; Hill et al., 2003). In either case, the presence of a duricrust caprock exerts a control upon slope development and hydrology and may significantly retard landscape denudation. The undercutting and subsequent collapse of caprocks may lead to the development of characteristic features such as 'breakaways' with the resulting slope surfaces mantled by duricrust-derived regolith.

Geochemical sedimentation may also play a more subtle but equally important role in preserving 'ephemeral' sediment bodies which would otherwise be highly susceptible to erosion and destruction. Calcium carbonate, gypsum or halite cementation of near-coastal and desert dune sands may, for example, significantly enhance their preservation potential once they are transformed to aeolianite (e.g. McKee, 1966; Gardner, 1998; McLaren and Gardner, 2004; see McLaren, Chapter 5). Similarly, the induration of fluvial terrace sediments through the development of pedogenic or groundwater calcretes may increase their resistance to erosion and reworking and hence preserve key palaeohydrological evidence (e.g. Candy et al., 2004a; see Wright, Chapter 2). In extreme cases, geochemical sedimentation may lead to the complete preservation of relict landforms, as, for example, in the case of the silica- and carbonate cemented palaeochannels described by Maizels (1987, 1990) from central Oman.

In addition to their geomorphological roles, chemical sediments of various types may act as important archives of palaeoenvironmental information. Even in the most arid deserts, where detailed hydrological or climatic data are often sparse, the occurrence of crusts such as calcrete or gypcrete at or near the land surface is a clear indication that the mobilisation and precipitation of minerals in the presence of water has occurred in the past. Evaporites in Death Valley, USA, for example, have been used to unravel sequences of regional climatic changes over the past 200,000 years (Lowenstein et al., 1999; see Chivas, Chapter 10). The accumulation of thick sequences of geochemical precipitates usually requires lengthy periods of landscape stability. As such, vast thicknesses of any fossil deposit may indicate relative tectonic, climatic and/or hydrological stability. However, it is essential that the morphological and geochemical characteristics of chemical sediments, as well as the environmental factors controlling their formation, are fully appreciated before they are used as evidence in palaeoenvironmental reconstruction. For example, when attempting to distinguish the significance of a calcrete within a sedimentary sequence, it is essential to determine whether it formed by pedogenic or non-pedogenic processes (see Wright, Chapter 2), since different processes of cementation may operate at different rates and represent different palaeohydrological conditions (e.g. Nash and Smith, 1998). This becomes even more critical when dealing with calcretes in the geological record (Pimentel et al., 1996) where fabrics may have been altered over time through processes of diagenesis and paragenesis.

Successful palaeoenvironmental interpretation is highly dependent upon the availability of representative and well-documented modern analogues. For many geochemical sediments, this is unproblematic as the processes involved in, and the controls upon, their formation are well understood. Studies of dripwater chemistry and environmental conditions within contemporary cave systems, for example, have greatly improved the hydrogeochemical interpretation of ancient speleothems (see Fairchild et al., 2006a,b; and Fairchild et al., Chapter 7). Similarly, Zhang et al. (2001) and Chen et al. (2004) have investigated the physico-chemical controls on contemporary carbonate precipitation at waterfalls, which has considerably enhanced our understanding of tufa and travertine formation (see Viles and Pentecost, Chapter 6). However, for materials such as silcrete (Nash and Ullyott, Chapter 4), there are virtually no representative modern equivalents, and debate continues over the precise environments under which they form (e.g. Summerfield, 1983b, 1986; Nash et al., 1994; Ullyott et al., 1998). Disagreements over the role of biological and physico-chemical mechanisms in the formation of rock varnish have also historically hindered their effective use as a palaeoenvironmental indicator, although recent developments will hopefully rectify this situation (see Liu, 2003; and Dorn, Chapter 8).

Geochemical sediments are increasingly being used as both relative and absolute age indicators. Duricrusts formed on palaeosurfaces as a result of pedogenic processes, for example, may represent important marker horizons and can, with considerable care, be used as a broad-scale correlative tool. However, it is the potential for absolute dating of geochemical sediments that is currently generating greatest interest. The dating of many $CaCO_3$-cemented sediments has long been considered inappropriate due to concerns over whether the carbonate-cementing environment could be viewed as geochemically 'closed'. Advances in the use of U-series dating mean that previously problematic materials such as calcrete can now be systematically dated (Kelly et al., 2000; Candy et al., 2004b, 2005). Similarly, the analysis and dating of microlaminations is permitting both palaeoenvironmental information and calibrated ages to be derived from rock varnish (see Liu et al., 2000; and Dorn, Chapter 8). These improvements may mean that such chemical sediments will, in the future, be used as routinely as speleothems (Fairchild et al., Chapter 7) and laminated lacustrine deposits (Verrecchia, Chapter 9) as chronometric and palaeoenvironmental tools.

Finally, many geochemical sediments are of major economic importance, both as sources of minerals and construction materials, and because of their potential impacts on human livelihoods through their influence upon soil properties and groundwater chemistry. In regions where alternative construction materials are scarce, geochemical sediments such as calcrete (Wright, Chapter 2), laterite (Widdowson, Chapter 3) and beachrock (Gischler, Chapter 11) may be used as building materials. For example, certain types of air-hardening laterite are widely employed as building bricks in Asia (Goudie, 1973), and calcrete was used as one of the main sources of road aggregate during the construction of the Trans-Kalahari Highway in Botswana in the late 1990s (Lawrance and Toole, 1984). In terms of mineral prospecting, evaporite sequences (Chivas, Chapter 10, and Goudie and Heslop, Chapter 12) provide economically significant sources of gypsum, nitrate, sulphate and borax, bauxite (Widdowson, Chapter 3) remains a key source of aluminium ore (e.g. Anand and Butt, 2003) and groundwater calcrete (Wright, Chapter 2) may contain significant concentrations of uranium (Carlisle et al., 1978; Carlisle, 1983).

Even where geochemical sediments are, in themselves, of little direct economic value, they may be of considerable utility in basin analysis, oil reservoir or aquifer characterisation and for locating economically important ore bodies (e.g. Smith et al., 1993; Abdel-Wahab et al., 1998; Butt et al., 2005). For example, chemical analyses of pedogenic calcretes are increasingly used as a gold-prospecting tool in southern Australia due to the preferential concentration of Au within profiles during bedrock weathering and cementation (Lintern et al., 1992); elevated levels of Au within the calcrete regolith may represent the near-surface expression of an area of concealed primary or secondary gold mineralisation (Lintern, 2002). Similarly, the upper ferruginous zone of lateritic profiles is frequently used as a sample medium for the detection of underlying Au ore bodies in southern and Western Australia (Butt et al., 2005). However, for these techniques to be successful, it is essential that sampling is undertaken with full regard to the local landform context (Craig, 2005), which requires that detailed regolith-landform mapping is carried out during the early phases of any mineral exploration programme (e.g. Hill et al., 2003). This ongoing work reinforces the premise behind this volume, namely that understanding the influence of landscape context upon the formation of any geochemical sediment is key to the successful exploitation of that precipitate or residual deposit. We are confident that as our understanding of the genesis of all geochemical sediments improves, then their economic value can only increase.

References

Abdel-Wahab, A., Salem, A.M.K. & McBride, E.F. (1998) Quartz cement of meteoric origin in silcrete and non-silcrete sandstones, Lower Carboniferous, western Sinai, Egypt. *Journal of African Earth Sciences* 27, 277–290.

Alley, N.F., Clarke, J.D.A., MacPhail, M. & Truswell, E.M. (1999) Sedimentary infillings and development of major Tertiary palaeodrainage systems of south-central Australia. In: Thiry, M. & Simon-Coinçon, R. (Eds) *Palaeoweathering, Palaeosurfaces and Related Continental Deposits*. International Association of Sedimentologists Special Publication 27. Oxford: Blackwell Science, pp. 337–366.

Anand, R.R. & Butt, C.R.M. (2003) Distribution and evolution of 'laterites' and lateritic weathering profiles, Darling Range, Western Australia. *Australian Geomechanics* 38, 41–58.

Barnes, L.C. & Pitt, G.M. (1976) The Mirackina Conglomerate. *Quarterly Geological Notes of the Geological Survey of South Australia* 59, 2–6.

Butt, C.R.M., Scott, K.M., Cornelius, M. & Robertson, I.D.M. (2005) Sample media. In: Butt, C.R.M., Robertson, I.D.M., Scott, K.M. & Cornelius, M. (Eds) *Regolith Expression of Australian Ore Systems*. Bentley, Western Australia: Cooperative Research Centre for Landscape Environments and Mineral Exploration, pp. 53–79.

Candy, I., Black, S. & Sellwood, B.W. (2004a) Complex response of a dryland river system to Late Quaternary climate change: implications for interpreting the climatic record of fluvial sequences. *Quaternary Science Reviews* 23, 2513–2523.

Candy, I., Black, S. & Sellwood, B.W. (2004b) Quantifying timescales of pedogenic calcrete formation using U-series disequilibria. *Sedimentary Geology* 170, 177–187.

Candy, I., Black, S. & Sellwood, B.W. (2005) U-series isochron dating of immature and mature calcretes as a basis for constructing Quaternary landform chronologies for the Sorbas Basin, southeast Spain. *Quaternary Research* 64, 100–111.

Carlisle, D. (1983) Concentration of uranium and vanadium in calcretes and gypcretes. In: Wilson, R.C.L. (Ed.) *Residual Deposits: Surface Related Weathering Processes and Materials*. Special Publication 11. London: Geological Society, pp. 185–195.

Carlisle, D., Merifield, P.M., Orme, A.R. & Kolker, O. (1978) *The distribution of calcretes and gypcretes in southwestern United States and their uranium favorability. Based on a study of deposits in Western Australia and South West Africa (Namibia)*. Open File Report 76-002-E. Los Angeles: University of California.

Chen, X.Y. & Roach, I.C. (Eds) (2005) *Calcrete: Characteristics, Distribution and Use in Mineral Exploration*. Perth, Western Australia: Cooperative Research Centre for Landscape Environments and Mineral Exploration.

Chen, J.A., Zhang, D.D., Wang, S.J., Xiao, T.F. & Huang, R.E. (2004) Factors controlling tufa deposition in natural waters at waterfall sites. *Sedimentary Geology* 166, 353–366.

Craig, M.A. (2005) Regolith-landform mapping, the path to best practice. In: Anand, R.R. & de Broekert, P. (Eds) *Regolith Landscape Evolution Across Australia*. Bentley, Western Australia: Cooperative Research Centre for Landscape Environments and Mineral Exploration, pp. 53–61.

Dorn, R.I. (1998) *Rock Coatings*. Amsterdam: Elsevier.

Fairchild, I.J., Smith, C.L., Baker, A., Fuller, L., Spötl, C., Mattey, D., McDermott, F. & E.I.M.F. (2006a) Modification and preservation of environmental signals in speleothems. *Earth-Science Reviews* 75, 105–153.

Fairchild, I.J., Tuckwell, G.W., Baker, A. & Tooth, A.F. (2006b) Modelling of dripwater hydrology and hydrogeochemistry in a weakly karstified aquifer (Bath,

UK): implications for climate change studies. *Journal of Hydrology* **321**, 213–231.

Gardner, R.A.M. (1988) Aeolianites and marine deposits of the Wahiba Sands: character and palaeoenvironments. *The Journal of Oman Studies Special Report* **3**, 75–95.

Goudie, A.S. (1973) *Duricrusts in Tropical and Subtropical Landscapes*. Oxford: Clarendon Press.

Goudie, A.S. (1984) Duricrusts and landforms. In: Richards, K.S., Arnett, R.R. & Ellis, S. (Eds) *Geomorphology and Soils*. London: Allen and Unwin, pp. 37–57.

Goudie, A.S. & Pye, K. (Eds) (1983a) *Chemical Sediments and Geomorphology*. London: Academic Press.

Goudie, A.S. & Pye, K. (1983b) Introduction. In: Goudie, A.S. & Pye, K. (Eds) *Chemical Sediments and Geomorphology*. London: Academic Press, pp. 1–5.

Hill, S.M., Eggleton, R.A. & Taylor, G. (2003) Neotectonic disruption of silicified palaeovalley systems in an intraplate, cratonic landscape: regolith and landscape evolution of the Mulculca range-front, Broken Hill Domain, New South Wales. *Australian Journal of Earth Sciences* **50**, 691–707.

Kelly, M., Black, S. & Rowan, J.S. (2000) A calcrete-based U/Th chronology for landform evolution in the Sorbas basin, southeast Spain. *Quaternary Science Reviews* **19**, 995–1010.

Lawrance C. J. & Toole T. (1984) *The location, selection and use of calcrete for bituminous road construction in Botswana*. Transport and Road Research Laboratory Report 1122. London: Department of Transport.

Lintern, M.J. (2002) Calcrete sampling for mineral exploration. In: Chen, X.Y. & Roach, I.C. (Eds) *Calcrete: Characteristics, Distribution and Use in Mineral Exploration*. Perth, Western Australia: Cooperative Research Centre for Landscape Environments and Mineral Exploration, pp. 31–109.

Lintern, M.J., Downes, P.M. & Butt, C.R.M. (1992) Bounty and Transvaal Au deposits, Western Australia. In: Butt, C.R.M. & Zeegers, H. (Eds) *Regolith Exploration Geochemistry in Tropical and Subtropical Terrains*. Amsterdam: Elsevier, pp. 351–355.

Liu, T. (2003) Blind testing of rock varnish microstratigraphy as a chronometric indicator: results on late Quaternary lava flows in the Mojave Desert, California. *Geomorphology* **53**, 209–234.

Liu, T., Broecker, W.S., Bell, J.W. & Mandeville, C. (2000) Terminal Pleistocene wet event recorded in rock varnish from the Las Vegas Valley, southern Nevada. *Palaeogeography, Palaeoclimatology, Palaeoecology* **161**, 423–433.

Lowenstein, T.K., Brown, C., Roberts, S.M., Ku, T.-L., Luo, S. & Yang, W. (1999) 200 k.y. paleoclimate record from Death Valley salt core. *Geology* **27**, 3–6.

Maizels, J. (1987) Plio-Pleistocene raised channel systems of the western Sharqiya (Wahiba), Oman. In: Frostick, L.E. & Reid, I. (Eds) *Desert Sediments, Ancient & Modern*. Special Publication 35. London: Geological Society, pp. 31–50.

Maizels, J. (1990) Raised channel systems as indicators of palaeohydrologic change: a case study from Oman. *Palaeogeography, Palaeoclimatology, Palaeoecology* **76**, 241–277.

Marker, M.E., McFarlane, M.J. & Wormald, R.J. (2002) A laterite profile near Albertinia, Southern Cape, South Africa: its significance in the evolution of the African Surface. *South African Journal of Geology* **105**, 67–74.

Martini, I.P. & Chesworth, W. (Eds) (1992) *Weathering, Soils and Palaeosols*. Developments in Earth Surface Processes 2. Amsterdam: Elsevier.

McKee, E.D. (1966) Structures of dunes at White Sands National Monument, New Mexico. *Sedimentology* 7, 1–69.

McLaren, S.J. & Gardner, R.A.M. (2004) Late Quaternary carbonate diagenesis in coastal and desert sands – a sound palaeoclimatic indicator? *Earth Surface Processes and Landforms* 29, 1441–1459.

Nash, D.J. & Smith, R.F. (1998) Multiple calcrete profiles in the Tabernas Basin, southeast Spain: their origins and geomorphic implications. *Earth Surface Processes and Landforms* 23, 1009–1029.

Nash, D.J., Thomas, D.S.G. & Shaw, P.A. (1994) Siliceous duricrusts as palaeoclimatic indicators: evidence from the Kalahari Desert of Botswana. *Palaeogeography, Palaeoclimatology, Palaeoecology* 112, 279–295.

Ollier, C.D. & Pain, C.F. (1996) *Regolith, Soils and Landforms*. Chichester: Wiley.

Pain, C.F. & Ollier, C.D. (1995) Inversion of relief – a component of landscape evolution. *Geomorphology* 12, 151–165.

Pimentel, N.L., Wright, V.P. & Azevedo, T.M. (1996) Distinguishing early groundwater alteration effects from pedogenesis in ancient alluvial basins: examples from the Palaeogene of southern Portugal. *Sedimentary Geology* 105, 1–10.

Smith, G.L., Byers, C.W. & Dott, R.H. (1993) Sequence stratigraphy of the Lower Ordovician Prairie Du-Chien Group on the Wisconsin Arch and in the Michigan Basin. *Bulletin of the American Association of Petroleum Geologists* 7, 49–67.

Summerfield, M.A. (1982) Distribution, nature and genesis of silcrete in arid and semi-arid southern Africa. *Catena, Supplement* 1, 37–65.

Summerfield, M.A. (1983a) Silcrete. In: Goudie, A.S. & Pye, K. (Eds) *Chemical Sediments and Geomorphology*. London: Academic Press, pp. 59–91.

Summerfield, M.A. (1983b) Silcrete as a palaeoclimatic indicator: evidence from southern Africa. *Palaeogeography, Palaeoclimatology, Palaeoecology* 41, 65–79.

Summerfield, M.A. (1986) Reply to discussion – silcrete as a palaeoclimatic indicator: evidence from southern Africa. *Palaeogeography, Palaeoclimatology, Palaeoecology* 52, 356–360.

Taylor, G. & Eggleton, R.A. (2001) *Regolith Geology and Geomorphology*. Chichester: Wiley.

Thiry, M. & Simon-Coinçon, R. (Eds) (1999) *Palaeoweathering, Palaeosurfaces and Related Continental Deposits*. International Association of Sedimentologists Special Publication Vol. 27. Oxford: Blackwell Science.

Ullyott, J.S., Nash, D.J. & Shaw, P.A. (1998) Recent advances in silcrete research and their implications for the origin and palaeoenvironmental significance of sarsens. *Proceedings of the Geologists' Association* 109, 255–270.

Wright, V.P. & Tucker, M.E. (Eds) (1991) *Calcretes*. International Association of Sedimentologists Reprint Series Vol. 2. Oxford: Blackwell Scientific.

Zhang, D.D., Zhang, Y., Zhu, A. & Cheng, X. (2001) Physical mechanisms of river waterfall tufa (travertines) formation. *Journal of Sedimentary Research* A71, 205–216.

Chapter Two

Calcrete

V. Paul Wright

2.1 Introduction: Nature and General Characteristics

Calcrete is a general term given to the near-surface, terrestrial accumulation of predominantly calcium carbonate, which occurs in a variety of forms from powdery to nodular, laminar and massive. It results from the cementation and displacive and replacive introduction of calcium carbonate into soil profiles, sediments and bedrock, in areas where vadose and shallow phreatic groundwaters are saturated with respect to calcium carbonate. This definition is modified from Wright and Tucker (1991) and based on that given by Goudie (1973) and Watts (1980). The term is effectively synonymous with caliche, which tends be more widely used in North America. Goudie (1973) provides a detailed review of the terminology used in various countries for this type of geochemical sediment. Calcrete is here used to mean not only highly indurated and massive accumulations but also unconsolidated accumulations. It could be argued that the above definition is too general and would include materials such as aeolianite or even beachrock. Where the calcium carbonate is introduced into a non-carbonate host its authigenic origin is clear, but calcretes can also develop in carbonate bedrock and sediments, including those formed around lake margins, seasonal wetlands (Chapter 9) and groundwater discharge zones, thus creating a spectrum of complex relationships (Tandon and Andrews, 2001; Alonso-Zarza, 2003). A common misconception is that calcretes are wholly pedogenic in origin, and arguably the term could be restricted to such occurrences, but in some semi-arid to arid regions extensive precipitation occurs in the shallow phreatic zone, producing large bodies of authigenic carbonate with many characteristics in common with pedogenic calcrete. These are termed groundwater calcretes and can display complex relationships with pedogenic forms and with some types of palustrine limestones (Figure 2.1).

There is a large, diverse and complex literature on calcretes, reflecting the interest in such material from not only pedologists and

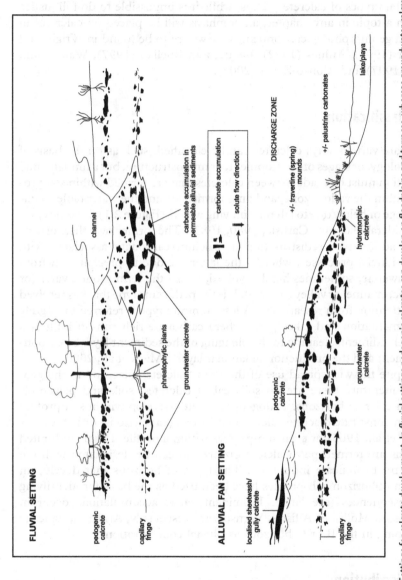

Figure 2.1 Settings for calcrete development. In fluvial settings pedogenic calcretes can develop on floodplains and terraces, whereas groundwater calcretes may form in channel deposits or around the capillary fringe and upper part of the phreatic zone in more permeable parts of the floodplain. In alluvial fans paired calcretes may develop on the fans, with hydromorphic calcretes near discharge zones.

geomorphologists but also geologists (as calcretes are particularly common in the stratigraphic record), geochemists, and increasingly from the fields of geobiology and geomicrobiology. Indeed, one of the key advances is that we are now more aware of the crucial role that vegetation plays in generating certain types of calcretes. Thus, while it is impossible to do full justice to such a topic in any chapter, an emphasis will be placed on calcretes in a broad geomorphological context. Reviews are to be found in Wright and Tucker (1991), Milnes (1992), Paquet and Ruellan (1997), Watson and Nash (1997) and Alonso-Zarza (2003).

2.2 Classification

There are various ways calcretes can be classified, such as on the basis of morphology, or stages of development, or microstructure, but a fundamental distinction must be made between calcretes that are formed within soil profiles, within the vadose zone, and ones formed around the water table–capillary fringe or below due to laterally moving waters (Figure 2.1), in some cases at considerable depth (Carlisle, 1980, 1983). The former owe their origins to the addition or redistribution of calcium carbonate associated with eluvial–illuvial processes, whereas the latter are due to precipitation from groundwaters, sometimes highly evolved, hence the term groundwater (or phreatic, channel or valley calcrete). It is the pedogenic type that has received most attention from researchers. A less common type is referred to as gully bed cementation and takes place where carbonate-rich run-off infiltrates channel sediments, leading to the plugging of the sediment layer by carbonate cement and the production of laminar layers (Mack et al., 2000).

Despite the widespread use of the term 'calcrete palaeosol' in the geological literature there are no soils called calcretes; soils and palaeosols contain calcrete horizons. Pedogenic calcretes develop within soil profiles and can constitute discrete calcic soil horizons (ca horizons, or K horizons of Gile et al., 1965) or even sub-profiles within a profile. Highly indurated horizons are termed petrocalcic horizons and can be designated with the suffix 'm' to indicate induration. The types of horizons that develop in these multihorizon sub-profiles have been used as the basis for identifying chronosequences (see later). Prominent calcic accumulations occur in Inceptisols, Mollisols, Alfisols, Vertisols and, especially, Aridisols; typically developing in the B or C horizons as illuvial concentrations.

2.3 Distribution

Yaalon (1988) has estimated that soils with calcic or petrocalcic horizons cover $20 \times 10^6 \, km^2$, or about 13% of the present land surface. The areal distribution of groundwater types is not known but they certainly cover many tens if not hundreds of thousands of square kilometres in Australia. In the stratigraphic record, pedogenic calcretes are common in many red-bed (dryland) successions, with individual formations containing many tens or even hundreds of calcrete-bearing palaeosols. Many shallow-marine carbonate successions, volumetrically so important in the stratigraphic record, also frequently contain large numbers of calcrete horizons at exposure surfaces (Wright, 1994). Ancient groundwater calcretes are also becoming more widely recognised (e.g. Colson and Cojan, 1996).

Calcic horizons develop in soils where there is a net moisture deficit, such that carbonate produced in a drier season is not leached away during a wetter season. Most present-day calcretes form in areas with warm to hot climates (mean annual temperature of 16–20°C) and low, seasonal rainfall (100–500 mm; Goudie, 1983). The upper boundary more likely spans the range 600–1000 mm (Mack and James, 1994). Royer (1999) has pointed out that from a large dataset of 1481 studies, carbonate-bearing soils correlate with a mean annual precipitation of <760 mm. Most of the dataset came from western USA and, as Retallack (2000) warns, there is great variability in the rainfall levels associated with the boundary between calcareous and non-calcareous soils in different parts of the world, or even as Birkeland (1999, p. 291) points out, there are marked local variations within the western USA.

Attempts have been made to use the depths at which carbonates accumulate to estimate annual rainfall (e.g. Retallack, 1994). However, the existence of a strong correlation between annual rainfall and depth to the top of the carbonate horizon has been questioned by Royer (1999). Retallack's dataset was based on a compilation from Inceptisols, Aridisols, Mollisols and Alfisols, whereas Royer's was based on a larger dataset including a spectrum of soil textures (Retallack, 2000). Stiles et al. (2001) recorded a strong linear relationship between mean annual precipitation and depth of carbonate-enriched horizons in Texan Vertisols; however, the moisture regime of many Vertisols is influenced by seasonal flood waters as well as local rainfall. Estimates of mean annual precipitation from the depth of carbonate accumulation can be compared with other values derived from the degree of chemical weathering (Sheldon et al., 2002; Sheldon and Retallack, 2004).

Local factors are clearly important, such as whether rainfall is mainly in summer or winter, or specific local drainage effects due to host-sediment permeability. An example is shown by the remarkable occurrence of a 4-m-thick Holocene carbonate-cemented zone with abundant root-related calcrete textures (Figure 2.2A,B), hosted in a late Pleistocene glacial gravel in North Yorkshire, hardly describable as a semi-arid region (Strong et al., 1992); as these highly porous carbonate-rich gravels occur many metres above the local water table, rapid free drainage occurs promoting high evapotranspiration.

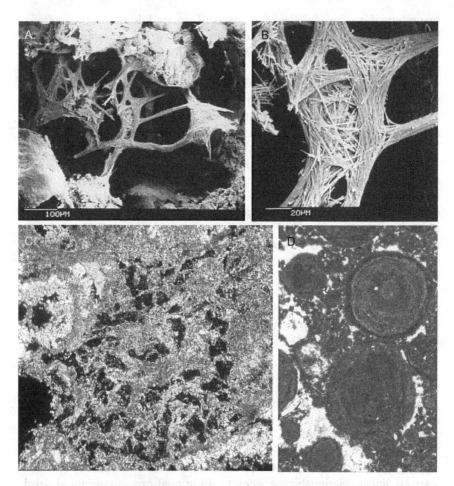

Figure 2.2 Calcrete microstructures. (A) Scanning electron microscopy micrograph of alveolar septal fabric from a Holocene gravel from North Yorkshire, UK (see Strong et al., 1992). (B) Enlargement of the centre of image in (A) showing needle fibre calcite. (C) Alveolar septal fabrics from Holocene soil, La Mora area, Tarragona, northeast Spain, showing septa made of aligned needle fibre calcite in a root mould; field of view is 2 mm wide. (D) Calcrete ooids from a Stage VI profile from near Carlsbad, New Mexico (see Wright et al., 1993); field of view is 2 mm wide.

Calcretes are also known from cold desert areas. Carbonate crusts have been identified in present day Arctic regions (Bunting and Christensen, 1980; Lauriol and Clarke, 1999). Vogt and Del Valle (1994) and Vogt and Corte (1996) have recorded calcretes from Pleistocene cold arid periods in various regions. Candy (2002) has described rhizogenic calcretes from deposits of the 450,000 year old Anglian glacial stage in Norfolk, UK, probably reflecting a period of climatic amelioration.

2.4 Calcretes in a Geomorphological Context

Pedogenic calcretes, especially those at a high degree of development (see below), require a lengthy residence time in the solum to accumulate, and as such are found at stable surfaces. These can range from classic loess settings, to prairies to river terraces, alluvial fans and marine terraces. The presence of highly developed forms indicates a lack of sediment input for extended periods, and such soils are developed where surfaces have become isolated as terraces or because climate changes have stabilised a formerly active surface such as an alluvial fan (more humid phases causing a stabilising vegetation cover or more stable channels), or a decrease in aeolian input on loessic-prairie surfaces (Gustavson and Holliday, 1999).

Pedogenic calcretes associated with alluvial fans are relatively well documented and can exhibit distribution patterns reflecting climate and or base-level related changes. The spatial variability of pedogenic calcrete development on an alluvial fan will ultimately reflect variations in the residence time of the solum in the zone of carbonate accumulation and the particle size of the host sediment. Wright and Alonso-Zarza (1990) and Wright (1992) offered models to explain the distribution of calcrete-bearing soils on alluvial fans based largely on the pedofacies concept (Bown and Kraus, 1987). The degree of calcic horizon development reflected the residence time, which is controlled by the rate of sedimentation. Fans can have highly complex patterns of stable and unstable surfaces depending on whether they are incised, on the type of incision, etc. Alonso-Zarza et al. (1998), in a study of the Pleistocene alluvial fan surfaces of the Campo de Cartegena–Mar Menor Basin in southeast Spain, have provided a more dynamic model, whereby profile development is an interaction of not simply sedimentation, but also erosion, reworking, surface hydrology and plant root activity, with complex feedback relationships. Marked differences in proximal and distal pedogenic calcrete development are present. In some alluvial systems, paired calcretes (Nash and Smith, 1998) (Figure 2.1) have been recorded, whereby alluvial deposits contain both pedogenic and groundwater calcrete (e.g. Kaemmerer and Revel, 1991); similar paired forms have also been recorded from Paleogene lake systems (Colson and Cojan, 1996).

Calcrete development can influence landscape development and as pedogenic calcretes become progressively more indurated they can produce a layer impervious to drainage. This may restrict rooting and the soil may become prone to erosion, especially on slopes. Some laminar calcretes capping impervious pedogenic calcrete horizons commonly show evidence of such soil stripping (see section 2.7). Likewise, such indurated zones may serve to armour the landscape against erosion, especially by wind action.

A common feature in some landscapes is inverted relief, usually of only a few metres, caused by the groundwater calcrete zones becoming exhumed and being more resistant to erosion than surrounding sediments (Reeves, 1983; Kaemmerer and Revel, 1991; Mann and Horowitz, 1979; Maizels, 1990). Groundwater calcretes act as cap rocks in some basins (Kaemmerer and Revel, 1991; Nash and Smith, 1998).

Tectonic effects influence landscape development and hydrology and so can also exert a strong influence on the nature and distribution of calcrete. For example, Mack et al. (2000), in a study of the Plio-Pleistocene Palomas Basin in New Mexico, found that 1.5–3-m-thick groundwater calcretes, producing sheet-like bodies, were preferentially developed in hanging-wall zones of half-grabens, whereas footwall fans were characterised by more dispersed sparry calcite cements. This partitioning was interpreted as reflecting the larger catchment areas of the hanging-wall fan, which is an order of magnitude larger than the footwall catchment area. Mack and Madoff (2005) have recently documented the lateral variability in the development of calcrete-bearing palaeosols in the same region, reflecting the effects of basin tilting and channel avulsion.

2.5 Macromorphological Characteristics

A set of distinctive horizon types has been recognised (Table 2.1) many of which can occur separately, or form profiles. The sequence given in Table 2.1 follows that seen in the typical chronosequence or stages of development (see below). Although the classic type of well-developed profile is commonly seen (Figure 2.3), there are a number of important exceptions to this pattern. For example, whereas in many well-developed profiles a continuous fine-grained, indurated layer or hardpan occurs (Figure 2.4A), some calcrete horizons are completely composed of dense calcified root mats producing thick, crudely laminar horizons, and in other examples the whole profile consists of dense masses of calcified root cells (Sanz and Wright, 1994; Wright et al., 1995).

Groundwater and pedogenic calcretes share many similarities in terms of the types of morphologies they exhibit (Table 2.1) but prismatic and pisolitic horizons are seemingly absent from the former. Phreatophytic plants, capable of reaching many metres or even tens of metres below the land surface, can produce rhizocretionary and even root-mat horizons in groundwater calcrete profiles (Semeniuk and Meagher, 1981). Pisolitic calcretes commonly occur above laminar calcretes but can form horizons over a metre thick at the bases of slopes, where the pisoids have grown during downslope movement (Figure 2.4B). Oolitic equivalents (<2 mm in diameter) are also prominent in association with root mats (Figure 2.5A) (Wright et al., 1995).

Table 2.1 Morphological types of calcrete horizons (in part sourced from Netterberg, 1967, 1980; Goudie, 1983; Esteban and Klappa, 1983; Wright, 1994; Alonso-Zarza, 2003)

Calcrete type	Characteristics
Calcareous soil	Very weakly cemented or uncemented soil with small carbonate accumulations as grain coatings, patches of powdery carbonate including needle-fibre calcite, carbonate-filled fractures and small nodules
Calcified soil	Friable to firmly cemented soil with scattered nodules; 10–50% carbonate
Chalky or powdery	Fine loose powder of calcium carbonate as a continuous body with little or no nodule development, consisting of micrite or microspar, with etched silicate grains, peloids and root and fungal-related microfeatures. Commonly is a transitional zone from calcareous soil to nodular horizons
Pedotubule or rhizocretionary	All or nearly all of the secondary carbonate occurs as root encrustations or calcifications, or around burrows, having a predominantly vertical structure
Nodular	Term synonymous with glaebular and refers to soft to highly indurated concretions of carbonate, or carbonate cemented host material. The margins may be gradational to sharp, and internally the nodules may be uniform, showing concentric laminae or septarian cracks or veins. The nodules can range in shape from spherical to elongate. The nodules typically consist of micrite or less commonly microsparite. Nodular calcrete develops primarily in siliciclastic host material
Honeycomb	Partial coalescence of nodules with softer interstitial areas produces a honeycomb-like effect
Mottled	Equivalent of nodular features where the host sediment is carbonate-dominated. It consists of irregular mottles of typically micritic carbonate that has cemented and replaced the original grains in the host (Wright, 1994)
Hardpan or massive	Consists of an indurated (petrocalcic) layer, which is sheet-like and typically has a sharp top and a gradational base into chalky or nodular calcrete. It may also be blocky or prismatic. Such layers may reach thicknesses of over a metre in pedogenic calcretes and many metres in groundwater forms

Table 2.1 *Continued*

Calcrete type	Characteristics
Platy	Consists of centimetre-thick plates, up to tens of centimetres in diameter, commonly found above hardpan or chalky layers. Plates may be tabular or wavy in form and may exhibit crude lamination. Some are fragmented calcified root mats. Distinguished from brecciated laminar calcrete by a lack of strong laminations
Laminar	Consists of sheets of laminated carbonate, in which the laminae are on a millimetre-scale. The term is synonymous with 'soilstone crust'. Some forms are dark in colour. They commonly occur capping hardpan layers but can also occur within chalky layers or in the host sediment or soil. They can be interlayered with pisolitic calcretes on a centimetre-scale. Most are only a few centimetres thick but some forms can reach 2 m and can be domal in form. This type of calcrete is strongly polygenetic in nature
Stringer	Closely related to laminar calcretes but not all forms are well laminated. These are sheets of carbonate, usually only a few centimetres thick, of sub-vertical to sub-horizontal form, that penetrate into carbonate-rich hosts and are related to root mats. They can occur singly or at multiple levels and varied orientations, extending for metres, and can cross-cut each other
Pisolitic	Millimetre- to centimetre-sized coated grains in layers typically only a few centimetres thick but up to metres thick at the bases of slopes. Can develop in a variety of hosts but are common on coarse-grained hosts. Laminae are generally micritic. Inverse grading is commonly seen. Commonly occur above laminar calcretes
Breccia or conglomeratic	Disrupted hardpans or other forms, with fracturing due to mechanical processes and roots and tree heave.

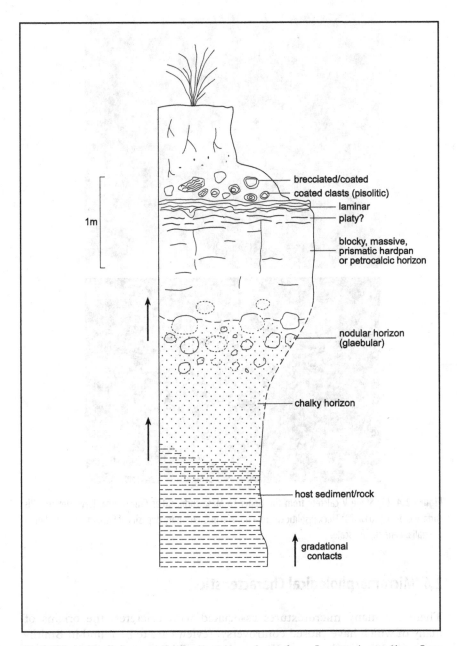

Figure 2.3 Idealised calcrete profile showing a range of macroforms. For examples see Alonso-Zarza et al. (1998).

Figure 2.4 (A) Stage V calcrete from the Upper La Mesa surface, Las Cruces area, New Mexico. This surface is over 400 ka. (B) Thick pisolitic calcrete horizon in Quaternary deposits of Fisherman's Bay, South Australia. Lens cap for scale.

2.6 Micromorphological Characteristics

There are many microtextures associated with calcretes, the origins of many of which have caused controversy; reviews are to be found in Braithwaite (1983), Esteban and Klappa (1983), Wright and Tucker (1991), Wright (1994) and Alonso-Zarza (2003). At a very general level, two end-member assemblages can be defined, with many calcretes showing varying proportions of each type but with pure end-member types

Figure 2.5 (A) Laminar calcrete (arrowed) overlain by an oolitic–pisolitic layer associated with a calcified root-mat layer, Holocene soil, La Mora area, Tarragona, northeast Spain (see Calvet and Julia, 1983). (B) Stage V–VI profile with hardpan layer overlain by pisolitic and brecciated level with a prominent calcified root–mat layer (arrowed), from San Miguel salinas area, Torrevieja, Alicante, southeast Spain. Such a profile would correspond to a thickened, polyphase profile as shown on the right of Figure 2.8.

occurring in nature (Wright and Tucker, 1991) (Figure 2.6). The alpha assemblages are predominantly composed of simple micritic to sparitic crystalline textures, and are more common in siliciclastic-rich hosts; for example, they are more prevalent in modern and ancient dryland settings. The common heterogeneity of crystal sizes and presence of microspar has led some workers to propose that recrystallisation has been an important process in creating these mosaics, but some studies assessing the

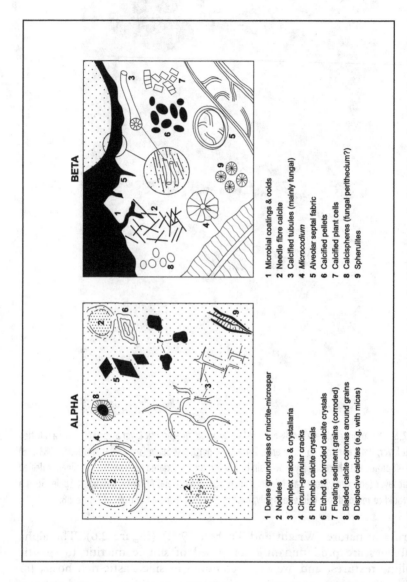

ALPHA

1 Dense groundmass of micrite–microspar
2 Nodules
3 Complex cracks & crystallaria
4 Circum-granular cracks
5 Rhombic calcite crystals
6 Etched & corroded calcite crystals
7 Floating sediment grains (corroded)
8 Bladed calcite coronas around grains
9 Displacive calcites (e.g. with micas)

BETA

1 Microbial coatings & ooids
2 Needle fibre calcite
3 Calcified tubules (mainly fungal)
4 *Microcodium*
5 Alveolar septal fabric
6 Calcified pellets
7 Calcified plant cells
8 Calcispheres (fungal perithecium?)
9 Spherulites

Figure 2.6 End-member types of calcrete microstructure. Alpha fabrics correspond to K fabrics or crystic plasmic fabrics of earlier authors. Corona structures are fringes of bladed or fibrous cement around sediment grains. Displacive structures (see text) are relatively rare. (Based on Wright and Tucker, 1991.)

role of recrystallisation have favoured processes of (over)growth and dissolution as the cause for the mosaics (Wright and Peeters, 1989; Deutz et al., 2002). Cathodoluminescence petrography is an important technique for distinguishing whether calcite microspar mosaics are the products of recrystallisation or complex growth histories (Budd et al., 2002). If the fabrics of alpha calcretes do represent significant aggrading neomorphism, the use of crystal sizes as an indicator of climate (e.g. Drees and Wilding, 1987), or for differentiating groundwater from pedogenic calcretes (Raghavan and Courty, 1987) is questionable, as is the validity of stable isotope analysis where values may have been diagenetically reset (Budd et al., 2002).

Beta assemblages, which are more common but not limited to carbonate-rich substrates, exhibit a very diverse range of textures of biogenic, mainly fungal and root-related, origins. Fungi certainly play a major role in fixing calcium carbonate in soils, and in producing a range of textures including alveolar septal fabrics and needle fibre calcite (Figure 2.2A–C) (see also Verrecchia and Verrecchia, 1994; Borsato et al., 2000). Calcification associated with roots is a major process (Klappa, 1980) and takes on a wide range of forms (Wright and Tucker, 1991; Alonso-Zarza, 1999; Kosir, 2004). The fixing of calcite in roots to produce cell-like crystal forms is a common process. *Microcodium* represents calcite precipitation in the cortical cells of roots and was the main contributor to enormous volumes of calcrete in the Cretaceous and Tertiary (Kosir, 2004). However, the subaerial parts of some trees can also contain cellular textures composed of calcium carbonate (Braissant et al., 2004). Even earthworms and slugs play a role in producing carbonate textures (Canti, 1998).

2.7 Laminar Calcretes

Laminar calcretes represent an important calcrete type requiring special consideration because of their polygenetic nature and their implications for elucidating landscape development. In the past they were considered to reflect abiogenic precipitation from ponded waters above impermeable hardpan layers or on bedrock. However, it has become clear that some forms develop at surfaces exposed to the atmosphere, caused by microbial activity, by lichens (Klappa, 1979), or possibly cyanobacteria (Verrecchia et al., 1995). Both the simple, abiogenically precipitated and the microbially influenced laminar calcretes are typically finely laminated and can resemble stromatolites. The latter can be regarded as terrestrial stromatolites and distinguishing them from their more common subaqueous equivalents is an issue for palaeoenvironmental analysis (Wright, 1989). Many laminar calcretes are found today buried beneath soil layers and if all laminar calcretes owed their origins to lichens and cyanobacteria (both groups requiring direct exposure to light), then such covered laminar calcretes above illuvial calcrete horizons must indicate that exhumation and reburial took place. In the case of a cyanobacterial origin, the crucial piece of evidence is seen as the presence of spherulites, regarded by Verrecchia et al. (1995) as of cyanobacterial origin. These are spherical structures about 100 μm in diameter showing a fibro-radial texture (Verrecchia et al., 1995) and tend to be more common at the tops of calcretes (Alonso-Zarza et al., 1998). However, identical structures can also be produced by bacteria (see discussion by Wright et al., 1996; Braissant et al., 2003), which would not require direct subaerial exposure. That said, where laminar calcretes show evidence of erosion (micro-unconformities) or influxes of detrital sediment, and spherulites, a strong case exists that exposure has played a role in their origin. Thus, some laminar calcretes are microprofiles that appear to record multiple phases of calcrete precipitation, erosion and exposure and burial (Verrecchia, 1987; Fedoroff et al., 1994), even constituting units 2 m thick (Sanz and Wright, 1994; Alonso-Zarza and Silva, 2002).

Most laminar calcretes, especially those many centimetres thick, are root related, produced by calcification in and around root mats, constituting rhizogenic or rhizolite calcretes (Wright et al., 1988; 1995). These can form as simple horizontal sheets or can be penetrative (Rossinsky and Wanless, 1992; Rossinsky et al., 1992) and associated with stringer calcretes. Such rhizolite calcretes typically show a somewhat less finely laminated fabric, with a millimetre-to-submillimetre microstructure of micritic laminae and tubular pores. Some rhizolite laminar calcretes can reach thicknesses of over 2 m and can constitute the only calcrete type in some successions (Wright et al., 1995). Rhizolite laminar calcretes can also occur in complex sequences with detrital intervals indicating possible periods of climate–vegetation change and landscape instability; such profiles are a prominent component of Quaternary profiles in mainland Spain and the Canary Islands (Alonso-Zarza et al., 1998; Alonso-Zarza and Silva, 2002) (Figure 2.5B).

Some thin laminar calcretes are associated with pisolitic–oolitic horizons and are produced along extensive root mats (Calvet and Julia, 1983; Wright et al., 1995). In some cases the ooids were formed effectively *in situ* by microbial processes (Calvet and Julia, 1983) (Figure 2.5A), but in other cases, especially where the ooids contain detrital grains, there may have been reworking (Alonso-Zarza and Silva, 2002). The occurrence of thick pisolitic horizons on the lower parts of slopes (Wright, 1994) (Figure 2.4B) is an indication that downslope transport may play a key role in the formation of such horizons. Thus, a laminar–pisolitic couplet does not have to have formed above a hardpan calcrete as is implied in earlier classifications based on chronosequences (see below), but can form as separate horizons as a result of root mats, with or without erosion and soil movement.

2.8 Mineralogy and Chemistry

The dominant mineralogy of calcretes, ancient and modern, seems to be low-Mg calcite (Wright and Tucker, 1991), although complex reactions should be expected as a result of microbial and evaporitic processes (Watts, 1980). Dolomite is commonly recorded and is probably primary in origin, but rarely constitutes the main mineral except in some groundwater dolocretes. This may be because dolomite forms when the Mg/Ca ratio is high, which is more likely to happen in evolved groundwaters where Ca has been removed as calcite (see below). This might also explain why dolomite occurs preferentially in the lower parts of some pedogenic profiles (see examples in Wright and Tucker, 1991). More complex carbonate mineralogies have been recorded from pedogenic carbonates associated with Mg-rich bedrock (e.g. Podwojewski, 1995). A variety of clays also occur with calcretes (see Wright and Tucker, 1991).

A very substantial body of knowledge has developed relating to C and O stable isotopes in pedogenic calcretes and a complex range of factors influencing the isotopic composition has been elucidated. These isotopes have been used to determine a range of palaeoenvironmental parameters, ranging from temperature, vegetation types and densities, and partial pressure of atmospheric carbon dioxide (Andrews et al., 1998; Cerling, 1999; Ekhart et al., 1999; Deutz et al., 2001; Royer et al., 2001). However, as Deutz et al. (2002) discuss, calcretes take prolonged periods of time to form, and as a result of overprinting (dissolution and overgrowth of crystals) and perhaps minor recrystallisation their chemical signatures are time averaged (as stressed by Kelly et al., 2000), and consequently the more developed the calcrete, the greater the heterogeneity in the C and O isotopic values. This severely limits the temporal resolution of C and O stable isotope interpretations, and they should only be made where the material shows a limited range of isotopic heterogeneity. In aggrading sedimentary systems, diagenetic overprinting by rising groundwaters may be a major process leading to recrystallisation and overgrowths (Budd et al., 2002). One little-used technique is where the isotopic tracers ($^{87}Sr/^{86}Sr$ and C) provenance the calcium and carbon found in soil carbonates (e.g. Quade et al., 1995). By using this technique it may be possible to distinguish the effects of local lithogenic sources of cations from atmospheric (long-transport dust) ones. The technique is not only of use for pedogenic calcretes, as Spotl and Wright (1992) used Sr isotopes to trace the likely source of cations in Triassic groundwater dolocretes from the Paris Basin.

Radiometric dating has been applied successfully to pedogenic Quaternary calcretes (e.g. Wang et al., 1996; Amundson et al., 1998; Rowe and Maher, 2000; Deutz et al., 2002; Ludwig and Paces, 2002; Candy et al., 2004). U/Th dating has been used on Quaternary calcretes by Kelly et al.

(2000), who stressed that such dates provide an apparent age which is the minimum for the commencement of calcrete formation. Rasbury et al. (2000) used U/Pb dating on Carboniferous calcretes and found that the uranium was fixed, effectively syndepositionally, in organic-rich rhizolite laminar calcretes.

2.9 Mechanisms of Formation of Pedogenic Calcretes

The fact that carbonate enrichment in soils is so widespread, even in soils not developed on carbonate parent materials, is clear proof that there are external sources for carbonate. The weight of evidence is that pedogenic calcretes are the products of illuvial concentrations of calcium carbonate (the 'per descensum' model of Goudie, 1983), whereby carbonate is moved mainly in solution (or possibly in colloidal form; Baghernejad and Dalrymple, 1993) from the upper soil layers to precipitate at a lower level. Rare examples of pedogenic carbonates fed from groundwater are known (Knuteson et al., 1989). The sources of carbonate are varied (Goudie, 1973, 1983), and besides any local lithogenic source, include rainfall, sea-spray, surface runoff, dust, calcareous fauna and even vegetation (e.g. Cailleau et al., 2004; Garvie, 2004). The primary source in pedogenic forms in semi-arid and arid areas must be the atmosphere, as dust and as rainfall (Monger and Gallegos, 2000). In the Las Cruces area of New Mexico, recent average atmospheric input of Ca expressed as $CaCO_3$ is $1.9\,\mathrm{g\,m^{-2}\,yr^{-1}}$. Of this only $0.3\,\mathrm{g}$ is directly from $CaCO_3$ in dust, $0.1\,\mathrm{g}$ as soluble Ca^{2+} in dust expressed as $CaCO_3$, and $1.5\,\mathrm{g}$ as Ca^{2+} expressed as $CaCO_3$ in rainfall (Monger and Gallegos, 2000). The route of the Ca^{2+} in the soil may be complex and influenced by vegetation, directly and indirectly; Garvie (2004) has estimated that in areas in the southwest USA with a high density of saguaro cacti, up to $2.4\,\mathrm{g\,m^{-2}\,yr^{-1}}$ of calcite are indirectly produced by the decay of the cacti.

Precipitation of carbonate is triggered by a number of processes (Wright and Tucker, 1991), including evaporation, evapotranspiration and degassing, but also biological processes play a crucial role. Calcite may be fixed intracellularly, particularly in the case of roots hosted on carbonate substrates, because the process enhances the production of protons, assisting the plant to acquire mineral nutrients in low-nutrient soils (McConnaughey and Wheelan, 1997; Kosir, 2004). Fixing calcite in the cells may also protect the plant from excessive Ca or bicarbonate concentrations (Kosir, 2004). Calcium oxalate monohydrate (whewellite) may be fixed in the cells and later converted by microbial decay processes, via calcium oxalate dihydrate (weddellite) to calcite (Verrecchia, 1990; Braissant et al., 2004). Weddellite can form as a primary mineral in saguaro cacti (Garvie, 2004), later converted on decay to calcite. Fungi also play a role in fixing of calcite directly or initially as Ca oxalates (Verrecchia et al., 1993). Cyanobacteria also trigger calcium carbonate precipitation (Verrecchia et al., 1995), as do bacteria (Loisy et al., 1999; Schmittner and Giresse, 1999; Braissant et al., 2003). Termites may even play a role in fixing calcite in soils (Monger and Gallegos, 2000), as can earthworms or slugs (Canti, 1998).

The newly formed calcium carbonate can have a passive role in filling pore space, or may be displacive (producing a series of macro- and micro-textures; reviewed by Wright and Tucker, 1991), or may be replacive of carbonate or silicate host grains. Displacive growth is more likely a process in non-carbonate hosts, as calcite is unable to form adhesive bonds with non-carbonate grains (Chadwick and Nettleton, 1990). The product of replacive and displacive growth is that many primary grains from the original host 'float' in the secondary calcrete matrix.

2.10 Profile Development

Profile development can be seen as a simple, progressive (linear) process or as a more dynamic one where carbonate accumulation can be interrupted by depositional and erosional events. In the simple progressive situation (Figure 2.7), three main developmental models can be identified. The simplest case is where progressive alteration takes place of a carbonate-rich parent material (Figure 2.7B). Here calcrete development on porous, carbonate-poor hosts mainly involves the downward translocation in solution of calcium carbonate as bicarbonate, but dissolution and reprecipitation are somewhat different in carbonate-rich hosts. In indurated carbonate hosts (Figure 2.7B), the accumulation of calcrete does not involve the classic stages of development so well documented from many Aridisols (Rabenhorst et al., 1991; see below), seemingly taking place along a downwardly migrating alteration front. The stages of alteration in porous carbonate hosts are reviewed by Wright (1994), where root mats and stringer calcretes play an important role.

Wright et al. (1995), based on studies of rhizogenic palaeosol calcrete profiles from southern Europe, proposed the rhizogenic model (Figure 2.7C). In Mesozoic palaeosols from northern Spain, they identified progressive profile development by calcified root mats from a few isolated laminar levels to thick (>2 m) dense carbonate sheets. In early Cenozoic successions in southern France they noted stages of development from scattered *Microcodium* remains to entire sheets composed almost wholly of this component.

The progressive illuviation model was devised by Gile et al. (1966) and Machette (1985), who produced the classic models for the development of calcrete profiles based on examples from the southwest USA (Figures 2.4A, 2.7A), whereby a series of time-dependent stages (chronosequences) are

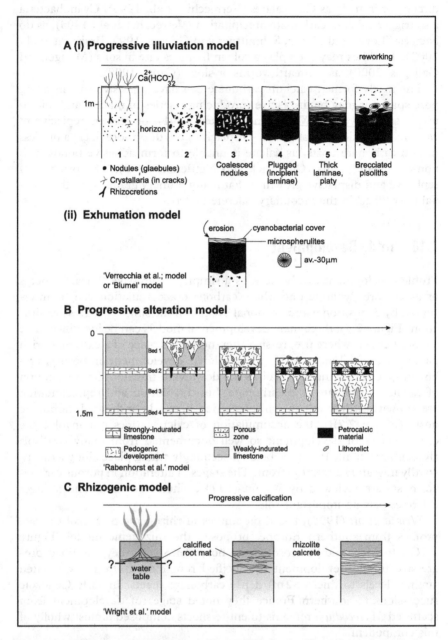

Figure 2.7 Models for pedogenic calcrete development: (A) (i) after Gile et al. (1966), and Machette (1985), Stages 1–5 as defined by Machette (1985); (ii) after Blumel (1982) and Verrecchia et al. (1995); (B) after Rabenhorst et al. (1991); (C) after Wright et al. (1995).

defined, although gradations exist between these stages. Such models can be tested with numerical simulations (McFadden et al., 1991). Klappa (1983) has also shown a broadly similar model emphasising diagenetic relationships. The developmental model envisages the progressive accumulation of carbonate through stages 1–4 (Figure 2.7A (i)), leading to the plugging of the profile, preventing free drainage, and to the ponding of waters over the impervious petrocalcic hardpan horizon; thus an important geomorphological and pedologic threshold is reached (Muhs, 1984). Carbonate precipitating from these ponded waters produces the laminar calcrete (Figure 2.7A (i), stages 4–5), although root mats forming at this level also produce laminar forms. Carbonate can also form around grains above the laminar layer to produce pisolitic calcrete. Presumably other salts also accumulate at this level. With time this laminar horizon may thicken or the soil cover may be eroded, leading to subaerial exposure and microbial colonisation of the exposed surface. Eventually the profile begins to break up to produce brecciated layers in which the fragments may also be coated by new carbonate (Figures 2.2D, 2.7A (i) stage 6). Calcretes associated with very old surfaces in the USA, such as the Ogallala calcretes (Machette, 1985), can reach many metres in thickness and have a highly brecciated appearance with evidence of multiple phases of brecciation and recementation. Such calcretes are rare and appear limited to major hiatal surfaces (Wright et al., 1993). The cause of the brecciation is unclear but root action must play a major role. If laminar calcretes owe their origin to precipitation by phototrophic microbial communities (section 2.7), as suggested by Verrecchia et al. (1995), and linked to the presence of microspherulites (see above), a phase of exhumation (Figure 2.7A (ii)) must have taken place followed by reburial by more soil to produce stage 4–5 profiles; Blumel (1982) also favoured laminar calcrete as being linked to phases of direct subaerial exposure.

The times required for the development of these stages are lengthy (Machette, 1985), and there is a very wide range of possible ages for individual stages (Wright, 1990). Such age estimates as are available were largely based on dating the surfaces at which the calcretes are found. In contrast, radiometric dating of calcretes tends to give minimum ages for the accumulation of carbonate (Kelly et al., 2000). The rate of supply of carbonate to the soil from the atmosphere, as well as the rates at which the carbonate is translocated and precipitated, are unlikely to have been constant or continuous over time. Perhaps the case of a dated (U/Th) nodule described by Kelly et al. (2000), which had a hiatus in its growth of possibly 90 kyr, is a glimpse of the episodic nature of calcrete formation? However, dating calcrete development has its pitfalls; Candy et al. (2004) have shown that to understand the complete growth history of a calcrete profile it may be more meaningful to date individual components whose relative ages are already established. Ludwig and Paces (2002) also show the benefits of fine-scale sampling to date carbonate formation. Over long periods of time rainfall levels also fluctuate and, as a result, carbonate already formed may become remobilised, causing upper layers to have undergone dissolution, and with phases of overprinting in the sense of Deutz et al. (2002).

Even if the rainfall and the atmospheric flux of carbonate remained constant (which is highly unlikely) the position of the land surface relative to the carbonate accumulation zone is unlikely to have remained constant. Studies have shown that at least some calcrete-bearing soils, of Quaternary (Alonso-Zarza et al., 1998) and even Palaeozoic age (Marriott and Wright, 1993), have developed through complex periods of pedogenesis, episodic sedimentation and erosion (Figure 2.8); although late Quaternary examples may have seen more extreme fluctuations in landscape stability than was the case throughout much of the geological record. Erosion not only causes reworking but also the lowering of the carbonate profile by leaching and translocation to lower levels (Elbersen, 1982). Depositional events, if introducing relatively small amounts of sediment, will produce cumulate profiles (Marriott and Wright, 1993; Alonso-Zarza et al., 1998). Considering the likely episodic nature of the calcrete process, many of these horizons or sub-profiles associated with present-day surfaces must be relict. The long periods represented by many calcrete-bearing soils mean that well developed calcretes must be polygenetic in that they are unlikely to be the product of continuous, linear and progressive pedogenesis. The reorganisation of carbonate due to fluctuations in the depths of leaching and precipitation may mask some of the complexity.

2.11 Groundwater Calcretes

The thickest calcrete bodies arise through the interstratal cementation, displacement and replacement of sediment bodies by carbonate produced from groundwater. Precipitation takes place in the capillary fringe or below in the phreatic zone (Figure 2.9). Groundwater calcretes are known from old drainage systems such as those of central Australia (Carlisle, 1983; Arakel, 1986; Morgan, 1993) or southern Africa (Carlisle, 1983; Nash et al., 1994), from alluvial fans (Pimentel et al., 1996; Nash and Smith, 1998; Mack et al., 2000) and playa-lake systems (Arakel, 1991; Colson and Cojan, 1996). In fluvial systems, the calcretes form ribbon-like bodies many tens or even hundreds of kilometres long, up to 10 km wide and many metres thick. In alluvial fans, they appear to form more sheet-like bodies (Mack et al., 2000) and may even constitute masses over 200 m thick (Maizels, 1987). Where they form in association with playa systems, precipitation may be triggered by the common ion effect, and the carbonate bodies may develop delta-like geometries or even halo-like patterns as in

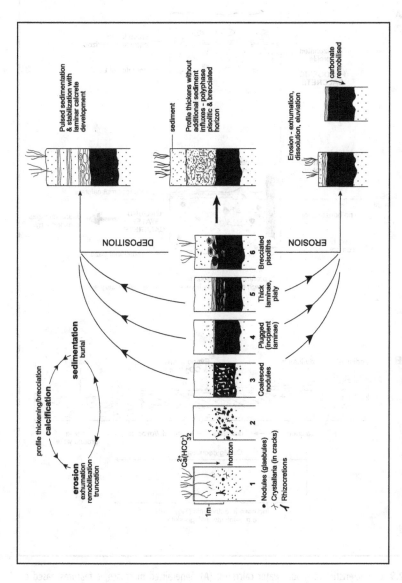

Figure 2.8 Dynamic model for pedogenic calcrete development based on Alonso-Zarza et al. (1998). Existing calcrete horizons of profiles may be remobilised following some degree of exhumation (Elbersen, 1982) or by changes in rainfall.

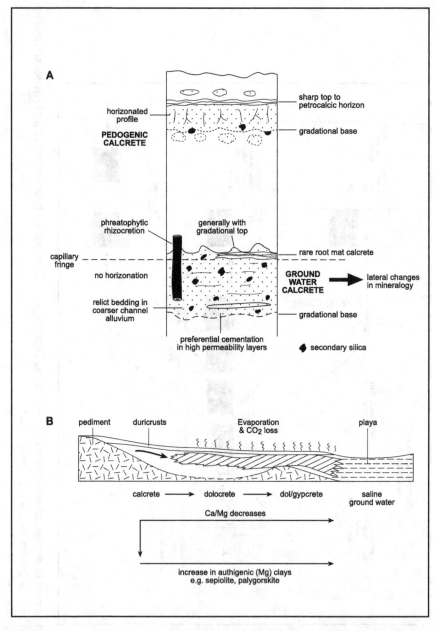

Figure 2.9 Characteristics of groundwater calcretes. (A) Generalised macroscopic features based on various sources. Some workers, such as Carlisle (1980), identify two zones in the massive phreatic unit: an upper earthy zone with remnant soil and alluvium, and a lower, dense 'porcellaneous' zone with abundant cracks and cavities. Phreatophytic plants may also produce features such as rhizocretions and laminar rhizolite crusts (Seminiuk and Meagher, 1981). (B) Generalised model for the evolution of groundwaters and their precipitates in semi-arid to arid alluvial systems; based on Arakel (1986).

the Danian continental deposits of Provence (Colson and Cojan, 1996) (Figure 2.10).

Groundwater calcretes in alluvial fans might be more likely in distal settings where water tables are shallower and more prone to evaporation, as in the Palomas Basin successions described by Mack et al. (2000). Near the toes of such fans, where they pass into finer grained playa-pan deposits, groundwaters are at shallower depths and there may be overlap between groundwater and vadose calcretes (Figure 2.1). This can result in overlap in the capillary fringe zone and even to calcrete developing in hydromorphic soils (Slate et al., 1996). Capillary zone calcretes may be characterised by thin, sharp-based zones of carbonate associated with an upper zone of nodules or tubules (Mack et al., 2000). Whereas groundwater calcretes form below the soil zone and usually lack biogenic features, phreatophytic plants are capable of forming laminar and rhizocretionary calcretes in the capillary fringe zone (Semeniuk and Meagher, 1981). Striking examples of phreatophyte rhizocretions have also been recorded from the pre-Quaternary record (Purvis and Wright, 1991).

As the groundwaters flow they evolve and the Mg/Ca ratio increases as calcite is first precipitated (Arakel, 1986; Morgan, 1993) (Figure 2.9B). This can result in elevated Mg/Ca and dolomite precipitation. As carbonate ions are depleted, gypsum and even Mg-rich clays such as palygorskite and sepiolite can precipitate, although the distribution of these minerals depends on local conditions (Pimentel, 2002). The distribution of groundwater dolomites in the Danian of Provence does not correspond to such a simple model, as dolomite precipitation was not limited to the more distal parts of the drainage system, and Colson and Cojan (1996) have proposed that dolomites developed where regional groundwaters mixed with saline lake waters.

Groundwater calcretes (and dolocretes) range from being thin developments of nodules to large bodies of massive carbonate (Arakel and McConchie, 1982). Alonso-Zarza (2003) has reviewed some of the diverse features recorded from Quaternary and pre-Quaternary examples. Criteria for their differentiation from pedogenic forms have been discussed by several authors, including Pimentel et al. (1996) and Mack et al. (2000), although as discussed by Alonso-Zarza (2003) groundwater carbonates can readily overprint and be overprinted by both pedogenic (including hydromorphic calcretes) and lacustrine–palustrine carbonates (Figure 2.1), and the tendency for such calcretes to become exhumed during drainage inversion means that they also undergo dissolution (Mann and Horwitz, 1979; Arakel, 1991; Spotl and Wright, 1992). Khadkikar et al. (1998) have emphasised the complex relationships between groundwater and pedogenic calcretes in the late Quaternary of Gujarat, India, especially the importance of pedogenic overprinting of groundwater forms. One of the most robust criteria for distinguishing pedogenic and groundwater calcretes is that the

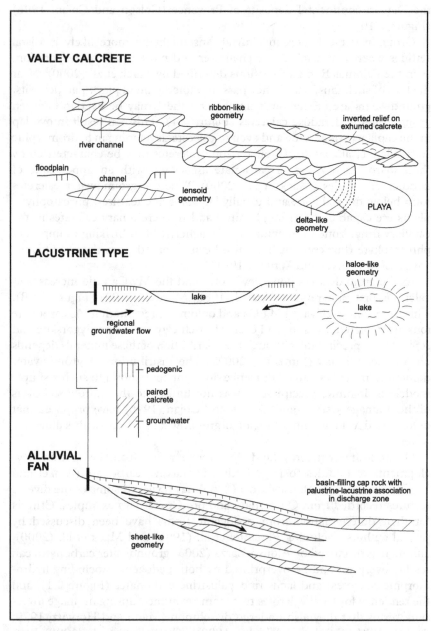

Figure 2.10 Geometries of groundwater calcretes and dolocretes. Linear, ribbon-like valley calcretes are known from the late Cenozoic of central and Western Australia, and the late Triassic of the Paris Basin (see text). Lacustrine groundwater calcretes or dolocretes forming halo-like masses are known from the Danian of Provence (see text). Alluvial fans are associated with sheet-like bodies, and are known from the Plio-Pleistocene of Oman and the Palomas Basin (New Mexico), and the Paleocene of the Lisbon and Sado basins (see text).

former tend to be associated with fine-grained substrates such as floodplain deposits, whereas the latter commonly occur in more permeable coarse, channel sediments. One commonly noted feature of some groundwater calcretes is the presence of more coarsely crystalline cements (spar) (Tandon and Naryan, 1981; Nash and Smith, 1998; Mack et al., 2000), although it seems that many if not most groundwater calcretes are micritic in grain size. Although pedogenic calcretes may be associated with silica concentrations in the lower parts of the profiles, silicification appears to be a more common feature of groundwater forms, and in Australia the zone of silicification occurs above zones of more hypersaline waters (Arakel et al., 1989).

If it appears that pedogenic calcretes have a potentially self-regulating stage where the profile becomes impermeable, do groundwater forms also have some similar phase? The degree of cementation a host undergoes must reach a point when the permeability is effectively reduced to zero; the flow paths must shift and cause cementation elsewhere. Any major lowering of the water table, such as caused by uplift or fluvial incision, must affect these calcretes (Nash and Smith, 1998). Such events would potentially isolate land surfaces on interfluves, favouring stable surfaces and pedogenic calcrete development. Groundwater calcretes and dolocretes have even been integrated into sequence stratigraphic models, based on the late Triassic reservoirs of the Paris Basin, France (Eschard et al., 1998; Goggin and Jacquin, 1998).

Little is known about the rates of groundwater calcrete formation, but since the supply of calcium is unlikely to be limited to atmospheric inputs, the rates will probably exceed those of pedogenic forms, with the proviso that there is likely to be a wide range of rates depending on the specific hydrogeological factors. Somewhat surprisingly, the isotopic composition of pedogenic and groundwater calcretes in the same basin is similar (Cerling and Hay, 1986; Spotl and Wright, 1992; Quade and Roe, 1999; Mack et al., 2000), although Slate et al. (1996) note that hydromorphic calcretes from the Plio-Pleistocene of Arizona do show some differences from vadose carbonates of the same age.

2.12 Palaeoenvironmental Significance

Calcretes have, as already discussed, many palaeoenvironmental applications from palaeotemperature and palaeoprecipitation determinations, to determining the C3/C4 of palaeovegetation. Not only can calcretes be used for identifying local or regional environmental factors but also for global ones such as estimating the partial pressure of CO_2 in ancient atmospheres, or even for identifying isotopic shifts due to massive scale disassociation of sedimentary methane hydrates (Magioncalda et al., 2004).

As discussed above, pedogenic calcretes can form in a variety of climates, especially as regards temperatures. Intriguing as the presence of calcretes are in temperate or even arctic settings, their main occurrence is in semi-arid to arid areas, and so when used with other criteria, they are good indicators of ancient climates. Even in those fortunate situations where the depth of the top of the carbonate zone can be estimated in fossil material, estimating the mean annual rainfall is not without its risks (see above). The use of pedogenic calcretes, especially their geochemical signatures, is hampered by our relatively poor understanding of the real nature of calcrete development. The issue relates to what is best regarded as a time resolution problem. Are the geochemical signals we record the products of long term, time averaging caused by overprinting? Alternatively, if calcrete development is not progressive but episodic, the geochemical signatures in the carbonate may reflect only a short, even unrepresentative fraction of the history of that profile. Diagenesis can re-set oxygen isotope values in calcretes, typically making them more negative, but carbon values can also be re-set, even early in their burial history (Budd et al., 2002). Indeed, calcrete nodules can even be intrusive, having been 'injected' into older buried soil horizons (Rowe and Maher, 2000). Exactly what we may be able to glean from the presence of ancient groundwater calcretes is unclear, but they are likely to be repositories of much information if we develop a better understanding of their origins.

2.13 Relationships to Other Terrestrial Geochemical Sediments

Calcretes, especially in the interior of Australia, are commonly associated with silcretes. However, only incipient silcrete development is known from present-day soils (Chapter 4), and pedogenic silcrete formation requires extensive leaching and a strongly seasonal climate (Thiry et al., 1999). The association of silcrete and calcrete in areas today is probably due to the former being relict. Pedogenic gypcretes are found where the mean annual rainfall is less than around 250 mm (Chapter 10), towards the lower range for calcrete formation. As gypsum is more soluble than calcite, if it occurs with calcrete it tends to form at lower levels in the profile than calcite, unless overprinted by groundwater effects. Unlike pedogenic calcretes, silcretes and gypcretes, which as stated above have slightly different climatic distributions, groundwater examples of each can occur along the same drainage system due to local variations in water chemistry (Arakel et al., 1989).

2.14 Directions for Future Research

Pedogenic calcretes form in soils and, of course, soils are the products of the interaction of the biosphere, atmosphere, hydrosphere and lithosphere; as these are complex, dynamic and interrelated systems we should expect calcretes to be equally complex. We will achieve a better appreciation of their origins and their potential uses for reconstructing the past when we know how continuous or episodic their growth has been. More accurate dating should provide this information. A well developed profile represents the sum of periods of carbonate accumulation, under varied rates of accumulation, periods of remobilisation and hiatuses, as climate, vegetation and the position of the soil surface have changed. We know that calcretes are most likely polygenetic and so we need to ask why do so many appear to follow such a simple developmental path? Perhaps there is a taphonomic dimension and that many profiles have been overprinted with a loss of detail? Despite the classic southwest USA model for calcrete development being entrenched in the collective psyche, not all pedogenic calcretes develop in that way and other models need to be identified. For example, we know surprisingly little about pedogenic calcrete catenas, and their study might provide new insights into how different calcretes actually develop.

It is becoming clear that vegetation plays a key role in concentrating carbonate and that some pedogenic calcretes are effectively rhizogenic in origin. It should be possible to identify specific calcrete types linked to specific vegetation types or indeed to specific metabolic strategies by specific plants. For example, the mysterious yet abundant *Microcodium*-maker created its own calcrete type. Groundwater calcretes warrant serious study because they are a major component of some dryland landscapes. Finally, calcretes are important as stores for CO_2 (Lal et al., 2000) and a fuller understanding of how they form is more important than ever.

Acknowledgements

I thank Adrijan Kosir for many helpful comments about biomediation in calcretes and for commenting on an earlier draft of this chapter. Thanks also to Ian Candy for his constructive comments. This is a contribution to the Cardiff Geobiology Initiative.

References

Alonso-Zarza, A.M. (1999) Initial stages of laminar calcrete formation by roots: examples from the Neogene of central Spain. *Sedimentary Geology* **126**, 177–191.

Alonso-Zarza, A.M. (2003) Palaeoenvironmental significance of palustrine carbonates and calcretes in the geological record. *Earth-Science Reviews* **60**, 261–298.

Alonso-Zarza, A.M. & Silva P.G. (2002) Quaternary laminar calcretes with bees nests: evidence of small-scale climatic fluctuations, Eastern Canary Islands, Spain. *Palaeogeography, Palaeoclimatology, Palaeoecology* **178**, 119–135.

Alonso-Zarza, A.M., Silva, P.G., Goy, J.L. & Zazo, C. (1998) Fan surface dynamics and biogenic calcrete development: interactions during ultimate phases of fan evolution in the semi-arid SE Spain (Murcia). *Geomorphology* 24, 147–167.

Amundson, R., Stern, L., Baisden, T. & Wang, Y. (1998) The isotopic composition of soil respired CO_2. *Geoderma* 82, 83–114.

Andrews, J.E., Singhvi, A.K., Kailath, A.J., Kuhn, R., Dennis, P.F., Tandon, S.K. & Dhir, R.P. (1998) Do stable isotope data from calcrete record Late Pleistocene monsoonal climate variation in the Thar Desert of India? *Quaternary Research* 50, 240–251.

Arakel, A.V. (1986). Evolution of calcrete in palaeodrainages of the Lake Napperby area, central Australia. *Paleogeography, Palaeoclimatology, Palaeoecology* 54, 282–303.

Arakel, A.V. (1991) Evolution of Quaternary duricrusts in Karinga Creek drainage system, central Australia groundwater discharge zone. *Australian Journal of Earth Sciences* 38, 333–343.

Arakel, A.V. & McConchie, D. (1982) Classification and genesis of calcrete and gypsite lithofacies in paleodrainage systems of inland Australia and their relationship to carnotite mineralisation. *Journal of Sedimentary Petrology* 52, 1149–1170.

Arakel, A.V., Jacobsen, G., Salehi, M. & Hill, C.M. (1989) Silicification of calcrete in palaeodrainage basins of the Australian arid zone. *Australian Journal of Earth Sciences* 36, 73–89.

Baghernejad, M. & Dalrymple, J.B. (1993) Colloidal suspensions of calcium carbonate in soils and their likely significance in the formation of calcic horizons. *Geoderma* 58, 17–41.

Birkeland, P.W. (1999) *Soils and Geomorphology*, 3rd edn. New York: Oxford University Press.

Blumel, W.D. (1982) Calcrete in Namibia and SE-Spain. Relations to substratum, soil formation and geomorphic factors. *Catena Supplement* 1, 67–82.

Borsato, A., Frisia, S., Jones, B. & Van der Borg, K. (2000) Calcite moonmilk: crystal morphology and environment of formation in caves in the Italian Alps. *Journal of Sedimentary Research* 70, 1171–1182.

Bown, T.M. & Kraus, M.J. (1987) Integration of channel and floodplain suites: 1 Developmental sequence and lateral relationships of alluvial paleosols. *Journal of Sedimentary Petrology* 57, 587–601.

Braissant, O., Guillaume, C., Dupraz, C. & Verrecchia, E.P. (2003) Bacterially induced mineralization of calcium carbonate in terrestrial environments: the role of exopolysaccharides and amino acids. *Journal of Sedimentary Research* 73, 485–490.

Braissant, O., Guillaume, C., Aragno, M. & Verrecchia, E.P. (2004) Biologically induced mineralization in the tree *Milicia excelsa* (Moraceae): its causes and consequences to the environment. *Geobiology* 2, 59–66.

Braithwaite, C.J.R. (1983) Calcrete and other soils in Quaternary limestones: structures, processes and applications. *Journal of the Geological Society of London* 140, 351–363.

Budd, D.A., Pack, S.M. & Fogel, M.L. (2002) The destruction of paleoclimatic isotopic signals in Pleistocene carbonate soil nodules of Western Australia. *Paleogeography, Palaeoclimatology, Palaeoecology* 188, 249–273.

Bunting, B.T. & Christensen, L. (1980) Micromorphology of calcareous crusts from the Canadian High Arctic. *Geologiska Föreningens i Stockholm Fordland* **100**, 361–367.

Cailleau, G., Braissant, O. & Verrecchia, E.P. (2004) Biomineralization in plants as a long term carbon sink. *Naturwissenschaften* **91**, 191–194.

Calvet, F. & Julia, R. (1983) Pisoids in the caliche profiles of Tarragona, northeast Spain. In: Peryt, T.M. (Ed.) *Coated Grains*. Berlin: Springer-Verlag, pp. 456–473.

Candy, I. (2002) Formation of a rhizogenic calcrete during a glacial stage (Oxygen Isotope Stage 12): its palaeoenvironmental stratigraphic significance. *Proceedings of the Geologists' Association* **113**, 259–270.

Candy, I., Black, S. & Sellwood, B.W. (2004) Quantifying time scales of pedogenic calcrete formation using U-series disequilibria. *Sedimentary Geology* **170**, 177–187.

Canti, M. (1998) Origin of calcium carbonate granules found in buried soils and Quaternary deposits. *Boreas* **27**, 275–288.

Carlisle, D. (1980) *Possible Variations in the Calcrete–Gypcrete Uranium Model.* US Department of Energy, Open File Report GJBX-53 (80), 38 pp.

Carlisle, D. (1983). Concentration of uranium and vanadium in calcretes and gypcretes. In: Wilson, R.C.L. (Ed.) *Residual Deposits*. Special Publication 11, Geological Society of London, pp. 185–195.

Cerling, T.E. (1999) Stable isotopes in palaeosol carbonates. In: Thiry, M. & Simon-Coinçon, R. (Eds) *Palaeoweathering, Palaeosurfaces and Related Continental Deposits*. Special Publication 27, International Association of Sedimentologists. Oxford: Blackwell Science, pp. 43–60.

Cerling, T.E. & Hay, H.L. (1986) An isotopic study of palaeosol carbonates from Olduvai Gorge. *Quaternary Research* **25**, 63–78.

Chadwick, O.A. & Nettleton, W.D. (1990) Micromorphological evidence of adhesive and cohesive forces in soil cementation. *Developments in Soil Science* **19**, 207–212.

Colson, J. & Cojan, I. (1996) Groundwater dolocretes ina lake-marginal environment: an alternative model for dolocrete formation in continental settings (Danian of the Provence Basin, France). *Sedimentology* **43**, 175–188.

Deutz, P., Montanez, I.P., Monger, H.C. & Morrison, J. (2001) Morphology and isotopic heterogeneiry of Late Quaternary pedogenic carbonates: implications for paleosol carbonates as paleoenvironmental proxies. *Palaeogeography, Palaeoclimatology, Palaeoecology* **166**, 293–317.

Deutz, P., Montanez, I.P. & Monger, H.C. (2002) Morphology and stable and radiogenic isotope composition of pedogenic carbonates in Late Quaternary relict soils, New Mexico, USA: an integrated record of pedogenic overprinting. *Journal of Sedimentary Research* **72**, 809–822.

Drees, L.R. & Wilding, L.P. (1987) Micromorphic record and interpretation of carbonate forms in the Rolling Plains of Texas. *Geoderma* **40**, 157–175.

Ekhart, D.D., Cerling, T.E., Montanez, I.P. & Tabor, N.J. (1999) A 400 million year carbon isotope record of pedogenic carbonate: implications for paleoatmospheric carbon dioxide. *American Journal of Science* **299**, 805–827.

Elbersen, G.W.W. (1982) *Mechanical Replacement Processes in Mobile Soft Calcic Horizons; their Role in Soil and Landscape Genesis in an Area near Merida, Spain.*

Wageningen: Centre for Agricultural Publishing and. Documentation, Agricultural Research Report, No. 919.

Eschard, R., Lemouzy, P., Bacchiana, C., Desaubliaux, G., Parpant, J. & Smart, B. (1998) Combining sequence stratigraphy, geostatistical simulations, and production data for modeling a fluvial reservoir in the Chaunoy Field (Triassic, France). *Bulletin of the American Association of Petroleum Geologists* **82**, 545–568.

Esteban, M. & Klappa, C.F. (1983) Subaerial exposure environment. In: Scholle, P.A., Bebout, D.G. & Moore, C.H. (Eds) *Carbonate Depositional Environments.* AAPG Memoir 33. Tulsa, OK: American Association of Petroleum Geologists, pp. 1–54.

Fedoroff, N., Courty, M.A., Lacroix, E. & Oleschko, K. (1994) Calcitic accretion on indurated volcanic materials (example of tepetate, Atliplano, Mexico). *Proceedings of the XVth World Congress of Soil Scientists*, Acapulco, Vol. 6A, pp. 459–472.

Garvie, L.A.J. (2004) Decay-induced biomineralization of the saguaro cactus (Carnegiea gigantean). *American Mineralogist* **88**, 1879–1888.

Gile, L.H., Peterson, F.F. & Grossman, R.B. (1965) The K horizon: a master soil horizon of carbonate accumulation. *Soil Science* **99**, 74–82.

Gile, L.H., Peterson, F.F. & Grossman, R.B. (1966) Morphological and genetic sequences of carbonate accumulation in desert soils. *Soil Science* **100**, 347–360.

Goggin, V. & Jacquin, T. (1998) A Sequence stratigraphic framework of the marine and continental Triassic series in the Paris Basin, France. In: de Graciansky, P-C., Hardenbol, J., Jacqin, T. & Vail, P.R. (Eds) *Mesozoic and Cenozoic Sequence Stratigraphy of European Basins.* Special Publication 60. Tulsa, OK: Society for Sedimentary Geology, pp. 667–690.

Goudie, A.S. (1973) *Duricrusts in Tropical and Subtropical Landscapes.* Oxford: Clarendon Press.

Goudie, A.S. (1983) Calcrete. In: Goudie, A.S. & Pye, K. (Eds) *Chemical Sediments and Geomorphology.* London: Academic Press, pp. 93–131.

Gustavson, T.C. & Holliday, V.T. (1999) Eolian sedimentation and soil development on a semi-arid to subhumid grassland, Tertiary Ogallala and Quaternary Blackwater Draw Formations, Texas and New Mexico High Plains. *Journal of Sedimentary Research* **69**, 622–634.

Kaemmerer, M. & Revel, J.C. (1991) Calcium carbonate accumulation in deep strata and calcrete in Quaternary alluvial formations of Morocco *Geoderma* **48**, 43–57.

Kelly, M., Black, S. & Rowna, J.S. (2000) A calcrete-based U/Th chronology for landform evolution in the Sorbas Basin, south east Spain. *Quaternary Science Reviews* **19**, 995–1010.

Khadkikar, A.S., Merh, S.S., Malik, J.N. & Chamyal, L.S. (1998) Calcretes in semi-arid alluvial systems: formative pathways and sinks. *Sedimentary Geology* **116**, 259–260.

Klappa, C.F. (1979) Lichen stromatolites: criterion for subaerial exposure and a mechanism for the formation of laminar calcrete (caliche). *Journal of Sedimentary Petrology* **49**, 387–400.

Klappa, C.F. (1980) Rhizoliths in terrestrial carbonates: classification, recognition, genesis and significance. *Sedimentology* **27**, 613–629.

Klappa, C.F. (1983) A process-response for the formation of pedogenic calcretes: model for calcrete formation. In: Wilson, R.C.L. (Ed.) *Residual Deposits*. Special Publication 11, Geological Society of London, pp. 211–220.

Knuteson, J.A., Richardson, J.L., Patterson, D.D. & Prunty, L. (1989) Pedogenic carbonates in a calciquoll associated with a recharge wetland. *Soil Science Society of America Journal* 53, 495–499.

Kosir, A. (2004) Microcodium revisited: root calcification products of terrestrial plants on carbonate-rich substrates. *Journal of Sedimentary Research* 74, 845–857.

Lal, R., Kimble, J.M., Eswaran, H. & Stewart, B.A. (Eds) (2000) *Global Climate Change and Pedogenic Carbonates*. Boca Raton: CRC Press/Lewis Publishers.

Lauriol, B. & Clarke, J. (1999) Fissure calcretes in the Arctic: a paleohydrologic indicator. *Applied Geochemistry* 14, 775–785.

Loisy, C., Verrecchia, E.P. & Dufour, P. (1999) Microbial origin for pedogenic micrite associated with a carbonate paleosol (Champagne, France). *Sedimentary Geology* 126, 193–204.

Ludwig, K.R. & Paces, J.B. (2002) Uranium-series dating of pedogenic silica and carbonate, Crater Lake, Nevada. *Geochimica et Cosmochimica Acta* 66, 487–506.

Machette, M.N. (1985). Calcic soils of the south-western United States. *Geological Society of America Special Paper* 203, pp. 1–21.

Mack, G.H. & James, W.C. (1994) Palaeoclimate and global distribution of paleosols. *Journal of Geology*, 102, 360–336.

Mack, G.H. & Madoff, R.D. (2005) A test of models of alluvial architecture and palaeosol development: Camp Rice Formation (Upper Pliocene-Lower Pleistocene), southern Rio Grande rift, New Mexico, USA. *Sedimentology* 52, 191–211.

Mack, G.H., Cole, D.R. & Trevino, L. (2000) The distribution and discrmination of shallow, authigenic carbonates in the Pliocene-Pleistocene Palomas Basin, southern Rio Grande Rift. *Bulletin of the Geological Society of America* 112, 643–656.

Maizels, J. (1987) Plio-Pleistocene raised channel systems of the western Sharqiya (Wahiba), Oman. In: Frostick, L.E. and Reid, I. (Eds) *Desert Sediments: Ancient and Modern*. Special Publication 35, Geological Society of London, pp. 31–50.

Maizels, J. (1990) Raised channel systems as indicators of palaeohydrologic change: a case study from Oman. *Palaeogeography, Palaeoclimatology, Palaeoecology* 76, 241–277.

Magioncalda, R., Dupuis, C., Smith, T., Steurbaut, E. & Gingerich, P.D. (2004) Paleocene-Eocene carbon isotope excursion in organic carbon and pedogenic carbonate: direct comparison in a continental stratigraphic section. *Geology* 32, 553–556.

Mann, A.W. & Horwitz, R.C. (1979) Groundwater calcrete deposits in Australia: some observations from Western Australia. *Journal of the Geological Society of Australia* 26, 293–303.

Marriott, S.B. & Wright, V.P. (1993) Palaeosols as indicator of geomorphic stability in two Old Red Sandstone alluvial suites, South Wales. *Journal of the Geological Society of London* 150, 1109–1120.

McConnaughey, T.A. & Whelan, J.F. (1997) Calcification generates protons for nutrient and bicarbonate uptake. *Earth-Science Reviews* **42**, 95–117.

McFadden, L.D., Amundson, R.G. & Chadwick, O.A. (1991) Numerical modeling, chemical and isotopic studies of carbonate accumulation in soils of arid regions. In: Nettleton, W.D. (Ed.) *Occurrence, Characteristics and Genesis of Carbonate, Gypsum and Silica Accumulations in Soils.* Special Publication 26. Madison, WI: Soil Science Society of America, pp. 17–35.

Milnes, A.R. (1992) Calcrete. In: Martini, I.P. & Chesworth, W. (Eds) *Weathering, Soils and Paleosols.* Amsterdam: Elsevier, pp. 309–347.

Monger, H.C. & Gallegos, R.A. (2000) Biotic and abiotic processes and rates of pedogenic carbonate accumulation in the southwestern United States – relationship to atmospheric CO_2 sequestration. In: Lal, R., Kimble, J.M., Eswaran, H. & Stewart, B.A. (Eds) *Global Climate Change and Pedogenic Carbonates.* Boca Raton: CRC Press/Lewis Publishers, pp. 273–289.

Morgan, K.H. (1993) Development, sedimentation and economic potential pf palaeoriver systems of the Yilgarn Craton of Western Australia. *Sedimentary Geology* **85**, 637–656.

Muhs, D.R. (1984) Intrinsic thresholds in soil systems. *Physical Geography* **5**, 99–110.

Nash, D.J. & Smith, R.F. (1998) Multiple calcrete profiles in the Tabernas Basin, SE Spain: their origins and geomorphic implications. *Earth Surface Processes and Landforms* **23**, 1009–1029.

Nash, D.J., Shaw, P.A. & Thomas, D.S.G. (1994) Duricrust development and valley evolution: process-landform links in the Kalahari. *Earth Surface Processes and Landforms* **19**, 299–317.

Netterberg, F. (1967) Some road making properties of South African calcretes. *Proceedings of the 4th Regional Conference of African Soil Mechanics and Foundation Engineers,* Cape Town, Vol. 1, pp. 77–81.

Netterberg, F. (1980). Geology of South African calcretes I: Terminology, description macrofeatures and classification. *Transactions of the Geological Society of South Africa,* **83**, 255–283.

Paquet, H. & Ruellan, A. (1997) Calcareous epigenetic replacement (epigenie) in soils and calcrete replacement. In: Paquet, H. & Clauer, N. (Eds) *Soils and Sediments: Mineralogy and Geochemistry.* Berlin: Springer-Verlag, pp. 21–48.

Pimentel, N.L.V. (2002) Pedogenic and early diagenetic processes in Palaeogene alluvial fan and lacustrine deposits from the Sado Basin (south Portugal). *Sedimentary Geology* **148**, 123–138.

Pimentel, N.L.V., Wright, V.P. & Azevedo, T.M. (1996) Distinguishing early groundwater alteration effects from pedogenesis in ancient alluvial basins: examples from the Palaeogene of southern Portugal. *Sedimentary Geology* **105**, 1–10.

Podwojewski, P. (1995) The occurrence and interpretation of carbonate and sulphate minerals in a sequence of Vertisols in New Caledonia. *Geoderma* **65**, 223–248.

Purvis, K. & Wright, V.P. (1991) Calcrete related to phreatophytic vegetation from the Middle Triassic Otter Sandstone of south west England. *Sedimentology* **38**, 539–551.

Quade, J. & Roe, L.J. (1999) The stable isotope composition of early groundwater cements from sandstones in paleoecological reconstruction. *Journal of Sedimentary Research* **69**, 667–674.

Quade, J., Chivas, A.R. & McCulloch, M.T. (1995) Strontium and carbon isotope tracers and the origins of soil carbonate in South Australia and Victoria. *Palaeogeography, Palaeoclimatology, Palaeoecology* **113**, 103–117.

Rabenhorst, M.C., West, L.T. & Wilding, L.P. (1991) Genesis of calcic and petrocalcic horizons in soils over carbonate rocks. In: Nettleton, W.D. (Ed.) *Occurrence, Characteristics and Genesis of Carbonate, Gypsum and Silica Accumulations in Soils*. Special Publication 26. Madison, WI: Soil Science Society of America, pp. 61–74.

Raghavan, H. & Courty, M.A. (1987) Holocene and Pleistocene environments in the Thar Desert (Didwana, India). In: Fedoroff, N. & Courty, M.A. (Eds) *Micromorphologie de Sols*. Paris: Association Francaise pour l'Etude du Sol, pp. 371–375.

Rasbury, E.T., Mayers, W.J., Hanson, G.N., Goldstein, R.H. & Saller, A.H. (2000) Relationship of uranium to petrography of caliche paleosols with application to precisely dating the time of sedimentation. *Journal of Sedimentary Research* **70**, 604–618.

Reeves, C.C. (1983) Pliocene channel calcrete suspenparallel drainage in West Texas. In: Wilson, R.C.L. (Ed.) *Residual Deposits*. Special Publication 11, Geological Society of London, pp. 179–183.

Retallack, G.J. (1994) The environmental approach to the interpretation of paleosols. In: Amundson, R., Harden, J. & Singer, M. (Eds) *Factors of Soil Formation: a Fiftieth Anniversary Retrospective*. Special Publication 33. Madison, WI: Soil Science Society of America, pp. 31–64.

Retallack, G.J. (2000) Depth to pedogenic carbonate horizon as a paleoprecipitation indicator: comment. *Geology* **28**, 572–573.

Rossinsky, V. Jr. & Wanless, H.R. (1992) Topographic and vegetation controls on calcrete formation: Turks and Caicos Islands, British West Indies. *Journal of Sedimentary Research* **62**, 84–98.

Rossinsky, V. Jr., Wanless, H.R. & Swart, P.K. (1992) Penetrative calcretes and their stratigraphic implications. *Geology* **20**, 331–334.

Rowe, P.J. & Maher, B.A. (2000) 'Cold' stage formation of calcrete nodules in the Chinese Loess Plateau: evidence from U-series dating and stable isotope analysis. *Palaeogeography, Palaeoclimatology, Palaeoecology* **157**, 109–125.

Royer, D.L. (1999) Depth to pedogenic carbonate horizons as a paleoprecipitation indicator? *Geology* **27**, 1123–1126.

Royer, D.L., Berner, R.A. & Beerling, D.J. (2001) Phanerozoic atmospheric CO_2 change: evaluating geochemical and paleobiological approaches. *Earth-Science Reviews* **54**, 349–392.

Sanz, M.E. & Wright, V.P. (1994) Modelo alternativo para el desarrollo de calcretas: un ejemplo del Plio-Cuaternario de la Cuenca de Madrid. *Geogaceta* **16**, 116–119.

Schmittner, K.-E. & Giresse, P. (1999) Micro-environmental controls on biomineralization: superficial processes of apatite and calcite precipitation in Quaternary soils, Roussillon, France. *Sedimentology* **46**, 463–476.

Semeniuk, V. & Meagher, T.D. (1981) Calcrete in Quaternary coastal dunes in south Western Australia: a capillary-rise phenomenon associated with plants. *Journal of Sedimentary Petrology* **51**, 47–68.

Sheldon, D.A. & Retallack, G.J. (2004) Regional paleoprecipitation records from the Late Eocene and Oligocene of North America. *Journal of Geology* **112**, 487–494.

Sheldon, N.D., Retallack, G.J. & Tanaka, S. (2002) Geochemical climofunction from North American soils and application to paleosols across the Eocene–Oligocene boundary in Oregon. *Journal of Geology* **110**, 687–696.

Slate, J.L., Smith, G.A., Wang, Y. & Cerling, T.E. (1996) Carbonate-paeosol genesis in the Plio-Pleistocene St. David Formation, southeastern Arizona. *Journal of Sedimentary Research* **66**, 85–94.

Spotl, C. & Wright, V.P. (1992) Groundwater dolocretes from the Upper Triassic of the Paris Basin, France: a case study of an arid, continental diagenetic facies. *Sedimentology* **39**, 1119–1136.

Stiles, C.A., Mora, C.I. & Driese, S.G. (2001) Pedogenic iron-manganese modules in Vertisols: a new proxy for paleoprecipitation? *Geology* **29**, 943–946.

Strong, G.E., Giles, J.R.A. & Wright, V.P. (1992) A Holocene calcrete from North Yorkshire, England: implications for interpreting palaeoclimates using calcretes. *Sedimentology* **39**, 333–347.

Tandon, S.K. & Andrews, J.E. (2001) Lithofacies associations and stable isotopes of palustrine and calcrete carbonates: examples from an Indian Maastrichtian regolith. *Sedimentology* **48**, 339–355.

Tandon, S.K. & Narayan, D. (1981) Calcrete conglomerate, case-hardened conglomerate and cornstone: a comparative account of pedogenic and non-pedogenic carbonates from the continental Siwalik Group, Punjab, India. *Sedimentology* **28**, 353–367.

Thiry, M., Schmitt, J-M. & Simon-Coinçon, R. (1999) Problems, progress and future research concerning palaeoweathering and palaeosurfaces. In: Thiry, M. & Simon-Coinçon, R. (Eds) *Palaeoweathering, Palaeosurfaces and Related Continental Deposits*. Special Publication 27, International Association of Sedimentologists. Oxford: Blackwell Science, pp. 3–17.

Verrecchia, E.P. (1987) Contexte morpho-dynamique des croutes calcaires: apport des analyses sequentielles a l'echelle microscopique. *Zeischrift für Geomorphologie* **31**, 179–193.

Verrecchia, E.P. (1990) Litho-diagenetic implications of the calcium oxalate-carbonate biochemical cycle in semi-arid calcretes, Nazareth, Israel. *Geomicrobiology* **8**, 87–99.

Verrecchia, E.P. & Verrecchia, K.E. (1994) Needle-fiber calcite: a critical review and a proposed classification. *Journal of Sedimentary Research* **A64**, 650–664.

Verrecchia, E.P., Dumont, J-L. & Verrecchia K. E. (1993) Role of calcium oxalate biomineralization by fungi in the formation of calcretes: a case study from Nazareth, Israel. *Journal of Sedimentary Research* **63**, 1000–1006.

Verrecchia, E.P., Freytet, P., Verrecchia, K.E. & Dumont, J.L. (1995) Spherulites in calcrete laminar crusts: biogenic CaCO₃ precipitation as a major contributor to crust formation. *Journal of Sedimentary Research* **65**, 690–700.

Vogt, T. & Corte, A.E. (1996) Secondary precipitates in Pleistocene and present cryogenic environments (Mendoza Precordillera, Argentina, Transbailalia, Siberia, and Seymour Island, Antarctica). *Sedimentology* **43**, 53–64.

Vogt, T. & Del Valle, H.F. (1994) Calcrets and cryogenic structures in the area of Puerto Madryn (Chubut, Patagonia, Argemtina). *Geografiska Annaler* **76A**, 57–75.

Wang, Y., McDonald, E., Amundson, R., McFadden, L. & Chadwick, O.A. (1996). An isotopic study of soils in chronological sequences of alluvial deposits, Providence mountains, California. *Bulletin of the Geological Society of America* **108**, 379–391.

Watson, A. & Nash, D.J. (1997) Desert crusts and varnishes. In: Thomas, D.S.G. (Ed.) *Arid Zone Geomorphology: Process, Form and Change in Drylands.* Chichester: John Wiley & Sons, pp. 69–107.

Watts, N.L. (1980) Quaternary pedogenic calcrete from the Kalahari (southern Africa): mineralogy, genesis and diagenesis. *Sedimentology* **27**, 661–686.

Wright, V.P. (1989) Terrestrial stromatolites: a review. *Sedimentary Geology* **65**, 1–13.

Wright, V.P. (1990) Estimating rates of calcrete formation and sediment accretion in ancient alluvial deposits. *Geological Magazine* **127**, 273–276.

Wright, V.P. (1992) Paleopedology: stratigraphic relationships and empirical models. In: Martini, I.P. & Chesworth, W. (Eds) *Weathering Soils and Paleosols.* Amsterdam: Elsevier, pp. 475–499.

Wright, V.P. (1994) Paleosols in shallow marine carbonate sequences. *Earth-Science Reviews* **35**, 367–39.

Wright, V.P. & Alonso-Zarza, A. (1990) Pedostratigraphic models for alluvial fan deposits: a tool for interpreting ancient sequences. *Journal of the Geological Society of London* **147**, 8–10.

Wright, V.P. & Peeters, C. (1989) Origins of some early Carboniferous calcrete fabrics revealed by cathodoluminescence: implications for interpreting the sites of calcrete formation. *Sedimentary Geology* **65**, 345–353.

Wright, V.P. & Tucker, M.E. (1991) Calcretes: an Introduction. In: Wright, V.P. & Tucker, M.E. (Eds) *Calcretes.* Reprint Series **2**, International Association of Sedimentologists. Oxford: Blackwell, pp. 1–22.

Wright, V.P., Platt, N.H. & Wimbledon, W.A. (1988) Biogenic laminar calcretes: evidence of root mats in paleosols. *Sedimentology* **35**, 603–620.

Wright, V.P., Turner, M.S., Andrews, J.E. & Spiro, B. (1993) Morphology and significance of super-mature calcretes from the Upper Old Red Sandstone of Scotland. *Journal of the Geological Society of London* **150**, 871–883.

Wright, V.P., Platt, N.H., Marriott, S.B. & Beck, V.H. (1995) A classification of rhizogenic (root-formed) calcretes, with examples from the Upper Jurassic–Lower Cretaceous of Spain and Upper Cretaceous of southern France. *Sedimentary Geology* **100**, 143–158.

Wright, V.P., Beck, V.H. & Sanz-Montero, M.E. (1996) Spherulites in calcrete laminar crusts: biogenic $CaCO_3$ precipitation as a major contributor to crust formation. Discussion. *Journal of Sedimentary Research* **66**, 1040–1041.

Yaalon, D.H. (1988) Calcic horizons and calcrete in Aridic soils and paleosols: progress in last twenty two years. *Soil Science Society of America Agronomy Abstracts.* (Cited in Wright and Tucker, 1991.)

Chapter Three

Laterite and Ferricrete

Mike Widdowson

3.1 Introduction

The term laterite, in its general sense, is given to a range of iron-rich, sub-aerial weathering products that develop as a result of intense, substrate alteration under tropical or sub-tropical climates. Physically, many laterites are rock-like, yet they cannot be easily placed into any of the major petrological groupings. Neither do they lend themselves readily to description as 'soils', other from the fact that they are the products of processes operating at the atmosphere – substrate interface. They are perhaps best considered to be metasomatic rock materials; i.e. rocks with chemical, mineralogical and physical characteristics that have been substantially changed by low-temperature and pressure alteration processes such as those operating under subaerial conditions. The conditions under which lateritic profiles form are primarily:

- a favourable climate, typified by seasonal, high annual rainfall (e.g. a monsoon-like climate) – high humidity and high mean annual temperatures further promote chemical weathering and mineral alteration;
- a favourable geomorphological environment, characterised by limited runoff and lack of aggressive erosion – the ingress of rainfall and/or the establishment of a water table may promote element enrichment and depletion processes;
- relative tectonic stability, characterised by minimal uplift and crustal deformation.

The manner in which ferricrete alteration profiles evolve differs from laterite weathering profiles in a number of ways: a genetic distinction between laterite and ferricrete is adopted here (Figure 3.1).

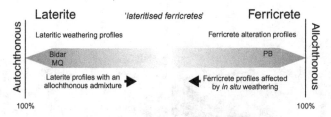

Figure 3.1 Schematic diagram showing the laterite–ferricrete genetic relationship, and the natural continuum between the autochthonous (i.e. *in situ* weathering profiles) and allochthonous end-members. Bidar and MQ (Merces Quarry) are two laterite weathering profiles from India, and PB (Palika Ba) is a ferricrete alteration profile from the Gambia (see section 3.3.2).

3.1.1 Nature, general characteristics and classification issues

Laterites *sensu stricto* comprise an important subset of the wider range of ferruginous and related aluminous weathering products (which also include mechanically resistant iron-cemented ferricretes and aluminium-rich bauxites). The material most commonly recognised as laterite typically forms the uppermost indurated surface layer of an extreme type of weathering profile developed under tropical or sub-tropical climates. Where preserved, this layer may extend over areas of thousands of square kilometres (Figure 3.2). Consequently, laterite is perhaps the most widespread of all duricrusts, with lateritic materials, or their eroded derivatives, covering about one-third of the emerged lands (Tardy, 1993).

The primary controls on lateritic profile development are the parent lithology, prevailing climate, and regional geomorphological and tectonic stability. Lateritic weathering profiles can develop from the chemical weathering of a variety of sedimentary, metamorphic and igneous parent (i.e. protolith) rocks. However, protoliths that have inherently higher iron contents are most suited to the formation of laterite (see section 3.5). By contrast, rock types such as limestone and sandstones less readily alter to laterite, although allochthonous ferruginous cementation can instead develop within these to form ferricretes. Climatic, tectonic and geomorphological evolution over periods of 10^6 years can also be important controls. Accordingly, many lateritic weathering profiles are not simply the result of equilibrium with a single set of prevailing environmental conditions; once formed, weathering profiles remain exposed to the environment and may continue evolving indefinitely.

There is a considerable and complex literature concerning the classification of laterite. These range from qualitative systems based upon physical properties (Buchanan, 1807; Sivarajasingham et al., 1962) or the envisaged processes of formation (e.g. pedogenetic and groundwater laterites;

Figure 3.2 Examples of mesa-like remnants of a Late Cretaceous lateritised palaeosurface developed on Deccan basalt from widely separated localities across the Maharashtra Plateau, western India. On the eastern reaches of the Deccan: (A) Bidar (17°55′N, 77°33′E); (B) Bidar area (17°53′N, 77°36′E); on the western reaches of the Deccan *c.* 400 km from Bidar, (C) Panchgani (17°56′N, 73°49′E); (D) Patan (17°23′N, 73°56′E). In all these areas, the mechanically resistant upper layers of the laterite profile typically form a protective capping to the less altered materials beneath and, following erosion, producing a characteristic cliff-like morphology.

McFarlane, 1976, 1983a), through to mineralogical and textural based schemes (e.g. Aleva, 1986; Bardossy and Aleva, 1990), to a system incorporated within soil taxonomy (Soil Survey Staff, 1975), as well as more quantitative, chemically based genetic classifications (Schellmann, 1982). It is not the intention of this chapter to discuss the merits of these schemes, nor to outline the others that have been proposed; this has been reviewed previously (e.g. McFarlane, 1976, 1983a; Aleva, 1994; Bourman and Ollier, 2002). Instead, a simple distinction is made between laterite and ferricrete (Aleva, 1994), and a chemical classification for defining lateritic materials (Schellmann, 1982, 1986) is adopted.

Following Aleva (1994), laterites *sensu stricto* are defined primarily as residual materials formed by rock breakdown *in situ*. They should not contain any significant allochthonous (i.e. externally derived) components. In this definition, laterites owe their composition to the relative enrichment of iron (and often aluminium) and other less mobile elements. The lateritisation process occurs because parent rock exposed at the surface is subject to weathering, but the degree of alteration diminishes with depth, so producing a laterite weathering profile (Figure 3.3). Accordingly, such profiles consist of an upward progression from unaltered protolith at the base to an indurated, iron-rich duricrust at the top (section 3.4).

Ferricretes do not achieve their high iron content through residual enrichment, but rather through the net input and absolute accumulation of allochthonous iron into a host rock or existing weathering profile. This pattern of enrichment and associated physical change is better termed an alteration profile. Elements such as Fe, Al and some trace metals can be redistributed or introduced into the evolving profile by groundwater, either in solution or as chelates (metallic ions attached to organic compounds). These can then be redeposited at specific levels within the alteration profile by reactions associated with variations in groundwater chemistry, Eh and pH, and/or water-table fluctuations. In effect, these iron-rich compounds invade the parent rock, then fill available pore spaces and/or replace the parent mineralogy in the profile. In the field, the distinction between laterite and ferricrete can often be subtle (Figure 3.1), and historically this has led to confusion in terms of description, nomenclature, and classification.

The division into laterite and ferricrete used in this chapter represents a useful process-based distinction, but the practicality of determining whether mineral components of a profile are allochthonous or autochthonous is problematic because many lateritic weathering profiles are subsequently modified by the introduction of allochthonous materials. Conversely, once formed, ferricretes can be subject to weathering processes *in situ* and evolve toward more lateritic-type profiles. Nevertheless, the distinction between dominantly autochthonous weathering profiles or allochthonous alteration profiles is an important one because it places constraints upon the processes operating during duricrust evolution, and also upon contemporaneous climatic and geomorphological conditions.

The Schellmann system (1982, 1986) is a quantitative approach which compares the extent of chemical alteration of a weathering product to that of its protolith in order to define increasing degrees of lateritisation (see section 3.5.4). Significantly, the scheme recognises the importance of the nature of the protolith upon the composition of the weathering product; it also requires that the weathering products are entirely the result of *in situ* alteration processes. Using a SiO_2, Fe_2O_3, Al_2O_3 ternary diagram, four stages of alteration are recognised in the lateritisation process: (a) kaolinitisation, (b) weak lateritisation, (c) moderate lateritisation, and (d) strong lateritisation. The relative positions of these stages on the ternary diagram vary according to the chemical composition of the protolith. As a result, the Schellmann scheme proves useful for determining the degree of weathering for laterites *sensu stricto*. However, it cannot be applied to ferricretes since these involve allochthonous inputs of material into an alteration profile (Bourman and Ollier, 2002; Schellmann, 2003).

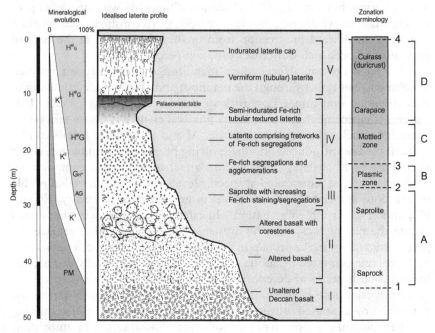

Figure 3.3 Generalised vertical section through the autochthonous Bidar laterite weathering profile (17°55′N, 77°32′E), illustrating the compositional and textural progression from the unaltered basalt protolith to indurated laterite. Different lateritic textures may be identified at various levels, with one texture merging into the next. A broad zonal distinction is as follows: (I) unaltered basalt progressing up-profile to material displaying limited alteration of primary minerals; (II) altered basalt containing 'unaltered' corestones within a soft saprolitic matrix, and progressing upward into (III) a saprolitic to weakly lateritised zone in which some primary lithological characteristics remain, but becoming increasingly obscured due to an up-profile increase of Fe-mottling and segregations; (IV) moderately lateritised zone comprising iron segregations and vermiform textures, and becoming increasingly indurated up-profile; (V) strongly indurated laterite with vermiform textures, becoming highly ferruginous in the uppermost levels of the profile. A similar alteration progression is commonly observed on other protolith substrates. The left-hand column shows the idealised vertical distribution of minerals through the profile (after Thomas, 1994): PM, parent minerals; K^1 to K^3, well-crystallised to poorly crystallised kaolinite; AG, amorphous goethite; G, goethite; $H^{al}G$, aluminous haematite and goethite. The symbol size illustrates the relative amount of each iron compound. The right-hand column shows the terminology of the various zones and alteration fronts within a typical laterite weathering profile (after Aleva, 1994): 1, *weathering front* – Saprolitic Zone (A), common relict parent rock fabrics, kaolinitic clay, and relict/etched quartz (saprock <20% primary minerals weathered; saprolite >20% primary minerals weathered); 2, *pedoplasmation front* – Plasmic Zone (B), absence of any parent rock fabrics, kaolinitic clay with relict/etched quartz (>90% primary minerals weathered); 3, *cementation front* – Mottled Zone (C), development of secondary structures and fabrics, with ferruginous (goethitic and haematitic) accumulations increasing in size and frequency upwards, set in a kaolinitic matrix; Lateritic Zone (D), increasing amalgamation of ferruginous accumulations (carapace) eventually coalescing to form an interconnected three-dimensional structure of mechanically resistant haematitic and goethitic masses in the uppermost levels (cuirass/duricrust) – any remaining tubes and voids are infilled with soft/friable Al-rich goethite or kaolinitic/gibbsitic material; 4, *surface erosional/weathering front*.

3.1.2 History of study

The properties of laterite, and its place within the wider classification of geomaterials, have been debated for 200 years. The term laterite (literally 'brick rock') was first given in 1807 by Francis Buchanan following a journey through western Peninsular India. He devised the term to describe a naturally occurring material that, once cut and allowed to harden, could be used as a building material. Buchanan had first observed the cutting of laterite bricks (c. 1800) at a quarry at Angadipuram (10°58′N, 76°13′E) in the Palghat region, but later noted that the practice was more widespread (i.e. throughout modern Kerala and Karnataka states). In fact, long before Buchanan's description, the Dutch East India Company had been using laterite to construct their fortresses and, similarly, the Moghuls and Mahrattas before them. However, its usage as a building stone (Figure 3.4A, B) and a source of smelting iron by native inhabitants in India and elsewhere considerably pre-dates these examples.

A more rigorous investigation of the lateritisation process was provided by Newbold (1844, 1846) from studies of laterite-capped plateaux of Deccan basalt near Bidar, India (17°54′N, 77°32′E). Newbold was the first to suggest that laterite developed as an *in situ* weathering product, but, by this time, Buchanan's description had become widely adopted as the 'type definition'. Buchanan's definition has ultimately proved inadequate and has caused considerable confusion throughout the 19th and 20th centuries. Problems of co-ordinating laterite and ferricrete description stem not only from investigation by a variety of different individuals and disciplines, but also from the development of extensive anglophone and francophone descriptive terminologies. The situation was not helped by the misinterpretation of the term 'ferricrete' (as used by Lamplugh, 1907) to describe 'an iron-rich crust'. Despite repeated attempts to define laterite and ferricrete, the terms have been used, often interchangeably, to describe a wide range of iron-rich weathering products and duricrusts. The genetic approach outlined earlier helps alleviate this problem.

With the rapid development of ideas in recent decades, particularly in geomorphology and geochemistry, greater consensus regarding the origin and composition of laterite has now emerged. Modern research is based upon the identification of common physical, chemical and mineralogical attributes and processes of formation. Field studies have also confirmed the genetic distinction between laterite and ferricrete. Interest in lateritisation was given renewed impetus as the result of the IGCP-129 *Lateritization Processes* (1975–83) and IGCP-317 *Paleoweathering Records and Paleosurfaces* programmes (cf. Widdowson, 1997a; Thiry and Simon-Coinçon, 1999), and the establishment of CORLat, a repository for type material (Moura, 1987). Modern analytical techniques (Chapter 13) have also provided a wealth of geochemical, isotopic and mineralogical data, which have

Figure 3.4 Examples of laterite and ferricrete profiles. (A) Merces Quarry near Panjim, Goa (15°29′N, 73°53′E), where an entire laterite weathering profile developed upon Proterozoic greywacke is exposed. (B) Laterite quarrying at Bidar (17°55′N, 77°33′E), showing recently cut sections (with 1 m rule for scale) where material from the upper parts of the profile (zone IV on Fig. 3) has been extracted for laterite bricks. The Bidar laterite localities were used by Newbold (1846) to argue for laterite as an *in situ* weathering product. (C) Ferricrete profile exposed in a road cutting to the south of Palika Ba (13°28′N, 15°14′W), near the Gambia River, Gambia, West Africa.

advanced our understanding of tropical weathering processes, and their response to, and effect upon, climate change.

3.2 Distribution, Field Occurrence and Geomorphological Relations

The global distribution of laterite and ferricrete is difficult to determine because few maps distinguish between laterites and ferricretes in the manner used in this chapter. However, iron-rich duricrusts have been documented on every continent except Antarctica (Goudie, 1973). In India, laterites and ferricretes are most widespread along the southwest coastal belt, beginning at Mumbai but particularly southwards from Ratnagiri (17°N), and also form prominent outcrops inland across the Maharashtra plateau (e.g. Foote, 1876; Fox, 1927, 1936; Prescott and Pendleton, 1952; Goudie,

1973; Sahasrabudhe and Deshmukh, 1981; Valeton, 1983; Ollier and Powar, 1985; Kumar, 1986; Brückner and Bruhn, 1992; Devaraju and Khanadali, 1993; Widdowson, 1997b; Kisakürek et al., 2004). Laterites and ferricretes in this region are commonly associated with the basalts of the Deccan Plateau, but also occur on other lithologies in the states of Goa, Karnataka and Kerala (Widdowson and Gunnell, 1999). Elsewhere in Asia, laterites and ferricretes have been documented in Cambodia, Indonesia, Malaysia, the Philippines, Sri Lanka and Thailand (e.g. Grubb, 1963; Eyles, 1967; Goudie, 1973; Debaveye and de Dapper, 1987). In Australia, laterite and ferricrete are well developed in the west, south and southeast of the continent, as well as in parts of Northern Territory and Queensland (e.g. Prescott and Pendleton, 1952; Connah and Hubble, 1960; Dury, 1967; Grant and Aitchison, 1970; Stephens, 1971; Gilkes et al., 1973; Alley, 1977; Twidale, 1983; Milnes et al., 1985; Bourman et al., 1987; Ollier, 1991; Bourman, 1993, 1995; Firman, 1994; Young et al., 1994; Benbow et al., 1995; Alley et al., 1999; Anand and de Broekert, 2005).

In Africa, iron-rich crusts are extensive on many of the palaeosurfaces developed on the continental mainland and offshore islands (e.g. Prescott and Pendleton, 1952; King, 1953; Goudie, 1973; McFarlane, 1976, 1983a,b). Lateritisation is most evident in a zone stretching from Senegal, Gambia, and Sierra Leone in the west, to Kenya, Tanzania and Uganda in the east, and has also been documented in Nigeria, Chad and the Sudan (de Swardt, 1964; McFarlane, 1971, 1976, 1983a; Grandin, 1976; Bowden, 1987, 1997; Gehring et al., 1992, 1994; Brown et al., 1994, 2003; Zeese et al., 1994; Zeese, 1996; Schwarz and Germann, 1999; Taylor and Howard, 1999; Gunnell, 2003). Further south, iron-rich duricrusts occur in the southern Democratic Republic of Congo, Angola, Zambia, Zimbabwe, Mozambique, South Africa and Madagascar (e.g. Alexandre and Alexandre-Pyre, 1987; Partridge and Maud, 1987; Marker and McFarlane, 1997; Marker et al., 2002).

Indurated laterite is uncommon in North America, but is widespread in Nicaragua and on plateau surfaces in Brazil, where it has been the focus of economic interest, and adjacent parts of Venezuela, Bolivia, Guyana, French Guiana, Surinam and Uruguay (e.g. Prescott and Pendleton, 1952; Valeton, 1983; LaBrecque and Schorin, 1987; O'Connor et al., 1987; Girard et al., 1997; Théveniaut and Freysinnet, 1999). Elsewhere, tropical-type deep weathering materials have been recognised as far north as Norway and Fennoscandia (Lidmar-Bergström et al., 1997, 1999; Whalley et al., 1997). Laterite *sensu stricto* has rarely been documented in Europe, with the exception of pockets of eroded lateritic materials which were stripped from a pre-Alpine etchplain formerly covering southern Germany (Borger, 2000; Borger and Widdowson, 2001).

The chemical and physical durability of lateritic duricrust has meant that it has played a prominent role in the evolution of tropical and sub-tropical landscapes (e.g. McFarlane, 1976; Bowden, 1987, 1997; Summerfield, 1991; Thomas, 1994; Widdowson and Cox, 1996; Widdowson, 1997c). A close association between palaeosurface formation and lateritic weathering profile development occurs primarily because both require protracted sub-aerial exposure, relative tectonic quiescence and a degree of climatic stability in order to develop (Widdowson, 1997c). Rapid uplift and/or climate change is likely to terminate the lateritisation process and lead to rapid erosion of the weathering profile. However, the development of thick, indurated laterites can also be dependent upon the slow incision of low-relief palaeosurfaces and may not, therefore, be coeval with palaeosurface formation (Thomas, 1994). Ferricretes, by contrast, develop preferentially in topographic depressions such as former lakes, swamps, valley floors, valley sides and peritidal zones or estuaries (Ollier, 1991; Bourman, 1993; Widdowson and Gunnell, 1999). This is because their formation is dependent upon the accumulation of iron-rich debris or solutes transported from higher landscape positions (Widdowson and Gunnell, 1999; see section 3.4). Where the erosion of less resistant material surrounding an indurated ferricrete occurs, the duricrust may act as a caprock and become preserved as inverted relief. In Australia, mesas linked together in a dendritic form represent inverted valley floor ferricretes (Ollier et al., 1988; Ollier and Pain, 1996).

The formation of lateritic materials has probably occurred since oxygenation of the atmosphere in the early Proterozoic. However, as laterite development is dependent upon the coincidence of conducive climatic, geomorphological and tectonic factors, it is difficult to predict periods when accelerated lateritisation might have occurred through examination of any one of these factors alone. Nevertheless, it is likely that global climate and atmospheric changes had a profound effect upon weathering rates in the geological past (Schmitt, 1999). During times of elevated CO_2, greater areas of the Earth's surface are likely to have been available for tropical-type deep weathering processes (Berner, 1991, 1994). Amongst the best known of these greenhouse episodes were those occuring during the Late Cretaceous and late Paleocene to early Eocene, when CO_2 levels were at least twice that of today; worldwide, there are extensive weathering profiles of this age (Bardossy, 1981; Valeton, 1983).

Conversely, periods of orogenesis are likely to curtail widespread laterite formation, since the associated deformation and uplift would promote mechanical erosion. Orogeny is also often associated with increasing continentality and the aridification of continental interiors, thereby limiting the effectiveness of lateritisation processes. The poor preservation potential of lateritic materials, due to the weaker nature of lower levels of the weathering profile (section 3.3), is also likely to have resulted in underrepresentation in the geological record. Periods of erosional stripping may mean that the only record of a once widespread laterite is the occurrence of eroded duricrust remnants (Borger and Widdowson, 2001). Accordingly, with the exception of the late Carboniferous and Early Permian when CO_2 levels were around half the present atmospheric levels (PAL), the paucity of laterite occurrences in the Mesozoic (twice to more than four times PAL) and Lower Palaeozoic (more than ten times PAL) may be an apparent rather than a real observation. The earliest examples of laterite are Proterozoic in age (e.g. Gamagara Formation, South Africa; Gutzmer and Beukes, 1998). These are important since the ferric iron they contain confirms an early oxygenation of the atmosphere. Prior to this, weathering horizons would have been characterised by the generation of ferrous rather than ferric iron, thus promoting a net loss of surficial iron and relative accumulation in the lowest levels of ancient profiles.

3.3 Laterite and Ferricrete at the Profile Scale

Many laterite and ferricrete profiles are >10–20 m in thickness. They have many structures and textures in common, and superficially can appear similar at outcrop. However, careful examination of their internal structure and arrangement can often aid in their discrimination.

3.3.1 Lateritic weathering profiles

Lateritic duricrusts are typically manifest as the uppermost layers of *in situ* tropical-type weathering profiles (Figure 3.3). Where these profiles are fully exposed, such as in western India, they typically display an uninterrupted series of zones from unaltered basal bedrock to highly indurated duricrust (Figure 3.5A–D). These zones are separated by interfaces which are commonly discernible in outcrop from observable changes in colour, mineralogy and texture.

The weathering front at the base of the profile is the junction between the unaltered bedrock and overlying chemically weathered materials (i.e. the base of the saprock zone). In many instances, this transition occurs over a narrow vertical interval (i.e. <1 m). A zone containing a mixture of weathered material and unweathered 'core stones' typically occurs immediately above the weathering front, passing upward into a saprolitic zone in which structures and crystal pseudomorphs from the protolith may still be recognised. Next, a microaggregated texture develops, known as the 'mottled zone', consisting of kaolinite particles along with crystals of iron oxyhydroxides. Such a texture promotes good vertical drainage and further accelerates the alteration process. Under alternating dry and wet conditions, this zone becomes characterised by a strong accumulation of iron oxides. The altered and iron-enriched zones begin with reddening due to the replacement of primary minerals by iron-rich secondary minerals and

Figure 3.5 Examples of weathering and lateritic textures at key horizons through the Merces Quarry lateritic weathering profile (starting at the base of the profile). (A) Slightly altered light grey greywacke, retaining original sedimentary structures (i.e. saprock). (B) Reddened greywacke, still retaining original sedimentary structures (i.e. saprolite). (C) Mottled zone with Fe concretions forming a nodular texture. (D) Indurated, vermiform laterite consisting of an interlocking framework of Fe and Al oxyhydroxides.

the development of minor iron segregations. These progress to nodular accumulations which may, up-profile, undergo further coalescence. This eventually culminates as an indurated 'vermiform' or 'tubular' laterite comprising a massive, interlocking fretwork of Fe- (and Al-) oxides and hydroxides at the top of the profile. The precise nature of the profile is, however, a response to climatic, geomorphological and geological control, and to long-term changes in external conditions. As such, this idealised profile can be complicated by rock structures, the groundwater regime, and the availability and accumulation of laterally derived solutes.

Laterite weathering profiles are best developed in landscapes character-ised by near-horizontal surfaces, such as those formed by long-term exogenic (erosional) or endogenetic processes (e.g. surfaces presented by areally extensive lavas; Widdowson, 1997c). These environments are char-acterised by limited runoff and erosion, and the long-term ingress of rainfall and/or the establishment of a water table may serve to promote the evacu-ation of more mobile constituents (section 3.4). Such conditions are typi-cally satisfied in regions of relative tectonic stability. In many weathering profiles, the deeper levels actually lie below the water table, even where this fluctuates seasonally, and alteration processes in this zone take place under saturated conditions. Since porosity and permeability are often higher in the more weathered parts of the profile, water movement, and the breakdown of primary minerals, may be significantly slower in the lowest levels comprising the saprock and saprolite.

Rather than simply regarding the lateritic weathering profile as a static system with its horizons defined by fixed water-table relationships, it is useful to consider it as a descending column in which horizons are trans-formed and reformed as both the ground surface and weathering front are lowered (Trendall, 1962; Bremer, 1981; Figure 3.6). The degree of surface lowering over time is difficult to quantify, but it is generally recognised that, whereas the transformation from unaltered rock to saprolite is essentially isovoluminous (i.e. preserving rock fabric and mineral structures), signifi-cant volume reductions can occur in the upper zones of the profile. These may be as much as 70%, largely due to chemical and mechanical loss of material liberated by the breakdown of primary minerals. If surface lower-ing, through volume collapse within the profile or erosional sheet-wash of the upper layers, keeps pace with the downward advance of the weathering front, then the landscape as a whole will continue to lower. However, these conditions may represent a special situation, and more probably the down-ward advancement of the weathering front will outpace surface lowering until it lies at the limit of penetration of active groundwaters (this typically being some metres below an established water table). Under such condi-tions, the downward progression of horizons will halt, until either surface erosion removes the upper layers and/or the onset of fluvial incision causes the water table to fall. Where the progression of the weathering front is

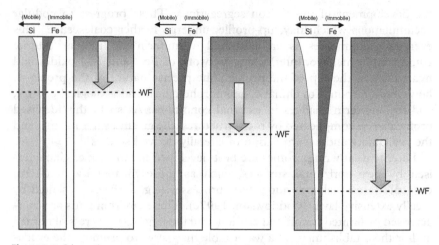

Figure 3.6 Schematic representation of the downward advancement of the weathering front, showing the relative changes in abundances (shaded) of the major lateritic components, Si and Fe, during profile evolution.

halted, the main modification to the profile will occur at or near the water table through the removal and introduction of allochthonous elements and compounds (Figure 3.7). Evidence of established palaeowater tables has been recorded in laterite profiles in India of different ages and developed upon different protoliths (e.g. Kisakürek et al., 2004), thus confirming the ubiquity of this process. In effect, this encompasses, in a single evolutionary model, the pedogenetic and groundwater laterite types identified by McFarlane (1976, 1983a).

Wherever lateritisation effects have become widespread, this 'descending column' model usefully couples laterite development with the long-term effects of landscape lowering, as required by models of tropical etchplanation (e.g. Büdel, 1980). However, this situation will end when incision proceeds beyond the rate at which the weathering front can keep pace with the falling water table. At this point, most lateritic profiles will become 'fossilised'. In such instances, the destruction and regeneration of neoformed minerals will end, preserving the chemical and mineralogical composition, and the ambient magnetic field will also become frozen in the once active parts of the profile. This behaviour has proved extremely useful in determining the ages of lateritic weathering profiles (section 3.8).

3.3.2 Ferricrete alteration profiles

In contrast to lateritic weathering profiles, ferricrete alteration profiles incorporate materials non-indigenous to the immediate locality of duricrust

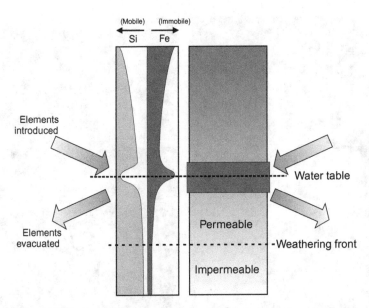

Figure 3.7 Schematic representation of changes in element abundances in a lateritic weathering profile affected by the establishment of a water table. This illustrates how an iron-enriched and silica depleted zone may develop within the weathering profile.

formation (Figure 3.4C). In mechanically derived ferricrete, transported materials can be readily identified as clasts derived from adjacent terrains or earlier generations of laterite or ferricrete. Importantly, the term ferricrete should also be extended to those materials whose constituents have been substantially augmented by the precipitation or capture of elements and compounds from allochthonous fluids. Although it is the allochthony of the ferricrete constituent materials which justify its appellation, determining whether the introduction of such fluids has taken place, and confirming their source, is often problematic. However, since ferricretes may develop as foot-slope accumulations, or within topographic depressions (Figure 3.8), they can often be distinguished by their obvious discordance with the underlying, relatively unaltered, substrate. In effect, they do not display the progressive weathering profile characteristic of many laterites (Figure 3.3).

3.3.3 The development of typical 'lateritic' textures

Typical to both ferricrete and lateritic duricrusts are 'tubiform' or 'vermiform' textures (McFarlane, 1976; Figures 3.3, 3.5D and 3.8B). These consist of a coherent indurated skeleton or 'fretwork' of primarily

Figure 3.8 Examples from the ferricrete alteration profile observed at outcrop at Palika Ba (13°28′N, 15°14′W), near the Gambia River, Gambia, West Africa. (A) Nodules of ferricrete developing within Quaternary alluvial sands and silts deposited by the Gambia River. Nodules consist of predominantly goethite with minor haematite (sample PG4 in Table 3.3). (B) Massive iron-cemented ferricrete horizon displaying characteristic tubes (i.e vermiform structure; sample PG2 in Table 3.3). Importantly, the iron cement is entirely due to the introduction of allochthonous iron into the pore spaces of the sediment.

secondary iron minerals. Vermiform tubes vary in size, both within and between duricrusts, but provide conduits for the through-flushing of rainfall. This often confers an edaphic aridity to many ferricrete- or laterite-dominated regions, making them useless for agriculture. The tube structure evolves in the upper part of lateritic profiles, and at the zone of ferralitisation within ferricrete profiles, through two processes. First, the increasing content of secondary iron minerals results in the growth of mottles, concretions and nodules that eventually coalesce into a 'skeleton'; those regions remaining interstitial to these tend to initially contain saprolitised remnants (in laterites) or patches of less altered host rock (in ferricretes). The second process arises because these remnants and patches are less resistant and so are removed or altered by further chemical or mechanical attack. This occurs especially in upper layers, where it is not uncommon to find tubes that have been evacuated of their original contents. Through-flushed constituents can migrate downwards into voids in the lower levels of the ferricrete or laterite profiles where they may accumulate into a clay-rich layer.

The other textures found in iron-rich duricrusts are pisoliths (or more strictly pisoids). These are more common in lower duricrust levels and seem to form at or near the water table, possibly as a result of fluctuations in the groundwater level or chemistry; their occurrence may thus be indicative of pronounced seasonality. Pisoids range in size from *c.* 3–30 mm, are generally well-rounded, and contain a core (e.g. primary mineral grains, or neo-formed clay agglomerates) surrounded by a cortex of fine, concentric colour-banded layers (Bardossy and Aleva, 1990). The chemical and mineralogical composition of the layers is variable, but typically consists of micro- or cryptocrystalline haematite/goethite, and kaolinite in more bauxitic examples. Pisolitic layers are not ubiquitous in all laterite and ferricrete profiles, suggesting that development is either an intermediary stage during the formation of the more robust duricrust skeleton, or else is controlled by conditions peculiar to certain landscape settings.

3.4 Mechanisms of Formation

Fundamental to the development of laterite and ferricrete is the formation of insoluble ferric iron derived from the breakdown of a parent mineralogy. In the case of laterites, it is the relative depletion of mobile elements liberated during weathering of the profile that produces the high iron and aluminium content. Clearly, those lithologies which contain quantities of Fe- or Al-bearing primary minerals will, under the correct weathering regime, most readily develop lateritic or bauxitic weathering residua. Accordingly, mafic igneous rocks and immature sediments lend themselves to lateritisation, whereas the weathering of granites or arkoses, rich in feldspars, would seem more conducive to the formation of bauxitic alteration profiles. In fact, the formation of bauxites is also dependent upon specific drainage conditions. In the case of ferricrete development, a donor source of iron is also required, but here an intervening stage where Fe becomes mobilised, transported, and then oxidised to the insoluble form is needed.

The most common rock minerals are metal silicates; in most lithologies these are either feldspars and/or ferromagnesian minerals. In the majority of weathering reactions, rain water or groundwater containing dissolved CO_2 (i.e. forming weak carbonic acid) are the main reactant that drives the breakdown of the protolith. The basic reactions that lead to the liberation of elements from the protolith minerals, and the subsequent loss or accumulation of elements in a weathering profile, can thus be expressed as a series of reactions involving water, CO_2 and/or carbonic acid. Plagioclase feldspars (e.g. calcic anorthite or sodic albite) breakdown by hydrolysis. In the case of albite, the highly mobile Na^+ ion is lost in solution, along with a proportion of the silica which is not then recombined to form kaolinite:

$$2NaAlSi_3O_5 + 3H_2O + CO_2 \rightarrow Al_2Si_2O_5(OH)_4 + 4SiO_2 + 2Na^+ + 2HCO_3^-$$
[albite + water + carbon dioxide → kaolinite + silica + Na and bicarbonate ions in solution]

In a similar manner, anorthite hydrolyses, liberating Ca^{2+}. The calcium ion may be flushed from the weathering profile or retained, depending on the rate of leaching and the availability of other ions in the system with which it may recombine:

$$CaAl_2Si_2O_5 + 3H_2O + CO_2 \rightarrow Al_2Si_2O_5(OH)_4 + Ca^{2+} + 2HCO_3^-$$
[anorthite + water + carbon dioxide → kaolinite + Ca and bicarbonate ions in solution]

Biotite may be also hydrolysed:

$$K_2Mg_6Si_6Al_2O_2O(OH)_4 + 14H^+ + 14HCO_3^- + H_2O$$
$$\rightarrow Al_2Si_2O_5(OH)_4 + 4Si(OH)_4° + 2K^+ + 6Mg_2 + 14HCO_3^-$$
[biotite + bicarbonate + water → kaolinite + silica + cations and bicarbonate ions in solution]

Any silica not combined as kaolinite typically goes into solution as silicic acid, and may then be lost from the system. However, precipitation of secondary silica is not uncommon in some laterite profiles, where it can become part of pore or void infillings.

$$SiO_2 + H_2O \rightarrow 4Si(OH)_4° \ (or \ H_4SiO_4)$$
[silica + water → silicic acid]

Under weakly acid conditions with sufficient water and free drainage, more silica may be removed, permitting gibbsite to form directly from kaolinite:

$$2Al_2Si_2O_5(OH)_4 + 105H_2O \rightarrow -42Al(OH)_3 + 42Si(OH)_4°$$
[kaolinite + water → gibbsite + silicic acid]

Similarly, gibbsite can form directly from plagioclase (albite):

$$2NaAlSi_3O_5 + 3H_2O + CO_2 \rightarrow Al(OH)_3 + 3H_4SiO_4 + Na^+ + OH^-$$
[albite + water + carbon dioxide → gibbsite + silicic acid + Na and hydroxyl ions]

Kaolinite is an important constituent of many lateritic profiles, and a common product of hydrolysis: it tends to be abundant where there is free-drainage. The reaction progressively separates silica in aqueous solution from aluminium, which remains in the solid phase (as kaolinite or gibbsite).

This process comprises dissolution of alumino-silicate minerals, and is the basis for much of the discussion concerning the development of saprolites and associated weathering products. Although conditions at the weathering front may initially favour the formation of micaceous clays (e.g. chlorite, vermiculite) and smectite, small quantities of gibbsite may also be formed, and, eventually, kaolinite can come to dominate those free-draining profiles where there is an adequate water supply.

Although it can develop direct from plagioclase feldspar, gibbsite formation is commonly regarded as the end-point of hydrolysis. It is thought to form more commonly from the breakdown of kaolinite, and hence is often the product of extreme weathering conditions. The distribution of kaolinite and gibbsite in lateritic and bauxitic terrains is far more complex than can be described by the general equations above (Tardy and Roquin, 1992). Rather, the occurrence of these particular Al-sesquioxides appears to be related to the activity of water, which is, in essence, a measure of its chemical potential to generate hydration and dehydration reactions. The activity of water is also controlled by the capillarity of the substrate, and so hydrological factors play an important role in determining the distribution of Al-sequioxide mineralogies in thick, lateritic weathering profiles. According to Kronberg et al. (1982), the formation of gibbsite (the principal mineral in bauxite) is accompanied by an increase in porosity due to the net loss of material during substrate alteration.

The presence of iron in the weathering reactions is fundamental to laterite development and also derives from the breakdown of iron silicates. One of the simplest ferromagnesian mineral structures is that of fayalite (Fe-rich olivine), which is a common silicate in many mafic rocks; according to Curtis (1976), ferric iron (Fe^{3+}) may be produced directly from a simple oxidation and hydration reaction:

$$2FeSiO_3 + (O) + 4H_2O \rightarrow Fe_2O_3 + 2Si(OH)_4^{\circ}$$
[iron + oxygen + water → ferric iron oxide + silica in solution]

However, in the presence of carbonic acid, Fe^{2+} is released into the weathering system by hydrolysis, and may then be oxidised to Fe_2O_3 (Krauskopf, 1967). In fact, reactions involving CO_2-rich rain- and groundwaters are likely to be the more common. Again, as an example, the effect of this acid upon a silicate such as fayalite can be used, but the principle also holds for Fe liberation from more complex silicates such as pyroxenes and amphiboles. As far as these silicate minerals are concerned, the two-stage weathering reaction can be summarised as:

$$Fe_2SiO_4 + 4H_2CO_3 \rightarrow 2Fe^{2+} + 4HCO_3^- + H_4SiO_4$$
[fayalite + carbonic acid → ferrous iron in solution + bicarbonate ions + silicic acid]

During the first stage of weathering, Fe^{2+} ions are liberated in solution, along with bicarbonate ions (HCO_3^-). Other elements present in these minerals, such as Na, can also liberate soluble ions. However, in the presence of dissolved oxygen, any Fe^{2+} ions are then immediately precipitated in the form of highly insoluble Fe^{3+} ions:

$$4Fe^{2+} + O_2 + 8HCO_3^- + 4H_2O \rightarrow 2Fe_2O_3 + 8H_2CO_3$$
[ferrous iron in solution + bicarbonate ions + water → insoluble ferric iron + carbonic acid]

The subsequent behaviour of iron in weathering systems is often complex because it is also affected by reduction and oxidation conditions, as well as by pH. Nevertheless, Fe is usually initially liberated during the breakdown of silicate minerals as soluble ferrous iron (Fe^{2+}). However, excess oxygen at the surface and within the weathering profile allows much of the iron to become quickly oxidised to the ferric form (Fe^{3+}) and become fixed, typically as haematite or goethite. The depth to which fixation penetrates depends on a number of factors. These include the partial pressure of CO_2 in the atmosphere and pore spaces, and the porosity and permeability of the protolith and its evolving weathering profile. The behaviour of iron within a profile is further complicated because CO_2 not only dissolves directly from the passage of rainwater through the atmosphere, but can also be delivered directly into groundwater through microbial and soil respiration processes. Whether or not the reaction is completed to generate ferric iron within the profile depends upon CO_2 and O_2 availability, and the availability and/or rate of supply of Fe, since these will control the rate of consumption of reactants. Thus, the balance between carbonic acid and oxygen in porewater and groundwater is of crucial importance in determining the pattern of iron distribution in lateritic profiles.

For instance, it may be more difficult for oxygen to diffuse into deeper parts of a profile and, with increasing depth, conditions may change to reducing, resulting in the removal of this ferrous iron from within deeper parts of the profile. Certainly, liberation and removal of iron takes place at some levels even within developing lateritic profiles. This can occur especially at or near the water-table level, where limited oxygen supply and/or microbial action permit some of the ferrous Fe generated there to be removed by groundwater. Such transportation can occur over considerable distances laterally, or over only a few metres laterally or vertically. Where this Fe-rich groundwater next meets oxygenated conditions, it then precipitates out into the ferric state. This seems to be the origin of the vertical redistribution of iron within some lateritic profiles, resulting in high Fe accumulations within the vadose zone. Importantly, it may also be the origin of many ferricrete horizons, where iron has been introduced into a receptor substrate.

There are many problems with these simplified representations of weathering reactions, not least because little is understood about the chemical reactions taking place at the crystal surface under field conditions. More-over, organic acids, derived from the microbial breakdown of vegetation colonising the upper parts of the profile, will also enhance silicate weather-ing processes via reactions more complex than those outlined above. Nevertheless, the reactions given offer a framework by which the gross mineralogical changes leading to the development of laterites and bauxites may be understood.

3.5 Mineralogy and Chemistry

The mineralogy of laterites, bauxites and ferricretes is complex, with nearly 200 minerals having now been identified in lateritic materials alone (e.g. Bardossy, 1979; Aleva, 1994). This range derives, in part, from the variety of protolith and/or host–rock compositions, and the numerous transforma-tions that can take place upon their mineral constituents (Tardy et al., 1973; Figure 3.9). Nevertheless, the majority of laterites and ferricretes consist of combinations of less than a dozen 'rock-forming' minerals; this assemblage is dominated by stable secondary minerals, and particularly iron and aluminium oxides and sesquioxides (Table 3.1).

3.5.1 Mineralogical variations within lateritic weathering profiles

Since laterites are primarily autochthonous residues, it is appropriate to first consider the range of available protolith mineralogies. Mafic rock types (e.g. gabbro) provide particularly suitable substrates for the development of laterites because a large proportion of their mineralogy consists of fer-romagnesian silicates and oxides (e.g. olivine, pyroxene and magnetite). Basaltic lavas and their intrusive equivalents also contain high proportions of feldspar, which is an important source of Al and readily converts to kaolinite and gibbsite (Figure 3.9). Other igneous rock types such as granite can provide a source of feldspar and associated minerals, although they contain significantly lower iron contents. Quartz is also an important con-stituent of many common igneous rocks, and may be dissolved during the development of laterite (Borger, 1993); relict quartz typically becomes increasingly sparse higher in the weathering profile. The complexity of the physico-chemical breakdown patterns of primary minerals has been docu-mented by Delvigne (1998). Immature sediments (e.g. greywackes and arkoses) can also provide suitable protoliths since they contain substantial proportions of the less stable minerals (e.g. pyroxenes, micas and feld-spars). However, chemically mature sediments (e.g. quartz-rich sandstones) or carbonates (e.g. limestones) are unlikely to contain sufficient iron or alumina to develop into laterite profiles *sensu stricto*. Those iron-rich

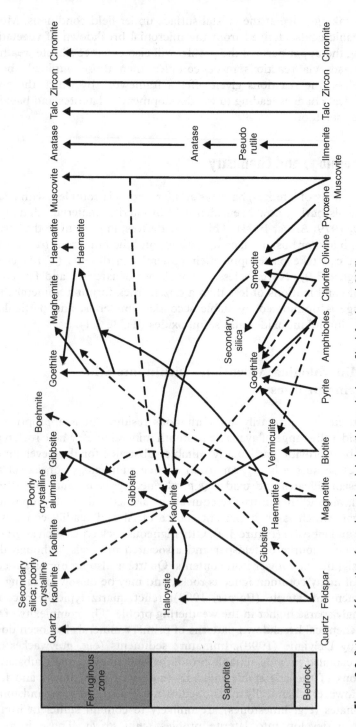

Figure 3.9 Pathways of formation of secondary minerals in lateritic weathering profiles (after Anand, 2005).

Table 3.1 Common alteration minerals found in laterites and bauxites (McFarlane, 1983)

Mineral	Formula
Haematite	Fe_2O_3
Goethite	$FeOOH$
Kaolinite	$Al_2Si_2O_5(OH)_4$
Gibbsite	$Al(OH)_3$
Boehmite	$AlOOH$
Diaspore	$AlOOH$
Corundum	Al_2O_3
Anatase	TiO_2
Rutile	TiO_2
Quartz	SiO_2

duricrusts that do become extensively developed in these rock types are more often ferricretes, since the high iron content requires introduction by transfer into the host rock.

Most of the common primary minerals and their associated relict textures do not survive beyond the saprolitisation stage, and only the most resistant types can remain as primary mineral remnants within the uppermost zones of the weathering profile (e.g. the extreme end members of this alteration sequence; Figures 3.3 and 3.9). Of these, it is disaggregated fragments of quartz that prove the most durable. Those primary minerals that are most resistant to alteration consist of accessory phases such as zircon, tourmaline and rutile, but these often occur in tiny quantities because they constitute a small proportion of the protolith. Accordingly, few, if any, of these minerals can be identified in most lateritic duricrusts. However, primary magnetite and ilmenite are an important constituent of mafic rocks (c. 3–8%) and prove highly resistant to alteration, and so can become preserved in all but the most highly altered laterite.

The most widespread of the secondary minerals formed during the development of a laterite weathering profile are iron and aluminium sesquioxides (Table 3.1). These may form either directly from the alteration of primary minerals, or else via a series of pathways involving the formation of intermediary sheet silicate minerals and clays (e.g. chlorite, illite, smectite, vermiculite and halloysite), which are then themselves broken down, stripped of their mobile ions and silica, and eventually converted to alumina and ferric oxyhydroxide residua (Figure 3.9). It is not possible to describe these mineral transformations in detail, but the key issue is that under tropical-type weathering conditions these transformation pathways lead to a dramatic increase in the abundance of goethite, haematite, kaolinite and gibbsite above the saprolite zone. These latter secondary minerals become increasingly important in the mottled zone and eventually form the bulk of the overlying laterite.

To illustrate the mineralogical changes and associated elemental variations that can occur through a lateritic weathering profile, it is useful to consider those observed in the classic laterite section at Bidar, India (Newbold, 1846; Borger and Widdowson, 2001; Kisakürek et al., 2004). At the base of the profile (Figure 3.3), the parent basalt is rapidly broken down and replaced initially by a mixture of clays in which kaolinite becomes increasingly dominant. With increasing alteration, much of the kaolinite is broken down, leaving a residua of kaolinite and gibbsite in voids in the laterite bulk (which consists of a mixture of goethite and haematite). Near the surface, through-flushing by rain water has often mechanically removed much of the softer Al-oxyhydroxides that originally filled the tubular fabric of the indurated laterite cap. In thin-section, apparently unaltered opaque phases (magnetite and ilmenite) can be identified throughout all levels of the weathering profile, except the topmost indurated laterite.

3.5.2 Mineralogy of ferricrete alteration profiles

The mineralogy of ferricrete alteration profiles can be complex and varied because of the incorporation of mechanically derived materials and the retained importance of host rock composition after the formation of secondary minerals. In general, ferricrete profiles do not display the progression of alteration minerals observed in laterites. Where ferricretes are formed by mechanical accumulation, they can lie disconformably above unaltered bedrock (Bowden, 1987, 1997). In these instances, the ferricrete mineral assemblage will be inherited, in part, from the derived materials, and in part from later cementation processes that involve remobilised iron and alumina deposited as neo-formed oxyhydroxides. In such examples, determining the sequence of mineralogical transformations becomes exceptionally difficult.

Where ferricretes develop through the precipitation of allochthonous iron and alumina, these elements form into Al- and Fe-oxyhydoxides, initially as microcrystalline aggregates within pore spaces; these can often appear as fine 'rust-spots' within the host rock. With further supply of allochthonous fluids, the frequency of these spots increases, they coalesce, and can progressively replace any existing silica or carbonate cements until regions of the host rock are largely cemented by Fe- and Al-rich secondary minerals. This process often becomes pronounced along bedding planes and/or joints, and can lead to patches and horizons which are preferentially affected (Figure 3.8). Accordingly, multiple or stacked ferruginous zones are not unusual in developing ferricrete profiles; although where the host rock is relatively homogeneous, concentration will occur at or near to the water table. At this stage, the altered profile becomes recognisable as a ferricrete and often becomes highly indurated. However, the parent mineralogy may still be retained and form the bulk of the altered host rock. Consequently, chemically unstable minerals may be found in some ferricrete materials, together with virtually unaltered quartz.

In the presence of iron, silica solubility can also increase (Morris and Fletcher, 1987), leading to an *in situ* alteration of host-rock silicates, together with etching, corrosion and, in extreme cases, the dissolution of any original quartz grains/crystals (Borger, 1993). Extreme etching of host-rock quartz grains and, in some instances, their replacement by goethite and haematite, is observed in the duricrust layers of ferricrete at Palika Ba, Gambia (Figs 3.4C and 3.8). The resulting secondary mineralogy generated by the introduction and replacement process is similar to that observed in laterite weathering profiles. However, the manner by which the secondary mineralisation progresses is controlled more by the invasion and infiltration of solutions than by the downward progression of a weathering front. Importantly, the three-dimensional relationship of quartz grains throughout the Palika Ba profile indicates that the ferralitisation process is essentially isovoluminous; this contrasts with the upper levels of lateritic weathering profiles where the removal of constituents can produce volume reduction and collapse.

3.5.3 Chemical variations within lateritic weathering profiles

When considering the distribution of elements within a lateritic weathering profile it is important to understand that no mineral is ever entirely unaffected by weathering, and that no element is immobile or ever entirely leached from any of the profile horizons. Many of the chemical characteristics of lateritic weathering profiles are related to the downward advancement of the weathering front (Figure 3.6), but processes such as the establishment of a stable water table may act to modify the pre-existing profile. The behaviour of major and minor elements during laterite development have been summarised for Australian examples by Butt et al. (2000), and relates to the leaching and retention of a range of elements in the principal horizons of the weathering profile. This work is applicable to laterite profiles developed elsewhere (e.g. Table 3.2A, B).

Weathering in the saprolite zone causes the destruction of feldspars and ferromagnesian minerals. Here, the mobile elements Na, Ca, K and Sr are leached and evacuated from the system (although K and Ba can be retained by adsorption to neo-formed clays); liberated Si and Al are partially retained by kaolinite and halloysite. In addition, K and Rb will be lost if hosted by orthoclase or biotite but, if present in muscovite, can be retained for as long as the mineral survives. The weathering of less stable ferromagnesian minerals in the saprolite produces Fe oxides with progressive loss of Mg and Si, except where retained in smectite (Mg, Si), kaolinite (Si) or quartz (Si); there may be partial retention of minor and trace elements such as Ni, Co, Cu, Mn and Ni. Paquet et al. (1987) detail the involvement of smectites as early hosts for selected trace elements (Zn, Mn, Co, Ni and Cu) during the development of the weathering profile.

The alteration of all but the most resistant primary minerals occurs in the mid- to upper saprolite zones; in addition, less stable secondary minerals such as smectite are also destroyed. Serpentine, magnetite, ilmenite and chlorite are progressively weathered through the zone. Ferromagnesian minerals are the principal hosts for transition metals such as Ni, Co, Cu and Zn in mafic and ultramafic rocks; they become leached from the upper horizons and reprecipitate with secondary Fe–Mn oxides in the mid- to lower-saprolite.

Most remnant primary minerals, except quartz, are finally destroyed in the mottled and ferruginous duricrust zones. Accordingly, these upper zones become dominated by Si, Al and Fe, resident in kaolinite, quartz, Fe oxides and, in places, gibbsite. The distribution of several minor and trace elements is controlled wholly or in part by the distribution of these major elements, due to substitution or co-precipitation. Thus, Cr, As, Ga, Sc and V tend to accumulate with Fe oxides; Cr, mainly derived from ferromagnesian minerals, is also associated with neo-formed kaolinite. Many residual and immobile elements tend to concentrate with clay and Fe oxides in the ferruginous zones. The distributions of Cr, K, Hf, Th, Nb, Ta, W, Sn, Rare Earth Elements, Ti and V within these relate wholly or in part to their relative inertness during continued weathering (e.g. V, Ti), or to the stability of primary and/or secondary host minerals (e.g. Zr and Hf in zircon; Ti in rutile and anatase; Cr in chromite; K in muscovite). The abundances of these trace elements tend to increase upwards through the profile due to the loss of other components, with marked accumulation in the lateritic residuum. Where there is significant chemical groundwater activity within a weathering profile there is also the potential to concentrate economically important trace elements such as U, V, Ag and Au (Mann, 1984).

As with section 3.5.1, the chemical changes than can occur through a lateritic weathering profile can be illustrated by comparing those observed in the classic laterite section at Bidar (Table 3.2A) with a similar profile of younger age developed upon Dharwar greywackes near Panjim (Table 3.2B). The observed mineralogical changes are reflected in the abundances of major elements, demonstrated by a rapid loss of the more mobile

Table 3.2   Geochemical analyses (by XRF) of autochthonous laterite profiles

(A) Developed on Deccan basalt exposed at Bidar, India (see Figures 3.2A and 3.3)

Type	Element	Sample/depth (m)								
		BB1 47.0	BB2 35.0	BB3 26.0	BB4 15.0	BB5 13.0	BB6 11.0	BB7 6.0	BB8 5.0	BB9 2.0
Major (wt %)	SiO_2	48.90	50.06	38.59	38.78	30.61	6.12	36.68	31.35	9.59
	TiO_2	2.16	2.29	5.11	4.78	5.76	1.40	2.44	2.33	2.03
	Al_2O_3	13.72	14.15	31.54	31.95	25.83	6.97	31.30	27.22	9.85
	Fe_2O_3	13.40	12.63	24.10	21.64	36.95	84.81	27.70	38.37	77.53
	MnO	0.19	0.22	0.11	0.06	0.06	0.03	0.33	0.07	0.23
	MgO	6.93	5.99	0.40	0.38	0.23	0.14	0.26	0.10	0.16
	CaO	10.99	11.45	0.19	1.91	0.07	0.00	0.05	0.00	0.04
	Na_2O	2.46	2.78	0.00	0.00	0.00	0.00	0.00	0.00	0.00
	K_2O	0.16	0.25	0.02	0.01	0.02	0.00	0.03	0.07	0.03
	P_2O_5	0.16	0.19	0.18	0.03	0.08	0.33	0.07	0.08	0.12
	Total	48.90	50.06	38.59	38.78	30.61	6.12	36.68	31.35	9.59
	LOI	2.16	2.29	5.11	4.78	5.76	1.40	2.44	2.33	2.03
Trace (ppm)	Rb	1	6	1	0	0	0	1	4	1
	Sr	210	222	13	23	4	41	15	7	10
	Y	31	35	693	8	3	14	5	4	4
	Zr	128	136	246	246	282	77	184	211	283
	Nb	10	11	20	19	23	5	15	17	15
	Ba	53	95	55	10	13	23	430	17	162
	Pb	1	1	2	4	10	6	45	20	37
	Th	1	1	2	2	1	1	4	6	7
	U	0	0	1	1	2	4	1	2	2
	Sc	38	37	70	49	15	154	20	24	190
	V	371	375	687	643	1846	975	671	967	2986
	Cr	148	139	201	213	250	766	257	737	692
	Ni	98	138	287	58	72	242	63	106	40
	Cu	177	185	394	173	452	852	211	183	581
	Zn	106	110	120	86	62	206	57	70	35

Table 3.2 Continued

(B) Developed on Proterozoic greywacke exposed at Merces Quarry near Panjim, Goa, India (see Figure 3.4A)

Type	Element	Sample/depth (m)												
		SQ2 34.0	SQ3 30.0	SQ4 25.5	SQ5 22.5	SQ6 15.0	SQ7 14.0	SQ8 13.5	SQ9 12.0	SQ10 8.5	SQ11 7.5	SQ12 3.5	SQ13 2.5	SQ14 0
Major (wt %)	SiO_2	67.77	69.67	79.50	64.21	75.27	71.19	67.46	67.86	17.21	15.20	19.62	17.50	13.63
	TiO_2	0.56	0.58	0.32	0.48	0.43	0.76	0.85	0.73	1.84	0.66	2.28	1.90	2.04
	Al_2O_3	14.65	13.90	11.41	16.38	13.43	18.62	19.62	18.95	30.52	13.58	40.89	34.57	34.15
	Fe_2O_3	6.08	5.58	2.43	7.63	3.38	7.64	10.25	10.78	49.64	68.33	36.42	44.95	48.48
	MnO	0.13	0.11	0.07	0.17	0.05	0.06	0.02	0.02	0.02	0.57	0.06	0.04	0.04
	MgO	3.07	2.88	1.29	4.13	1.61	0.51	0.32	0.30	0.10	0.13	0.15	0.18	0.19
	CaO	0.95	0.65	0.55	0.43	0.43	0.05	0.05	0.04	0.05	0.05	0.06	0.06	0.06
	Na_2O	3.31	3.24	3.54	2.76	3.50	0.09	0.08	0.07	0.64	0.07	0.12	0.21	0.13
	K_2O	3.68	3.20	1.63	3.95	2.44	1.76	1.43	1.35	0.28	0.51	0.65	0.74	0.72
	P_2O_5	0.12	0.11	0.07	0.05	0.08	0.08	0.04	0.06	0.16	0.93	0.25	0.27	0.70
	Total	100.33	99.93	100.79	100.20	100.63	100.75	100.14	100.15	100.45	100.02	100.50	100.43	100.14
	LOI	1.34	1.52	1.29	2.52	1.80	6.59	7.23	7.28	12.55	12.15	17.22	15.37	16.39
Trace (ppm)	Rb	118	100	43	132	73	37	30	26	7	15	20	24	24
	Sr	114	108	100	81	109	7	9	7	19	36	38	35	38
	Y	15	25	13	14	36	12	11	8	15	25	21	21	23
	Zr	194	228	92	138	134	265	261	219	257	99	339	296	310
	Nb	13	14	7	12	10	15	17	16	20	11	28	24	26
	Ba	721	630	431	677	576	674	583	573	61	458	179	175	164
	Pb	12	11	13	12	12	24	15	13	26	32	47	36	38
	Th	16	19	7	14	14	25	22	21	21	11	24	17	20
	U	6	8	3	4	4	6	7	6	4	7	9	5	13
	Sc	14	12	5	10	8	17	20	23	29	28	31	47	80
	V	92	98	44	88	65	124	151	157	780	60	626	837	765
	Cr	108	107	44	81	66	91	130	149	839	153	908	1156	924
	Ni	38	43	15	41	22	67	42	41	59	119	58	52	53
	Cu	41	31	51	33	42	36	33	30	37	65	38	55	59
	Zn	109	108	32	137	60	39	32	24	19	92	25	27	32

elements (e.g. Ca, Na, K, Mg) in the earliest stages of weathering (i.e. within the saprolite), and are accompanied by an escalating loss of silica content. Silica loss results initially from the breakdown of primary minerals, and subsequently from the breakdown of neo-formed clays. These losses result in a concomitant relative increase in the concentration of the less mobile elements (Fe, Al and Ti) within the developing laterite profile, which are then considered as being 'residual'.

3.5.4 A quantitative estimation of the degree of lateritisation

Having adopted a process-based distinction between laterite and ferricrete, it is appropriate to consider a quantitative method of determining the degree of alteration within a lateritic weathering profile. A distinction between the lower saprolitic and mottled zones, and the uppermost lateritic levels, may be achieved using the chemical method of Schellmann (1984, 1986). This uses major oxide data for the weathering products and compares them with protolith composition through a SiO_2, Al_2O_3, Fe_2O_3 ternary plot. The method determines three levels of lateritisation (weak, moderate and strong) that lie beyond a 'limit of kaolinitisation'. This limit is calculated by assuming that all the Al available in the protolith is first converted to kaolinite, and that further weathering beyond this condition marks the stage where the material can be called laterite *sensu stricto*.

The principles and limitations of the Schellmann technique are illustrated using two examples of Indian laterite weathering profiles, from Bidar and Merces Quarry near Panjim, Goa. The limit of kaolinitisation on the ternary plot for the protolith of the Bidar (i.e. basalt) and Merces Quarry (greywacke) profiles have been determined at 43% and 57% SiO_2 respectively, and the fields of increasing degree of lateritisation are given in Figure 3.10A, B (see Schellmann (1986) for details of the calculation procedure). If these profiles had been formed through a simple downward progression of the weathering front and an associated expansion of the saprolitic, mottled and lateritic zones, then samples taken from increasingly higher levels within the weathering profile should plot sequentially away from an origin at the protolith composition, and toward the strongly lateritised field.

Considering first the Bidar profile (Table 3.2A). Here samples BB1 and 2 are from the unaltered basalt protolith and immediately overlying saprock, respectively, and display little evidence of chemical change. Samples BB3 and 4 are from the lower and upper mottled zone, and have been stripped of mobile elements. The progressive break down of primary silicates has resulted in the loss of Si, driving composition into the weakly lateritised field. Samples BB7 and 8 are from the upper carapace level and plot in the weakly and moderately lateritised zones, respectively. These show increasing concentrations of Fe and a concomitant loss of Si and Al. The top of the profile is capped by a resistant iron-rich duricrust, BB9, which plots in the strongly lateritised field. Although the chemical composition of the samples is consistent with an up-profile increase in weathering, samples BB5 and BB6 do not follow this progression. These samples are from the lower and middle zones of a palaeowater-table level (Mason et al., 2000; Kisakürek et al., 2004), and display high Fe, Cr and V concentrations which have been generated by allochthonous input of these elements. Importantly, since the composition of these samples cannot be considered as entirely due to alteration caused by *in situ* weathering processes, they cannot fulfil the conditions required by Schellmann for this approach.

The chemistry of the Merces Quarry profile (Table 3.2B) follows a similar pattern to that of Bidar. Here, natural variation in the protolith presents a more variable starting composition than Deccan basalt. Sample SQ2 is unaltered basement, and SQ3 and 4 lie at progressively higher levels within a thick saprock zone. Sample SQ5 lies at the base of the saprolite zone proper. This has also undergone some depletion in mobile elements but displays Fe enrichment and silica depletion, placing it in the kaolinitised field. Sample SQ6 lies near to a quartz vein that passes through the laterite profile (Figure 3.5B), which may explain its elevated Si content. Sample SQ7 lies at the top of the reddened saprolite and displays marked depletion of Ca, Na and Sr. Primary mineralogies are largely retained up to this level; their persistence may be because the protolith is a derived sediment and contains a higher proportion of stable minerals. Above the saprolite, the porosity of the profile increases markedly, and Fe-rich mottling begins to dominate. Samples SQ8 and 9 are both more depleted in mobile elements and enriched in Al and Fe relative to the samples beneath. Above these, Fe-mottling increases rapidly and the weathering profile passes into massive, vermiform laterite; samples SQ12–14 represent a progression toward increasingly Fe-dominated secondary minerals. These samples define a trend from the moderately to the strongly lateritised fields of the tri-plot. In the uppermost levels, the remaining inherited quartz grains become progressively etched and disaggregated; these are eventually lost in the duricrust layer. In a similar fashion to the Bidar profile, these samples also define an upward trend of increasing lateritisation. However, samples SQ10 and 11 do not follow this progression. Isotopic studies have confirmed that these too lie at the level of a palaeowater-table (Wimpenny et al., 2007), and their chemistry confirms the high Fe concentrations which have been generated by allochthonous input.

A number of principles are illustrated by these two lateritic weathering profiles and by the pattern in which the constituent samples appear on Figure 3.10. The ubiquity of the lateritisation process is confirmed by the fact that comparable patterns of element behaviour occur in two profiles that differ

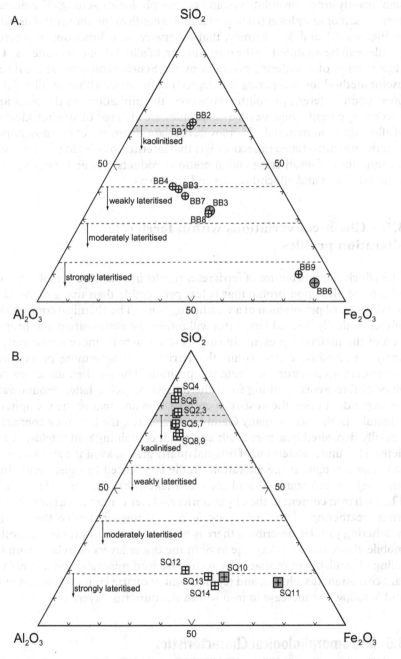

Figure 3.10 Ternary or tri-plots (SiO₂, Fe₂O₃, Al₂O₃) of (a) Bidar and (b) Merces Quarry data. Limits of kaolinitisation are determined according to the calculation of Schellmann (1986). The shaded band at the top of each tri-plot represents the natural chemical range of the protolith. The shaded symbols are those samples which correspond to the level of a palaeowater table. See text for discussion.

significantly in age, protolith type and geomorphological setting. The element enrichment of samples at palaeowater tables confirm the concepts illustrated in Figures 3.1 and 3.7, namely, that otherwise autochthonous weathering profiles can be modified by the introduction of allochthonous elements. The classification of weathering products by the Schellmann scheme provides a useful method for comparing the degree of alteration within profiles developed upon different protoliths. However, the limitations of the technique become apparent when weathering profiles are affected by the introduction of allochthonous materials. Moreover, the data affirm one of the most important themes introduced in section 3.1; that laterites and ferricretes represent a continuum of weathering and alteration products that lie between purely autochthonous and allochthonous end-members.

3.5.5 Chemical variations within ferricrete alteration profiles

The allochthonous nature of ferricretes results in a distribution of elements within the alteration profile that is less predictable than that generated by the downward progression of a weathering front. The chemical composition of mechanically derived ferricretes will reflect the composition and proportion of the materials present. In such cases, it may be more appropriate to analyse individual clasts within the ferricrete to determine provenance, rather than investigating the bulk composition. The geochemical characteristics of ferricretes resulting from the ingress of solute-laden groundwater are dependent upon the host-rock composition and that of the precipitated minerals. In the case of many Gambian ferricretes, the host rock consists of fluvially deposited quartz-rich silt and sand, containing minor feldspar and detrital kaolinite. Retention of original quartz grains, kaolinite and accessory minerals throughout the alteration profile is reflected by consistently high silica values and static Al and trace-element concentrations (Table 3.3). The high iron content in the upper duricrust layer cannot have been derived from weathering of the host rock *in situ*. Moreover, unlike the lateritic weathering profiles described, there is no evidence of up-profile removal of mobile elements. The decrease in Si in the upper layers correlates with the filling of available pore spaces by secondary iron minerals, and a concomitant corrosion, dissolution and replacement of quartz grains. Increases in Cr and V follow the increase in iron within the duricrust layer.

3.6 Micromorphological Characteristics

For lateritic profiles, routine thin-section analysis can readily be achieved on those materials that represent the lower degrees of alteration

Table 3.3 Geochemical analyses (by XRF) of the ferricrete alteration profile exposed at Palika Ba, Gambia, West Africa. (see Figs 3.4C and 3.8)

Type	Element	Sample/depth (m)					
		PG7 16.7	PG6 13.5	PG5 10	PG4 6.1	PG2 2	PG3 0.2
Major	SiO_2	88.42	86.39	83.55	79.45	47.21	40.91
(wt %)	TiO_2	0.94	1.02	1.18	1.40	1.06	1.13
	Al_2O_3	9.44	11.31	12.60	14.66	14.79	14.95
	Fe_2O_3	1.12	1.40	2.59	4.71	37.00	43.21
	MnO	0.01	0.01	0.01	0.01	0.01	0.01
	MgO	0.04	0.05	0.05	0.04	0.02	0.01
	CaO	0.01	0.02	0.02	0.02	0.01	0.02
	Na_2O	0.05	0.15	0.22	0.11	0.04	0.19
	K_2O	0.05	0.04	0.05	0.02	0.02	0.03
	P_2O_5	0.03	0.03	0.02	0.02	0.06	0.06
	Total	100.13	100.41	100.30	100.44	100.24	100.52
	LOI	3.64	4.24	4.73	5.49	11.48	13.38
Trace	Rb	1.4	2.2	2.8	1.1	0.7	1.8
(ppm)	Sr	35.4	18.2	16.9	16.8	14.4	13.5
	Y	12.8	12.8	15.8	16.6	11.4	12.3
	Zr	434	455	572	613	341	400
	Nb	15.7	18.3	21.4	26	18.7	21.5
	Ba	102	65	52	47	17	12
	Pb	13	10	9	10	21	26
	Th	7	7	10	10	10	16
	U	1	2	2	2	3	5
	Sc	3	4	5	8	8	8
	V	32	36	63	94	n/d	n/d
	Cr	48	49	58	86	89	167
	Ni	12	15	17	14	7	9
	Cu	4	4	5	6	33	55
	Zn	9	10	9	10	14	15

(e.g. saprock, saprolite), since the primary mineralogy and mineral relationships remain recognisable. In these, the progressive breakdown of primary minerals may be documented and can offer a link to observed chemical changes. With increased weathering (i.e. higher in the profile), mechanically resistant minerals, such as quartz, tend to become fractured and disaggregated, with dissolution progressing along fracture joints or as pitting at sites of crystal defects. Once attacked, olivine is particularly susceptible to rapid destruction. The cleavage of many of the more complex silicates also provides pathways for alteration; feldspars, pyroxenes and amphiboles typically crumble into polyhedral fragments, and the secondary mineral growth that occurs along mica cleavages tends to 'burst apart' crystal structures. The physical breakdown of the common rock-forming minerals is illustrated in Figure 3.11.

Many of the key micromorphological features produced within lateritic weathering profiles can be illustrated with reference to samples collected from a profile from the coastal belt of Western India (Figure 3.12A–D): extensive pseudomorphing of primary minerals by neo-formed kaolinite, together with disaggregation of remnant fragments makes thin-section analysis increasingly difficult when applied to materials from the mottled zone and above. Here, increasing quantities of semi-opaque secondary iron minerals effectively obscure mineralogical detail. However, because the evolution of fabrics during the development of both lateritic and ferricrete duricrusts is largely the result of increasing dominance of secondary iron-rich oxides and oxyhydroxides, a sequential series of micromorphological 'textures' can be recognised instead (Figure 3.13). These arise from the processes of iron replacement and void filling, and apply equally to the development of laterite and ferricrete duricrusts. Ferricrete alteration profiles lend themselves more readily to thin-section analysis since much of the primary mineralogy can remain recognisable throughout the profile. The main observable changes are the increasing quantity of secondary iron minerals and associated void-filling, and the replacement, etching and dissolution of quartz grains.

3.7 Dating and Palaeoenvironmental Significance of Lateritic Weathering Profiles

One of the perennial problems of laterite study is the determination of their age (Bourman, 1993). Until recently, stratigraphical techniques have provided the only method. However, since the protolith may be considerably older than any laterite profile developed upon it, the value of such techniques is limited. It is clear that many lateritic duricrusts are of considerable antiquity because their current distribution places them in climatic, geomorphological or tectonic settings that otherwise would have been inimical to their formation. Establishing when the lateritisation process that formed them began, or the point at which they ceased to form, remain polemic issues, but a number of recent studies have provided some insights.

Palaeomagnetic dating techniques, which rely upon measuring the chemical remnant magnetism (CRM) preserved by secondary iron minerals within the duricrust, have proved relatively successful. This approach effectively records the cessation of the mineralogical transformations that cause

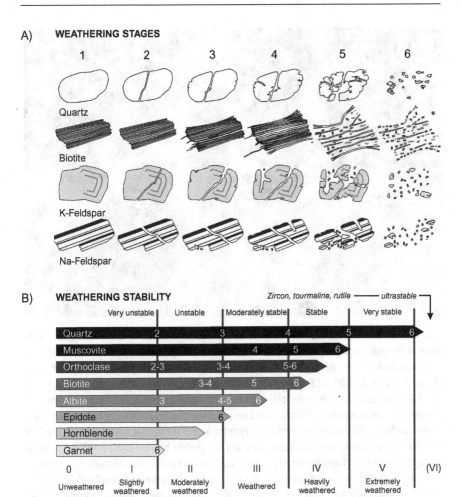

Figure 3.11 (A) Weathering stages of quartz, biotite, K-feldspar and Na-feldspar: (1) fresh and unweathered mineral; (2) grain splitting; (3) marginal etching of quartz and K-feldspar, incipient kaolinisation of Na-feldspar, mobilisation of Fe from biotite causing fanning; (4) deep corrosion of quartz, K- and Na-feldspar, bleaching, widening, and onset of kaolinisation of biotite; (5) fragmentation with very strong widening and kaolinisation of biotite; (6) complete dissolution of primary minerals. (B) Weathering resistance and degree of weathering in humid tropical environments: numbers within the horizontal bars correspond with the weathering stages of quartz shown in (A); the vertical lines match with the weathering stages of quartz and are also used for both the classification of the weathering resistance of other minerals as well as the degree of weathering of any weathered sample as a whole (after Borger, 2000).

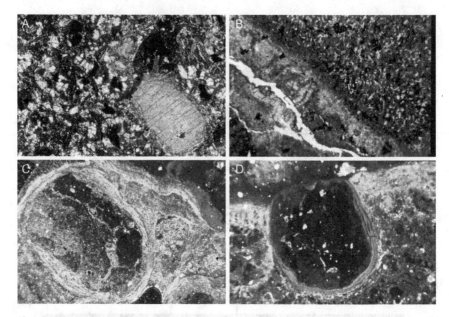

Figure 3.12 Photomicrographs illustrating the micromorphology through a low-level coastal laterite profile of Neogene age developed within Deccan basalt from Guhagar, western India (field of view is *c*. 5 × 3 mm for each). The stated 'degree of lateritisation' has been derived using the chemical classification of Schellmann (1982, 1986). (A) Kaolinitised basalt from the saprolitic zone: on the right, a large pyroxene crystal has developed an alteration rind of clay minerals; groundmass pyroxene has been broken down into iron-rich clays; lamellar-twinned feldspars have been pseudomorphed by kaolinite; opaque primary oxides remain unaffected. (B) Weakly lateritised basalt from the base of the mottled zone: on the left, a vein of kaolinitic clays containing microcrystallising haematite and goethite bisects a highly weathered pyroxene crystal; the basalt groundmass texture, in the upper right, has been completely replaced by haematite and kaolinite. (C) Moderately lateritised basalt (carapace): on the left a finely layered cutane (pisoid) of goethite encloses a haematite-rich zone, formerly a pyroxene; all protolith mineralogy textures have been obscured by the growth of anastomosing haemetitic/goethitic veinlets and infillings; tiny, bright, unaltered disaggregated crystal fragments and kaolinite pseudomorphs are scattered throughout. (D) Strongly lateritised basalt (cuirasse): a thin goethitic cutane surrounds a haematised pisoid within a highly ferruginous haematitic zone; coaslescence of haematite patches and zones has produced a resistant skeleton penetrated by voids and tubes (not shown); a few tiny, disaggregated crystal fragments and kaolinite patches occur scattered throughout.

iron accumulation within the profile. The derived CRM palaeomagnetic poles are then used to ascertain weathering ages by comparing them with the trajectories of palaeomagnetic poles of known age. The method has proved particularly useful for sites that have altered their latitudinal position as a result of plate tectonic motion (e.g. Australia and India during the Tertiary; Schmidt and Embleton, 1976; Schmidt et al., 1983; Kumar,

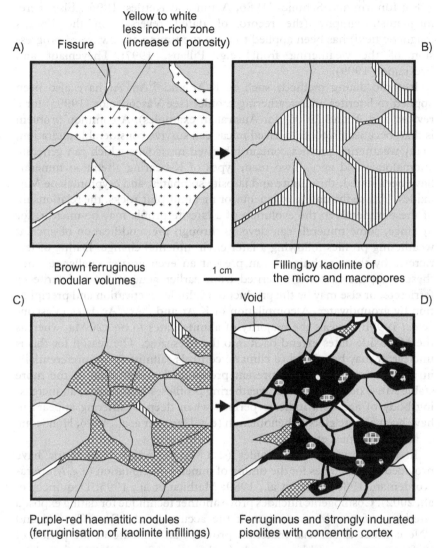

A)

Fissure

Yellow to white
less iron-rich zone
(increase of porosity)

B)

Brown ferruginous
nodular volumes

1 cm

Filling by kaolinite of
the micro and macropores

C)

Void

D)

Purple-red haematitic nodules
(ferruginisation of kaolinite infillings)

Ferruginous and strongly indurated
pisolites with concentric cortex

Figure 3.13 Schematic illustration of the formation and evolution of successive laterite facies. (A) Mottled clay layer with ferruginised nodules and bleached zones. (B) Secondary fillings of kaolinite in bleaching zones. (C) Ferruginisation of kaolinite and the formation of pseudo-conglomeratic iron crust. (D) Evolution of haematite nodules into pisolites with the formation of pisolitic crust (after Nahon, 1986; Thomas, 1994).

1986; Idnurm and Schmidt, 1986; Acton and Kettles, 1996). Elsewhere, magnetostratigraphy (the record of dated reversals in the Earth's magnetic field) has been applied to trace the timing of downward progression of the weathering front (e.g. Pillans, 1997; Théveniaut and Freyssinet, 1999).

Isotopic dating methods such as K/Ar and $^{40}Ar/^{39}Ar$ have also been applied to laterites and weathering profiles (see Vasconcelos (1999a) for a review, and Anand (2005) for Australian examples). A common problem is that, because protolith-derived micas can survive considerable alteration, many weathering profiles contain inherited muscovite which can generate anomalously old ages. Two main types of K-bearing alteration minerals have been dated; the alunite and jarosite sulphates, and cryptomelane Mn-oxides. Unlike the dating of igneous or metamorphic rocks, the relationship of these minerals to the evolution of a laterite profile may be unclear; for instance, some minerals can develop through the modification of ancient weathering profiles following a later environmental change. Dating of ferricretes by these techniques can present an even greater challenge since these comprise material inherited from earlier generations of laterite or ferricrete, or else may be the product of multiple reactivation and precipitation by groundwater. A compilation of K/Ar and $^{40}Ar/^{39}Ar$ dates (Vasconcelos, 1999b) reveals the majority of alunite dates to be <20 Ma, whereas the Mn-oxide dates extend back into the Mesozoic. The reason for this is unclear; it may be related to climatic controls (alunite forms preferentially in arid climates) or may represent preferential preservation of the more stable Mn-oxide within older weathering profiles. Nevertheless, an increasing body of age data indicate periods when deep weathering appears to have been a more global phenomenon (e.g. Dammer et al., 1996; Henocque et al., 1998; Spier et al., 2006).

In younger weathering profiles (<<1 Ma), U-series disequilibria have provided chronologies for the timing of mineral precipitation (e.g. Moreira-Nordeman, 1980; Short et al., 1989; Mathieu et al., 1995; Dequincey et al., 2002). Cosmogenic nuclides prove another technique for dating exposed lateritic surfaces. Measurement of the accumulated ^{10}Be, ^{36}Cl, ^{26}Al and ^{23}Ne content of surficial materials provides an estimate of the period of surface exposure, and has proved of value in producing denudation chronologies (Lal, 1991). The technique also has been applied to constrain the mechanisms of weathering profile development (Braucher et al., 1998a, b, 2000), and has revealed information about the erosion and burial of weathering products and lateritic profiles (Brown et al., 1994; Heimsath et al., 2000).

Advances in isotope geoscience have also provided a number of environmental tracers. Bird and Chivas (1988, 1989) were amongst the first to recognise the potential of oxygen isotopes in weathering studies, not only as a technique for establishing palaeoenvironmental conditions during the formation of weathering profiles, but also as a chronometer for dating residual kaolinite. They noted that pedogenic carbonates from different stages of the Mesozoic and Tertiary displayed different $\delta^{18}O$ values, and were able to calibrate this isotopic change over time. Using $\delta^{18}O$ analyses of numerous weathering profiles, Bird and Chivas (1993) concluded that a considerable proportion of the Australian landscape was of a greater antiquity than previously had been supposed.

Other isotope systems that have been applied to the study of laterite include those of strontium (i.e. $^{87}Sr/^{86}Sr$ ratio), neodymium ($^{143}Nd/^{144}Nd$ ratio; reported as ϵ_{Nd}), and lithium (i.e. $^{7}Li/^{6}Li$ ratio; reported as $\delta^{7}Li$). For example, Mason et al. (2000) revealed a marked upward increase in $^{87}Sr/^{86}Sr$ ratio (and associated decreases in the $^{143}Nd/^{144}Nd$ ratio) in early Paleogene lateritic profiles from the Deccan region of India, concluding that allochthonous aeolian material had been introduced into upper levels of the laterite profile. Viers and Wasserburg (2004) reported a similar isotopic shift in a Neogene lateritic profile from Cameroon, and suggested that aeolian inputs further controlled the isotopic variations of groundwater that passed through the weathering profile. Lithium isotopes have also been used successfully (Pistiner and Henderson, 2003; Rudnick et al., 2004). For example, Kisakürek et al. (2004) analysed Li behaviour in the Bidar laterite profile and determined the position of a palaeowater table on the basis of measured $\delta^{7}Li$ values: the technique also revealed the importance of an aeolian input to the surface levels of the laterite.

Rhenium–osmium isotope systematics have been applied to laterite profiles (Sharma et al., 1998; Wimpenny et al., 2007), although the interpretation of such data remains at a preliminary stage. Nevertheless, some first-order inferences may be drawn; rhemium is preferentially mobilised and leached under oxidising conditions, whereas osmium is retained in upper horizons through its affinity with Fe–Al–Mn oxyhydroxides. In these upper levels, the $^{187}Os/^{188}Os$ ratio is found to become significantly less radiogenic. Accordingly, if regional stripping of laterite profiles was to occur, such an event could contribute high levels of non-radiogenic Os to the oceans and have consequences for the record of isotopic change in marine sediments (Puecker-Ehrenbrink and Ravizza, 2000). Sharma et al. (1998) also used the $^{187}Os/^{187}Os$ ratio to generate 'model ages' as a method for dating the evolution of different weathering zones within a profile. However, this approach relies on the laterite becoming a closed system for Re and Os at an early stage in the development of the weathering profile; any allochthonous aeolian input would serve to obfuscate meaningful age data.

Finally, since the formation of laterite profiles depends upon prevailing climate, the distribution of laterites, and to a lesser extent ferricretes, of different ages have become increasingly used as proxies for palaeoclimate reconstruction. However, because the precise climatic conditions needed for lateritisation remain uncertain (e.g. Bardossy and Aleva, 1990; Tardy et al., 1991; Taylor et al., 1992), this has led to differences in climatic interpretation. Nevertheless, since lateritisation generally requires consistently warm temperatures and high amounts of seasonal precipitation, quantitative, as well as qualitative, climatic information can now be inferred from the occurrence of laterites (e.g. Bardossy, 1981; Thomas, 1994; Tsekhovski et al., 1995; Price et al., 1997; Tardy and Roquin, 1998).

3.8 Relationship to other Terrestrial Geochemical Sediments

Laterites and ferricretes are sometimes found in association with silcretes (Alley, 1977; Twidale, 1983; Partridge and Maud, 1987; Firman, 1994; Benbow et al., 1995; see Chapter 4), which has generated considerable debate over possible interlinkages between the origins of the three materials. Given that potentially large volumes of silica are released during the development of a lateritic weathering profile, it is unsurprising that studies have suggested a genetic link between lateritisation and silcrete formation (e.g. Stephens (1971) and Marker et al. (2002) in Australia and South Africa respectively). However, as our understanding of the factors controlling the formation of silica- and iron-cemented duricrusts has improved, this link has been challenged (Young, 1985; Taylor and Ruxton, 1987). More rarely, ferricretes have been recorded in association with calcretes. For example, Nash et al. (1994) note ferricretes that have developed above a relict calcrete profile in the southeast Kalahari, presumably as a result of the ingress of iron-rich groundwater into overlying sediments. Laterites have also been documented in southeastern Australia in catenary relationships with Mn-rich duricrusts (Taylor and Ruxton, 1987), the 'manganocretes' forming as a result of the absolute accumulation of Mn released during the lateritisation of basalts at higher elevations.

3.9 Directions for Future Research

Amongst the most urgent issues for future research is the need for a consensus over the usage of the terms laterite and ferricrete. A process-based distinction has been adopted here and has provided a useful and workable framework for this chapter. Although the concept and usage of laterite and ferricrete terminology will inevitably continue to vary between authors, any advancement in the understanding of the weathering and alteration processes involved will be hampered without such a consensus. Sampling, and the chemical and mineralogical analysis of laterite and ferricrete profiles, should, where possible, be systematised, as this will help enormously with comparisons between localities, types of occurrence, and between studies by different investigators. For more information, see Aleva (1994). Further, any chemical analysis of samples should be conducted with reference to appropriate standards; laterite standards VL1, VL2 and SLB are recommended (Schorin and LaBreque, 1986; LaBreque and Schorin, 1987; Schorin and Carias, 1987).

An improvement in the documentation of laterite and ferricrete occurrences is needed to produce a global distribution map. Creating such a map is problematic, not least because of the acknowledged terminological issues. However, it will prove essential for evaluating landscape stability and landform evolution, understanding the influence of tropical weathering on geochemical cycles, as well as in the evaluation of mineral and agricultural resources.

Further work is required to better constrain the environmental conditions under which laterite and ferricretes form (e.g. Bremer, 1995). A reliable series of climatic parameters, quantified on the basis of the degree and type of secondary mineral assemblages, for instance, would be both desirable and useful. Such data would improve the reliability of using remnants of lateritic weathering profiles as palaeoenvironmental indicators, and aid in realising their value as monitors of past climate change. Likewise, a rigorous understanding of the tectonic and topographic requirements for laterite formation is needed. It is often assumed that laterite formed as a uniform blanket, so that present occurrences are erosional remnants of fossil laterite. Where this assumption is correct, these remnants can prove invaluable in deciphering landscape chronologies. However, uniform development may not necessarily occur, since variations in local topography, hydrology or lithology can cause patchy laterite or ferricrete development.

Determining the age of laterite and the weathering processes responsible for its formation are crucial to understanding the manner in which landscapes evolve, and for evaluating environmental records provided by palaeoweathering materials. Attempts at utilising palaeomagnetism and other techniques have met with varying success. Some radiometric dating methods seem promising, but the relationship of the dateable secondary minerals to the long-term evolution of the weathering profile requires further investigation. Strontium, lithium, samarium – neodymium and rhenium – osmium analyses, and their associated isotopic studies, have great potential as 'tracers' for illustrating the role that secondary redistribution processes play in shaping laterite and ferricrete composition. The value of such techniques, when coupled with appropriate mineralogical and geochemical analyses, cannot be underestimated. Since tropical weathering products constitute a major component of global fluvial fluxes, investigations into the development and erosion of laterite profiles will also have an impact upon marine geochemical and isotopic records. Isotopic and chemical approaches also have the potential to reveal information about past climatic and atmospheric conditions, and may ultimately aid in identifying important periods of climate perturbation and change.

References

Acton, G.D. & Kettles, W.A. (1996) Geologic and palaeomagnetic constraints on the formation of weathered profiles near Inverell, Eastern Australia. *Palaeogeography, Palaeoclimatology, Palaeoecology* **126**, 211–225.

Aleva, G.J.J. (1986) Classification of laterites and their textures. In: Banerji, P.K. (Ed.) Lateritisation processes. *Memoirs of the Geological Survey of India* **120**, 8–28.

Aleva, G.J.J. (1994) *Laterites: Concepts, Geology, Morphology and Chemistry*. Wageningen (Netherlands): International Soil Reference and Information Centre.

Alexandre, J. & Alexandre-Pyre, S. (1987) La reconstitution à l'aide des cuirasses lateritiques de l'histoire geomorphologique du Haut-Shaba. *Zeitschrift für Geomorphologie N.F., Supplement Band* **64**, 119–131.

Alley, N.F. (1977) Age and origin of laterite and silcrete duricrusts and their relationship to episodic tectonism in the Mid North of South Australia. *Journal of the Geological Society of Australia* **24**, 107–116.

Alley, N.F., Clarke, J.D.A., MacPhail, M. & Truswell, E.M. (1999) Sedimentary infillings and development of major Tertiary palaeodrainage systems of south-central Australia. In: Thiry, M. & Simon-Coinçon, R. (Eds) *Palaeoweathering, Palaeosurfaces and Related Continental Deposits*. Special Publication 27, International Association of Sedimentologists. Oxford: Blackwell, pp. 337–366.

Anand, R.R. (2005) Weathering history, landscape evolution and implications for exploration. In: Anand, R.R. & de Broekert, P. (Eds) (2005) *Regolith Landscape Evolution Across Australia*. Bentley, Western Australia: Cooperative Research Centre for Landscape Environments and Mineral Exploration, pp. 2–40.

Anand, R.R. & de Broekert, P. (Eds) (2005) *Regolith Landscape Evolution Across Australia*. Bentley, Western Australia: Cooperative Research Centre for Landscape Environments and Mineral Exploration.

Bardossy, G. (1979) Growing significance of bauxites. *Episodes* **2**, 22–25.

Bardossy, G. (1981) Paleoenvironments of laterites and lateritic bauxites – effect of global tectonism on bauxite formation. In: *International Seminar on Lateritisation Processes (Trivandrum, India)*. Rotterdam: Balkema, pp. 284–297.

Bardossy, G. & Aleva, G.J.J. (1990) *Lateritic Bauxites*. Developments in Economic Geology 27. Amsterdam, Oxford: Elsevier, 624 pp.

Benbow, M.C., Callen, R.A., Bourman, R.P. & Alley, N.F. (1995) Deep weathering, ferricrete and silcrete. In: Drexel, J.F. & Preiss, W.V. (Eds) *The Geology of South Australia, Volume 2: the Phanerozoic*. Bulletin 54, Geological Survey of South Australia, pp. 201–207.

Berner, R.A. (1991) A model for atmospheric CO_2 over Phanerozoic time. *American Journal of Science* **291**, 339–376.

Berner, R.A. (1994) GEOCARB II: a revised model of atmospheric CO_2 over Phanerozoic time. *American Journal of Science* **294**, 56–91.

Bird, M.I. & Chivas, A.R. (1989) Stable isotope geochronology of the Australian regolith. *Geochimica et Cosmochimica Acta* **53**, 3239–3256.

Bird, M.I. & Chivas, A.R. (1993) Geomorphic and palaeoclimatic implications of an oxygen-isotope chronology for Australian deeply weathered profiles. regolith. *Australian Journal of Earth Sciences* **40**, 345–348.

Borger, H. (1993) Penetration of solutions into quartz grains – a scale for weathering intensity. In: Ford, D., McCann, B. & Vajoczki, S. (Eds) *Volume of Abstracts, Third International Geomorphology Conference,* 23–28 August 1993, Hamilton, Canada, p. 106.

Borger, H. (2000) *Mikromorphologie und Paläoenvironment: Die Mineralverwitterung als Zeugnis der cretazisch-tertiären Umwelt in Süddeutschland.* Relief, Boden, Paläoklima 15. Stuttgart: E. Schweizerbart'sche Science Publishers.

Borger, H. & Widdowson, M. (2001) Indian laterites, and lateritous residues of southern Germany: a petrographic, mineralogical, and geochemical comparison. *Zeitschrift für Geomorphologie N.F.* **45,** 177–200.

Bourman, R.P. (1993) Modes of ferricrete genesis: evidence from southeastern Australia. *Zeitschrift für Geomorphologie N.F.* **37,** 77–101.

Bourman, R.P. (1995) A review of laterite studies in southern South Australia. *Transactions of the Royal Society of South Australia* **199,** 1–28.

Bourman, R.P. & Ollier, C.D. (2002) A critique of the Schellmann classification of laterite. *Catena* **47,** 117–131.

Bourman, R.P., Milnes, A.R. & Oades, J.M. (1987) Investigations of ferricretes and related surficial materials in parts of southern and eastern Australia. *Zeitschrift für Geomorphologie N.F., Supplement Band* **64,** 1–24.

Bowden, D.J. (1987) On the composition and fabric of the footslope laterites (duricrust) of Sierra Leone, West Africa, and their geomorphological significance. *Zeitschrift für Geomorphologie N.F., Supplement Band* **64,** 39–53.

Bowden, D.J. (1997) The geochemistry and development of lateritized footslope benches, the Kasewe Hills, Sierra Leone. In: Widdowson, M. (Ed.) *Palaeosurfaces, Recognition, Reconstruction and Palaeoenvironmental Interpretation.* Special Publication 120. Bath: Geological Society Publishing House, pp. 295–306.

Braucher, R., Bourles, D.L., Colin, F., Brown, E.T. & Boulange, B. (1998a) Brazilian laterite dynamics using *in-situ* produced ^{10}Be. *Earth and Planetary Science Letters,* **163,** 197–205.

Braucher, R., Colin, F., Brown, E.T., Bourles, D.L., Bamba, O., Raisbeck, G.M., Yiou, F. & Koud, J.M. (1998b) African laterite dynamics using *in situ*-produced ^{10}Be. *Geochimica et Cosmochimica Acta* **62,** 1501–1507.

Braucher, R., Bourles, D.L., Brown, E.T., Colin, F., Muller, J.-P., Braun, J.-J., Delaune, M., Edou Minko, A., Lescouet, C., Raisbeck, G.M. & Yiou. F. (2000) Application of *in situ*-produced cosmogenic ^{10}Be and ^{26}Al to the study of lateritic soil development in tropical forest: theory and examples from Cameroon and Gabon. *Chemical Geology* **170,** 95–111.

Bremer, H. (1981) Relieformen und reliefbildende Prozesse in Sri Lanka. In: *Relief Boden Paläoklima* (Zur Morphogenese in den feuchten Tropen. Verwitterung und Reliefbildung am Beispiel von Sri Lanka), Vol. 1. Berlin: Borntraeger, pp. 7–183.

Bremer, H. (1995) *Boden und Relief in den Tropen: Grundvorstellungen und Datenbank.* Berlin: Gerb. Borntraeger.

Brown, D.J., Helmke, P.A. & Clayton, M.K. (2003) Robust geochemical indices for redox and weathering on a granitic laterite landscape in Central Uganda. *Geochimica et Cosmochimica Acta* **67,** 2711–2723.

Brown, E.T., Bourlès, D.L., Colin, F., Sanfo, Z., Raisbeck, G.M. & Yiou, F. (1994) The development of iron crust lateritic systems in Burkino Faso, West

Africa examined with *in situ* produced cosmogenic nuclides. *Earth and Planetary Science Letters* **124**, 19–33.

Brückner, H. & Bruhn, N. (1992) Aspects of weathering and peneplanation in southern India. *Zeitschrift für Geomorphologie, Supplement Band* **91**, 43–66.

Buchanan, F. (1807) *A Journey from Madras through the Countries of Mysore, Kanara, and Malabar*, Vols 2 & 3. London: East India Company.

Büdel, J. (1980) Climate and climatomorphic geomorphology. *Zeitschrift für Geomorphologie N.F., Supplement Band* **36**, 1–8.

Butt, C.R.M., Lintern, M.J. & Anand, R.R. (2000) Evolution of regoliths and landscapes in deeply weathered terrain – implications for geochemical exploration. *Ore Geology Reviews* **16**, 167–183.

Connah, T.H. & Hubble, G.D. (1960) Laterites in Queensland. *Journal of the Geological Society of Australia* **7**, 373–386.

Curtis, C.D. (1976) Chemistry of rock weathering: fundamental reactions and controls. In: Derbyshire, E. (Ed.) *Geomorphology and Climate*. New York: Wiley, pp. 25–57.

Dammer, D., McDougall, I. & Chivas, A.R. (1999) Timing of weathering-induced alteration of manganese deposits in Western Australia; evidence from K/Ar and $^{40}Ar/^{39}Ar$ dating. *Economic Geology* **94**, 87–108.

De Swardt, A.M.J. (1964) Lateritisation and landscape development in parts of equatorial Africa. *Zeitschrift für Geomorphologie N.F.* **8**, 313–333.

Debaveye, J. & de Dapper, M. (1987) Laterite, soil and landform development in Kedah, Peninsular Malaysia. *Zeitschrift für Geomorphologie N.F., Supplement Band* **64**, 145–161.

Delvigne, J.E. (1998) *Atlas of Micromorphology of Mineral Alteration and Weathering*. Canadian Mineralogist Special Publication 3. Ottawa: Mineralogical Association of Canada.

DeQuincey, O., Chabaux, F., Clauer, N., Sigmarsson, O., Liewig, N. & Leprun, J-C. (2002) Chemical mobilizations in laterites: evidence from trace elements and $^{238}U-^{234}U-^{230}Th$ disequilibria. *Geochimica et Cosmochimica Acta* **66**, 1197–1210.

Devaraju, T.C. & Khanadali, S.D. (1993) Lateritic bauxite profiles of South-Western and Southern India – characteristics and tectonic significance. *Current Science* **64**, 919–920.

Dury, G.H. (1967) An introduction to the geomorphology of Australia. In: Dury, G.H. & Logan, H.I. (Eds) *Studies in Australian Geography*. London: Heinemann, pp. 1–36.

Eyles, R.J. (1967) Laterite at Kerdau, Pahang, Malaya. *Journal of Tropical Geography* **25**, 18–23.

Firman, J.B. (1994) Paleosols in laterite and silcrete profiles: evidence from the south east margin of the Australian Precambrian Shield. *Earth-Science Reviews* **36**, 149–179.

Foote, R.B. (1876) Geological features of the south Maratta County and adjacent districts. *Memoir of the Geological Survey of India* **12**, 1–268.

Fox, C.S. (1923) The bauxite and aluminous laterite occurrences of India. *Memoir of the Geological Survey of India* **49**, 1–287.

Fox, C.S. (1936) Buchanan's laterite of Malabar and Kanara. *Records of the Geological Survey of India* **69**, 389–422.

Gehring, A.U., Keller, P. & Heller, F. (1992) Magnetic evidence for the origin of lateritic duricrusts in southern Mali. *Palaeogeography, Palaeoclimatology, Palaeoecology* **95**, 33–40.

Gehring, A.U., Langer, M.R. & Gehring, C.A. (1994) Ferriferous bacterial encrustations in lateritic duricrusts from southern Mali (West Africa). *Geoderma* **61**, 213–222.

Gilkes, R.J., Scholz, G. & Dimmock, G.M. (1973) Lateritic deep weathering of granite. *Journal of Soil Science* **24**, 523–536.

Girard, J.P., Razanadronorosoa, D. & Freyssinet, Poh. (1997) Laser oxygen isotope analysis of weathering goethite from the lateritic profile of Yaou, French Guiana: paleoweathering and paleoclimatic implications. *Applied Geochemistry* **12**, 163–174.

Goudie, A.S. (1973) *Duricrusts in Tropical and Subtropical Landscapes*. Oxford: Clarendon Press.

Grandin, G. (1976) *Aplanissements Cuirassés et Enrichements des Gisements de Manganèse dan Quelques Régions d'Afrique de l'Ouest*. Paris: Mémoire de L'ORSTOM 82.

Grant, K. & Aitchison, G.D. (1970) The engineering significance of silcretes and ferricretes in Australia. *Engineering Geology* **4**, 93–120.

Grubb, P.L.C. (1963) Critical factors in the genesis, extent and grade of some residual bauxite deposits. *Economic Geology* **58**, 1267–1277.

Gunnell, Y. (2003) Radiometric ages of laterites and constraints on long-term denudation rates in West Africa: *Geology* **31**, 131–134.

Gutzmer, J. & Beukes, N.J. (1998) Earliest laterites and possible evidence for terrestrial vegetation in the Early Proterozoic. *Geology* **26**, 263–266.

Heimsath, A.M., Chappell, J., Dietrich, W.E., Nishiizumi, K. & Finkel, R.C. (2000) Soil production on a retreating escarpment in southeastern Australia: *Geology* **28**, 787–790.

Henocque, O., Ruffet, G., Colin, F & Feraud, G. (1998) ^{40}Ar/^{39}Ar dating of West African lateritic cryptomelanes. *Geochimica et Cosmochimica Acta* **62**, 2739–2756.

Idnurm, M. & Schmidt, P.W. (1986) Palaeomagnetic dating of weathered profiles. *Memoirs of the Geological Survey of India* **120**, 79–89.

King, L.C. (1953) Canons of landscape evolution. *Bulletin of the Geological Society of America* **64**, 721–752.

Kisakürek, B., Widdowson, M. & James, R.H. (2004) Behaviour of Li isotopes during continental weathering: the Bidar laterite profile, India. *Chemical Geology* **212**, 27–44.

Krauskopf, K.B. (1967) *Introduction to Geochemistry*. New York: McGraw-Hill.

Kronberg, B.I., Fyfe, W.S., McKinnon, B.J., Couston, J.F., Filho, B.S. & Nash, R.A. (1982) Model for bauxite formation: Paragominas (Brazil). *Chemical Geology* **35**, 311–320.

Kumar, A. (1986) Palaeolatitudes and the age of Indian laterites. *Palaeogeography, Palaeoclimatology, Palaeoecology* **53**, 231–237.

LaBrecque, J.J. & Schorin, H. (1987) Some statistical parameters for selected trace elements in VL-1. *Zeitschrift für Geomorphologie N.F., Supplement Band* **64**, 33–38.

Lal, D. (1991) Cosmic ray labeling of erosion surfaces: *in situ* nuclide production rates and erosion models. *Earth and Planetary Science Letters* **104**, 424–439.

Lamplugh, G.W. (1907) The geology of the Zambezi Basin around the Batoka Gorge (Rhodesia). *Quarterly Journal of the Geological Society of London* **63**, 162–216.

Lidmar-Bergström, K., Olsson, S. & Olvmo, M. (1997) Palaeosurfaces and associated saprolites in southern Sweden. In: Widdowson, M. (Ed.) *Palaeosurfaces: Recognition, Reconstruction and Palaeoenvironmental Interpretation*. Special Publication 20. Bath: Geological Society Publishing House, pp. 95–124.

Lidmar-Bergström, K., Olsson, S. & Roaldset, E. (1999) Relief features and palaeoweathering remnants in formerly glaciated Scandinavian basement areas. In: Thiry, M. & Simon-Coinçon, R. (Eds) *Palaeoweathering, Palaeosurfaces and Related Continental Deposits*. Special Publication 27, International Association of Sedimentologists. Oxford: Blackwell Science, pp. 275–301.

Mann, A.W. (1984) Mobility of gold and silver in lateritic weathering profiles; some observations from Western Australia. *Economic Geology* **79**, 38–49.

Marker, M.E. & McFarlane, M.J. (1997) Cartographic analysis of the African Surface complex between Albertinia and Mossel Bay, southern Cape, South Africa. *South African Journal of Geology* **100**, 185–194.

Marker, M.E., McFarlane, M.J. & Wormald, R.J. (2002) A laterite profile near Albertinia, Southern Cape, South Africa: its significance in the evolution of the African Surface. *South African Journal of Geology* **105**, 67–74.

Mason, T.F.D., Widdowson, M., Ellam, R.M. Oxburgh, R. (2000) Isotopic variability of Sr and Nd in lateritic deposits from the Deccan Traps, India: Evidence for an input of Aeolian material to the laterites. *Journal of Conference Abstracts* **5**, 674.

Mathieu, D., Bernat, M. & Nahon, D. (1995) Short-lived U and Th isotope distribution in a laterite derived from granite (Pitinga river basin, Amazouia, Brazil): Application to assessment of weathering rate. *Earth and Planetary Science Letters* **136**, 703–714.

McFarlane, M.J. (1971) Lateritization and landscape development in Kyagwe, Uganda. *Quarterly Journal of the Geological Society of London* **126**, 501–539.

McFarlane, M.J. (1976) *Laterite and Landscape*. London: Academic Press.

McFarlane, M.J. (1983a) Laterites. In: Goudie, A.S. & Pye, K. (Eds) *Chemical Sediments and Geomorphology*. London: Academic Press, pp. 7–58.

McFarlane, M.J. (1983b) A low level laterite profile from Uganda and its relevance to the question of parent material influence on the chemical composition of laterites. In: Wilson, R.C.L. (Ed.) *Residual Deposits: Surface Related Weathering Processes and Materials*. Special Publication 11, Geological Society of London, pp. 69–76.

Milnes, A.R., Bourman, R.P. & Northcote, K.H. (1985) Field relationships of ferricretes and weathered zones in southern South Australia: a contribution to 'laterite' studies in Australia. *Australian Journal of Soil Research* **23**, 441–465.

Moreira-Nordeman, L.M. (1980) Use of $^{234}U^{238}U$ disequilibrium in measuring chemical weathering rate of rocks. *Geochimica et Cosmochimica Acta* **44**, 103–108.

Morris, B.A. & Fletcher, I.A. (1987) Increased solubility of quartz following ferrous-ferric iron reactions. *Nature* **330**, 558–561.

Moura, M.L. (1987) The establishment of an international interdisciplinary collection of reference laterite profiles. *Zeitschrift für Geomorphologie N.F., Supplement Band* **64**, 111–118.

Nahon, D. (1986) Evolution of iron crusts in tropical landscapes. In: Colman, S.M. & Dethier, D.P. (Eds) *Rates of Chemical Weathering of Rocks and Minerals.* New York: Academic Press, pp. 169–191.

Nash, D.J., Shaw, P.A. & Thomas, D.S.G. (1994) Duricrust development and valley evolution – process – landform links in the Kalahari. *Earth Surface Processes and Landforms* **19**, 299–317.

Newbold, J.T. (1844) Notes, chiefly geological, across the Peninsula from Masulipatam to Goa, comprising remarks on the Regur and Laterite: occurrence of manganese veins in the latter, and certain traces of aqueous denudation on the surface of Southern India. *Journal of the Asiatic Society of Bengal* **13**(II), 984–1004.

Newbold, J.T. (1846) Summary of the geology of Southern India. VI: Laterite. *Journal of the Royal Asiatic Society of Great Britain & Northern Ireland* 227–240.

O'Connor, E.A., Pitfield, P.E. & Litherland, M. (1987) Landscape and Landsat over the Eastern Bolivian Shield. *Zeitschrift für Geomorphologie N.F., Supplement Band* **64**, 97–109.

Ollier, C.D. (1991) Laterite profiles, ferricrete and landscape evolution. *Zeitschrift für Geomorphologie N.F.* **35**, 165–173.

Ollier, C.D. & Pain, C.F. (1996) *Regolith, Soils and Landforms.* Chichester: John Wiley.

Ollier, C.D. & Powar, K.B. (1985) The Western Ghats and the morphotectonics of Peninsular India. *Zeitschrift für Geomorphologie Supplementband* **54**, 57–69.

Ollier, C.D., Chan, R.A., Craig, M.A. & Gibson, D.L. (1988) Aspects of landscape history and regolith in the Kalgoorlie region, Western Australia. *BMR Journal of Australian Geology and Geophysics* **10**, 309–321.

Partridge, T.C. & Maud, R.R. (1987) Geomorphic evolution of southern Africa since the Mesozoic. *South African Journal of Geology* **90**, 179–208.

Paquet, H., Colin, F., Duplay, J., Nahon, K. & Millot, G. (1987) Ni, Mn, Zn, Cr-smectites, early and effective traps for transition elements in supergene ore deposits. In: Rodrigez-Clemente, R. & Tardy, Y. (Eds). *Geochemistry and Mineral Formation in the Earth Surface.* Madrid: Consejo Superior de Investigaciones Cientificas, pp. 221–229.

Pillans, B. (1997) Soil development at snail's pace: evidence from a 6 Ma soil chronosequence on basalt in north Queensland, Australia. *Geoderma* **80**, 117–128.

Pistiner, J.S. & Henderson, G.M. (2003) Lithium-isotope fractionation during continental weathering processes. *Earth and Planetary Science Letters* **214**, 327–339.

Prescott, J.A. & Pendleton, R.L. (1952) *Laterite and Lateritic Soils.* Technical Communication 47. Farnham Royal: Commonwealth Bureau of Soil Science.

Price, G.D., Valdes, P.J. & Sellwood, B.W. (1997) Prediction of modern bauxite occurrence: implications for climate reconstruction. *Paleogeography, Paleoclimatology, Palaeoecology* **131**, 1–13.

Puecker-Ehrenbrink, B. & Ravizza, G. (2000) The marine osmium isotope record. *Terra Nova* **12**, 205–219.

Rudnick, R.L., Tomascak, P.B., Njo, H.B. & Gardner, L.R. (2004) Extreme lithium isotopic fractionation during continental weathering revealed in saprolites from South California. *Chemical Geology* **212**, 45–57.

Sahasrabudhe, Y.S. & Deshmukh, S.S. (1981) The laterites of the Maharashtra State. *Lateritisation Processes: Proceedings of the International Seminar on Lateritisation Processes, Trivandrum, 1979*. Rotterdam: Balkema, pp. 209–220.

Schellmann, W. (1982) Eine neue Lateritdefinition. *Geologische Jahrbuch* **D58**, 31–47.

Schellmann, W. (1986) A new definition of laterite. In: Banerji, P.K. (Ed.) Lateritisation processes. *Memoir of the Geological Survey of India* **120**, 1–7.

Schellmann, W. (2003) Discussion of 'A critique of the Schellmann definition and classification of laterite' by R.P. Bourman and C.D. Ollier (Catena 47, 117–131). *Catena* **52**, 77–79.

Schmidt, P.W. & Embleton, B.J.J. (1976) Palaeomagnetic results from sediments of the Perth Basin, Western Australia, and their bearing on the timing of regional lateritisation. *Palaeogeography, Palaeoclimatology, Palaeoecology* **19**, 257–273.

Schmidt, P.W., Prasad, V. & Ramam, P.K. (1983) Magnetic ages of some Indian laterites. *Palaeogeography, Palaeoclimatology, Palaeoecology* **44**, 185–102.

Schmitt, J.-M. (1999) Weathering, rainwater and atmospheric chemistry: example and modelling of granite weathering in present conditions in a CO_2-rich, and in an anoxic palaeoatmosphere. In: Thiry, M. & Simon-Coinçon, R. (Eds) *Palaeoweathering, Palaeosurfaces and Related Continental Deposits*. International Association of Sedimentologists Special Publication 27. Oxford: Blackwell, 21–41.

Schorin, H. & Carias, O. (1987) Analysis of natural and beneficiated ferruginous bauxite by both X-ray diffraction and X-ray fluorescence. *Chemical Geology* **60**, 19–204.

Schorin, H. & LaBreque, J.J. (1986) Three Laterite Standard Reference materials from Venezuela. *Memoirs of the Geological Survey of India* **120**, 89–101.

Schwarz, T. & Germann, K. (1999) Weathering surfaces, laterite-derived sediments and associated mineral deposits in north-east Africa. In: Thiry, M. & Simon-Coinçon, R. (Eds) *Palaeoweathering, Palaeosurfaces and Related Continental Deposits*. International Association of Sedimentologists Special Publication 27. Oxford: Blackwell, pp. 367–390.

Sharma, M., Clauer, N. & Toulkeridis, T. (1998). Rhemium-omsium systematics of an ancient laterite profile. Abstract: Goldschmidt Conference, Toulouse, 1998. *Mineralogical Magazine* **62A**, 1373–1374.

Short, S.A., Lowson, R.T., Ellis, J. & Price, D.M. (1989) Thorium-uranium disequilibrium dating of Late-Quaternary ferruginous concretions and rinds. *Geochimica et Cosmochimica Acta* **53**, 1379–1389.

Sivarajasingham, S., Alexander, L.T., Cady, J.G., & Cline, M.G. (1962): Laterite. *Advances in Agronomy* **14**, 1–60.

Soil Survey Staff (1975) *Soil Taxonomy*. USDA Handbook 436. Washington, DC: United States Department of Agriculture.

Spier, C.A., Vasconcelos, P.M. & Oliviera, M.B.S. (2006) $^{40}Ar/^{39}Ar$ geochronological constraints on the evolution of lateritic iron deposits in the Quadrilatero Ferrifero, Minas Gerais, Brazil. *Chemical Geology* **234**, 79–104.

Stephens, C.G. (1971) Laterite and silcrete in Australia: a study of the genetic relationships of laterite and silcrete and their companion materials, and their collective significance in the formation of the weathered mantle, soils, relief and drainage of the Australian continent. *Geoderma* **5**, 5–52.

Summerfield, M.A. (1991). *Global Geomorphology*. Harlow: Longman.

Tardy, Y. (1993*) Pétrologie des latérites et des soils tropicaux*. Masson, Paris, France.

Tardy, Y. & Roquin, C. (1992) Geochemistry and evolution of lateritic landscapes. In: Martini, I.P. and Chesworth, W. (Eds) *Weathering, Soils and Palaeosols*. Developments in Earth Surface Processes 2. Amsterdam: Elsevier, pp. 407–443.

Tardy, Y., Bocquier, G., Paquet, H. & Millot, G. (1973) Formation of clay from granite and its distribution in relation to climate and topography. *Geoderma* 10, 271–284.

Tardy, Y. Kobilsek, B. & Paquet, H. (1991) Mineralogical composition and geographical distribution of African and Brazilian periatlantic laterites. The influence of continental drift and tropical paleoclimates during the past 150 million years and implications for India and Australia. *Journal of African Earth Sciences* 12, 283–295.

Taylor, R.G. & Howard, K.W.F. (1999) The influence of tectonic setting on the hydrological characteristics of deeply weathered terrains: evidence from Uganda. *Journal of Hydrology* 218, 44–71.

Taylor, G.R. & Ruxton, B.P. (1987) A duricrust catena in South-East Australia. *Zeitschrift für Geomorphologie, N.F.* 31, 385–410.

Taylor, G.R., Eggleton, R.A.S., Holzhauer, C.C., Maconachie, L.A., Gordon, M., Brown, M.C. & McQueen, K.G. (1992) Cool climate lateritic and bauxitic weathering. *Journal of Geology* 100, 669–677.

Théveniaut, H. & Freysinnet, Ph. (1999) Paleomagnetism applied to lateritic profiles to assess saprolite and duricrust formation processes: the example of Mont Baduel profile (French Guiana). *Palaeogeography, Palaeoclimatology, Palaeoecology* 148, 209–231.

Thiry, M. & Simon-Coinçon, R. (Eds) (1997) *Palaeoweathering, Palaeosurfaces and Related Continental Deposits*. Special Publication 27, International Association of Sedimentologists. Oxford: Blackwell.

Thomas, M.F. (1994) *Geomorphology in the Tropics. A Study of Weathering and Denudation in Low Latitudes*. Chichester: John Wiley.

Trendall, A.F.J. (1962) The formation of apparent peneplains by a process of combined lateritisation and surface wash. *Zeitschrift für Geomorphologie, N.F.* 6, 183–197.

Tsekhovskii, Yu. G., Shchipakina, I.G. & Khramtsov, I.N. (1995). Lateritic eluvium and its redeposition products as indicators of Aptian–Turonian climate. *Stratigraphy and Geological Correlation* 3, 285–294.

Twidale, C.R. (1983) Australian laterites and silcretes: ages and significance. *Revue de Geologie Dynamique et de Geographie Physique* 24, 35–45.

Valeton, I. (1983) Palaeoenvironment of lateritic bauxites with vertical and lateral differentiation. In: Wilson, R.C.L. (Ed.) *Residual Deposits: Surface Related Weathering Processes and Materials*. Special Publication 11, Geological Society of London, pp. 77–90.

Vasconcelos, P.M. (1999a) K–Ar and $^{40}Ar/^{39}Ar$ geochronology of weathering processes. *Annual Review of Earth & Planetary Sciences* 27, 183–229.

Vasconcelos, P.M. (1999b) $^{40}Ar/^{39}Ar$ geochronology of supergene processes in ore deposits. *Economic Geology* 12, 73–113.

Viers, J. & Wasserburg, G.J. (2004) Behavior of Sm and Nd in a lateritic profile. *Geochimica et Cosmochimica Acta* 68, 2043–2054.

Whalley, W.B., Rea, B.R., Rainey, M.M. & McAlister, J.J. (1997) Rock weathering in blockfields: some preliminary data from mountain plateaus in North Norway. In: Widdowson, M. (Ed.) *Palaeosurfaces: Recognition, Reconstruction and Palaeoenvironmental Interpretation.* Geological Society Special Publication 120. Bath: Geological Society Publishing House, pp. 133–145.

Widdowson, M. (Ed.) (1997a) *Palaeosurfaces: Recognition, Reconstruction and Palaeoenvironmental Interpretation.* Geological Society Special Publication 120. Bath: Geological Society Publishing House.

Widdowson, M. (1997b) Tertiary palaeosurfaces of the SW Deccan, Western India: implications for passive margin uplift. In: Widdowson, M. (Ed.) *Palaeosurfaces: Recognition, Reconstruction and Palaeoenvironmental Interpretation.* Geological Society Special Publication 120. Bath: Geological Society Publishing House, pp. 221–248.

Widdowson, M. (1997c) The geomorphological and geological importance of palaeosurfaces. In: Widdowson, M. (Ed.) *Palaeosurfaces: Recognition, Reconstruction and Palaeoenvironmental Interpretation.* Special Publication 120. Bath: Geological Society Publishing House, pp. 1–12.

Widdowson, M. & Cox, K.G. (1996) Uplift and erosional history of the Deccan Traps, India: Evidence from laterites and drainage patterns of the Western Ghats and Konkan Coast. *Earth and Planetary Science Letters* 137, 57–69.

Widdowson, M. & Gunnell, Y. (1999) Lateritization, geomorphology and geodynamics of a passive continental margin: the Konkan and Kanara coastal lowlands of western Peninsula India. In: Thiry, M. & Simon-Coinçon, R. (Eds) *Palaeoweathering, Palaeosurfaces and Related Continental Deposits.* Special Publication 27, International Association of Sedimentologists. Oxford: Blackwell, pp. 245–274.

Wimpenny, J.B., Gannoun, A., Burton, K.W., Widdowson, M., James, R.H. & Gislason, S. (2007) Rhenium and osmium isotope and elemental behaviour accompanying laterite formation in the Deccan region of India. *Earth and Planetary Science Letters*, doi: 10.1016/j.epsl.2007.06.028.

Young, R.W. (1985) Silcrete distribution in eastern Australia. *Zeitschrift für Geomorphologie, N.F.* 29, 21–36.

Young, R.W., Short, S.A., Price, D.M., Bryant, E.A., Nanson, G.C., Gardiner, B.H. & Wray, R.A.L. (1994) Ferruginous weathering under cool temperate climates during the Late Pleistocene in southeastern Australia. *Zeitschrift für Geomorphologie N.F.* 38, 45–57.

Zeese, R. (1996) Tertiary weathering profiles in central Nigeria as indicators of paleoenvironmental conditions. *Geomorphology* 6, 61–70.

Zeese, R., Schwertmann, U., Tietz, G.F. & Jux, U. (1994) Mineralogy and stratigraphy of three deep lateritic profiles of the Jos Plateau (Central Nigeria). *Catena* 21, 195–214.

Chapter Four

Silcrete

David J. Nash and J. Stewart Ullyott

4.1 Introduction: Nature and General Characteristics

Silcrete is a term first used by Lamplugh (1902) to describe the products of near-surface processes by which silica accumulates in and/or replaces a soil, sediment, rock or weathered material to form an indurated mass (Watson and Nash, 1997). Silcretes are defined as containing >85 wt. % SiO_2, with many comprising >95 wt.% SiO_2 (Summerfield, 1983a). They are most widespread in Australia, southern Africa and western Europe, with localised occurrences elsewhere (section 4.2). With the exception of biogenic silcretes in Botswana (Shaw et al., 1990), dorbanks in South Africa (Ellis and Schloms, 1982), and duripans in North America (Flach et al., 1969; Chadwick et al., 1989; Dubroeucq and Thiry, 1994), most silcretes are relict features. The majority require stable geomorphological conditions to develop, although the genesis of some silcretes may be related to actively evolving landscapes (Thiry, 1999).

Silcretes exhibit a wide variety of forms (Figure 4.1), but commonly consist of brittle masses or nodules of hard, silica-cemented quartzose sand with a conchoidal or sub-conchoidal fracture. Very hard, fine-grained 'porcellanitic' and cherty varieties also occur (Peterson and von der Borch, 1965; Wopfner, 1983; Schubel and Simonson, 1990; Mišík, 1996), with less well-cemented silcretes common in mid-latitude settings (Summerfield and Goudie, 1980; Thiry et al., 1988a). The silcrete cement (or matrix) can contain a range of silica minerals, of which opal, chalcedony, cryptocrystalline silica and quartz are the most widely documented (section 4.4). The presence of other minerals may influence the silcrete colour, with grey, white, buff, brown, red and green varieties reported. Root casts (Goudie, 1973; Milnes and Twidale, 1983), preserved plant material (Callender, 1978; Ambrose and Flint, 1981; Callen, 1983; Milnes and Twidale, 1983; Lindqvist, 1990; Taylor and Eggleton, 2001; Campbell, 2002), and ant or termite burrows (Milnes et al., 1991; Benbow et al., 1995; Thiry et al., 1995, 2006) are also documented.

Figure 4.1 Silcretes in the landscape: (A) Silcrete outcrop near Albertinia, Western Cape, South Africa, with eroded remnants of the deeply-weathered African palaeosurface in the background; (B) Silcrete caprock overlying basalt at the eastern margin of the Kalahari Group sediments, Sesase Hill, Botswana; (C) Undulating surface of a silcrete-capped mesa north of Tibooburra, New South Wales, Australia; (D) Silcrete outcrop on the flanks of the Moshaweng Valley, south of Letlhakeng, Botswana.

Attempts to classify silcretes have come full circle since the work of Goudie (1973), who was the first to subdivide duricrusts according to their genesis (Table 4.1). Smale (1973) suggested classification on the basis of macroscale characteristics (Table 4.2), whereas Wopfner (1978) proposed a scheme based on mineralogy and macromorphology (Table 4.3). However, application of both classifications has been problematic. Summerfield (1983a) suggested a simpler micromorphological classification (Table 4.4), which has been widely used and applied to other duricrusts. All of these schemes were superceded by Milnes and Thiry (1992) who, echoing Goudie (1973), proposed the first workable genetic classification. This divided silcretes into two groups on the basis of micromorphology; pedogenic (or complex) and groundwater (or simple), with Thiry (1999) adding a third category of silicification associated with evaporites (Table 4.5). The authors, however, consider this classification to be too limited and propose that shown in Figure 4.2. This scheme subdivides silcretes into pedogenic (Figure 4.3) and 'non-pedogenic' varieties, with the latter category grouped on the basis of the geomorphological context of silicification into groundwater (Figure 4.4), drainage-line and pan/

Table 4.1 Genetic classification of duricrusts (after Goudie, 1973)

1. Fluvial	Deposition or precipitation in valleys and channels, or deposition or alteration by sheet-flood action
2. Lacustrine	Formation in, or marginal to lakes, pans or playas
3. *In situ* models	Relative accumulation by removal of other constituents by weathering
4. Capillary rise from, or fluctuation of, groundwater	Evaporation draws up solute; or precipitation in zone of intermittent saturation
5. Pedogenic models (a) *per ascensum*	Precipitation by surface evaporation, or on mixing with descending low pH or saline solutions
(b) *per descensum*	Precipitation after leaching, or down-washing of plant accumulated, or aeolian derived material
6. Detrital types of secondary origin	Transport, deposition, and recementation of fragments of duricrust material

Table 4.2 Morphological classification of silcrete (after Smale, 1973)

Silcrete type	Silcrete characteristics
Terrazzo	Framework of *c.* 60% quartz grains, with solution cavities, in a cryptocrystalline quartz or opaline matrix with colloform banding; conchoidal fracture typical
Conglomeratic	Conglomeratic fabric with abundant pebbles of terrazzo silcrete or other material
Albertinia	Detrital quartz absent or rare, cryptocrystalline quartz matrix of terrazzo type
Opaline/fine-grained massive	No detrital component, massive, comprising opaline, chalcedonic or cryptocrystalline silica
Quartzitic	Grain supported fabric cemented by overgrowths on detrital quartz

Table 4.3 Classification of silcrete according to matrix (cement) type and macromorphology (Wopfner, 1978, 1983)

Group I silcretes	Matrix:	crystalline quartz
		(1) irregular crystalline
		(2) optically continuous overgrowths
	Habit:	(a) blocky
		(b) bulbous/pillowy
	Retention of host-rock texture in 1a and 2a	
	Profile:	bleached or kaolinised, below zone of pedogenesis
Group II silcretes	Matrix:	cryptocrystalline quartz
		(1) massive
		(2) pisolitic/pseudopebbly
		(3) laminated
	Habit:	(a) columnar
		(b) platy
		(c) botryoidal
		(d) pillowy
	Host-rock textures obliterated	
	Profile:	within intensely kaolinised palaeosol
Group III silcretes	Matrix:	amorphous (opaline)
	Habit:	(a) breccia
		(b) conglomeratic
		(c) replacement/infill
	Perfect preservation of host-rock texture	
	Profile	lack of distinct profile but associated with gypsum and alunite

Table 4.4 Micromorphological classification of silcrete (after Summerfield, 1983c)

Fabric type	Fabric characteristics
GS-fabric	*Grain supported fabric* – skeletal grains (i.e. grains >30 µm diameter) constitute a self-supporting framework. Subdivided by cement type: (a) optically continuous quartz overgrowths (b) chalcedonic overgrowths (c) microquartz/cryptocrystalline/opaline silica in-fill
F-fabric	*Floating fabric* – skeletal grains comprise >5% but float in matrix, grain solution or fretting common. Sub-types: (a) massive (glaebules absent) (b) glaebular (glaebules present)
M-fabric	*Matrix fabric* – skeletal grains comprise <5%. Sub-types: (a) massive (glaebules absent) (b) glaebular (glaebules present)
C-fabric	*Conglomeratic fabric* – skeletal grains include fractured bedrock, gravel or duricrust fragments >4 mm

Table 4.5 Genetic classification of silcrete (after Thiry, 1999)

Genetic type	Macromorphology	Host material	Fabric/chemistry	Palaeoenvironment
Pedogenic silcrete	Columnar jointing, vertical profile organization	Kaolinitic sediments	Host features destroyed; microquartz dominant, Ti enriched, low Al and Fe, complex micromorphology with illuvial features	Soil profile; tropical/sub-tropical alternating wet and dry, silicification in soil profile
			Host features destroyed; opal dominant, low Ti, clay and Fe rich, complex micromorphology with illuvial features	Soil profile; tropical/sub-tropical alternating wet and dry/evaporitic
Groundwater silcrete	Massive, lenticular/tabular	Sands and pebbly sands	Host features preserved, simple micromorphology illuvial features absent	Water table marginal to valleys; temperate–tropical
	Irregular/nodular	Carbonates or clay-rich sediments	Host features preserved, simple micromorphology illuvial features absent	Related to karst circulation or water table fluctuation; temperate–tropical
Silicification with evaporites	Nodular	Lacustrine/playa sediments	Simple, chert like	Marginal to evaporitic depressions, arid/semi-arid

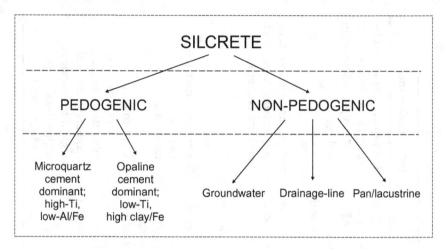

Figure 4.2 Geomorphological classification of silcretes.

Figure 4.3 Pedogenic silcrete profiles. (A) *In situ* silcrete profile developed within deeply weathered bedrock, near Inniskillin, Western Cape, South Africa. (B) Cambered silcrete profile in the Mt Wood Hills area near Tibooburra, New South Wales, Australia, showing columnar structures and 'candle-wax' drips. (C) Columnar silcrete profile, Montagny Lencoup, Paris Basin, France.

Figure 4.4 Groundwater silcrete outcrops. (A) Superposed groundwater silcrete lenses developed within Fontainebleau Sand at Bonnevault Quarry, Paris Basin, France (with Médard Thiry for scale). (B) Groundwater silcrete developed within Red Bluff Sand underlying Newer Volcanics basalt at Taylor Creek, north of Melbourne, Australia (with John Webb for scale). (C) Groundwater silcrete boulder (sarsen) 'train' within a valley floor at Clatford Bottom, Wiltshire, UK.

lacustrine types (Figure 4.5). This classification is used in the remainder of this chapter.

4.2 Distribution, Field Occurrence and Geomorphological Relations

Silcretes have been documented in every continent except Antarctica (Table 4.6), but are most widespread in the interior, west and southeast of Australia. Silcrete is also common in the Kalahari and Cape coastal areas of southern Africa, and has been described in various parts of North Africa. In Europe, silcretes occur in Belgium, France, Germany, Italy, The Netherlands, Slovakia, Spain, Portugal and the UK. Silcretes have also been noted in parts of North and South America and Asia. In the geological record, silcrete has been identified from the Precambrian to Quaternary (Cros and Freytet, 1981; Twidale, 1983; Mustard and Donaldson, 1990;

Figure 4.5 (A) Massive drainage-line silcrete in the floor of the Boteti River, Botswana, at Samedupe Drift (see Figures 4.7A, B for microscopic thin-sections and 4.10A for associated sedimentary cores). (B) Close-up of a partially silicified non-pedogenic calcrete beneath the floor of Kang Pan, near Kang, Botswana – light coloured parts of the profile are cemented by calcium carbonate, with darker silicified patches occurring along joint-faces as well as towards the centre of joint-bounded blocks (see Figure 4.7C for microscopic thin-section). (C) Sheet-like pan/lacustrine silcrete developed through the desiccation of formerly floating colonies of the silica-fixing cyanobacteria *Chloriflexus* at Sua Pan, Botswana.

Table 4.6 Examples of regional investigations of silcrete

Continent	Location	Key investigations
Australasia	Australia	Stephens (1964, 1966, 1971); Brückner (1966); Grant and Aitchison (1970); Langford-Smith (1978); Watts (1977, 1978a,b); Wopfner (1978); Callen (1983); Milnes and Twidale (1983); van der Graaff (1983); Wopfner (1985); Taylor and Ruxton (1987); Ollier (1991); Thiry and Milnes (1991); Simon-Coinçon et al. (1996); Webb and Golding (1998); Hill et al. (2003); Alexandre et al. (2004); Lee and Gilkes (2005); Thiry et al. (2006)
	New Zealand	Lindqvist (1990); Youngson (2005)
Africa	Kalahari	Goudie (1973); Summerfield (1982, 1983c); Shaw and de Vries (1985); Shaw et al. (1990); Nash et al. (1994a,b, 2004); Nash and Shaw (1998); Shaw and Nash (1998); Ringrose et al. (2002, 2005); Nash and Hopkinson (2004)
	Southern Cape	Frankel and Kent (1938); Bosazza (1939); Mountain (1946, 1951, 1980); Frankel (1952); Smale (1973); Summerfield (1981, 1983a–d, 1984); Roberts et al. (1997)
	Namibia	Storz (1926)
	East Africa	Cecioni (1941); Dixey (1948); Bassett (1954)
	North Africa	Hume (1925); Auzel and Cailleux (1949); Alimen and Deicha (1958); Millot (1964); Thiry and Ben Brahim (1990, 1997)
Asia	Arabia	Khalaf (1988)
	Karakum	Fersmann and Wlodawetz (1926)

Table 4.6 *Continued*

Continent	Location	Key investigations
Europe	Belgium	Demoulin (1989, 1990)
	France	Thiry (1978, 1981, 1988, 1989, 1999); Thiry and Turland (1985); Thiry et al. (1988a,b); Thiry and Simon-Coinçon (1996); Roulin et al. (1986); Roulin (1987); Simon-Coinçon et al. (2000); Thiry and Marechal (2001); Basile-Doelsch et al. (2005)
	Germany	Wopfner (1983); Henningsen (1986)
	Italy	Cros and Freytet (1981)
	Netherlands	Van der Broek and van der Vaals (1967)
	Slovakia	Curlík and Forgác (1996); Mišík (1996)
	Spain	Bustillo and Bustillo (1993, 2000); Rodas et al. (1994); Armenteros et al. (1995); Ballesteros et al. (1997); Gomez-Gras et al. (2000); Bustillo (2001); Parcerisa et al. (2001); Bustillo and García Romero (2003).
	Portugal	Meyer and Pena dos Reis (1985)
	UK	Davies and Baines (1953); Kerr (1955); Clark et al. (1967); Isaac (1979, 1983); Summerfield (1979); Summerfield and Goudie (1980); Summerfield and Whalley (1980); Hepworth (1998); Nash et al. (1998); Ullyott et al. (1998, 2004); Hancock (2000); Ullyott and Nash (2006)
North America	Canada	Ross and Chiarenzelli (1985); Leckie and Cheel (1990)
	Mexico	Dubroeucq and Thiry (1994)
	USA	Dury and Knox (1971); Dury and Haberman (1978); Gassaway (1988, 1990); Chadwick et al. (1989); Bromley (1992); Terry and Evans (1994)
South America	Brazil	King (1967)

Dubroeucq and Thiry, 1994; Martini, 1994), with the majority of occurrences being of Tertiary age (Thiry, 1999).

Silcretes frequently crop out in inverted relief as caprocks on residual hills, mesas or escarpments (Summerfield, 1983a; Twidale and Milnes, 1983; Ballesteros et al., 1997; Hill et al., 2003; Figure 4.1), and may be extensive where formation occurred on a palaeosurface (Figure 4.3). Spatially limited silcrete bodies associated with palaeochannels, on the other hand, may form elongate sinuous ridges (Pain and Ollier, 1995) or more localised areas of inverted topography (Hill et al., 2003). Undercut slopes often occur at the edges of such features, leading to the development of 'breakaways', which may be mantled by silcrete 'gibber plains' (Mountain, 1946; Twidale and Milnes 1983). Silcretes also outcrop as layers or lenses part way up slopes (Mountain, 1952; Young, 1978; Thiry et al., 1988a; Milnes et al., 1991). Such occurrences may represent exhumed silcretes that formed at earlier stages of landscape evolution, although it is important to distinguish these from recemented silcrete talus found in similar settings (Frankel and Kent, 1938; Ruxton and Taylor, 1982). Silcretes in less erosional contexts occur in, or marginal to, valleys or ephemeral lakes (Summerfield, 1982; Young, 1985; Taylor and Ruxton, 1987; Thiry et al., 1988a; Nash et al., 1994a,b, 2004; Shaw and Nash, 1998; Ringrose et al., 2005; Figures 4.4 and 4.5).

Silcrete may form within, or by the replacement of, a range of host material types. Many Australian and South African silcretes have developed within deeply kaolinised or bleached regolith (Summerfield, 1978; 1983b–d; Wopfner, 1983; Milnes and Thiry, 1992; Thiry et al., 2006). Formation can, however, occur in (or on) unweathered bedrock or otherwise unaltered and unconsolidated sediments. Silcrete development is also frequently associated with the replacement of carbonate sediments or calcretes (Cros and Freytet, 1981; Twidale and Milnes, 1983; Ribet and Thiry, 1990; Armenteros et al., 1995; Nash and Shaw, 1998; Thiry and Ribet, 1999; Nash et al., 2004).

4.3 Macromorphological Characteristics

Individual silcrete masses are usually 1–3 m thick, although profiles of >15 m have been recorded (Thiry and Simon-Coinçon, 1996). Multiple horizons are common, with surface layers underlain by lenses at depths of up to 100 m (Mountain, 1946, 1952; Frankel, 1952; Senior and Senior, 1972; Summerfield, 1978, 1981; Thiry et al., 1988a; Basile-Doelsch et al., 2005). This may reflect the operation of different surface and sub-surface silicification mechanisms, as in the Eromanga Basin of Australia (Milnes et al., 1991; Thiry et al., 2006) and the Apt region of southeast France (Basile-Doelsch et al., 2005), where multiple groundwater silcrete layers occur at depths of 4–10 m and 50–100 m, respectively, beneath surficial pedogenic silcretes. Alternatively, superposed lenses may relate to successive stages of landscape dissection, as in the Paris Basin (Thiry et al., 1988a). The lateral continuity of silcrete bodies is often hard to ascertain, although pedogenic silcretes are usually more extensive than other types (Thiry and Milnes, 1991). In Australia, for example, widespread silcrete sheets occur, especially related to the Cordillo palaeosurface (Langford-Smith, 1978a; Wopfner, 1978; Alley, 1998). In other areas, silcretes are only localised (Frankel, 1952; Wopfner, 1983; Wopfner, 1985; Taylor and Ruxton, 1987; Thiry et al., 1988a; Shaw and Nash, 1998). Lateral variations in thickness within individual silcrete bodies are also typical (Mountain, 1952; Thiry, 1978; Summerfield, 1981).

The morphology of individual silcrete profiles reflects the mode of formation (section 4.7) and nature of the host materials (Summerfield, 1983a; Milnes and Thiry, 1992). However, appearances vary, even between profiles developed by similar mechanisms, to the extent that it is difficult to distinguish silcrete types on the basis of macromorphology alone. For example, groundwater silcretes at Stuart Creek, Australia, appear as amoeboid opalite masses where silicification occurred in Cretaceous shales but as massive or tuberous bodies within Tertiary fluvial sands (Thiry and Milnes, 1991). Some generalisations can be made. Strongly developed vertical jointing, giving a columnar, rodded or even phallic appearance, is common in pedogenic silcretes (Goudie, 1973; Wopfner, 1978; Callen 1983; van der Graaff, 1983; Thiry, 1988, 1999; Dubroeucq and Thiry, 1994; Ballesteros et al., 1997; Thiry et al., 2006; Figure 4.3), and often occurs in conjunction with macroscopic geopetal features such as laminated drapes, 'candle wax drips' (Thiry, 1978; Summerfield, 1981; Webb and Golding, 1998) and cap-like forms (Callen, 1983; van de Graaff, 1983). Less regular blocky, vertical or horizontal jointing and pseudo-bedding has also been reported (Mountain, 1952; Summerfield, 1981, 1984). The degree of cementation of pedogenic profiles commonly decreases with depth, in contrast to groundwater, drainage-line and pan/lacustrine silcretes, which are more uniformly massive (Thiry, 1999) or less systematically jointed (Thiry and Millot, 1987; Figures 4.4 and 4.5). Grooved, ropy or fluted surfaces are widely documented, as are nodular, concretory or pisolitic horizons, and tabular, lenticular, pillowy, botryoidal and mammiform shapes (Summerfield, 1983a). Exhumed silcretes may have a fretted appearance as a result of surface or sub-surface weathering (Wopfner, 1978; Twidale and Milnes, 1983; Thiry and Millot, 1987; Milnes et al., 1991), and dissolution pipes and basins are common.

4.4 Mineralogy and Chemistry

Silcretes are chemically simple (Summerfield, 1983a,d), usually comprising >90% silica with minor amounts of titanium, iron and aluminium oxides (Table 4.7). As with macromorphology, it is difficult to identify

Table 4.7 Bulk chemistry of world silcretes (analyses by X-ray fluorescence)

Region		SiO_2	TiO_2	Al_2O_3	Fe_2O_3	MnO	MgO	CaO	Na_2O	K_2O	P_2O_5	SO_3	LOI
Inland Australia[*]	Mean	94.81	2.09	0.50	0.76	0.01	0.10	0.18	0.07	0.08	0.05	No data	1.22
n = 72	SD	5.06	4.06	0.77	1.50	0.01	0.09	0.20	0.16	0.09	0.08	No data	1.18
Eastern Australia[†]	Mean	97.58	0.42	0.30	0.79	0.02	0.10	0.02	0.02	0.03	0.02	0.004	0.47
n = 63	SD	1.76	0.36	0.37	1.28	0.05	0.14	0.02	0.02	0.06	0.02	0.001	0.44
Kalahari[‡]	Mean	91.63	0.12	1.69	0.86	0.01	0.98	0.85	0.36	0.95	0.01	No data	2.80
n = 48	SD	4.04	0.06	0.98	0.54	0.00	0.66	1.51	0.35	0.86	0.00	No data	1.51
Cape Coastal[§]	Mean	95.04	1.79	0.61	1.28	0.01	0.28	0.13	No data	0.05	0.04	No data	0.99
n = 66	SD	2.54	0.58	0.46	1.70	0.00	0.20	0.37	No data	0.12	0.03	No data	0.67
Europe[¶]	Mean	96.98	0.53	0.68	0.43	0.01	0.09	0.10	0.10	0.07	0.02	0.100	0.48
n = 96	SD	2.46	0.89	1.42	0.47	0.00	0.13	0.13	0.09	0.13	0.02	0.000	0.31

[*]Calculated from analyses in Hutton et al. (1972), Senior and Senior (1972), Watts (1977), Callender (1978), Senior (1978), Wopfner (1978), O'Neill (1984), Collins (1985), Thiry and Milnes (1991), van Dijk and Beckmann (1978), Tait (1998) and Webb and Golding (1998).

[†]Calculated from analyses in Taylor and Smith (1975), O'Neill (1984), Collins (1985) and Webb and Golding (1998).

[‡]Calculated from analyses in Summerfield (1982), Nash and Shaw (1998) and Nash et al. (1994b, 2004).

[§]Calculated from analyses in Frankel and Kent (1938), Bosazza (1939), Frankel (1952) and Summerfield (1983d).

[¶]Calculated from analyses in Kerr (1955), Summerfield (1978), Thiry (1978), Isaac (1983), Ullyott et al. (2004) and Ullyott and Nash (2006).

n, number; SD, standard deviation; LOI, loss on ignition.

mineralogical or chemical criteria to distinguish silcrete types. Silcrete mineralogy reflects the properties of the host material as well as the conditions under which silica precipitation occurred (section 4.6). The detrital component of silcretes is usually quartzose. The matrix may, however, contain a range of silica species (or polymorphs), of which the most widely recognised are, in order of increasing crystal organisation, opal, chalcedony and quartz. Opal may occur both within the matrix and as cavity linings (Bustillo and Bustillo, 2000), and may be present as non-crystalline opal-A and opal-AG or in near-amorphous to microcrystalline forms such as opal-T, opal-C or opal-CT (Flörke et al., 1991). Chalcedony, a microcrystalline fibrous form of silica comprising intergrowths of quartz and the silica polymorph moganite (Flörke et al., 1984; Heaney, 1993, 1995), is documented as a cement, void-lining and void-fill (Summerfield, 1983a,c; Nash and Shaw, 1998). It can occur in optically length-fast or -slow forms (Graetsch, 1994). The length-fast variety may develop in a range of environments, but length-slow chalcedony is frequently, but not exclusively, associated with formation in evaporitic settings (Folk and Pittman, 1971). The presence of enhanced levels of moganite (>20 wt.%) may also provide an indication of precipitation under an evaporitic regime (Heaney, 1995). Quartz may be present within the matrix as microquartz or as well-developed crystals of authigenic α-quartz (or megaquartz). Microquartz (grain size <20 μm) is present both as a cement and void-lining, whilst megaquartz (>20 μm) commonly occurs as intergrown microscopic grains, overgrowths and as subhedral quartz crystals towards the centre of cavities (Nash and Hopkinson, 2004).

There is considerable evidence for systematic organisation of silica cements. Thiry and Millot (1987), for example, identified a sequence of silicification starting with amorphous opal and progressing through chalcedony to microquartz and megaquartz. This sequence has been widely documented within pedogenic silcretes, where more ordered silica is typically found toward the top of the profile (e.g. Thiry, 1978, 1988; Ballesteros et al., 1997), but is uncommon or absent in non-pedogenic types. The sequence is also recognised within vugh fills, where increasingly ordered forms of silica occur typically toward the centre of voids (Summerfield, 1982; Thiry and Milnes, 1991; Nash et al., 1994a; Rodas et al., 1994), although occasionally this order is reversed (Abdel-Wahab et al., 1998). The increasing organisation of silica minerals towards the centres of pore spaces and voids may be accompanied by increasing fractionation of Si isotopes (Basile-Doelsch, 2006), with Basile-Doelsch et al. (2005) reporting $\delta^{30}Si$ values as low as −5.7‰ for megaquartz cements within groundwater silcretes from southeast France. The nature of the host material may also influence silica mineralogy (Taylor and Smith, 1975; Watts, 1978a,b; Summerfield, 1982; Wopfner, 1983; Webb and Golding, 1998), with opal prevalent in argillaceous substrates and microquartz more common in porous terrigenous sediments or carbonates (Callen, 1983; Thiry and Ben Brahim, 1990; Benbow et al., 1995).

A range of other minerals may also be present. There has been controversy over the significance of titania, particularly where TiO_2 levels have been used as an indicator of palaeoenvironmental conditions (section 4.8). Titanium is usually present as anatase and may be disseminated throughout opaline cements or concentrated within illuvial features (Thiry, 1978). TiO_2 may be enriched (to >1%) in some silcretes (Summerfield, 1979, 1983d; Thiry and Millot, 1987) and exceed 20% in silcrete 'skins' (Hutton et al., 1972). TiO_2 content appears to be related to host-material chemistry and the ratio of matrix to detrital grains, as well as to the environment of silicification (Summerfield, 1979, 1983a; Nash et al., 1994b; Webb and Golding, 1998). Late aeolian inputs of Ti may be significant (McFarlane et al., 1992), although this suggestion is at odds with TiO_2 distribution in pedogenic profiles where levels are often highest near the base (Thiry, 1978) and may be elevated in the substrate (Frankel and Kent, 1938). The mode of TiO_2 emplacement has also been debated. Relative accumulation of titania was favoured by Hutton et al. (1972, 1978), and titanium mobility during silcrete development is suggested by the distribution of anatase both in silcrete skins (Milnes and Hutton, 1974) and colloform structures (Frankel and Kent, 1938; Summerfield, 1978; Milnes and Twidale, 1983; section 4.5.3). However, analyses of Australian silcrete geochemistry by Webb and Golding (1998) indicate limited Ti mobility during silicification. An interplay of relative and absolute accumulation is more likely, as TiO_2 levels in some silcretes are too great to be accounted for by relative accumulation alone (Thiry, 1978). Other resistate trace elements may also be enriched in silcretes (Twidale and Hutton, 1986).

The alumina content of silcrete is usually <1% and reflects in-washed, authigenic or inherited clay minerals (Thiry, 1978; Summerfield, 1983a; Roulin, 1987). However, under certain conditions, significant amounts (>10%) of clay minerals may be retained during silicification (Meyer and Pena dos Reis, 1985; Thiry and Turland, 1985; Bustillo and Bustillo, 1993; Rodas et al., 1994; Armenteros et al., 1995; Ballesteros et al., 1997). In such cases, silcretes may grade into Al-rich siliceous duricrusts. Iron minerals, such as goethite and haematite may also be retained (Wopfner 1978; Meyer and Pena dos Reis 1985; Thiry and Turland, 1985; Ballesteros et al., 1997) and may account for 5–10% of the bulk chemistry. The K- and Fe-rich aluminosilicate mineral glauconite has been reported in some Kalahari silcretes (Summerfield, 1982; Shaw and Nash, 1998) and may indicate sub-oxic, partially reducing groundwater conditions during silicification (Nash et al., 2004).

4.5 Micromorphological Characteristics

The examination of silcretes in thin-section and by techniques such as scanning electron microscopy (Figures 4.6 and 4.7) can provide important

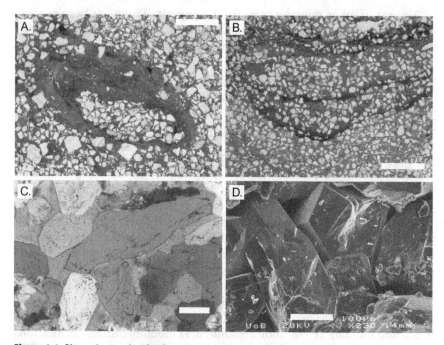

Figure 4.6 Photomicrographs of pedogenic and groundwater silcretes. (A) Thin-section view of a grain-supported to floating fabric glaebular pedogenic silcrete from Stuart Creek, South Australia, consisting of quartz grains surrounded by a microquartz and opal matrix (plain polarised light; scale bar 2 mm; micrograph courtesy of John Webb); (B) Thin-section view of a grain-supported to floating fabric pedogenic silcrete from Stuart Creek, South Australia, consisting of quartz grains surrounded by a microquartz and opal matrix and exhibiting anatase-rich geopetal laminations (plain polarised light; scale bar 2 mm; micrograph courtesy of John Webb). (C) Thin-section view of a grain-supported fabric groundwater silcrete from Chyngton, South Downs, UK, consisting of quartz grains cemented by optically continuous quartz overgrowths (cross-polarised light; scale bar 0.1 mm). (D) Scanning election microscope view of quartz overgrowth cements within a grain-supported fabric groundwater silcrete from Stanmer, South Downs, UK (scale bar 0.1 mm).

clues to their mode of development and subsequent diagenetic history. When viewed at this scale, silcretes can be seen to comprise varying proportions of detrital minerals, silica cements and void spaces, which may be partially or completely filled with secondary silica or other minerals. The nature and variety of these components reflect not only the diagenetic processes operating during and after formation, but may, in part, be inherited from the host sediment, regolith or bedrock.

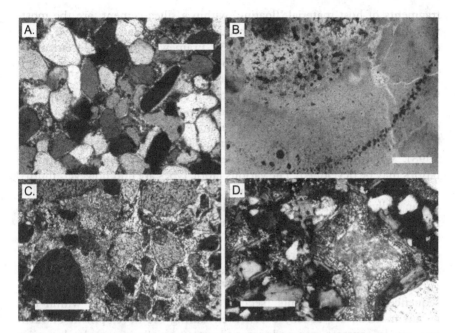

Figure 4.7 Photomicrographs of drainage-line and pan/lacustrine silcretes. (A) Thin-section view of a grain-supported fabric drainage-line silcrete from the floor of the Boteti River at Samedupe, Botswana, consisting of quartz grains cemented by microquartz and cryptocrystalline silica and exhibiting length-fast chalcedonic silica grain-coating cements (cross-polarised light, scale bar 0.5 mm). (B) Thin-section view of a pisolithic matrix fabric drainage-line silcrete from beneath the floor of the Boteti River at Samedupe, Botswana, consisting of alternating layers of disordered chalcedony, cryptocrystalline silica and sparry calcite (cross-polarised light, scale bar 0.5 mm). See Figures 4.5A and 4.10A for geomorphological and stratigraphic context. (C) Thin-section view of a partially silicified non-pedogenic calcrete from a depth of 4.0 m beneath the floor of Kang Pan, near Kang, Botswana, showing patchy replacement of carbonate cement (left of view) by chalcedonic and cryptocrystalline silica (right of view), with sharp boundary between carbonate and silica cements (cross-polarised light, scale bar 0.5 mm). See Fig. 4.5B for profile context. (D) Thin-section view of a complex floating fabric pan/lacustrine silcrete from Sua Pan, Botswana, consisting of quartz grains cemented by chalcedonic and cryptocrystalline silica which is partially replaced by micritic calcite. The sample also contains voids that are lined by opal and chalcedony and infilled with late-stage calcite (cross-polarised light, scale bar 0.5 mm).

4.5.1 Silcrete microfabrics

The classification proposed by Summerfield (1983a; Table 4.4) is the most widely used for describing silcrete microfabrics, which are defined according to the proportion and size of any detrital grains which may be present (Figures 4.6 and 4.7). Floating (or F) fabric silcretes, which consist of fragments of host material 'floating' within a silica cement, are the most widespread and commonly form through the replacement of mixed clay–silt/sand parent materials (Smale, 1973). Framework grain replacement or expansion during silicification has also been proposed to account for such fabrics (Watts, 1978a; Summerfield, 1983c). Matrix (M) fabric silcretes, which contain only minor detrital material, commonly occur within silicified clay-rich or carbonate hosts. Grain-supported (GS) fabrics are typical of silcretes developed in arenaceous sediments, where cementation occurs by void-filling between grains (Summerfield, 1983a; Shaw and Nash, 1998). Where clasts of >4mm diameter are present, conglomeratic (C) fabrics may be developed. The various fabric types are not, however, mutually exclusive. For example, F-fabrics may grade into GS- or M- fabrics within individual profiles or even single samples (Ullyott et al., 2004). All four fabrics occur across the range of silcrete types. In order to distinguish between pedogenic and non-pedogenic silcretes it is necessary to identify diagnostic features. According to Thiry (1999), pedogenic silcretes normally contain structures indicative of soil-forming processes (e.g. nodules, cutans, caps and eluvial structures; Figure 4.6A, B) and exhibit vertical sequences of silica cements (section 4.4). Textures present within the parent material are usually overwritten during the development of such features. In contrast, non-pedogenic silcretes partially or completely maintain inherited structures and typically lack systematic vertical organisation (Milnes and Thiry, 1992; Shaw and Nash, 1998; Ullyott and Nash, 2006).

4.5.2 Silcrete cements

Silcretes may be cemented by a range of silica and other minerals. Neither specific microfabrics nor silcrete types are associated with particular cementing agents, although some generalisations can be made. The matrix of F- and M-fabric silcretes usually consists of microquartz, cryptocrystalline or opaline silica and may include silt- or clay-sized detrital quartz as well as disseminated or colloform anatase, iron oxides and clay minerals. Diffuse or fretted boundaries are typical on host grains and may reflect replacement and dissolution during formation (Frankel and Kent, 1938; Smale, 1973; Taylor, 1978; Thiry, 1978; Watts, 1978a). GS-fabric silcretes commonly exhibit microquartz or cryptocrystalline silica cements. Syntaxial optically continuous quartz overgrowth cements (Figure 4.6C, D) may form, providing there is a lack of intergranular matrix or grain coatings (Heald and Larese, 1974; Dewars and Ortoleva, 1991) and porewaters are relatively dilute so that silica precipitation can occur slowly (e.g. Thiry and Millot, 1987). Rare chalcedonic overgrowths have been documented (Taylor and

Smith, 1975; Shaw and Nash, 1998; Webb and Golding, 1998; Figure 4.7A). Gradations between overgrowth and less well-ordered silica cements may also occur (Ullyott et al., 2004; Ullyott and Nash, 2006).

The variety of silica species within a silcrete matrix depends not only upon the polymorph initially precipitated, but also the subsequent diagenetic history of the material. Silica polymorphs may, for example, transform over time by dissolution and recrystallisation into other silica species. In general, the most soluble polymorph will transform in a stepwise manner to less soluble species until the most stable form remains (Dove and Rimstidt, 1994). Normally, amorphous opal-A transforms to near-amorphous opal-CT, better-ordered opal-CT, cryptocrystalline quartz or chalcedony and finally microcrystalline quartz (Nash and Hopkinson, 2004). The precise sequence and extent of paragenesis is determined by a range of processes which influence the degree of supersaturation of solutions present within, or passing through, a silcrete. These include silica complexation, adsorption by clays, and the neoformation of clays, zeolites and other silicates (Williams et al., 1985). Transformation may be accompanied by textural changes, but the original silcrete fabric and textures can also be retained (Williams et al., 1985). This can lead to confusion when identifying minerals in thin-section. Studies by Heaney (1995) of materials known to have formed in evaporitic environments showed that, despite the identification of optically length-slow silica in thin-section, the samples contained little or no moganite. X-ray diffraction analyses indicated that the materials were pure quartz, suggesting that moganite had altered to quartz over time.

4.5.3 Geopetal features

A range of geopetal structures have been described within silcretes, especially in pedogenic varieties (Figure 4.6A, B). The most common are illuvial 'colloform' features. These are cusp-like structures containing alternating lamellae of silica and TiO_2, or silica and iron or manganese oxides (e.g. Frankel and Kent, 1938; Callender, 1978; Taylor, 1978; Thiry, 1978, 1988; Terry and Evans, 1994; Curlík and Forgác, 1996; Ballesteros et al., 1997; Thiry et al., 2006). They frequently occur as vertically stacked (or laterally extended) concave-upward cusps and may occupy tubular structures (Summerfield, 1983a; Taylor and Ruxton, 1987). Colloform features are more common in M- and F-fabrics (Frankel and Kent, 1938), but can occur in matrix-dominated patches in GS-fabrics (Taylor and Ruxton, 1987). Their origin is unclear, but appears to be related to gravitational movement of solutions under alternating wet and dry conditions. Laminated conical or cap-like structures have also been identified on top of host-sediment clasts, silcrete fragments or lining dissolution pipes (Callen, 1983; van der Graaff, 1983; Thiry, 1978, 1981, 1988; Thiry and Milnes, 1991). These often occur in conjunction with geopetal void fills and drip-like structures. Unlaminated, but sometimes graded, anatase-rich 'caps' have been documented on clasts in non-pedogenic silcretes from eastern Australia (Taylor and Ruxton, 1987). Geopetal patches or 'beards' of titaniferous cryptocrystalline silica (Watts, 1978a) and iron-stained opal (Callender, 1978) also occur on the underside of grains and glaebules in some F-fabric silcretes in Australia (Milnes et al., 1991). Care must be taken when identifying geopetal features. For example, Ullyott et al. (2004) and Ullyott and Nash (2006) describe cap-like structures in UK groundwater silcretes which developed through the illuviation of fine-grained sediment into the silcrete host material prior to silicification.

4.5.4 Glaebular and nodular features

'Glaebule' is a pedological term for a concretionary or nodular structure which encloses greater amounts of certain constituents, or has a different fabric, or a distinct boundary with the surrounding matrix (Brewer, 1964). Such features have been widely documented within F- and M-fabric silcretes (Summerfield, 1983a,d) of presumed pedogenic origin (Figure 4.6A). They can usually be distinguished due to darker colours caused by concentrations of anatase or iron oxides, and have either a concentrically zoned or undifferentiated internal fabric. Nodular, granular (Thiry, 1978, 1988, 1989), pisolitic (Frankel, 1952; Wopfner 1978, 1983; van der Graaff, 1983) and 'ped-like' structures (Milnes and Twidale, 1983) have also been described, which may be equivalent to glaebules. Their origin is uncertain but many appear to have developed as a result of discrete phases of silica accretion with differing concentration of impurities during pedogenesis (Milnes and Thiry, 1992). Glaebules should not, however, be regarded as uniquely diagnostic features of pedogenic silcretes. Some glaebular structures within Kalahari silcretes (Figure 4.7B), for example, formed by silica accumulation at or near the groundwater table (Shaw and Nash, 1998) or were inherited by silica replacement of pedogenic calcretes (Nash et al., 1994b). Care is also needed to ascertain whether glaebules formed *in situ* during silicification or have a detrital origin (Hill et al., 2003).

4.5.5 Vugh-fills

Void-spaces within silcretes may be partially or completely infilled with a variety of silica and other minerals as a result of porewater movement during late-stage diagenesis (Figure 4.7D). Vugh-fills range from simple to complex and provide an important indication of porewater chemistry. The most widely documented sequence of silica void-infilling is for massive or laminated opal to line void margins and, in turn, be overlain by increasingly well-ordered forms of silica in the order chalcedony–microquartz–mega-quartz (Summerfield, 1982; Thiry and Milnes, 1991; Rodas et al., 1994). This sequence reflects a range of factors including the presence of impurities around the void margin, changes to porewater chemistry over time, reductions in flow rates as porosity is reduced and water becomes trapped in cavities, and decreasing rates of silica precipitation (Summerfield, 1983a,c). Void fills of other minerals such as calcite (Summerfield, 1982, 1983c; Nash and Shaw, 1998), iron oxides or zeolites (Terry and Evans, 1994) are also reported.

4.6 Silica Sources, Transfers and Precipitation Mechanisms

Any attempt to understand silcrete genesis must take into account three key factors: (a) potential sources of silica; (b) routes by which this silica is transported to the site of silicification; and (c) factors leading to the precipitation of the various silica species.

4.6.1 Silica sources

The silica required for silcrete genesis may be derived locally from within the soil, sediment or rock profile or be transported from distant sites. Weathering of silicate minerals provides the ultimate source for the majority of silica within surface and subsurface waters, the atmosphere, and plants and animals (Summerfield, 1983a). Enhanced silica availability may arise for a variety of reasons, but one of the most important controls is environmental pH. Silica solubility is relatively stable under weakly acidic to neutral pH, but increases rapidly above pH 9 (Lindsay, 1979; Dove and Rimstidt, 1994; Figure 4.8A). Such pH levels are not uncommon in arid and semi-arid environments, where alkaline conditions may occur as a result of evaporation (Chadwick et al., 1989). Enhanced silica concentrations in soils also occur at very low pH (Figure 4.8B), primarily due to the breakdown of aluminosilicate (clay) minerals under increasingly acidic conditions (Beckwith and Reeve, 1964; Taylor and Eggleton, 2001). Low pH levels may arise, for example, within soils or sediments sustaining high organic productivity or as a result of sulphide oxidation (Summerfield, 1979), with high silica availability also associated with lateritic weathering environments (Stephens, 1971).

Of the silica released by weathering, only a small proportion comes from direct dissolution of quartz (except under intense leaching), although

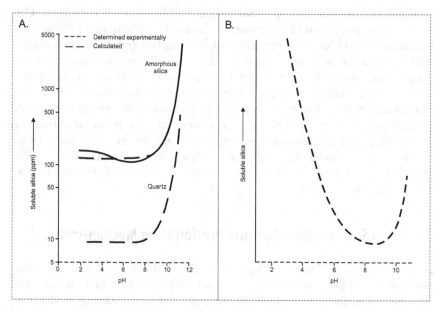

Figure 4.8 (A) Variations in silica solubility with pH (after Williams and Crerar, 1985; Dove and Rimstidt, 1994). (B) The release and sorption of monosilicic acid by a black earth soil under varying pH (after Beckwith and Reeve, 1964).

metamorphic quartz, abraded quartz grains and quartzose aeolian dust have higher solubilities (Dove and Rimstidt, 1994). As such, previous suggestions that the silica within silcrete cements may have been derived from water percolating through quartzose cover sediments (Watts, 1978a) or by diurnal dewfall (Summerfield, 1982) seem unlikely. Despite their relatively low silica mineral content, rocks rich in ferromagnesian minerals and feldspars readily release silica as they weather. Similarly, volcanic ash containing poorly ordered silicate glass is susceptible to rapid breakdown (Terry and Evans, 1994). Kaolinisation of feldspars and clay mineral diagenesis are also important silica sources (Wopfner, 1978, 1983). For example, the transformation of montmorillonite to illite, or montmorillonite or illite to kaolinite releases silica (Birkeland, 1974), as does the breakdown of kaolinite to form gibbsite or alunite (Summerfield, 1978; McArthur et al., 1991; Thiry et al., 1995). Such reactions are generally favoured by low pH environments, related to the oxidation of sulphide minerals (Thiry et al., 1995), high organic productivity (Summerfield, 1983a) or the precipitation of ferric oxyhydroxides (McArthur et al., 1991). The direct breakdown of clays under acidic leaching conditions is also widely identified (Summerfield, 1983d, 1984; Thiry, 1978, 1981; Roulin et al., 1986; Thiry and Millot, 1987; Borger et al., 2004). Alteration of kaolinite to opaline silica has been reported (Thiry, 1978, 1981; Milnes and Thiry, 1992; Rayot et al., 1992), as has the breakdown of halloysite (Dubroeucq and Thiry, 1994), palygorskite (Meyer and Pena dos Reis 1985; Thiry and Ben Brahim, 1990), smectite and illite (Roulin et al., 1986; Roulin, 1987).

In addition, there are a number of biological silica sources (Clarke, 2003). Many plants, notably grasses, reeds, horsetails, palms and some hardwoods, concentrate amorphous silica within their tissues (Goudie, 1973; Summerfield, 1978), which may be readily released upon decay or temporarily stored as phytoliths (Meunier et al., 1999). Lichens also release silica during biological weathering (Chen et al., 2000). Diatoms, common in many lacustrine environments, extract silica from surface waters for their tests. These may subsequently become a silica source upon death (Passarge, 1904; Du Toit, 1954). In fact, Knauth (1994) regards diatoms as one of the most important silica sources in the near-surface environment during the past 50 Myr. Biocorrosional and/or bioerosional processes may also actively release silica from silicate minerals. Experimental work by Brehm et al. (2005) demonstrates that heterotrophic bacteria and cyano-bacteria biofilm growth can significantly increase the breakdown of silica in amorphous, sub-crystalline and crystalline forms of quartz, with diatoms also able to etch quartz and glass surfaces to extract silica.

4.6.2 Silica transfer mechanisms

Silica may be transferred to the site of silcrete development by wind and/or water. Quartz dust, for example, may be transported considerable distances from desert areas by the wind (Goudie and Middleton, 2001). Similarly, plant phytoliths, sponge spicules and diatoms can be subject to aeolian transport (Clarke, 2003). All other transfers rely upon silica transport in solution as undissociated monosilicic acid, either as the monomer H_4SiO_4 or the dimer $H_6Si_2O_7$ (Dove and Rimstidt, 1994). Organic or inorganic complexes may also be formed.

Silica transfer mechanisms can, for simplicity, be divided into those involving lateral or vertical movement (Summerfield, 1983a). It should, however, be noted that silcrete genesis invariably involves both in combination. For example, the development of sub-basaltic silcretes in eastern Australia involved the downward and then lateral movement of silica-rich solutions liberated from weathered basalt to the site of silicification (Young, 1978; Gunn and Galloway, 1978; Ruxton and Taylor, 1982; Webb and Golding, 1998). The most widely documented lateral transfer mechanisms involve silica movement within surface- and groundwater, over scales ranging from the local to sub-continental (Stephens, 1971; Hutton et al., 1972, 1978). Silica transport in river water, most commonly during periodic flood events, is a key control upon the development of drainage-line silcretes (section 4.7.3). The average silica content of river water is around 13 ppm, but this varies with catchment geology and climate (Goudie, 1973), with higher levels in the tropics (e.g. 23.2 ppm average for Africa). Similarly, the formation of pan/lacustrine silcretes is dependent upon the transfer of silica-bearing waters into topographic lows to promote evaporation and mixing with saline solutions (section 4.7.4). Sub-surface water transfers are vital for groundwater silcretes (section 4.7.2) but may also provide silica for the development of other silcrete types.

Vertical transfer mechanisms involve movements of silica-bearing soil- or groundwater within the vadose zone, and are often referred to as *per ascensum* and *per descensum* models (Goudie, 1973). *Per ascensum* mechanisms involve upward movement by capillary rise, with silica precipitation promoted by evaporation or the mixing of upward-migrating solutions with downward percolating water (Frankel and Kent, 1938; Smale, 1973). However, since silica deposition inhibits further capillary rise, such mechanisms are unlikely to generate thick crusts (Summerfield 1982, 1983a). *Per descensum,* or downward percolation mechanisms, suffer none of the drawbacks of *per ascensum* models (Summerfield, 1983a). Silica may be supplied from surface waters, leaching of overlying sediments (Watts, 1978a,b; Marker et al., 2002), or aeolian or biological sources (Goudie, 1973; Summerfield, 1978) and be washed down through the soil or sediment profile. Silica precipitation may be triggered by a range of processes (section 4.6.3), thereby reducing subsequent percolation (Summerfield, 1983a). In the case of pedogenic silcretes, downward-moving silica may be supplied by the progressive dissolution of silica within the developing silcrete profile, with mobilisation and precipitation promoted by wetting and drying respectively (section 4.7.1; Thiry, 1999).

4.6.3 Silica precipitation

Silica precipitation is controlled by the concentration of silica in solution, itself related to silica availability and, to a lesser extent, the duration of wetting/drying cycles (Knauth, 1994). The most soluble silica species will be precipitated first from any supersaturated solution (Millot, 1960, 1970). This is usually amorphous silica, which is both poorly ordered and has a higher solubility under neutral conditions (60–130 ppm) compared with opal-C (20–30 ppm) and quartz (6–10 ppm) (Williams et al., 1985). Silica monomers polymerise and aggregate to form colloids from solutions, and these colloids may precipitate to form opal-A (Iler, 1979; Williams et al., 1985). If solutions are supersaturated with respect to quartz then precipitation can occur in the absence of other changes, providing solutions are slow moving (Summerfield, 1982). Under such circumstances, precipitation is much more likely if there is a suitable template available (e.g. a crystalline quartz grain) to initiate silica deposition. Precipitation from fast-moving silica-undersaturated waters has also been proposed (Thiry et al., 1988a). Crystal size and order are controlled by the speed of precipitation, itself influenced by the host material permeability and evaporation rate.

Silica precipitation is also influenced by pH, Eh, temperature, pressure, evaporation, the presence of other dissolved constituents, and life processes (Nash and Hopkinson, 2004). The effects of fluctuating temperature and pressure are negligible in near-surface environments due to the slow associated rate of change in silica solubility. Evaporation is significant at or very near the surface, for example in the development of the upper parts of pedogenic silcrete profiles (van der Graaff, 1983; Thiry and Millot, 1987; Webb and Golding, 1998) and in highly evaporitic settings (Stephens, 1971; Wopfner, 1978, 1983; Milnes et al., 1991). Biological influences upon silica precipitation have been proposed and may be of local significance. In Botswana, Shaw et al. (1990) have described the role of cyanobacteria in silcrete development, whilst McCarthy and Ellery (1995) identify transpiration from aquatic grasses as a factor in the precipitation of amorphous silica from groundwater.

The importance of pH for silica solubility has been noted in section 4.6.1, with solubility increasing rapidly above pH9 (Dove and Rimstidt, 1994; Figure 4.8A). However, the work of Beckwith and Reeve (1964) suggests that neutralisation of an alkaline solution to below pH9, other than in highly evaporative settings, may not directly trigger silica precipitation unless the pH change is intimately associated with clay dissolution. Environmental Eh also has a complex effect. Quartz placed in solutions containing ferrous iron has been shown to become more soluble following exposure to oxidising conditions (Morris and Fletcher, 1987). The presence of other dissolved constituents affects silica solubility in a number of ways. Solubility is reduced if adsorption onto iron or aluminium oxides occurs. In contrast, the presence of NaCl may enhance solubility due to the ionic strength effect (Dove and Rimstidt, 1994). A range of solutes, including Fe^{3+}, UO_2^+, Mg^{2+}, Ca^{2+}, Na^{2+} and F^-, react with dissolved silica to form complexes, thereby increasing silica solubility (Dove and Rimstidt, 1994). Some dissolved organic compounds have a similar effect (Bennett et al., 1988; Bennett and Siegel, 1989), and the destruction of these complexes, for example following oxidation (Thiry, 1999), may promote silica precipitation. Where significant quantities of salts are present in solution, silica colloids may aggregate, leading to the formation of hydrated opaline silica gels (Iler, 1979). Under high salinity, it is possible for chalcedony to precipitate directly (Heaney, 1993).

4.7 Models of Silcrete Formation

4.7.1 Pedogenic silcretes

Pedogenic silcretes (Figure 4.3) are widely regarded as forming over long periods of time (>10^6 years), although, given the difficulties in determining absolute ages for silcrete cements (section 4.8.), this has yet to be confirmed. There is, however, general agreement that such silcretes form within soil profiles developed on stable basement or basin-marginal areas (Thiry, 1978; Callen, 1983; Milnes and Thiry, 1992), and indicate a hiatus or unconformity. Most are relict features and date from the Paleogene or early Neogene (Thiry, 1999), although some have been attributed to the Mesozoic (Langford-Smith, 1978b; Wopfner, 1978; Ballesteros et al., 1997), late Neogene and Quaternary (Taylor, 1978; Dubroeucq and Thiry, 1994; Čurlík and Forgác, 1996). They develop through the downward percolation of silica-bearing water in cycles of flushing and precipitation, possibly under strongly seasonal climates with high rates of evaporation (Webb and Golding, 1998), with a progressive evolution of the profile down into the landscape (Milnes and Thiry, 1992). Consequently, a diagnostic feature is a horizon of eluviation overlying one of illuviation (Thiry and Ben Brahim, 1990; Kendrick and Graham, 2004). The most obvious manifestations of this transfer process are geopetal colloform structures developed along water pathways, together with glaebules, rootlets and evidence of bioturbation (Summerfield, 1983d; Thiry et al., 1991, 1995, 2006; Terry and Evans, 1994). In rare cases, silicified roots exhibiting clear cellular structures may be present (Lee and Gilkes, 2005), suggesting either active silica fixation by biological processes during pedogenic silcrete development or the petrification of plant materials after death. Lateral transfers of silica may also be important (Hutton et al., 1972; Thiry and Millot, 1987; Terry and Evans, 1994; Simon-Coinçon et al., 1996) producing macromorphological variations according to palaeolandscape position (van der Graaff, 1983; Thiry and Turland, 1985). Thiry (1999) distinguishes two principal types of pedogenic silcrete: those with abundant microquartz and Ti-enrichment but lacking clay and iron oxide; and predominantly opaline 'hardpans' or 'duripans' in which clay and iron are retained.

Complex profile organisation is diagnostic of pedogenic silcretes, with macro- and micromorphology varying consistently throughout (Figure 4.9A). A 'typical' pedogenic profile can be differentiated into two sections. Upper parts exhibit a columnar structure with well-ordered silica cements and evidence of interplay of dissolution and percolation (Milnes and Thiry, 1992). This may be surmounted by a pseudo-brecciated or, more rarely, nodular zone containing evidence for dissolution and eluviation. Lower parts of the profile are weakly cemented by poorly ordered silica and often

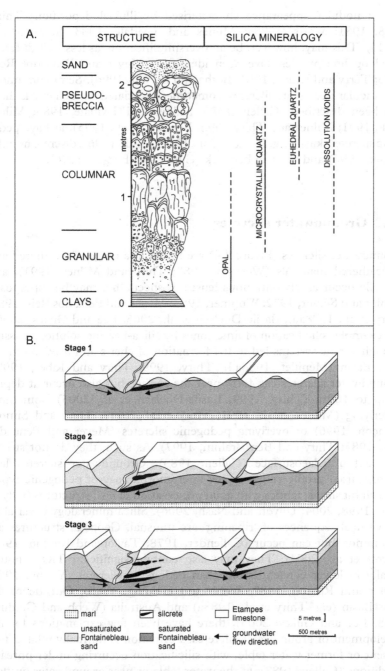

Figure 4.9 (A) Schematic representation of a 'typical' pedogenic silcrete profile, showing silcrete structures and the vertical arrangement of silica cements (after Milnes and Thiry, 1992; Thiry 1999). (B) Model of groundwater silcrete development in the Paris Basin (after Thiry and Bertrand-Ayrault, 1988) showing the development of superposed silcrete lenses within bedrock adjacent to a progressively incising valley system.

have a nodular appearance characterised by illuvial deposition (Thiry, 1978, 1981; Watts, 1978a; Milnes and Twidale, 1983; Summerfield, 1983d). This may, however, be an oversimplification, as less well-differentiated opaline profiles have been identified (Meyer and Pena dos Reis, 1985; Terry and Evans, 1994; Botha and de Wit, 1996). Semi-continuous or lenticular pedogenic silcretes, sometimes associated with channels, have also been described (Callen, 1983; Milnes and Twidale, 1983; Milnes et al., 1991; Milnes and Thiry 1992; Benbow et al., 1995), as have pedogenic silcrete 'skins' (e.g. Hutton et al., 1972; Mišík, 1996), discrete nodules (Mišík, 1996) and veins in bedrock (Čurlík and Forgác, 1996).

4.7.2 Groundwater silcretes

Groundwater silcretes (Figure 4.4) are found in a range of weathered and unweathered materials (Wopfner, 1983; Thiry and Milnes, 1991), and typically occur as discontinuous lenses or sheets that may be superposed (Senior and Senior, 1972; Wopfner, 1983; Meyer and Pena dos Reis, 1985; Thiry et al., 1988a,b; Basile-Doelsch et al., 2005; Lee and Gilkes, 2005). For example, silicification of limestones by silica-bearing solutions passing through karstic systems led to the formation of cherts and *meulières* in the Paris Basin (Ménillet, 1988a,b; Thiry, 1999; Thiry and Ribet, 1999). Groundwater silcretes can form near the surface but also occur at depths of up to 100m (Thiry, 1999; Basile-Doelsch et al., 2005), sometimes underlying (Milnes et al., 1991; Thiry et al., 1995; Thiry and Simon-Coinçon, 1996) or overlying pedogenic silcretes (Meyer and Pena dos Reis, 1985; Thiry and Ben Brahim, 1997). As such, they do not strictly represent a palaeosurface (Callen, 1983). Groundwater silcretes lack the organised profile and complex micromorphology of pedogenic types, and exhibit simple fabrics with good preservation of host structures (Ullyott et al., 1998, 2004; Ullyott and Nash, 2006). Silica mineralogy is variable, but vertical sequences of 'ripening' are unusual. Geopetal structures are uncommon but can occur (Callender, 1978; Taylor and Ruxton, 1987; Ullyott et al., 2004). They may also contain significant TiO_2, though usually in lower concentrations than pedogenic silcretes (Young, 1985; Taylor and Ruxton, 1987; Webb and Golding, 1998). Outside of the Paris Basin (e.g. Thiry et al., 1988a) and Australia (Webb and Golding, 1988; Lee and Gilkes, 2005), there have been few explanations for the development of groundwater silcretes. Formation is normally related to a present or former water table, with silicification occurring under phreatic conditions (Callen, 1983), at the water table or near groundwater outflow zones (Thiry et al., 1988a,b; Demoulin, 1989; Nash, 1997). Water table fluctuations have also been proposed as a control (Taylor and Ruxton,

1987; Demoulin, 1990; Milnes et al., 1991; Thiry and Milnes 1991; Rodas et al., 1994).

The main genetic models for groundwater silcrete development (Figure 4.9B) arise from work by Médard Thiry and colleagues in the Paris Basin, where silcretes are found both in Paleogene arenaceous formations, most notably the Fontainebleau Sand (Thiry and Bertrand-Ayrault, 1988; Thiry et al., 1988a,b; Thiry, 1999; Thiry and Marechal, 2001), and in Neogene and Paleogene lacustrine limestones (Ribet and Thiry, 1990; Thiry and Ribet, 1999). Silcrete formation was initiated during Plio-Quaternary landscape incision (Thiry, 1999). In arenaceous host sediments, extensive bleaching and silica release from clay minerals and poorly ordered silica phases is suggested to have occurred due to acidic conditions generated by the oxidation of glauconite-, organic- and pyrite-rich sand bodies (Thiry et al., 1988a). Silica moved downwards to the water table and then laterally towards zones of groundwater outflow within newly eroded valley systems, where physico-chemical reactions triggered silica deposition at the water table. The precise mechanisms involved are unclear, but may have included oxidation of dissolved organic compounds during meteoric and groundwater mixing (Thiry, 1999). Silica precipitated as a series of successive envelopes around an initially silicified core, leading to the progressive formation of silcrete lenses. Development was relatively rapid, in the order of 30 kyr for each lens (Thiry et al., 1988a). The process was repeated during subsequent phases of landscape incision to create a series of superposed silcrete lenses, with older (highest) lenses partially dissolved and reworked during the development of later layers (Figure 4.4A). Similar processes have been proposed for the development of groundwater silcretes in arenaceous sediments in Australia (Milnes et al., 1991; Thiry and Milnes, 1991; Simon-Coinçon et al., 1996; Lee and Gilkes, 2005) and other parts of Europe (Riezebos, 1974; Wopfner, 1983; Demoulin, 1989, 1990; Ullyott et al., 1998, 2004; Basile-Doelsch et al., 2005; Ullyott and Nash, 2006).

For carbonate-hosted groundwater silcretes in the Paris Basin (Ribet and Thiry, 1990; Thiry and Ribet, 1999), silicification appears to have been directly related to high-porosity zones, reflecting either primary fabric or karst systems in the limestone. Silica was derived from overlying sands and soils and transported through the limestone by groundwater, with silicification proceeding both by void infilling and carbonate replacement. The resulting silcretes comprise 3–5% of the host rock and occur as discordant irregular bodies. As with groundwater silcretes developed in sand formations, modelling suggests relatively rapid formation under high groundwater flow rates at timescales of 10 to 100 kyr (Thiry and Ribet, 1999). Similar processes have been suggested for the silicification of limestones in Morocco (Thiry and Ben Brahim, 1997) and calcretes in the Kalahari (Nash et al., 2004).

4.7.3 Drainage-line silcretes

Drainage-line silcretes (Figures 4.5A, B and 4.10A, B) are closely related to groundwater varieties but develop *within* alluvial fills in current or former fluvial systems (Summerfield, 1982; van de Graaff, 1983; Young, 1978, 1985; Leckie and Cheel, 1990; Nash et al., 1994a, 1998; Shaw and Nash, 1998), as opposed to within bedrock *marginal* to valley systems (as is the case with many groundwater silcretes in France and Australia). In such settings, silcretes may develop in zones of water table fluctuation (Taylor and Ruxton, 1987) or at sites marginal to drainage courses that are subject to seasonal wetting and drying (McCarthy and Ellery, 1995). One of the most extensive examples is the Mirackina Conglomerate in South Australia. This represents the silicified remnants of a 200-km-long palaeodrainage system (Barnes and Pitt, 1976; McNally and Wilson, 1995; Figure 4.10B), and formed as a product of silica-bearing groundwater moving through channel alluvium (although it is possible that pedogenic silicification may also have occurred in sections of the channel; Benbow et al., 1995). Few models for the formation of drainage-line silcrete exist. This is, in part, due to a lack of drainage-line silcretes cropping out in the geomorphological context in which they formed. Shaw and Nash (1998) have proposed mechanisms for the development of silcretes within the Boteti River, Botswana (Figures 4.5A and 4.10A). Silcretes exposed in the ephemeral river bed (Figure 4.7A) are suggested to have formed by the accumulation of clastic and phytolithic silica from annual floodwaters, along with precipitation of amorphous silica from shallow groundwater. Silica precipitation was induced by near-surface evaporation and transpiration from stands of aquatic grasses within seasonal pools in the channel (McCarthy and Ellery, 1995). In contrast, massive and pisolitic silcrete layers developed deep within the channel alluvium (Figure 4.7B) in response to salinity shifts associated with vertical and lateral movements of the wetting front during flood events (Figure 4.10A). During groundwater recharge by floodwaters, conditions beneath the channel floor shifted from unsaturated to saturated, with associated decreases in salinity due to dilution effects. Silica precipitation subsequently occurred as the water table dropped.

4.7.4 Pan/Lacustrine silcretes

Pan/lacustrine silcretes (Figure 4.5B, C) most commonly develop within, or adjacent to, ephemeral lakes, pans or playas within endoreic basins (Goudie, 1973; Summerfield, 1982; Figure 4.10C). In the majority of modern evaporitic lacustrine environments, silica precipitation is driven by changes in pH and salt concentration (Thiry, 1999), both of which can

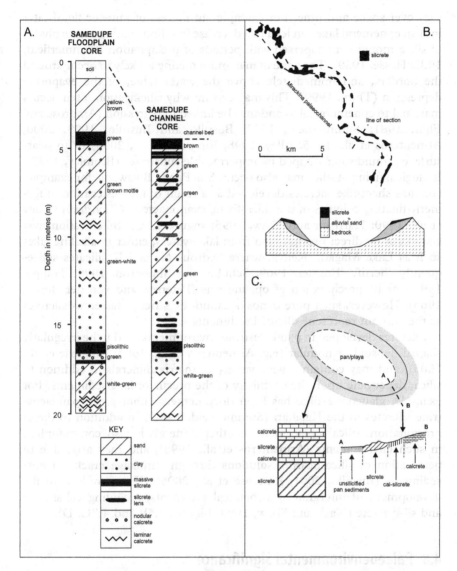

Figure 4.10 (A) Cores extracted from the bed of the Boteti River at Samedupe Drift, Botswana, showing a range of geochemical sediments developed beneath the channel floor, including massive and pisolithic drainage-line silcretes (after Shaw and Nash, 1998). (B) Section of silcretes in the Mirackina palaeochannel, South Australia (after Ollier and Pain, 1996), showing the planform distribution of silcrete outcrops along the palaeochannel course and the relationship of silcrete to underlying non-silicified units. (C) Schematic representation of geochemical sedimentation patterns in the vicinity of a pan or playa (after Summerfield, 1982).

vary over space and time. For example, an ingress of rain- or floodwater into an ephemeral lake can lead to a decrease in salinity and generate phases of silica mobility interspersed with periods of precipitation (Summerfield, 1982; Hesse, 1989). The zone of maximum mixing is likely to occur around the borders, and immediately above the water table, of the evaporitic depression (Thiry, 1999). This may explain why silicification often occurs marginal to ephemeral lakes and may be linked to regression (Ambrose and Flint, 1981; Summerfield, 1982; Bustillo and Bustillo, 1993, 2000; Armenteros et al., 1995; Alley, 1998; Ringrose et al., 2005) or to near-surface groundwater trapped by impermeable substrates (Bromley, 1992). Biological fixing of silica may also occur. Sua Pan in Botswana, for example, contains sheet-like silcretes developed as a result of the desiccation of formerly floating colonies of the silica-fixing cyanobacteria *Chloriflexus* (Shaw et al., 1990; Harrison and Shaw, 1995; Figure 4.5C). Silicification may also occur by direct precipitation from lake waters, either in alkaline lakes such as Lake Magadi, Kenya, where hydrous Na and K silicates subsequently chertify (Eugster, 1969; Schubel and Simonson, 1990; Trappe, 1992), or by precipitation of opaline gels (Peterson and von der Borch, 1965). However, such pure deposits cannot be strictly classed as silcretes as they do not represent silicified sediments.

Silcretes developed in pan/lacustrine environments tend to be irregularly shaped, lensoid or nodular (e.g. Armenteros et al., 1995; Ringrose et al., 2005), and may contain a wide variety of matrix minerals in addition to silica, depending upon the chemistry of the precursor lake sediments. For example, glauconite–illite has been documented within green pan/lacustrine silcretes in the Kalahari (Summerfield, 1982). In addition to direct precipitation, silica replacement of other minerals is also commonplace in such environments (Armenteros et al., 1995), and may arise due to penetration of silica-bearing solutions through shrinkage cracks in host sediments (Bustillo, 2001; Ringrose et al., 2005). This may lead to the development of intergrade geochemical sediments, including cal-silcrete and sil-calcrete (Nash and Shaw, 1998; Figures 4.5B and 4.7C, D).

4.8 Palaeoenvironmental Significance

The environment of silcrete formation, and hence the use of relict silcretes as palaeoenvironmental indicators, has been the subject of considerable debate (Summerfield, 1983b, 1986; Twidale and Hutton, 1986; Nash et al., 1994b). Suggested environments of formation have ranged from arid (e.g. Beetz, 1926; Kaiser, 1926; Prescott and Pendleton, 1952) to humid tropical (e.g. Whitehouse, 1940; Van den Broek and Van der Vaals, 1967; Wopfner, 1978; Young, 1978). However, as shown above, silcrete genesis depends upon a complex interplay between climate and silica supply. Compared with other geochemical sediments, there are virtually no examples of contemporary silcretes, and those that do exist are atypical. This makes the identification of environmental parameters for silcrete formation highly problematic.

A further issue is the difficulty of applying dating techniques to silcretes (Watchman and Twidale, 2002), not least because of the impacts of paragenesis and the occurrence of multiphase cements. Methods such as electron-spin resonance have been tested on Australian silcretes (Radtke and Brückner, 1991), but the use of bulk samples in such examples renders the resulting ages almost meaningless. The majority of studies, however, use relative dating approaches, where the ages of silcretes are determined from their stratigraphic position. This concept has been employed in France (Thiry, 1999), and in Australia where the presence of plant fossils (Taylor and Eggleton, 2001) and the relationship between silcretes and basalts of known age has been used to age-constrain outcrops (e.g. Wopfner et al., 1974; Twidale, 1983; Webb and Golding, 1998).

Early attempts to determine environments of silcrete formation were based upon the identification of micromorphological and geochemical criteria. Summerfield (1983b), for example, proposed that silcretes in southern Africa could be divided into two groups. 'Weathering profile silcretes' in the Cape coastal zone occurred in deep weathering profiles, whereas 'non weathering profile silcretes' in the Kalahari were developed in unweathered sediments. Summerfield argued that the TiO_2-rich (>1%) colloform features typical of 'weathering profile silcretes' formed under a humid tropical climate. Conversely, low TiO_2 contents and the association of 'non weathering profile silcretes' with pan and aeolian sediments suggested silicification in a much drier climate (Summerfield, 1982, 1983b,c). In Australia, Wopfner (1978, 1983) reached similar conclusions, envisaging humid tropical conditions for his Type II silcretes and an arid palaeoclimate for Type III silcretes.

Summerfield's ideas have been contested on several grounds. It soon became clear that 'non weathering profile silcretes' were not restricted to arid or semi-arid environments (Young, 1978, 1985; Thiry et al., 1988a). The criteria used to distinguish palaeoclimates were also disputed, with colloform structures reported in Australian 'non-weathering profile' silcretes (Callender, 1978; Taylor and Ruxton, 1987). Further, TiO_2 patterns were shown to be reversed in Australia (Young, 1985), with higher TiO_2 contents in inland silcretes formed under seasonally arid conditions (Wopfner, 1978, 1983; Ambrose and Flint, 1981; Thiry and Milnes, 1991) and lower TiO_2 in eastern silcretes generated under humid conditions (Taylor and Smith, 1975; Young, 1978, 1985; Taylor and Ruxton, 1987; Webb and Golding, 1998). TiO_2-enriched silcretes originating under different climatic regimes were also recognised within the same deepweathered profiles (Thiry and Milnes, 1991). Finally, the TiO_2 contents of silcrete host materials were suggested to be more influential than palaeoenvironmental conditions (Taylor and Smith 1975; Hutton et al., 1978; Nash et al., 1994b), with Webb and Golding (1998) identifying TiO_2 losses during the silicification of sandy hosts and gains in clay-rich sediments.

In many of these discussions, silcretes were considered as a whole with little recognition of the nuances between different silcrete types. Recent advances in our understanding of silcrete genesis may, however, provide a starting point for re-evaluating palaeoenvironmental signals. We know, for example, that pedogenic silicification is directly influenced by near-surface conditions, making pedogenic silcretes potentially useful palaeoclimatic indicators. Pedogenic silicification has been attributed to climates comprising alternating wet/humid and dry/arid seasons or periods (Thiry, 1981, 1989; Callen, 1983; Milnes and Twidale, 1983; Meyer and Pena dos Reis, 1985; Thiry and Milnes, 1991; Čurlík and Forgác, 1996). Under such conditions, wetter phases favour leaching and drier episodes promote silica precipitation. The significance of evaporation has also been confirmed both by micromorphological (van der Graaff, 1983) and ^{18}O analyses (Webb and Golding, 1998). A low pH environment is envisaged in many cases, although this may be due either to high organic activity (Summerfield, 1979, 1983a) or oxidation of sulphides (Thiry and Milnes, 1991). However, Terry and Evans (1994) attribute silicification to high pH conditions under a Mediterranean palaeoclimate.

Groundwater, drainage-line and, to a degree, pan/lacustrine silcrete formation is less directly dependent on climate but is strongly influenced by (palaeo)topography. Silicification is normally related to the water table or groundwater flow within sediments, or to external fluvial inputs of silica. The development of groundwater and drainage-line silcretes has been attributed to a range of palaeoenvironments. Low pH conditions have been suggested, arising from high organic productivity in a humid, possibly tropical climate (Wopfner, 1978, 1983, Young, 1985; Taylor and Ruxton, 1987; Lindqvist, 1990), or from oxidation of organic matter and sulphides within host sediments (Meyer and Pena dos Reis, 1985; Thiry et al., 1988a,b; Thiry and Milnes, 1991). This may occur in arid or even temperate climates. In contrast, many non-pedogenic silcretes in arid regions are the product of high pH environments (Summerfield, 1982; Meyer and Pena dos Reis, 1985; Leckie and Cheel, 1990; Nash and Shaw, 1998). However, recent work by Alexandre et al. (2004), which combines stable isotope, X-ray diffraction and Fourier transform infrared spectroscopic analyses of a single inland Australian groundwater silcrete sample, suggests that silicification may have occurred under cooler and wetter conditions than present. There are also arguments for development both in oxidising (Meyer and Pena dos Reis, 1985; Thiry et al., 1988a) and reducing conditions (Callen, 1983; Wopfner, 1983; Lindqvist, 1990), although an interplay of the two processes may be indicated (Dove and Rimstidt, 1994).

Silicification is, in the majority of cases, demonstrably *not* contemporaneous with deposition or formation of the host material. Pedogenic silcretes typically transect various ages and types of bedrock (Frankel and Kent, 1938; Langford-Smith, 1978b; Thiry and Ben Brahim, 1990), and the development of the weathered profiles within which many silcretes occur usually pre-dates silicification (Thiry et al., 1995). Groundwater, drainage-line and pan/lacustrine silicification can be intraformational, but typically occurs long after the deposition of the host sediment (Milnes and Thiry, 1992). Moreover, the superposition of groundwater silcrete lenses may indicate a reversed sequence of ages as the oldest is frequently uppermost (Thiry et al., 1988a; Thiry and Milnes, 1991).

4.9   Relationships to other Terrestrial Geochemical Sediments

Although the preceding discussion suggests four broad models of silcrete formation, the situation may be more complex. Many silcretes develop over long time periods under a range of environmental conditions, and may be multiphase and/or polygenetic (Thiry, 1981; Ullyott et al., 1998). Earlier phases of silicification may, for example, be reworked into, or overprinted by, one or more later phases. Pedogenic profiles can become superimposed on earlier groundwater silcretes during the course of landscape evolution (Milnes and Thiry, 1992; Thiry and Simon-Coinçon, 1996). Intergrade duricrusts which contain cementing agents in addition to silica may also develop, reflecting either changing conditions or the replacement of calcrete, dolo-crete or ferricrete by silcrete or *vice versa* (Ambrose and Flint, 1981; Sum-merfield, 1982; Arakel et al., 1989; Nash et al., 1994a,b, 2004; Rodas et al., 1994; Armenteros et al., 1995; Nash and Shaw, 1998; Ringrose et al., 2005). Complex catenary relationships are reported (Taylor and Ruxton, 1987; Germann et al., 1994), with silcretes forming complementary to other duri-crusts, sometimes with lateral gradation into calcrete, dolocrete (Summer-field, 1982; Armenteros et al., 1995) or laterite (Twidale, 1983; van der Graaff, 1983; Alley, 1998). Vertical gradations also occur, with calcrete, fer-ricrete or bauxite over- or underlying silcrete (Summerfield, 1983d; Nash et al., 1994a, 2004; Mišík, 1996; Nash and Shaw, 1998). Co-precipitation of silcrete with calcrete (Nash and Shaw, 1998) or ferricrete (Frankel and Kent, 1938) has also been suggested. The formation of laterite may also lead to the release of silica, which may be transported laterally (Stephens, 1971) or verti-cally (Marker et al., 2002) to a site of silcrete accumulation.

4.10   Directions for Future Research

Despite the leaps in silcrete research over the past 20 years, work is still needed to fully understand the variations in silcrete micromorphology, mechanisms of formation (especially in non-pedogenic varieties), and timing of development on a global scale. Where possible, detailed micromorphological investigations of *in situ* profiles should be undertaken, similar to those employed by Thiry and colleagues in France and Australia. The application of approaches such as oxygen isotope (Webb and Golding, 1998) and Raman-FTIR analyses (Alexandre et al., 2004; Nash and Hopkinson, 2004) also offers scope for identifying the conditions under which silicification occurred. Absolute dating of silcretes remains a problem, although the development of techniques for dating diagenetic events (McNaughton et al., 1999) and K–Mn oxides in associated weathered profiles (Bird et al., 1990; Vasconcelos, 1999; Vasconcelos and Conroy, 2003) gives some grounds for optimism. As a footnote, Webb and Golding (1998, p. 992) suggested 'it is impossible to generalise that the presence of silcrete is indicative of a particular climate'. We feel, however, that with improvements in dating and the identification of macro- and micromorphological criteria for distinguishing silcrete types, it may soon be possible to use silcretes as palaeoenvironmental indicators with increasing confidence and predict their surface or subsurface extent, thereby aiding both aquifer and oil reservoir characterisation.

Acknowledgements

The authors would like to thank the researchers whose input and discussion over the years has contributed to the development of the ideas included within this chapter. Particular mention must go to Andrew Goudie, John Hepworth, Steve Hill, Laurence Hopkinson, Marty McFarlane, Cliff Ollier, Paul Shaw, Mike Summerfield, Graham Taylor, Médard Thiry, John Webb and Brian Whalley, as well as to all members of the BGRG Terrestrial Geochemical Sediments and Geomorphology Working Group.

References

Abdel-Wahab, A., Salem, A.M.K. & McBride, E.F. (1998) Quartz cement of meteoric origin in silcrete and non-silcrete sandstones, Lower Carboniferous, western Sinai, Egypt. *Journal of African Earth Sciences* 27, 277–290.

Alexandre, A., Meunier, J.-D., Llorens, E., Hill, S.M. & Savin, S.M. (2004) Methodological improvements for investigating silcrete formation: petrography, FT-IR and oxygen isotope ratio of silcrete quartz cement, Lake Eyre Basin, Australia. *Chemical Geology* 211, 261–274.

Alimen, H. & Deicha, G. (1958) Observations pétrographiques sur les meulières pliocènes. *Bulletin de la Societé Géologique de France, Séries 6* 8, 77–90.

Alley, N.F. (1998) Cainozoic stratigraphy, palaeoenvironments and geological evolution of the Lake Eyre Basin. *Palaeogeography, Palaeoclimatology, Palaeoecology* 144, 239–263.

Ambrose, G.J. & Flint, R.B. (1981) A regressive Miocene lake system and silicified strandlines in northern South Australia; implications for regional stratigraphy and silcrete genesis. *Journal of the Geological Society of Australia* **28**, 81–94.

Arakel, A.V., Jacobson, G., Salehi, M. & Hill, C.M. (1989) Silicification of calcrete in palaeodrainage basins of the Australian arid zone. *Australian Journal of Earth Sciences* **36**, 73–89.

Armenteros, I., Bustillo, M.A. & Blanco, J.A. (1995) Pedogenic and groundwater processes in a closed Miocene basin (northern Spain). *Sedimentary Geology* **99**, 17–36.

Auzel, M. & Cailleux, A. (1949) Silicifications nord-saharieenes. *Bulletin de la Société géologique de France, Series 5* **19**, 553–559.

Ballesteros, E.M., Talegón, J.G. & Hernández, M.A. (1997) Palaeoweathering profiles developed on the Iberian hercynian basement and their relationship to the oldest Tertiary surface in central and western Spain. In: Widdowson, M. (Ed.) *Palaeosurfaces: Recognition, Reconstruction and Palaeoenvironmental Interpretation.* Special Publication 120. Bath: Geological Society Publishing House; pp. 175–185.

Barnes, L.C. & Pitt, G.M. (1976) The Mirackina Conglomerate. *Quarterly Geological Notes of the Geological Survey of South Australia* **59**, 2–6.

Basile-Doelsch, I. (2006) Si stable isotopes in the Earth's surface: a review. *Journal of Geochemical Exploration* **88**, 252–256.

Basile-Doelsch, I., Meunier, J.D. & Parron, C. (2005) Another continental pool in the terrestrial silicon cycle. *Nature* **433**, 399–402.

Bassett, H. (1954) Silicification of rocks by surface waters. *American Journal of Science*, **252**, 733–735.

Beckwith, R.S. & Reeve, R. (1964) Studies on soluble silica in soils. II. The release of monosilicic acid from soils. *Australian Journal of Soil Research* **2**, 33–45.

Beetz, W. (1926) Die Tertiärablegerungen der Küstennamib. In: Kaiser, E. (Ed.) *Die Diamantenwüste Südwest-Afrikas*, Vol. 2. Berlin: Deitrich Riemer, pp. 1–54.

Benbow, M.C., Callen, R.A., Bourman, R.P. & Alley, N.F. (1995) Deep weathering, ferricrete and silcrete. In: Drexel, J.F. & Preiss, W.V. (Eds) *Bulletin 54: The Geology of South Australia, Volume 2: the Phanerozoic.* Geological Survey of South Australia, pp. 201–207.

Bennett, P.C. & Siegel, D.I. (1989) Silica organic complexes and enhanced quartz dissolution in water by organic acids. In: Miles, D.L. (Ed.) *Proceedings, 6th International Symposium on Water-Rock Interaction,* Rotterdam: Balkema, pp. 69–72.

Bennett, P.C., Melcer, M.E., Siegel, D.I. & Hassett, J.P. (1988) The dissolution of quartz in dilute aqueous solutions at 25°C. *Geochimica Cosmochimica Acta* **52**, 1521–1530.

Birkeland, P.W. (1974) *Pedology, Weathering and Geomorphological Research.* New York: Oxford University Press.

Borger, H., McFarlane, M.J. & Ringrose, S. (2004) Processes of silicate karstification associated with pan formation in the Darwin-Koolpinya area of Northern Australia. *Earth Surface Processes and Landforms* **29**, 359–371.

Bosazza, V.L. (1939) The silcrete and clays of the Riversdale-Mossel Bay area. *Fulcrum, Johannesburg* **32**, 17–29.

Botha, G.A. & de Wit, M.C.J. (1996) Post-Gondwanan continental sedimentation, Limpopo region, southeastern Africa. *Journal of African Earth Science* **23**, 167–187.

Brehm, U., Gorbushina, A. & Mottershead, D. (2005) The role of microorganisms and biofilms in the breakdown and dissolution of quartz and glass. *Palaeogeography, Palaeoclimatology, Palaeoecology* **219**, 117–129.

Brewer, R.C. (1964) *Fabric and Mineral Analysis of Soils*. New York: Robert E. Kreiger.

Bromley, M. (1992) Topographic inversion of early interdune deposits, Navajo Sandstone (Lower Jurassic), Colorado Plateau, USA. *Sedimentary Geology* **80**, 1–25.

Brückner, W.D. (1966) Origin of silcretes in Central Australia. *Nature* **209**, 496–497.

Bustillo, M.A. (2001) Cherts with moganite in continental Mg-clay deposits: an example of 'false' Magadi-type cherts, Madrid Basin, Spain. *Journal of Sedimentary Research* **A71**, 436–443.

Bustillo, M.A. & Bustillo, M. (1993) Rhythmic lacustrine sequences with silcretes from the Madrid Basin, Spain: Geochemical trends. *Chemical Geology* **107**, 229–232.

Bustillo, M.A. & Bustillo, M. (2000) Miocene silcretes in argillaceous playa deposits, Madrid Basin, Spain: petrological and geochemical features. *Sedimentology* **47**, 1023–1037.

Bustillo, M.A. & García Romero, E. (2003) Arcillas fibrosas anómalas en encostramientos y sedimentos superficiales: características y génesis (Esquivias, Cuenca de Madrid). *Boletin de la Sociedad Española de Cerámica y Vidrio* **42**, 289–297.

Callen, R.A. (1983) Late Tertiary 'grey billy' and the age and origins of surficial silicifications (silcrete) in South Australia. *Journal of the Geological Society of Australia* **30**, 393–410.

Callender, J.H. (1978) A study of the silcretes near Marulan and Milton, New South Wales. In: Langford-Smith, T. (Ed.) *Silcrete in Australia*. Armidale: University of New England Press, pp. 209–221.

Campbell, J.D. (2002) Angiosperm fruit and leaf fossils from Miocene silcrete, Landslip Hill, northern Southland, New Zealand. *Journal of the Royal Society of New Zealand* **32**, 149–154.

Cecioni, G. (1941) Il crostone selcioso di Bur-Uen in Somalia. *Bollettino Società geologica italiana* **58**, 235–241.

Chadwick, O.A., Hendricks, D.M. & Nettleton, W.D. (1989) Silicification of Holocene soils in Northern Monitor Valley, Nevada. *Soil Science Society of America Journal* **53**, 158–164.

Chen, J., Blume, H.-P. & Beyer, L. (2000) Weathering of rocks by lichen colonisation – a review. *Catena* **39**, 121–146.

Clark, M.J., Lewin, J. & Small, R.J. (1967) The sarsen stones of the Marlborough Downs and their geomorphological implications. *Southampton Research Series in Geography* **4**, 3–40.

Clarke, J. (2003) The occurrence and significance of biogenic opal in the regolith. *Earth-Science Reviews* **60**, 175–194.

Collins, N. (1985) *Sub-basaltic silcrete at Morrisons, central Victoria*. Unpublished Honours Thesis, Department of Earth Sciences, University of Melbourne, Australia.

Cros, P. & Freytet, P. (1981) Paleogeographical importance of *in situ* and reworked calcretes and silcretes in the Carnian of southern Italian Alps. *Comptes Rendus de l'Academie des Sciences Series II* **292**, 737–740.

Čurlík, J. & Forgác, J. (1996) Mineral forms and silica diagenesis in weathering silcretes of volcanic rocks in Slovakia. *Geologica Carpathica* **47**, 107–118.

David, T.W.E. & Browne, W.R. (1950) *The Geology of the Commonwealth of Australia*, Vols 1–3. London.

Davies, A.M. & Baines, A.H.J. (1953) A preliminary survey of the sarsen and puddingstone blocks of the Chilterns. *Proceedings of the Geologists' Association* **64**, 1–9.

Demoulin, A. (1989) Silcretes and ferricretes in the Oligocene sands of Belgium (Haute-Belgique). *Zeitschrift für Geomorphologie* **33**, 103–118.

Demoulin, A. (1990) Les silicifications tertiaires de la bordure nord de l'Ardenne et du Limbourg méridional (Europe NO). *Zeitschrift für Geomorphologie* **34**, 179–197.

Dewers, T. and Ortoleva, P. (1991) Influences of clay-minerals on sandstone cementation and pressure solution. *Geology* **19**, 1045–1048.

Dixey, F. (1948) *Geology of Northern Kenya*. Report 15, Geological Society of Kenya.

Dove, P.M. & Rimstidt, J.D. (1994) Silica-water interactions. In: Heaney, P.J., Prewitt, C.T. & Gibbs, G.V. (Eds) *Silica: Physical Behaviour, Geochemistry and Materials Applications*. Reviews in Mineralogy 29. Washington: Mineralogical Society of America, pp. 259–308.

Du Toit, A.L. (1954) *The Geology of South Africa*, 3rd edn. Edinburgh: Oliver & Boyd.

Dubroeucq, D. & Thiry, M. (1994) Indurations siliceuses dans des sols volcaniques. Comparison avec des silcrètes anciens. *Transactions of the 15th World Congress of Soil Sciences*, Acapulco, Mexico, 10–16 July, Vol. 6a, pp. 445–459.

Dury, G.H. & Habermann, G.M. (1978) Australian silcretes and northern-hemisphere correlatives. In: Langford-Smith, T. (Ed.) *Silcrete in Australia*. Armidale: University of New England Press, pp. 223–259.

Dury, G.H. & Knox, J.C. (1971) Duricrusts and deep-weathering profiles in southwestern Wisconsin. *Science* **174**, 291–292.

Ellis, F. & Schloms, B.H.A. (1982) A note on the Dorbanks (duripans) of South Africa. *Palaeoecology of Africa* **15**, 149–157.

Eugster, H.P. (1969) Inorganic bedded cherts from Magadi area, Kenya. *Contributions to Mineralogy and Petrology* **22**, 1–31.

Fersmann, A. & Wlodawetz, N. (1926) Uber die Erscheinungen der Silicifierung in der Mittelaisatischen Wuste Karakum. *Comptes Rendus de l'Acadèmie des Sciences de l'U.S.S.R.* 145–8.

Folk, R.L. & Pittman, J.S. (1971) Length-slow chalcedony: a new testament for vanished evaporites. *Journal of Sedimentary Petrology* **41**, 1045–1058.

Flach, K.W., Nettleton, W.D., Gile, L.H. & Cady, J.C. (1969) Pedocementation: induration by silica, carbonate and sesquioxides in the Quaternary. *Soil Science* **107**, 442–453.

Flörke, O.W., Flörke, U. & Giese, U. (1984) Moganite: a new microcrystalline silica-mineral. *Neues Jahrbuch für Mineralogie, Abhandlungen* 149: 325–226.

Flörke, O.W., Graetsch, B., Martin, B., Röller, K. & Wirth, R. (1991) Nomenclature of micro- and non-crystalline silica minerals, based on structure and microstructure. *Neues Jahrbuch für Mineralogie, Abhandlungen* 163: 19–42.

Frankel, J.J. (1952) Silcrete near Albertinia, Cape Province. *South African Journal of Science* 49, 173–182.

Frankel J.J. & Kent L.E. (1938) Grahamstown surface quartzites (silcretes). *Transactions of the Geological Society of South Africa* 15, 1–42.

Gassaway, J.S. 1988. Silcrete – Paleocene marker beds in the western interior. *Geological Society of America, Abstracts with Programs*, Abstract No. 7068.

Gassaway, J.S. 1990. Silcrete of Wyoming. *Geological Society of America, Abstracts with Programs*, Abstract No. 16,300.

Germann, K., Schwarz, T. & Wipki, M. (1994) Mineral-deposit formation in Phanerozoic sedimentary basins of northeast Africa – the contribution of weathering. *Geologische Rundschau* 83, 787–798.

Gomez-Gras, D., Parcerisa D., Bitzer K., Calvet F., Roca E. & Thiry M. (2000) Hydrogeochemistry and diagenesis of Miocene sandstones at Montjuic, Barcelona (Spain). *Journal of Geochemical Exploration* 69, 177–182.

Goudie, A.S. (1973) *Duricrusts in Tropical and Subtropical Landscapes*. Oxford: Clarendon Press.

Goudie, A.S. & Middleton N.J. (2001) Saharan dust storms: nature and consequences. *Earth-Science Reviews* 56, 179–204.

Graetsch, H. (1994) Structural characteristics of opaline and microcrystalline silica minerals. In: Heaney, P.J., Prewitt, C.T. & Gibbs, G.V. (Eds) *Silica: Physical Behaviour, Geochemistry and Materials Applications*. Reviews in Mineralogy 29, Washington: Mineralogical Society of America, pp. 209–232.

Grant, K. & Aitchison, G.D. (1970) The engineering significance of silcretes and ferricretes in Australia. *Engineering Geology* 4, 93–120.

Gunn, R.H. & Galloway, R.W. (1978) Silcretes in south-central Queensland. In: Langford-Smith, T. (Ed.) *Silcrete in Australia*. Armidale: University of New England Press, pp. 51–71.

Hancock, J.M. (2000) The Gribun Formation: clues to the latest Cretaceous history of western Scotland. *Scottish Journal of Geology* 36, 137–141.

Harrison, C.C. & Shaw, P.A. (1995) Bacterial involvement in the production of silcretes? *Bulletin de l'Institute océanographie, Monaco, No. Spécial* 14, 291–295.

Heald, M.T. & Larese, R.E. (1974) Influence of coatings on quartz cementation. *Journal of Sedimentary Petrology* 44, 1269–1274.

Heaney, P.J. (1993) A proposed mechanism for the growth of chalcedony. *Contributions to Mineralogy and Petrology* 115, 66–74.

Heaney, P.J. (1995) Moganite as an indicator for vanished evaporites: a testament reborn? *Journal of Sedimentary Research* A65, 633–638.

Henningsen, D. (1986) *Einführung in die Geologie der Bundesrepublik Deutschland*. Stuttgart: Ferdinand Enke Verlag.

Hesse, R. (1989) Silica diagenesis: origin or inorganic and replacement cherts. *Earth-Science Reviews* 26, 253–284.

Hill, S.M., Eggleton, R.A. & Taylor, G. (2003) Neotectonic disruption of silicified palaeovalley systems in an intraplate, cratonic landscape: regolith and landscape evolution of the Mulculca range-front, Broken Hill Domain, New South Wales. *Australian Journal of Earth Sciences* 50, 691–707.

Hume, W.G. (1925) *Geology of Egypt. Volume 1.* Government Press, Cairo.

Hutton, J.T., Twidale, C.R., Milnes, A.R. & Rosser, H. (1972) Composition and genesis of silcretes and silcrete skins from the Beda valley, southern Arcoona plateau, South Australia. *Journal of the Geological Society of Australia* **19**, 31–39.

Hutton, J.T., Twidale, C.R. & Milnes, A.R. (1978) Characteristics and origin of some Australian silcretes. In: Langford-Smith, T. (Ed.) *Silcrete in Australia.* Armidale: University of New England Press, pp. 19–39.

Iler, R. (1979) *The Chemistry of Silica: Solubility, Polymerization, Colloid and Surface Properties and Biochemistry.* New York: Wiley.

Isaac, K.P. (1979) Tertiary silcretes of the Sidmouth area, east Devon. *Proceedings of the Ussher Society* **4**, 41–54.

Isaac, K.P. (1983) Silica diagenesis of Palaeogene residual deposits in Devon, England. *Proceedings of the Geologists' Association* **94**, 181–186.

Kaiser, E. (1926) *Die Diamantenwüste Südwest-Afrikas,* Vol. 2. Berlin: Dietrich Reimer.

Kendrick, K.J. & Graham, R.C. (2004) Pedogenic silica accumulation in chronosequence soils, Southern California. *Soil Science Society of America Journal* **68**, 1295–1303.

Kerr, M.H. (1955) On the origin of silcretes in southern England. *Proceedings of the Leeds Philosophical and Literary Society (Scientific Section)* **6**, 328–337.

Khalaf, F.I. (1988) Petrography and diagenesis of silcrete from Kuwait, Arabian Gulf. *Journal of Sedimentary Petrology* **58**, 1014–1022.

King, L.C. (1967) *The Morphology of the Earth.* Edinburgh. Oliver & Boyd.

Knauth, L.P. (1994) Petrogenesis of chert. In: Heaney, P.J., Prewitt, C.T. & Gibbs, G.V. (Eds) *Silica: Physical Behaviour, Geochemistry and Materials Applications.* Reviews in Mineralogy 29. Washington: Mineralogical Society of America, pp. 233–258.

Lamplugh, G.W. (1902) Calcrete. *Geological Magazine* **9**, 75.

Langford-Smith, T. (Ed.) (1978a) *Silcrete in Australia.* Armidale: University of New England Press.

Langford-Smith, T. (1978b) A select review of silcrete research in Australia. In: Langford-Smith, T. (Ed.) *Silcrete in Australia.* Armidale: University of New England Press, pp. 1–11.

Leckie, D.A. & Cheel, R.J. (1990) Nodular silcretes of the Cypress Hills Formation (Upper Eocene to Middle Miocene) of Southern Saskatchewan, Canada. *Sedimentology* **37**, 445–454.

Lee, S.Y. & Gilkes, R.J. (2005) Groundwater geochemistry and composition of hardpans in southwestern Australian regolith. *Geoderma* **126**, 59–84.

Lindqvist, J.K. (1990) Deposition and diagenesis of Landslip Hill silcrete, Gore Lignite Measures (Miocene), New Zealand. *New Zealand Journal of Geology and Geophysics* **33**, 137–150.

Lindsay, W.L. (1979) *Chemical Equilibria in Soils.* New York: John Wiley.

Marker, M.E., McFarlane, M.J. & Wormald, R.J. (2002) A laterite profile near Albertinia, Southern Cape, South Africa: its significance in the evolution of the African Surface. *South African Journal of Geology* **105**, 67–74.

McArthur, J.M., Turner, J.V., Lyons, W.B., Osborn, A.O. & Thirlwall, M.F. (1991) Hydrochemistry on the Yilgarn Block, Western Australia – ferrolysis and mineralization in acidic brines. *Geochimica et Cosmochimica Acta* **55**, 1273–1288.

McCarthy, T.S. & Ellery, W.N. (1995) Sedimentation on the distal reaches of the Okavango Fan, Botswana, and its bearing on calcrete and silcrete (gannister) formation, *Journal of Sedimentary Research* **A65**, 77–90.

McFarlane, M.J., Borger, H. & Twidale, C.R. (1992) Towards an understanding of the origin of titanium skins on silcrete in the Beda Valley, South Australia. In: Schmitt, J.-C. & Gall, Q. (Eds) *Mineralogical and Geochemical Records of Palaeoweathering. ENSMP Mémoire Science de la Terre, École des Mines de Paris* **18**, 39–46.

McNally, G.H. & Wilson, I.RE. (1995) Silcretes of the Mirackina Palaeochannel, Arckaringa, South Australia. *Australian Geological Survey Organisation Journal of Australian Geology and Geophysics* **16**, 295–301.

McNaughton, N.J., Rasmussen B. & Fletcher I.R. (1999) SHRIMP uranium-lead dating of diagenetic xenotime in siliciclastic sedimentary rocks. *Science* **285**, 78–80.

Martini, J.E.J. (1994) A late Archean-Palaeoproterozoic (2.6 Ga) paleosol on ultramafics in the eastern Transvaal, South-Africa. *Precambrian Research* **67**, 159–180.

Ménillet, F. (1988a) Désilicifications et silicifications au Plioquaternaire dans le karsts de calcaires tertiares du Bassin de Paris. *Bulletin d'Information des géologiques du Bassin de Paris* **25**, 81–91.

Ménillet, F. (1988b) Les accident siliceux des calcaires continentaux à lacustres du Tertiare du Bassin de Paris. *Bulletin d'Information des géologiques du Bassin de Paris* **25**, 57–70.

Meunier, J.D., Colin, F. & Alarcon, C. (1999) Biogenic silica storage in soils. *Geology* **27**, 835–838.

Meyer, R. & Pena dos Reis, R.B. (1985) Palaeosols and alunite silcretes in continental Cenozoic of Western Portugal. *Journal of Sedimentary Petrology* **55**, 76–85.

Millot, G. (1960) Silice, silex, silicifications et croissance des cristaux. *Bulletin de Service Carte Geologique, Alsace Lorraine* **13**, 129–146.

Millot, G. (1964) *Géologie des Argiles.* Paris: Masson.

Millot, G. (1970) *Geology of Clays: Weathering, Sedimentology, Geochemistry* (translated by Farrand, W.R. & Paquet, H.). New York: Springer-Verlag.

Milnes, A,R. & Hutton, J.T. (1974) The nature of microcryptocrystalline titania in 'silcrete' skins from Beda Hill area of South Australia. *Search* **5**, 153–154.

Milnes, A.R. & Thiry, M. (1992) Silcretes. In: Martini, I.P. and Chesworth, W. (Eds) *Weathering, Soils and Palaeosols.* Developments in Earth Surface Processes 2. Amsterdam: Elsevier, pp. 349–377.

Milnes, A.R. & Twidale, C.R. (1983) An overview of silicification in Cainozoic landscapes of arid central and southern Australia. *Australian Journal of Soil Research* **21**, 387–410.

Milnes, A.R., Wright, M.J. & Thiry, M. (1991) Silica accumulations in saprolites and soils in South Australia. In: Nettleton, W.D. (Ed.) *Occurrence, Characteristics and Genesis of Carbonate, Gypsum, and Silica Accumulations in Soils.* Special Publication 26. Madison, WI: Soil Science Society of America, pp. 121–149.

Mišík, M. (1996) Silica spherulites and fossil silcretes in carbonate rocks of the Western Carpathians. *Geologica Carpathica* **47**, 91–105.

Morris, B.A. & Fletcher, I.A. (1987) Increased solubility of quartz following ferrous-ferric iron reactions. *Nature* **330**, 558–561.

Mountain, E.D. (1946) *The Geology of the Area East of Grahamstown.* Pretoria: Department of Mines.

Mountain, E.D. (1952) The origin of silcrete. *South African Journal of Science* **48**, 201–204.

Mountain, E.D. (1980) Grahamstown peneplain. *Transactions of the Geological Society of South Africa* **83**, 47–53.

Mustard, P.S. & Donaldson, J.A. (1990) Palaeokarst breccias, calcretes, silcretes and fault talus breccias at the base of Upper Proterozoic 'Windermere' strata, northern Canadian Cordillera. *Journal of Sedimentary Petrology* **60**, 525–539.

Nash, D.J. (1997) Groundwater as a geomorphological agent in drylands. In: Thomas, D.S.G. (Ed.) *Arid Zone Geomorphology: Process, Form and Change in Drylands.* Chichester: John Wiley & Sons, pp. 319–348.

Nash, D.J. & Hopkinson, L. (2004) A reconnaissance Laser Raman and Fourier Transform Infrared survey of silcretes from the Kalahari Desert, Botswana. *Earth Surface Processes & Landforms* **29**, 1541–1558.

Nash, D.J. & Shaw, P.A. (1998) Silica and carbonate relationships in silcrete-calcrete intergrade duricrusts from the Kalahari Desert of Botswana and Namibia. *Journal of African Earth Sciences* **27**, 11–25.

Nash, D.J., Shaw, P.A. & Thomas, D.S.G. (1994a) Duricrust development and valley evolution – process–landform links in the Kalahari. *Earth Surface Processes and Landforms* **19**, 299–317.

Nash, D.J., Thomas, D.S.G. & Shaw, P.A. (1994b) Siliceous duricrusts as palaeo-climatic indicators: evidence from the Kalahari Desert of Botswana. *Palaeogeography, Palaeoclimatology, Palaeoecology* **112**, 279–295.

Nash, D.J., Shaw, P.A. & Ullyott, J.S. (1998) Drainage-line silcretes of the Middle Kalahari: an analogue for Cenozoic sarsen trains? *Proceedings of the Geologists' Association* **109**, 241–254.

Nash, D.J. McLaren, S.J. & Webb, J.A. (2004) Petrology, geochemistry and environmental significance of silcrete-calcrete intergrade duricrusts at Kang Pan and Tswaane, central Kalahari, Botswana. *Earth Surface Processes and Landforms* **29**, 1559–1586.

Ollier, C.D. (1991) Aspects of silcrete formation in Australia. *Zeitschrift für Geomorphologie* **35**, 151–163.

Ollier, C.D. & Pain, C.F. (1996) *Regolith, Soils and Landforms.* Chichester: John Wiley.

O'Neill, G. (1984) *Geochemistry of silcrete in the Sunbury area.* Unpublished Honours Thesis, Department of Earth Sciences, University of Melbourne, Australia.

Pain, C.F. & Ollier, C.D. (1995) Inversion of relief – a component of landscape evolution. *Geomorphology* **12**, 151–165.

Parcerisa, D., Thiry M., Gomez-Gras D. & Calvet F. (2001) Tentative model for the silicification in Neogene Montjuic sandstones, Barcelona (Spain): authigenic minerals, geochemical environment and fluid flow. *Bulletin de la Societe Geologique de France* **172**, 751–764.

Passarge, S. (1904) *Die Kalahari.* Berlin: Dietrich Reimer.

Peterson, M.N.A. & von der Borch, C.C. (1965) Chert: modern inorganic deposition in a carbonate-precipitating locality. *Science* **149**, 1501–1503.

Prescott, J.A. & Pendleton, R.L. (1952) Laterite and lateritic soils. *Technical Communications of the Commonwealth Bureau of Soil Science* **47**.

Radtke, U. & Brückner, H. (1991) Investigation on age and genesis of silcretes in Queensland (Australia) – preliminary results. *Earth Surface Processes & Landforms* **16**, 547–554.

Rayot, V., Self, P. & Thiry, M. (1992) Transition of clay minerals to opal-CT during groundwater silicification. In: Schmitt, J.-C. & Gall, Q. (Eds) *Mineralogical and Geochemical Records of Palaeoweathering. ENSMP Mémoire Science de la Terre, École des Mines de Paris* **18**, 47–59.

Ribet, I. & Thiry, M. (1990) Quartz growth in limestone: example from water-table silicification in the Paris Basin. In: Noack, Y. & Nahon, D. (Eds) *Geochemistry of the Earth's Surface and of Mineral Formation, 2nd International Symposium*, 2–8 July, Aix en Provence, France. *Chemical Geology* **84**, 316–319.

Riezebos, P.A. (1974) Scanning electron microscopical observations on weakly cemented Miocene sands. *Geologie en Mijnbouw* **53**, 109–122.

Ringrose, S., Kampunzu, A.B., Vink, B.W., Matheson, W. & Downey, W.S. (2002) Origin and palaeo-environments of calcareous sediments in the Moshaweng dry valley, southeast Botswana. *Earth Surface Processes and Landforms* **27**, 591–611.

Ringrose S., Huntsman-Mapila, P., Kampunzu A.B., Downey W.S., Coetzee, S., Vink B.W., Matheson W. & Vanderpost, C. (2005) Sedimentological and geochemical evidence for palaeo-environmental change in the Makgadikgadi subbasin, in relation to the MOZ rift depression, Botswana. *Palaeogeography, Palaeoclimatology, Palaeoecology* **217**, 265–287.

Roberts, D.L., Bamford, M. & Millsteed, B. (1997) Permo-Triassic macro-plant fossils in the Fort Grey silcrete, East London. *South African Journal of Geology* **100**, 157–168.

Rodas, M., Luque, F.J., Mas, R. & Garzon, M.G. (1994) Calcretes, palycretes and silcretes in the Paleogene detrital sediments of the Duero and Tajo Basins, Central Spain. *Clay Minerals* **29**, 273–285.

Ross, G.M. & Chiarenzelli, J.R. (1985) Paleoclimatic significance of widespread Proterozoic silcretes in the Bear and Churchill provinces of the northwestern Canadian Shield. *Journal of Sedimentary Petrology* **55**, 196–204.

Roulin, F. (1987) Geodynamic, climatic and geochemical evolution in a tertiairy continental basin – detrital deposits, silcretes, calcretes and associated clays – the Eocene Apt Basin (Vaucluse, France). *Comptes Rendus de l'Academie des Sciences Series II* **305**, 121–125.

Roulin, F., Boudeulle, M. & Truc, G. (1986) Clay-opal transformations in Eocene silcretes of the Bassin-d'Apt (Vaucluse). *Bulletin de Mineralogie* **109**, 349–357.

Ruxton, B.P. & Taylor, G. (1982) The Cainozoic geology of the Middle Shoalhaven Plain. *Journal of the Geological Society of Australia* **29**, 239–246.

Schubel, K.A. & Simonson, B.M. (1990) Petrography of cherts from Lake Magadi, Kenya. *Journal of Sedimentary Petrology* **60**, 761–776.

Senior, B.R. (1978) Silcrete and chemically weathered sediments in southwest Queensland. In: Langford-Smith, T. (Ed.) *Silcrete in Australia*. Armidale: University of New England Press, pp. 41–50.

Senior, B.R. & Senior, D.A. (1972) Silcrete in southwest Queensland, *Bulletin of the Bureau of Mineral Resources, Geology and Geophysics of Australia* **125**, 23–28.

Shaw, P.A. & de Vries, J.J. (1988) Duricrust, groundwater and valley development in the Kalahari of south-east Botswana. *Journal of Arid Environments* **14**, 245–254.

Shaw, P.A. & Nash, D.J. (1998) Dual mechanisms for the formation of fluvial silcretes in the distal reaches of the Okavango Delta Fan, Botswana. *Earth Surface Processes and Landforms* **23**, 705–714.

Shaw, P.A., Cooke, H.J. & Perry, C.C. (1990) Microbialitic silcretes in highly alkaline environments: some observations from Sua Pan, Botswana. *South African Journal of Geology* **93**, 803–808.

Simon-Coinçon, R., Milnes, A.R., Thiry, M. & Wright, M.J. (1996) Evolution of landscapes in northern South Australia in relation to the distribution and formation of silcretes. *Journal of the Geological Society, London* **153**, 467–480.

Simon-Coinçon, R., Thiry, M. & Quesnel, F. (2000) Siderolithic palaeolandscapes and palaeoenvironments in the northern Massif Central (France). *Comptes Rendus de l'Academie des Sciences Series II* **330**, 693–700.

Smale, D. (1973) Silcretes and associated silica diagenesis in southern Africa and Australia, *Journal of Sedimentary Petrology* **43**, 1077–1089.

Stephens, C.G. (1964) Silcretes of central Australia. *Nature* **203**, 1407.

Stephens, C.G. (1966) Origin of silcrete in central Australia. *Nature* **209**, 497.

Stephens, C.G. (1971) Laterite and silcrete in Australia: a study of the genetic relationships of laterite and silcrete and their companion materials, and their collective significance in the formation of the weathered mantle, soils, relief and drainage of the Australian continent. *Geoderma* **5**, 5–52.

Storz, M. (1926) Zur petrogenesis der sekundären Kieselgesteine in der südlichen Namib. In: Kaiser, E. (Ed.) *Die Diamantenwüste Südwest-Afrikas*, Vol. 5. Berlin: Dietrich Reimer, pp. 254–282.

Summerfield, M.A. (1978) *The nature and origin of silcrete with particular reference to Southern Africa.* Unpublished PhD Thesis, University of Oxford.

Summerfield, M.A. (1979) Origin and palaeoenvironmental interpretation of sarsens. *Nature* **281**, 137–9.

Summerfield, M.A. (1981) *The Nature and Occurrence of Silcrete, Southern Cape Province, South Africa.* School of Geography Research Paper 28, University of Oxford.

Summerfield, M.A. (1982) Distribution, nature and genesis of silcrete in arid and semi-arid southern Africa. *Catena, Supplement* **1**, 37–65.

Summerfield, M.A. (1983a) Silcrete. In: Goudie, A.S. & Pye, K. (Eds) *Chemical Sediments and Geomorphology.* London: Academic Press, pp. 59–91.

Summerfield, M.A. (1983b) Silcrete as a palaeoclimatic indicator: evidence from southern Africa. *Palaeogeography, Palaeoclimatology, Palaeoecology* **41**, 65–79.

Summerfield, M.A. (1983c) Petrography and diagenesis of silcrete from the Kalahari Basin and Cape Coastal Zone, southern Africa. *Journal of Sedimentary Petrology* **53**, 895–909.

Summerfield, M.A. (1983d) Geochemistry of weathering profile silcretes, southern Cape Province, South Africa. In: Wilson, R.C.L. (Ed.) *Residual Deposits: Surface*

Related Weathering Processes and Materials. Special Publication 11, Geological Society of London, pp. 167–178.

Summerfield, M.A. (1984) Isovolumetric weathering and silcrete formation, Southern Cape Province, South Africa. *Earth Surface Processes and Landforms* **9**, 135–141.

Summerfield, M.A. (1986) Reply to discussion – silcrete as a palaeoclimatic indicator: evidence from southern Africa. *Palaeogeography, Palaeoclimatology, Palaeoecology* **52**, 356–360.

Summerfield, M.A. & Goudie, A.S. (1980) The sarsens of southern England: their palaeoenvironmental interpretation with reference to other silcretes. In: Jones, D.K.C. (Ed.) *The Shaping of Southern England.* London: Academic Press, pp. 71–100.

Summerfield, M.A. & Whalley, W.B. (1980) Petrographic investigations of sarsens (Cainozoic silcretes) from southern England. *Geologie en Mijnbouw* **59**, 145–53.

Tait, M. (1998) *Geology and landscape evolution of the Mt Wood Hills area, near Tibooburra, northwestern New South Wales.* Unpublished Honours Thesis, Department of Earth Sciences, La Trobe University, Australia.

Taylor, G. (1978) Silcretes in the Walgett-Cumborah region of New South Wales. In: Langford-Smith, T. (Ed.) *Silcrete in Australia.* Armidale: University of New England Press, pp. 187–193.

Taylor, G. & Eggleton, R.A. (2001) *Regolith Geology and Geomorphology.* Chichester: John Wiley.

Taylor, G. & Ruxton, B.P. (1987) A duricrust catena in South-east Australia. *Zeitschrift für Geomorphologie* **31**, 385–410.

Taylor, G. & Smith, I.E. (1975) The genesis of sub-basaltic silcretes from the Monaro, New South Wales. *Journal of the Geological Society of Australia* **22**, 377–385.

Terry, D.O. & Evans, J.E. (1994) Pedogenesis and paleoclimatic implications of the Chamberlain Pass Formation, Basal White River Group, Badlands of South Dakota. *Palaeogeography, Palaeoclimatology, Palaeoecology* **110**, 197–215.

Thiry, M. (1978) Silicification des sédiments sablo-argileux de l'Yprésien du sud-est du bassin de Paris. Genèse et évolution des dalles quartitiques et silcrètes. *Bulletin du B.R.G.M.* (deuxième serie) **1**, 19–46.

Thiry, M. (1981) Sédimentation continentale et altérations associées: calcitisations, ferruginisations et silicifications. *Les Argiles Plastiques de Sparnacien du Bassin de Paris.* Mémoire 64. Sciences Géologiques.

Thiry, M. (1988) Les Grès lustrés de l'Éocène du Bassin de Paris: des silcrètes pédologiques. *Bulletin d'Information des Géologues du Bassin de Paris* **25**, 15–24.

Thiry, M. (1989) Geochemical evolution and palaeoenvironments of the Eocene continental deposits in the Paris basin. *Palaeogeography, Palaeoclimatology, Palaeoecology* **70**, 153–163.

Thiry, M. (1999) Diversity of continental silicification features: examples from the Cenozoic deposits in the Paris Basin and neighbouring basement. In: Thiry, M. & Simon-Coinçon, R. (Eds) *Palaeoweathering, Palaeosurfaces and Related Continental Deposits.* Special Publication 27, International Association of Sedimentologists. Oxford: Blackwell Science, pp. 87–127.

Thiry, M. & Ben Brahim, M. (1990) Silicifications pédogénétiques dans les dépôts hamadiens du piémont de boudenib (Maroc). *Geodinamica Acta* 4, 237–251.

Thiry, M. & Ben Brahim, M. (1997) Ground-water silicifications in the calcareous facies of the Tertiary piedmont deposits of the Atlas Mountain (Hamada du Guir, Morocco). *Geodinamica Acta* 10, 12–29.

Thiry, M. & Bertrand-Ayrault, M. (1988) Les grès de Fontainebleau: Genèse par écoulement de nappes phréatiques lors de l'entaille des vallées durant le Plio-Quaternaire et phénomènes connexes. *Bulletin d'Information des géologues du Bassin de Paris* 25, 25–40.

Thiry, M. & Marechal, B. (2001) Development of tightly cemented sandstone lenses in uncemented sand: example of the Fontainebleau sand (Oligocene) in the Paris Basin. *Journal of Sedimentary Research* A71, 473–483.

Thiry, M. & Millot, G. (1987) Mineralogical forms of silica and their sequence of formation in silcretes. *Journal of Sedimentary Petrology* 57, 343–352.

Thiry, M. & Milnes, A.R. (1991) Pedogenic and groundwater silcretes at Stuart Creek opal field, South Australia. *Journal of Sedimentary Petrology* 61, 111–127.

Thiry M. & Ribet I. (1999) Groundwater silicification in Paris Basin limestones: fabrics, mechanisms, and modeling. *Journal of Sedimentary Research* 69, 171–183.

Thiry, M. & Simon-Coinçon, R. (1996) Tertiary palaeoweatherings and silcretes in the southern Paris Basin. *Catena* 26, 1–26.

Thiry, M. & Turland, M. (1985) Paléotoposéquences de sols ferrugineux et de cuirassements siliceux dans le Sidérolithique du nord du Massif Central. *Géologie de France* 2, 175–192.

Thiry, M., Bertrand-Ayrault, M. & Grisoni, J.-C. (1988a) Ground-water silicification and leaching in sands: example of the Fontainebleau Sand (Oligocene) in the Paris Basin. *Bulletin of the Geological Society of America* 100, 1283–1290.

Thiry, M., Bertrand-Ayrault, M., Grisoni, J.-C., Menillet, F. & Schmitt, J.-M. (1988b) Les grès de Fontainebleau: silicification de nappes liées à l'évolution géomorphologique du Bassin de Paris durant le Plio-Quaternaire. *Bulletin de Société géologique de France* 8, 419–430.

Thiry, M., Schmitt, J.-M. & Milnes, A.R. (1991) *Silcretes: Structures, Micromorphology, Mineralogy, and Their Interpretation*, IGCP 317 Workshop, Fontainebleau, 25–27 November.

Thiry, M., Schmitt, J.M., Rayot, V. & Milnes, A.R. (1995) Geochemistry of the bleached profiles of the Tertiary regolith of inland Australia. *Comptes Rendus de l'Academie des Sciences Series II* 320, 279–285.

Thiry, M., Milnes, A.R., Rayot, V. & Simon-Coinçon, R. (2006) Interpretation of palaeoweathering features and successive silicifications in the Tertiary regolith of inland Australia. *Journal of the Geological Society, London* 163, 723–736.

Trappe, J. (1992) Synsedimentary silicified stromatolites in Palaeocene playa deposits of the western Basin of Ouarzazate, Morocco. *Neues Jahrbuch für Geologie und Paläontologie. Monatshefte* 8, 458–468.

Twidale, C.R. (1983) Australian laterites and silcretes: ages and significance. *Revue de Geologie Dynamique et de Geographie Physique* 24, 35–45.

Twidale, C.R. & Campbell, E.M. (1995) Pre-Quaternary landforms in the low-latitude context – the example of Australia. *Geomorphology* **12**, 17–35.

Twidale, C.R. & Hutton, J.T. (1986) Silcrete as a palaeoclimatic indicator: Discussion. *Palaeogeography, Palaeoclimatology, Palaeoecology* **52**, 351–356.

Twidale, C.R. & Milnes, A.R. (1983) Aspects of the distribution and disintegration of siliceous duricrusts in arid Australia. *Geologie en Mijnbouw* **62**, 373–382.

Ullyott, J.S. (2002) *The distribution and petrology of sarsens on the eastern South Downs and their relationship to Palaeogene and Neogene sediments and palaeoenvironments.* Unpublished PhD Thesis, University of Brighton.

Ullyott, J.S. & Nash, D.J. (2006) Micromorphology and geochemistry of groundwater silcretes in the eastern South Downs, UK. *Sedimentology* **53**, 387–412.

Ullyott, J.S., Nash, D.J. & Shaw, P.A. (1998) Recent advances in silcrete research and their implications for the origin and palaeoenvironmental significance of sarsens. *Proceedings of the Geologists' Association* **109**, 255–270.

Ullyott, J.S., Nash, D.J., Whiteman, C.A. & Mortimore, R. (2004) Distribution, petrology and mode of development of silcretes (sarsens and puddingstones) on the eastern South Downs, UK. *Earth Surface Processes and Landforms* **29**, 1509–1539.

Van den Broek, J.M.M. & van der Waals, L. (1967) The late Tertiary peneplain of south Limburg (The Netherlands): Silicifications and fossil soils; a geological and pedological investigation. *Geologie en Mijnbouw* **46**, 318–332.

Van der Graaff, W.J.E. (1983) Silcrete in Western Australia: geomorphological settings, textures, structures, and their possible genetic implications. In: Wilson, R.C.L. (Ed.) *Residual Deposits: Surface Related Weathering Processes and Materials.* Special Publication 11, Geological Society of London, pp. 159–166.

Van Dijk, D.C. & Beckmann, G.G. (1978) The Yuleba Hardpan, and its relationship to soil-geomorphic history, in the Yuleba-Tara region, Southeast Queensland. In: Langford-Smith, T. (Ed.) *Silcrete in Australia.* Armidale: University of New England Press, pp. 73–91.

Vasconcelos, P.M. (1999) K–Ar and ^{40}Ar/^{39}Ar geochronology of weathering processes. *Annual Review of Earth and Planetary Sciences* **27**, 183–229.

Vasconcelos, P.M. & Conroy, M. (2003) Geochronology of weathering and landscape evolution, Dugald River valley, NW Queensland, Australia. *Geochimica et Cosmochimica Acta* **67**, 2913–2930.

Watchman, A.L. & Twidale, C.R. (2002) Relative and 'absolute' dating of land surfaces. *Earth-Science Reviews* **58**, 1–49.

Watson, A. & Nash, D.J. (1997) Desert crusts and varnishes. In: Thomas, D.S.G. (Ed.) *Arid Zone Geomorphology: Process, Form and Change in Drylands.* Chichester: John Wiley & Sons, pp. 69–107.

Watts, S.H. (1977) Major element geochemistry of silcrete from a portion of inland Australia. *Geochimica et Cosmochimica Acta* **41**, 1164–7.

Watts, S.H. (1978a) The nature and occurrence of silcrete in the Tibooburra area of northwestern New South Wales. In: Langford-Smith, T. (Ed.) *Silcrete in Australia.* Armidale: University of New England Press, pp. 167–85.

Watts, S.H. (1978b) A petrographic study of silcrete from inland Australia. *Journal of Sedimentary Petrology* **48**, 987–94.

Webb, J.A. & Golding, S.D. (1998) Geochemical mass-balance and oxygen-isotope constraints on silcrete formation and its palaeoclimatic implications in southern Australia. *Journal of Sedimentary Research* **68**, 981–993.

Whitehouse, F.W. (1940) Studies in the late geological history of Queensland. *Publications of the University of Sydney Geology Department* **2**(1).

Williams, L.A. & Crerar, D.A. (1985) Silica diagenesis, II. General mechanisms. *Journal of Sedimentary Petrology* **55**, 312–321.

Williams, L.A., Parks, G.A. & Crerar, D.A. (1985) Silica diagenesis, I. Solubility controls. *Journal of Sedimentary Petrology* **55**, 301–311.

Wopfner, H. (1978) Silcretes of northern South Australia and adjacent regions. In: Langford-Smith, T. (Ed.) *Silcrete in Australia*. Armidale: University of New England Press, pp. 93–141.

Wopfner, H. (1983) Environment of silcrete formation: a comparison of examples from Australia and the Cologne Embayment, West Germany. In: Wilson, R.C.L. (Ed.) *Residual Deposits: Surface Related Weathering Processes and Materials*. Special Publication 11, Geological Society of London, pp. 151–157.

Wopfner, H., Callen, R. & Harris, W.K. (1974) The lower Tertiary Formation of the south-western Great Artesian Basin. *Journal of the Geological Society of Australia* **21**, 17–51.

Young, R.W. (1978) Silcrete in a humid landscape: the Shoalhaven Valley and adjacent coastal plains of southern New South Wales. In: Langford-Smith, T. (Ed.) *Silcrete in Australia*. Armidale: University of New England Press, pp. 195–207.

Young, R.W. (1985) Silcrete distribution in eastern Australia. *Zeitschrift für Geomorphologie* **29**, 21–36.

Youngson, J.H. (2005) Diagenetic silcrete and formation of silcrete ventifacts and aeolian gold placers in Central Otago, New Zealand. *New Zealand Journal of Geology and Geophysics* **48**, 247–263.

Chapter Five

Aeolianite

Sue J. McLaren

5.1 Introduction: Nature and General Characteristics

Aeolianites (or eolianites) were originally described as consolidated aeolian sediments by Sayles in 1931. However, many variations on the definition of aeolianite exist in the literature. For example, Livingstone and Warren (1996, p. 141) describe aeolianites as 'cemented coastal dune sands, generally on semi-arid, tropical and high energy shorelines', yet this ignores those found in inland drylands (see section 5.2). James and Choquette (1984) argue that 'the most important intrinsic factor is the original mineralogy' (p. 161), as carbonate deposits respond differently to subaerial exposure when compared with siliclastic deposits (Dravis, 1996). This has led to a number of people attempting to make a distinction between siliclastic and carbonate aeolianites. For example, Brooke (2001) defines aeolianites as comprising mostly reworked shallow-marine biogenic carbonate sediment, although as a strict definition this ignores the abundant record of terrigenous sand-dominated aeolianites such as those in southeast Australia (e.g. Bryant et al., 1990, 1992, 1994; Hunter et al., 1993, 1994). The definition therefore needs to be broad, and aeolianites *sensu stricto* should encompass all aeolian sands (carbonate and siliclastic) that are partially cemented by calcium carbonate under subaerial conditions (Gardner, 1983; McLaren, 2003).

The genesis of aeolianites involves both the accumulation of wind-derived deposits on the ground and their subsequent alteration by various geochemical processes and changes. These processes are collectively known as diagenesis and may involve changes in the mineralogy, geochemistry, texture and fabric of the sediment. The mineralogical changes associated with the diagenesis of carbonate-rich aeolian deposits include the alteration of unstable aragonite and high-Mg calcite and precipitation of relatively stable low-Mg calcite. Geochemical changes are associated with the mineralogical changes and result in the loss of Mg and Sr ions. Formation of cement and changes in porosity values (primary and secondary) comprise the main textural and fabric changes.

There are a number of factors that can affect the processes of subaerial lithification (Gardner and McLaren, 1994), including sediment type, time and climate (e.g. Ward, 1973; Gardner and McLaren, 1993; McLaren and Gardner, 2004) at the macroscale; the effect of rhizoliths (e.g. Klappa, 1980, 1983; Jones and Kwok Choi, 1988; McLaren 1995a), sea spray (e.g. Bruckner, 1986; McLaren, 1995b, 2001), proximity to the surface (e.g. Yaalon, 1967) and texture at the mesoscale; and finally, the rate, amount and chemistry of pore waters at the microscale (e.g. Berner, 1978; Chafetz et al., 1985).

5.2 Distribution, Field Occurrence and Geomorphological Relations

Aeolianites are found in a wide variety of climatic zones, from arid through subtropical to temperate. Cemented dune sands (Figure 5.1) are largely late Pleistocene in age (although many older deposits do occur; see section 5.3) and their existence in the landscape today may be a function of conditions favouring preservation rather than formation (Gardner, 1983). Aeolianites form most commonly in coastal locations (see Brooke, 2001), although they have also been found in deserts such as Iran (Thomas et al., 1997), Wadi Araba Desert, Jordan (Saqqa and Atallah, 2004), the Thar (Goudie and Sperling, 1977), the Ubari Sand Sea, southern Libya (Mattingly et al., 1998; McLaren and Gardner, 2004), and the Wahiba Sands (Gardner, 1988; Preusser et al., 2002; McLaren and Gardner, 2004).

Three of the most extensive deposits of aeolianites can be found in coastal Western and South Australia (e.g. Playford et al., 1976; Playford, 1988; Price et al., 2001), South Africa (e.g. Malan, 1987; Illenberger, 1996) and the Wahiba Sands in Oman (Gardner, 1988; Glennie and Singhvi, 2002; Preusser et al., 2002). The longest temporal record of aeolianite deposition is along the Coroong coastal plain in Southern Australia (Brooke, 2001). The total thickness of aeolianite deposits varies between < 1 m to in excess of 100 m (e.g. South Africa, Bateman et al., 2004). In coastal locations, aeolianite often forms cliffs at the back of beaches, for example on Isla Cancun, Mexico (McLaren and Gardner, 2000) and southeast Sushastra in India (Khadkikar and Basavaiah, 2004). Barrier dune ridges, sometimes extending miles inland, dominate some coastal locations such as the Mediterranean coast of northwest Egypt (El-Asmar and Wood, 2000), Israel (e.g. Gvirtzman et al., 1984) and in South Africa (Illenberger and Verhagen, 1990; Illenberger, 1996). A number of tropical islands are made up largely of aeolianite, the most well known being the islands of Bermuda, the Bahamas, Lord Howe Island and Rottnest Island

Figure 5.1 Aeolianite from North Point, San Salvador, The Bahamas.

(e.g. Land et al., 1967; Hearty et al., 1992; Hearty and Kindler, 1993; Carew and Mylroie, 1995; Vacher et al., 1995; Vacher and Rowe, 1997; Hearty and Kaufman, 2000; Price et al., 2001; Hearty, 2002, 2003).

5.3 Macro- and Micromorphological Characteristics

All dune forms have the potential to become cemented, which increases their preservation potential as they become more resistant to erosion. As most aeolianites are found along coastal shorelines, transverse, barrier (Illenberger, 1996), oblique and parabolic (e.g. Kindler and Mazzolini, 2001) are the most common dune forms found, but climbing or cliff-front (e.g. Clemmensen et al., 1997), barchan (Gardner, 1988; Frechen et al., 2002), nebkha (Saqqa and Atallah, 2004) and linear dunes are also known (Gardner, 1988; Kindler and Mazzolini, 2001; Saqqa and Atallah, 2004).

5.3.1 Internal structures

Internal sedimentary structures within aeolian dunes (Figure 5.2) are a function of the depositional nature of entrained sand grains. According to Livingstone and Warren (1996), there are three key processes that result in the accumulation of distinct laminae and depositional surfaces. First of all, avalanching slip faces result in sandflow cross-strata developing on lee slopes often creating high-angle beds with dips of 30–34°. Accretion of bedload grains moved by creep occurs on lower slopes where angles are less than 15°. Sorting of grain sizes often results in distinct pin-stripe

Figure 5.2 Internal sedimentary structures, Wahiba Sands, Oman.

bedding (Fryberger and Schenk, 1988). Inverse size grading of sediments occurs as a result of the migration and climbing of aeolian ripples and dunes (see Hunter (1977) for more details). Most other aeolian deposition occurs in the zone of flow separation, leeward of the dune crests. Larger, lighter allochems and smaller, heavier terrigenous clasts are often sorted into different laminae (e.g. Gardner, 1988).

Junctions between the accumulation of older and younger dune sands have been called bounding surfaces (Brookfield, 1977), and Kocurek (1988, 1991) has identified a three-order hierarchy of these features. First-order surfaces develop as a result of the movement of similar sized dunes over one another. Second-order bounding surfaces form by the passage of smaller dunes over larger ones. Surfaces created as a result of a change in an extrinsic control, such as direction of dune movement, make up the third-order surfaces. Major breaks in aeolian sedimentation are known as supersurfaces (Kocurek, 1991). Stokes surfaces also exist; these are wind-deflated surfaces that have been eroded down to the groundwater table (Stokes, 1968).

5.3.2 Texture

A sand dune's texture comprises the particle sizes and morphology, together with their sorting and packing, and the primary sedimentary structures. Aeolian dunes are predominantly made up of subangular to well-rounded grains, between 63 and 1000 μm, that are moderately through to well sorted and may have become reddened (Thomas, 1997). Processes of saltation can transport larger carbonate grains, which have a lower specific gravity and are often platy in nature, more readily than similar sized terrigenous clasts that generally have a greater bulk density. Tucker and Vacher (1980) note, however, that these characteristics are not unique to aeolian sands. Texture affects both the porosity and permeability of a deposit. Porosity, permeability, original grain sizes and the packing of grains are controlled by a number of factors, including the physical processes of sedimentation, the environmental history of the grain prior to deposition, and by post-depositional diagenetic processes. The diagenetic events create as well as destroy pore spaces. Porosity development progresses both in time and space and therefore is four-dimensional (Schroeder, 1988).

The importance of porosity and permeability in diagenesis has been emphasised by a number of researchers, including Evans and Ginsburg (1987). They proposed that precipitation and dissolution of calcium carbonate that is controlled or guided by depositional facies, must be related to the rate and path of water movement, itself determined by the porosity and permeability of the sediment and availability of nuclei for precipitation. Harris and Matthews (1968) found that during large-scale solution-reprecipitation (other factors being equal) the relative rates of percolating water movement determine the extent of calcium carbonate mobility.

Tightly bound films of water surrounding particles result in higher porosities in carbonate sediments by separating the particles (Enos and Sawatsky, 1981). It may be possible that this aids diagenesis, as the waters surrounding these grains are thought to be sites of early rim and meniscus cements. Therefore, porosity and permeability have been proposed to be important in vadose diagenesis as they influence the hydraulic behaviour of the water (see Morrow, 1971; Schroeder, 1988). However, experiments by Sippel and Glover (1964) led them to suggest that flow transport is relatively unimportant as a device for inducing cementation or dissolution in the pore system.

The textural combination of grain size and sorting together are thought to be influential in the vadose diagenetic zone. Morrow (1971) studied fluid motion in sediments composed of various mixtures of fine and coarse grains. Where clustering of fine clasts occurred, fluid was retained, but where grains were more evenly distributed the sediment was drained. Thus, sorting and packing heterogeneities also influence the hydraulic behaviour. This observation was also made by other authors (e.g. Buddemeier and Oberndorfer, 1986; Strasser and Davaud, 1986), including Gardner (1988), who recognised that calcite cement occurs preferentially in the finer grained and more densely packed laminae of aeolianites in the Wahiba Sands.

Even though texture is thought to be an important control in vadose diagenesis, exactly how it influences cementation over time during diagenetic alteration is still uncertain. Pittman (1974) found that porosity decreases primarily with the onset of diagenesis because of the growth of sparry calcite in voids, whereas permeability increases because of changes

Figure 5.3 Differentially cemented laminae (chisel for scale). The finer grained laminae are better cemented and more resistant to weathering than the coarser-grained laminae. (Note the slumped laminae in the centre of the figure.)

in pore aperture size that result from fabric alterations associated with dissolution of aragonite and precipitation of low-Mg calcite. Schmalz (1967) noted that with the onset of cementation there is a reduction in pore size and thus narrower diffusion pathways, which results in a slowing down of diagenetic changes. Martin et al. (1986) and Evans and Ginsburg (1987), working on the Pleistocene Miami Limestone, also found that texture acts as an important control on diagenesis, with finer grained, well-sorted strata being better cemented than coarser grained beds (Figure 5.3). Gardner (1986) did not record a textural control in the Kanya Kumari aeolianite series in southern India. She found that it was the type and amount of allochems that controlled the degree of cementation.

Variations between laminae are often visible in exposures after differential weathering (Figure 5.2). Fryberger and Schenk (1988, p. 1) found that the textural segregation associated with the deposition of laminae in ancient aeolian sediments 'in most cases leads to early cementation along and near the finest sand and silt comprising the pin stripe laminations'. Roberts et al. (1973, p. 98) also found that cement often 'follows the dipping accretion planes of the original dune structure rather than being generally horizontal' in Holocene dunes from Luskentyre, northern Scotland. Caputo (1993, p. 256) suggested that laminae become 'selectively well cemented because of fine grain size, close packing and small pores, and high retention of cement generating pore water'. The variability is probably also because permeability values are higher parallel to laminae than normal to them, and so fluids tend to flow along parallel to laminae. Agnew (1988) found in modern dunes from Oman that subsurface dune bedding affects infiltration rates and that drainage moves along subsurface bedding and not vertically downwards. This is supported by research carried out by Gardner and McLaren (1999) on coastal sand dunes at Studland, Dorset, UK (Figure 5.4). Therefore, it is likely that cementation will also develop along laminae; and production of cement will, moreover, then retard fluid movement across laminae and hinder even further cement growth across laminae.

5.3.3 Micromorphology

In the vadose zone, both water and air fill pore spaces, giving an uneven cement distribution. Pendant or gravity cements result from calcite precipitation from saturated waters held on the underside of grains, due to the force of gravity, but this cement type is uncommon in the vadose zone (McLaren, 1993). Meniscus, rim and pore-filling cements are common in the freshwater vadose zone. Rim cementation (Figure 5.5) is rarely complete and the cement crystals tend to decrease in size around the grain, disappearing in air-filled pore spaces. Average sizes of crystals often range from about 5 to 50 μm in diameter and they commonly increase away from the initial substrate, towards the centre of the pores. The rate of crystal growth affects crystal size; for example, rapid precipitation (favoured where saturation is high and nuclei are abundant) may result in fine crystal sizes of less than 5 μm (Given and Wilkinson, 1985).

Cement crystals are commonly bladed, equant or syntaxial in shape. Syntaxial overgrowths are precipitated in optical continuity with the substrate

Figure 5.4 Alternating darker wet and lighter dry layers in a modern-day dune, Studland, UK. The central wetter layer is about 15 cm thick.

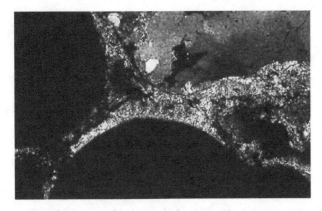

Figure 5.5 Rim cements developed in an aeolianite from Cabo de Gato, southern Spain (cross-polarised light, ×10 magnification).

crystal. The development of this cement type is dependent upon the nature and abundance of contained allochems, thus syntaxial rims may not be present in all vadose-cemented deposits. Pore-filling cements are often patchy in distribution, with some occluded pores next to pores devoid of any pore-filling cement. Gardner and McLaren (1993) provide evidence that there are no distinct temporal cement generations as indicated by Land et al. (1967).

During dissolution of aragonite, grains and cement may be entirely dissolved, with moulds filled in later by calcite cement. Dissolution of aragonitic shells and ooids often results in fossil outlines being preserved by micritic envelopes (Figure 5.6). According to Knox (1977), planar and concavo-convex grain-to-grain contacts indicate vadose dissolution–compaction. Corrosion embayments on quartz grains are a result of the peripheral replacement by calcite (Frechen et al., 2002).

Aeolianites that become submerged below the groundwater table may become cemented in the phreatic environment. In this zone, pore spaces between sand grains are completely filled with water; and any cements derived from the interstitial waters are often isopachous in nature (Muller, 1971). Large solution volumes and longer residence times in the phreatic zone can result in coarser spar compared with the vadose zone. Increases in temperature, degree of supersaturation and NaCl content appear to result in larger cement crystal sizes (Badiozamani et al., 1977).

5.4 Chemistry and Mineralogy

The sand grains that make up aeolian dunes can include detrital biogenic carbonate grains (e.g. Ward, 1973, 1975), ooids (e.g. Budd, 1988) and

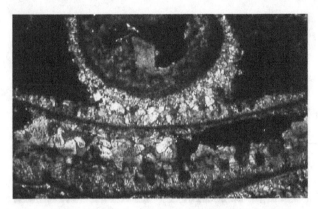

Figure 5.6 Micritic envelopes developed around a former shell fragment that has undergone dissolution and has been partially replaced by secondary porosity and neomorphic spar, Campo de Tiro, Mallorca (cross-polarised light, ×10 magnification).

non-carbonate grains (e.g. Gardner, 1986; McLaren, 1995b). Carbonate-rich dunes are common along tropical and sub-tropical coastal banks and shelves because there is greater production of carbonates, including biogenic, oolitic and peloidal materials in the warm seas. Lithic-rich dunes are more common along temperate coastlines, in the Mediterranean and in deserts. In coastal dryland environments, the ratio of quartz to carbonate often increases in a landward direction. For example, Hadley et al. (1998) recorded that around the Arabian Gulf the siliclastic content of dunes increases from 16% near the coast to as much as 89% eighty kilometres inland. In deserts, one occasionally finds the preservation of evaporite cementing agents such as halite (e.g. Gardner, 1988) or gypsum (e.g. McKee, 1966) along with calcium carbonate. Halite and dolomite crystals can also be found in coastal locations exposed to sea spray (Muller and Tietz, 1975; McLaren, 1995b, 2001).

In coastal settings, numerous types of allochems originating from marine environments are moved on land where they are relatively unstable under subaerial conditions. Meteoric waters are thought of as being aggressive toward sedimentary carbonate minerals in the vadose environment. Due to accessibility of meteoric waters to huge reservoirs of CO_2 present in the atmosphere and the ability of waters to dissolve large quantities of CO_2 present in both the atmosphere and in soils (Moore, 1989), pCO_2 of soils can be two orders of magnitude higher than atmospheric pCO_2 (Matthews, 1974). The dominant carbonate mineral in the meteoric diagenetic environment is low-Mg calcite, as Mg/Ca ratios are usually very low in vadose waters and low-Mg calcite is far more stable at surface temperatures and pressures than aragonite and high-Mg calcite. Diagenesis may result in a number of processes occurring, including the dissolution of aragonite and reprecipitation of low-Mg calcite cement, calcitisation of aragonite *in situ*, and loss of Mg^{2+} from within the lattice of high-Mg calcite (during the transformation from magnesian calcite to calcite, Mg^{2+} is leached from the crystal and the initial grain or cement architecture remains unaffected; Scoffin, 1987). The presence of Mg^{2+} and other ions in precipitating waters may act as either inhibitors or promoters of precipitation of one or more of the $CaCO_3$ polymorphs because they affect the mineral equilibria. If there is a high amount of Na^+ or Mg^{2+}, then aragonite or high-Mg calcite is preferentially developed, such as would occur in a marine environment. Trace element composition of vadose cements is usually low, and iron and manganese are generally absent from this oxidising environment.

Substrate mineralogy, crystal size and orientation, and degree of cleanliness of surfaces control nucleation (Bathurst, 1971). The earliest cements are in lattice-continuity with pre-existing crystals in the original free surface (Dickson, 1983). As growth continues, competition results in the survival of the more favourably orientated crystals; others are stunted by the obstruction or overgrowth of more successful crystals. Alternatively, a compromise boundary (Sippel and Glover, 1965) may be created where adjacent crystals meet but continue to grow, maintaining contact along a plane interface. Cement crystal development is retarded if the substrate is dirty or if it is a different mineralogy to the cement (Bathurst, 1971).

Past environmental and climatic conditions can be obtained from cements by measuring $\delta^{13}C$ and $\delta^{18}O$ values; $\delta^{13}C$ values allow an interpretation of the source of CO_2 within the precipitating waters and $\delta^{18}O$ values reflect the water temperature and isotopic composition of the pore waters at the time of precipitation. Depleted $\delta^{13}C$ whole dune rock values largely result from the concentration of $\delta^{12}C$ associated with evaporation, the extension of plant roots into the sands and pedogenic processes. During subaerial lithification of carbonate sands there is a progressive negative increase in the stable isotope value of carbon. This change reflects the alteration and loss of high-Mg calcite and aragonite carbonate grains and their replacement with low-Mg calcite cements. Isotope studies are discussed in more detail in Chapter 13.

5.5 Mechanisms of Formation or Accumulation

5.5.1 Dune formation

Wind-blown sand dunes are accumulations of sediment laid down by aeolian processes and fashioned into bedforms by deflation and deposition (Livingstone and Warren, 1996). A supply of sand grains, either marine carbonate material or terrigenous lithoclasts that can be entrained by the wind is necessary. Deposition of dune sands implies a reduction in the sediment transport rate, which can be caused by a number of factors, including variations in microtopography and changes in surface roughness that result in the decrease in shear stress, or convergence of streamlines in the lee of obstacles (Lancaster, 1995). Net sedimentation must occur if dunes are to be preserved in the landscape, and long-term persistence is aided by cementation. Factors that affect the accumulation of dune sands in coastal environments include onshore winds, sediment supply, sea level, ocean currents, the type of continental margin/platform geometry, level of carbonate production, vegetation and climate (precipitation, temperature and potential evaporation). In drylands, a sand supply, dry, vegetation-free surfaces and turbulent winds may aid the entrainment of grains into the wind velocity profile. Unstabilised dunes are prone to reworking of the sand grains and their movement downwind (Livingstone and Warren, 1996). According to Kocurek and Havholm (1993), the conditions necessary for preservation of wind-blown sediments are not always the same as those needed for their accumulation. Preservation of aeolian deposits need not always be by cementation but can occur as a result of subsidence, burial, rise in the water table (Kocurek, 1999) or, in the case of coastal dunes, a rise in sea level during sediment accumulation.

5.5.2 Subaerial diagenesis

After the accumulation of the wind-derived deposits, the next set of processes in the formation of aeolianites involves their diagenetic alteration in the vadose zone. Because meteoric fluids are commonly undersaturated with regard to calcium carbonate, dissolution initially occurs. The rate of dissolution of the more soluble (aragonite) phase greatly exceeds the rate of precipitation of the less soluble (calcite) phase (Schmalz, 1967). Continued dissolution may lead to supersaturation with respect to calcium carbonate, resulting in the precipitation of cement. Supersaturation can occur by a number of processes, including the loss of CO_2 by degassing or photosynthetic extraction (Phillips et al., 1987; Semeniuk and Meagher, 1981), warming increasing solubility, mixing of high- and low-salinity solutions (e.g. in the sea-spray zone; Land et al., 1967; McLaren, 1995b, 2001), increasing concentration by evaporation (James, 1974), or pressure reduction, especially along local pressure gradients adjacent to grain contacts (Berner, 1980).

Precipitation of cement in all environments is controlled by a number of factors, including the presence of saturated solutions, the degree of supersaturation of pore fluids (which affects the amount of cement precipitated as it controls both the crystal fabric developed and the amount of CaO available for precipitation), the composition of the solution, the rate of pore-water movement and the chemistry of the substrate.

It is now generally accepted that diagenetic variability in the vadose zone of Late Quaternary aeolianites is a characteristic feature (Schroeder, 1988). Some Late Pleistocene and Holocene aeolianites have undergone substantial diagenetic alteration (e.g. Gavish and Friedman, 1969; Halley and Harris, 1979; Budd, 1988). Dravis (1996) studied subaerially exposed oolitic sands that had become case-hardened in the upper 10–20 cm in less than 10 years as a result of 'fabric selective dissolution and associated low-Mg calcite cementation in the freshwater vadose zone' (p. 8.). This indicated that the cements were derived locally from the dissolution of the aragonitic clasts making up the sands. Other deposits of similar age remain extremely friable (e.g. Magaritz et al., 1979; Gardner, 1988; McLaren and Gardner, 2004). Indeed, time is not always a dominant control on diagenesis. For example, parts of the Australian Gambier limestone that have been in the vadose zone for over ten million years, remain virtually unlithified (James and Bone, 1989).

A number of researchers have argued that one of the key extrinsic controls on diagenesis is climate (Ward, 1973; James and Choquette, 1984). McLaren and Gardner (2004) attempted to assess whether a macroscale palaeoclimatic control on vadose diagenesis can be identified. Calcarenites were studied from modern-day arid to sub-humid settings. McLaren and Gardner (2004) found that a palaeoclimatic signal could be recognised in some deposits. Examples include the Bahamas archipelago where precipitation rates decrease from $1500 \, mm \, yr^{-1}$ in the northwest to $700 \, mm \, yr^{-1}$ in the southeast. As a consequence, meteoric diagenesis on Joulters Cay, northern Bahamas, is fast and widespread (Halley and Harris, 1979), whereas on Hogsty reef, southern Bahamas, it is negligible (Pierson and Shinn, 1985). Brooke (2001, p. 138) contends that 'rates of lithification of carbonate sand may have been reduced in areas that experienced arid glacial conditions'. Hussein (2002), working in the arid Gulf Coastal Province in eastern Saudi Arabia, studied Quaternary coastal aeolianites that were weakly cemented and still in their original depositional framework. However, McLaren and Gardner (2004) found that sandy deposits from climatically wet regions did not always contain the most cement; instead it was those sediments that were located in zones of accelerated diagenesis, such as the sea-spray zone, close to the groundwater table or near to a palaeosurface. A range of intrinsic and extrinsic controls that affect diagenetic processes can complicate palaeoclimatic interpretations.

Meso-scale controls on vadose diagenesis include the concentration of ions in water within dune sand bodies and surficial soils as a result of the processes of evapotranspiration. Klappa was one of the first to conduct a detailed study on the important role that biota can have on early carbonate diagenesis. Organo-sedimentary structures are created by mineral precipitation encrusting, impregnating or replacing organic root tissues (Klappa, 1980, 1983). Microflora and microfauna around roots (Figure 5.7) encourage diagenesis and result in the development of calcified filaments (Jones

Figure 5.7 Aeolianite from Cap Blanc, Tunisia. Evidence of the original dune structure has been lost due to the growth and later induration of root structures in the dune sands.

and Kwok Choi, 1988). Needle-fibre meshes of acicular (Figure 5.8), randomly oriented calcite needles and shorter, fatter, ropy grain-coating needle mats are common in the root zone where there is strong evapotranspiration (McLaren, 1995a).

Moisture conditions close to or at the land surface are subject to large variations due to the processes of drainage and evaporation, particularly in the upper 2m (Friedman, 1964). Vadose waters that are highly charged with calcium and carbonate ions move up and down through deposits. Rapid fluctuations in fluid composition as a result of alternating periods of rainfall followed by longer periods of evaporation can lead to enhanced cementation. In coastal locations, sea spray can result in accelerated rates of diagenesis at palaeoerosion and other exposure surfaces, such as coastal cliffs (McLaren, 1995b, 2001).

5.5.3 Cement sources

Sources for cement are far more easily explained for carbonate aeolianites in comparison to siliclastic ones. Dissolution of contained bioclastic grains, ooids and peloids can provide a local source, as Dravis (1996), for example, has shown. Because aragonite is more dense than calcite, Bathurst (1971) has calculated that there is an increase in volume of 8% when calcite is precipitated from dissolved aragonite. Other minor intrinsic sources may come from the breakdown of lithoclasts that contain calcium, such as garnet and calcic-feldspars. Leaching of overlying sands (Gardner, 1983, 1986) or dissolution of bedrock may supply $CaCO_3$ for cement, and weathering of decalcified sands and limestones are known to result in the development of

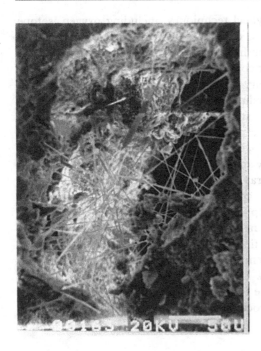

Figure 5.8 Scanning electron microscopy image showing needle fibre cement developed in a root mould, Campo de Tiro, Mallorca. Scale bar is 50 μm in length.

red soils such as on Bermuda (Land et al., 1967; Vacher and Rowe, 1997). Other extrinsic cement sources include groundwater, overland flow and aerosols (both terrigenous dust and sea spray; McLaren, 1995b, 2001).

5.6 Palaeoenvironmental Significance

Studying aeolianites in terms of their mineralogy (indicating the provenance or source area of the sediments), bedding structures, palaeosol horizons and post-depositional modification, particularly at or near palaeosurfaces, should potentially allow aeolianites to be used as aids in palaeoenvironmental reconstruction (see Curran and White, 1995). However, up until now, the overall usefulness of aeolianites in reconstructing former environments has been somewhat limited.

In terms of providing information on past climates, dune development under arid or semi-arid conditions is broadly indicative of rainfall conditions being less than 250 mm yr^{-1} (Thomas, 1997) at the time of their formation. In coastal locations, dunes form under a wide range of climatic regimes but do require strong onshore winds. The diagenetic variability in aeolianites, as a result of the range of different controls that can affect lithification, makes the identification of the role of palaeoclimate on vadose diagenesis difficult but not impossible (see McLaren and Gardner, 2004).

Using a wide range of analytical techniques, including oxygen isotope studies, Rose et al. (1999) have derived a detailed palaeoenvironmental history of landscape development from marine oxygen isotope stages (MIS) 6 through to 2 on the northeast coast of the island of Mallorca. Rose et al. (1999) have dated late Quaternary stacked sequences of soils and terrestrial sediments (including aeolian sediments, some of which are lithified) and argue that warmer climates with stable land surfaces and soil development were associated with high sea levels during MIS 5e, 5c, 5a and 3. Colder climatic episodes during MIS 5d, 5b, 4, 3–2 and 2 were characterised by low sea levels, aeolian and fluvial activity and open vegetation (Rose et al., 1999).

Studies of the diagenetic alteration of palaeodunes can help to determine post-depositional environmental conditions. For example, different cement types and patterns often develop in the phreatic and vadose zones, which may allow the identification of the height of ancient water tables. The ion-enriched chemistry of cements that have formed when exposed to significant sea spray (McLaren, 2001) may indicate higher sea levels in aeolianites inland from current shorelines. Evidence for higher shorelines or storm events may also come from the interdigitation of coastal aeolianites and raised beaches or raised beachrock (e.g. McLaren and Gardner, 2000).

5.6.1 Dating aeolianites

There are a range of dating techniques that have been used to try and establish either relative or numerical ages of dunes and aeolianites. Until the advent of luminescence dating, it was extremely difficult to determine the age of aeolianites, especially those older than the last glacial. Initially, relative techniques applying lithostratigraphic relationships between aeolianites and palaeoshorelines (e.g. Butzer, 1962) or the presence of human artefacts were used (Kallweit and Hellyer, 2003).

Uranium-series dating on shells from coastal deposits is known to be problematic, as shells often behave as geochemically open systems and display considerable variation in isotopic uptake both within different parts of one shell as well as between different shells of the same age (see McLaren and Rowe, 1996). Conventional radiocarbon dating requires organic matter to be present in the dunes (which is often not the case, especially in dryland dunes), can only be used back to about 50,000 years ago and the dates need to be calibrated. However, the technique has shown some level of success for obtaining ages for aeolianites. For example, McLaren and Gardner (2000) have obtained nine radiocarbon dates from shells collected within Holocene cemented dune and beach sands and from a midden on top of the aeolianite on Isla Cancun, Mexico. The radiocarbon chronology was self-consistent through the sedimentary profile. These dates provided evidence to suggest that the dunes were active from less than 5000 years ago and that dune formation ceased about 2500 years ago. After stabilisation, the dunes then became rapidly cemented in the vadose zone (McLaren and Gardner, 2000). However, a number of people, unable to obtain enough material for radiocarbon dating, have dated whole-rock samples (e.g. Ward, 1975). Such dates obtained from aeolianites and beach deposits are fairly meaningless as they are derived from a mix of relatively young cements and older marine carbonate grains that may have been reworked a number of times.

Amino acid racemisation techniques can be used on fossil samples by utilising the systematic changes undergone by various amino acids in fossils after the death of the animal. This technique is only a relative age indicator but has been used successfully to date aeolianites when the results can be calibrated (e.g. Hearty, 2002, 2003). El Asmar (1994) studied a sequence of aeolianites and palaeosols along the northwest coast of Egypt. Using $\delta^{18}O$ and $\delta^{13}C$ and amino acid racemisation, he established that phases of dune accumulation occurred during relatively drier phases in aminozones A, C, E and G and these periods were separated by relatively more humid times when soils accumulated.

Luminescence dating of terrigenous grains such as quartz, feldspar and zircon has revolutionised aeolianite chronostratigraphies, as the technique obtains ages from the actual constituent grains and can be used to date deposits that range in age from a few decades to several hundred thousand years. The usefulness of luminescence dating of aeolianites and dune deposits has been comprehensively reviewed by Prescott and Robinson (1997), Stokes (1997) and Singhvi and Wintle (1999). Perhaps the best-established luminescence-derived chronology for aeolianites and Holocene dune deposits has been along the coasts of Western Australia, Queensland and eastern South Australia (e.g. Bryant et al., 1990, 1992, 1994; Hunter et al., 1993, 1994). Hunter et al. (1993, 1994) have dated aeolianites from south of Adelaide back to 500,000 years.

A large dating programme on aeolianites, using a combination of luminescence, radiocarbon and archaeological techniques, has also been carried out in Israel. Palaeosols containing Mousterian archaeological tools have been found within aeolianites along the Carmel, Givat Olga and Sharon Coastal Plains (Frechen et al., 2001, 2002). The dunes at Habonim have been luminescence dated by Tsatskin and Ronen (1999) to glacial phases around 30.7 ± 7 ka and 160 ± 40 ka. The dates indicate that the palaeosols developed through the last interglacial stage. These deposits reflect a succession of wetter soil-forming episodes separated by drier, windier stages of coastal instability. Frechen et al. (2001, 2002) have identified phases of sand dune accumulation between 65 and 50, 7 and 5 and 5 and 0.2 ka along the Sharon Coastal Plain, and before 50 ka at Givat Olga. At the latter site, there are eight palaeosols, five of which contain Lower Palaeolithic and Mesolithic artefacts. Sapropels sampled from the Mediterranean

Sea suggest that the sand dunes correlate with periods of strongly increased African monsoon activity (Frechen et al., 2002). Godfrey-Smith et al. (2003) also applied luminescence chronology to culture-bearing aeolianites in coastal Israel. Although there was abundant mammalian bone in the deposits, it was too degraded to be dated using radiocarbon methods. The aeolianite was found to be 42.7 ± 6.3 ka. Human occupation occurred between 21.0 and 14.0 ka (i.e. during the Last Glacial Maximum), when conditions are thought to have been cold and dry and sea levels were about 60 to 120 m lower than today, meaning that the coastline would have been several kilometres further westwards.

5.6.2 Relationship to sea level

Over recent years, one of the key debates has been on how dune formation correlates with Quaternary glacial–interglacial cycles. Up until about 10 years ago, the timing of dune development was uncertain due to the lack of numerical dates on aeolianites. Eustatic changes in sea level are important controls on the timing of the deposition of large-scale coastal carbonate dunes (Brooke, 2001). It now appears to be the case that coastal dunes can form at any time during a transgression–regression cycle. In environments where there are long ramping continental shelves, such as in the southern Arabian Gulf (Williams and Walkden, 2001), development of dunes can occur at any stage of the sea-level cycle. According to Carew and Mylroie (1995), dune and island building can occur during early transgression but the features formed would be subject to later destruction by the same continued transgression. Dune units of glacial age probably reflect aeolian reworking following marine regression (Brooke, 2001). Clemmensen et al. (1997) and Nielsen et al. (2004) have all found evidence for dunes forming at times of Mid- and Late Pleistocene glacial low sea levels on the island of Mallorca. However, dune sands cannot accumulate on isolated, steep-rimmed platforms such as Bermuda during low sea-level stands as the supply of sediment is cut off because the shoreline has dropped below the platform edge (Brooke, 2001). Terrestrial weathering, karstification or diagenetic processes may dominate during lowstands in such environments.

Price et al. (2001) thermoluminescence-dated aeolianites on both Lord Howe and Rottnest islands, off the coast of Western Australia (a lack of suitable crystalline minerals meant that attempts to date aeolianites on Norfolk Island were unsuccessful). They found that emplacement occurred at different times in different places, leading Price et al. (2001) to suggest that sediment availability and the type of offshore shelf play important roles in addition to sea level. On Lord Howe Island, the numerical dates obtained by Price et al. (2001) indicated that dunes were deposited only during high sea stands, when the platform around the island was flooded. On Rottnest Island, Price et al. (2001) argue that there was a more continuous supply of sediment, allowing dunes to form at both low and high sea levels. In contrast, Hearty (2003) argues against the findings of Price et al. (2001) in his study of amino acid racemisation rates on Rottnest Island. Hearty found that deposition of dune sands began in MIS 5e, continued through MIS 5c and peaked in MIS 5a, approximately 70–80 ka ago. He purports that most of the dunes built up during Aminozone C, when sea levels were close to present-day, and that these aeolianites represent the landward facies of interglacial sandy shorelines, not dunes, that have migrated across an exposed shelf. Hearty goes on to state (p. 220) that the 'whole rock A/I ratios reflect stratigraphic order, internal and incremental consistency in all cases and show no concordance against the TL ages from many of the same deposits that were obtained by Price et al. (2001)'.

Working along the southern Cape Coast in South Africa, Bateman et al. (2004), using optically stimulated luminescence (OSL) techniques, have dated aeolian and barrier construction over the past two glacial–interglacial cycles. Numerical dates for the aeolianites group together at 68–75, 89–95, 104–122, 178–188 and >200 ka ago. The ages suggest to Bateman et al. (2004) that sand dunes were deposited when sea levels were either rapidly transgressing or regressing. In the mid- to late Holocene on Isla Cancun, Mexico, McLaren and Gardner (2000) have found that dune sands were emplaced when sea level was transgressing and approaching a similar height to its current level. In the Bahamas, Hearty and Neumann (2001) argue that bank margin ooids are moved onshore into island dune and beach ridges as sea level falls. By far the largest volume of aeolianite is formed during and emplaced after a prolonged highstand (Carew and Mylroie, 1995). As sea level falls these sediments are swept backward and are preserved by vadose cementation.

5.6.3 Aeolianites in the stratigraphic record

Despite the range of diagnostic sedimentary and diagenetic features recognised within Quaternary aeolianites (Hunter, 1977; Kocurek and Dott, 1981; Gardner and McLaren, 1994), relatively few examples can be found in Tertiary and older stratigraphic sequences (although an edited book by Abegg et al. (2001) addresses carbonate aeolianites in the rock record). There are three main reasons that may help to explain this lack of recognition: non-deposition, non-preservation or misinterpretation.

The abundance of coastal aeolianites of late Quaternary age may be a reflection of climate and sea level change associated with the last ice age. The lack of evidence of aeolianites outside of ice age conditions may be because, during warmer phases, rising sea levels resulted in significant reworking of the aeolian dune deposits. Barrell (1906) and Rust (1990) have suggested that the probability of preservation of widespread aeolian sands in the geological record is low because they are largely destroyed by subsequent marine transgressions. However, a slow rising sea level may simply push a coastal dune field landwards, whereas a fast rise in sea level may drown the dunes and actually protect them below the erosive action of the wave base. For example, Martin and Flemming (1987) found off-shore aeolianite ridges of Pleistocene age at depths of 40, 50–55 and 65–75 m in southern Africa, and Hearty (2003) reports aeolian deposits submerged off the coast of Western Australia. Eriksson and Simpson (1998) have studied the distribution of Precambrian aeolian sandstones and have found that many are preserved, not in former coastal environments but in continental rift systems and compressional basins.

Preservation problems may arise from two main factors. First of all, the balance between leaching and lime production is an important control on the preservation of Quaternary aeolianites (Goudie, 1991). Early decalci-fication of the sands would only be of any real significance in dune sands with a high carbonate content; quartz-rich dune sands would be relatively unaffected. Rain water absorbs carbon dioxide from the atmosphere and vegetation, which makes it slightly acidic. The water travels through the dune sands and dissolves and removes $CaCO_3$ from the upper layers of the sand. Dissolution of the sands would be most active under conditions of high rainfall, rapid drainage and acid vegetation.

Secondly, post-formation erosion may be significant. After the stabilisa-tion of a dune deposit, a number of processes can occur that may result in either the loss or preservation of the sediment. A change in climate to wetter conditions may introduce more water onto and into the dune. Surface water may transport sand grains down towards the interdunal areas, which, over time, may result in a decline in the slope angle of the dune and loss of dune form. However, Rust (1990, p. 299) has remarked that 'coastal dune systems appear to be remarkably resistant to destruction by erosion, especially after they have made the transition from active dune fields to vegetated or fixed fields, and after the unconsolidated dune sand has been transformed into aeolianite'. Perhaps more of a problem would be erosion by marine processes, particularly, as mentioned earlier, as a result of long-term, large-scale transgressions and regressions.

Thirdly, misinterpretation of the diagenetic fabric and/or sedimentary environment may occur. The Permian Cedar Mesa Sandstone (Colorado Plateau, southwest USA) 'has been interpreted by several recent workers as shallow-marine' even though this deposit contains abundant rhizoliths (Loope, 1984, p. 563). Diagenetically, aeolianites are highly variable in nature (Gardner and McLaren, 1994), and it is possible that some of the indurated aeolianites containing abundant pore filling cements have been misinterpreted in the past.

There are many deposits that may have been initially cemented by low-Mg calcite, but over time have been moved out of the vadose zone and their original cements either lost or masked by other diagenetic processes. For example, as a result of the burial of sediments, diagenesis may occur at depth and totally change the cementation and alteration patterns of a former aeolianite. This will result in the loss of evidence of early vadose diagenesis and will hinder identification in the geological record. Glennie et al. (1978) studied the Permian Rotliegendes sandstones in the southern North Sea and found that the aeolian sands are cemented by halite, gypsum and later pore-destroying clay, feldspar, quartz, dolomite and anhydrite cements. Any of the early calcite cements were rapidly dolomitised, creating some secondary porosity, shortly after burial below water.

5.7 Relationships to other Terrestrial Geochemical Processes and Sediments

Cemented dunes form stable surfaces that are prone to subaerial weathering and carbonate solution that can produce karst topography. Karstification of aeolianite is commonplace and includes solution pipes (e.g. Vacher and Rowe, 1997), caves and dolines (Mylroie and Carew, 1995). Diagenesis of aeolianites at or near to palaeosurfaces often results in the formation of a pedogenic or laminar calcrete horizon (Wright et al., 1988) that generates an indurated resistant capping to the aeolianite (see Chapter 2).

Insoluble residues may develop into an incipient soil. Fossilised Quaternary dunes from Bermuda are capped by *Terra rossa* palaeosols, which are thought to have formed during minor breaks within and between interglacial substages and glacial conditions at times of low sea level (Vacher et al., 1995), and represent periods of extensive soil formation and development of solutional unconformities (Morse and Mackenzie, 1990). In Mallorca, different phases of Quaternary dune activity are separated from one another by unconformities, microkarst and soil development. More extensive soil formation can lead to silica diagenetic alteration of dune sands (see Gardner and Hendry, 1995).

Khadkikar and Basavaiah (2004) have recognised five types of *Terrae rossae*, ranging from karstified limestone to soil development, that have developed as a result of leaching and residual accumulation of limestone as well as the input of aerosols. The palaeosols commonly have a sharp upper contact but a diffuse lower boundary and residual lumps of weathered aeolianite are present.

Soils that developed both in and on aeolianites in Pleistocene interglacial periods in the Mediterranean are usually reddish and, according to Nielsen et al. (2004), have a high magnetic susceptibility and negative $\delta^{18}O$ values, suggesting that the climate was warm and moist. In general, palaeosols associated with aeolianites show signs of oxidation (giving the red colouration), although in interdune areas, gleyed horizons, silty clays and salt-rich crusts are indicative of wetter conditions (e.g. in the Woody Cape aeolianite in South Africa; Smuts and Rust, 1988). Here, interdune drying-up sequences have been identified and consist of waterlain ripples, followed by organic debris, adhesion ripples and laminae and finally, wind ripples or cross-strata at the top of the sequence (Smuts and Rust, 1988).

Many coastal aeolianites are found in close (either adjacent to or inter-digitating with) association with raised beaches and beachrocks, and represent the palaeojunction between land and sea. The texture of all three deposits can be very similar. Distinguishing between aeolianites and beachrock can be accomplished through a study of their distinct cement types and cement mineralogies (see Chapter 11). Differentiating between sandy beach and dune deposits is far more difficult, as they are often both cemented in the vadose zone and thus have similar low-Mg calcite cement types. Steeply dipping (30–34°) sandflow cross-strata, however, are unique to aeolian dunes.

5.8 Directions for Future Research

General improvements in dating techniques should allow the palaeoenvironmental records preserved within aeolianites to be fully utilised and permit a better understanding of the timing of major dune depositional events throughout the Quaternary. Luminescence dating has proven to be one of the best tools for establishing aeolianite chronologies (as is evident from recent research conducted in Australia and South Africa e.g. Bryant et al., 1994; Hunter et al., 1994; Bateman et al., 2004), and continuing refinement and application of the technique will represent an important future research direction. In addition, further pairing of luminescence dating with amino acid racemisation (cf. Rose et al., 1999) will be of considerable benefit in establishing long-term records and substantiating the chronological control for aeolianites and their associated deposits.

In terms of post-depositional processes leading to the generation of aeolianites, a better understanding of the rates of lithification away from zones of accelerated diagenesis is required. This will be aided by isotopic analyses of aeolianite cements (as opposed to whole rock samples) which are needed to provide information on the environmental conditions under which diagenesis occurred.

Many studies on aeolianites have been conducted in coastal locations in a wide range of climatic environments, but, as yet, there have been few detailed studies on the diagenetic processes operating under dryland conditions. Finally, models of diagenesis in the vadose zone need to be refined/improved as they do not properly reflect the complexity of processes operating, and the changes that occur to deposits, during the alteration of dune sands into aeolianites *sensu stricto*.

References

Abegg, F.E., Harris, P.M. & Loope, D.B. (Eds) (2001) *Modern and Ancient Carbonate Eolianites: Sedimentology, Sequence Stratigraphy and Diagenesis*. Special Publication 71. Tulsa, OK: Society of Economic Paleontologists and Mineralogists, 207 pp.

Agnew, C.T. (1988) Soil hydrology in the Wahiba Sands. *Journal of Oman Studies Special Report* 3, 191–200.

Badiozamani, K., Mackenzie, F.T. & Thorstenson, D.C. (1977) Experimental carbonate cementation: salinity, temperature and vadose-phreatic effects. *Journal of Sedimentary Petrology* 47, 529–542.

Barrell, J. (1906) Relative geological importance of continental, littoral and marine sediments 2. *Journal of Geology* 14, 430–457.

Bateman, M.D., Holmes, P.J., Carr, A.S., Horton, B.P. & Jaiswal, M.K. (2004) Aeolianite and barrier dune construction spanning the last two glacial-interglacial cycles from the southern Cape coast, South Africa. *Quaternary Science Reviews* 23, 1681–1698.

Bathurst, R.G.C. (1971) *Carbonate Cements and their Diagenesis. Developments in Sedimentology* 12, Amsterdam: Elsevier.

Berner, R.A. (1978) Rate control of mineral dissolution under earth surface conditions. *American Journal of Science* 278, 1235–1252.

Berner, R.A. (1980) *Early Diagenesis: a Theoretical Approach*. Princeton: Princeton University Press, 241 pp.

Brooke, B. (2001) The distribution of carbonate eolianite. *Earth-Science Reviews* 55, 135–164.

Brookfield, M.E. (1977) The origin of bounding surfaces in ancient eolian sandstones. *Sedimentology* 24, 303–332.

Bryant, E.A., Young, R.W., Price, D.M. & Short, S.A. (1990) Thermoluminescence and uranium-thorium chronologies of Pleistocene coastal landforms of Illawarra region, New South Wales. *Australian Geographer* 21, 101–112.

Bryant, E.A., Young, R.W., Price, D.M. & Short, S.A. (1992) Evidence for Pleistocene and Holocene raised marine deposits, Sandon Point, New South Wales. *Australian Journal of Earth Sciences* 39, 481–493.

Bryant, E.A., Young, R.W., Price, D.M. & Short, S.A. (1994) Late Pleistocene dune chronology: near coastal New South Wales and eastern Australia. *Quaternary Science Reviews* 13, 209–223.

Bruckner, H. (1986) Stratigraphy, evolution and age of Quaternary marine terraces in Morocco and Spain. *Zeitschrift für Geomorphologie Supplement Band* 62, 83–103.

Budd, D.A. (1988) Petrographic products of freshwater diagenesis in Holocene ooid sands, Schooner Cays, The Bahamas. *Carbonates and Evaporites* 3, 143–163.

Buddemeier, R.W. & Oberndorfer, J.A. (1986) Internal hydrology and geochemistry of coral reefs and atoll islands: key to diagenetic variations. In: Schroeder, J.H. Purser, B.H. (Eds) *Reef Diagenesis*. Berlin: Springer-Verlag, pp. 91–111.

Butzer, K.W. (1962) Pleistocene stratigraphy and prehistory in Egypt. *Quaternaria* **VI**, 451–478.

Caputo, M.V. (1993) Eolian structures and textures in oolitic-skeletal calcarenites from the Quaternary of San Salvador Island, Bahamas: a new perspective on eolian limestones. In: Keith, B.D. & Zuppann, C.W. (Eds) *Mississippian Oolites and Modern Analogs*. Studies in Geology 35. Tulsa, OK: American Association of Petroleum Geologists, pp. 243–259.

Carew, J.L. & Mylroie, J.E. (1995) Depositional model and stratigraphy for the Quaternary geology of the Bahamas islands. In: Curran, H.A. & White B. (Eds) *Terrestrial and Shallow Marine Geology of the Bahamas and Bermuda*. Special Publication 300. Boulder, CO: Geological Society of America, pp. 5–32.

Chafetz, H.S., Wilkinson, B.H. & Love, K.M. (1985) Morphology and composition of non-marine carbonate cements in near surface settings. In: Schneidermann, N. & Harris, P. (Eds) *Carbonate Cements*. Special Publication 36. Tulsa, OK: Society of Economic Paleontologists and Mineralogists, pp. 337–348.

Clemmensen, L.B., Fornós, J.J. & Rodríguez-Perea, A. (1997) Morphology and architecture of a late Pleistocene cliff-front dune, Mallorca, Western Mediterranean. *Terra Nova* **9**, 251–254.

Curran, H.A. & White, B. (1995) *Terrestrial and Shallow Marine Geology of the Bahamas and Bermuda*. Special Publication 300. Boulder, CO: Geological Society of America, 335 pp.

Dickson, J.A.D. (1983) Graphical modelling of crystal aggregates and its relevance to cement diagenesis. *Philosophical Transactions of the Royal Society* **A309**, 465–502.

Dravis, J.J. (1996) Rapidity of freshwater calcite cementation – implications for carbonate diagenesis and sequence stratigraphy. *Sedimentary Geology* **107**, 1–10.

El Asmar, H.M. (1994) Aeolianite sedimentation along the north-western coast of Egypt: evidence for middle to late Quaternary aridity. *Quaternary Science Reviews* **13**, 699–708.

El Asmar, H.M. & Wood, P. (2000) Quaternary shoreline development: the north-western coast of Egypt. *Quaternary Science Reviews* **19**, 1137–1149.

Enos, P. & Sawatsky, L.H. (1981) Pore networks in Holocene carbonate sediments. *Journal of Sedimentary Petrology* **51**, 961–985.

Eriksson, K.A. & Simpson, E.L. (1998) Controls on spatial and temporal distribution of Precambrian eolianites. *Sedimentary Geology* **120**, 275–294.

Evans, C.C. & Ginsburg, R.N. (1987) Fabric selective diagenesis in the late Pleistocene Miami Limestone. *Journal of Sedimentary Petrology* **57**, 311–318.

Frechen, M., Dermann, B., Boenigk, W. & Ronen, A. (2001) Luminescence chronology of aeolianites from the section at Givat Olga – coastal plain of Israel. *Quaternary Science Reviews* **20**, 805–809.

Frechen, M., Neber, A., Dermann, B., Tsatskin, A., Boenigk, W. & Ronen, A. (2002) Chronostratigraphy of aeolianites from the Sharon coastal plain of Israel. *Quaternary International* **89**, 31–44.

Friedman, G.M. (1964) Early diagenesis and lithification in carbonate sediments. *Journal of Sedimentary Petrology* **34**, 777–813.

Fryberger, S.G. & Schenk, C.J. (1988) Pin stripe lamination: a distinctive feature of modern and ancient eolian sediments. *Sedimentary Geology* **55**, 1–15.

Gardner, R.A.M. (1983) Aeolianite. In: Goudie, A.S. & Pye, K. (Eds) *Chemical Sediments and Geomorphology*. London: Academic Press, pp. 263–301.

Gardner, R.A.M. (1986) Quaternary coastal sediments and stratigraphy, south-east India. *Man and Environment* **X**, 51–72.

Gardner, R.A.M. (1988) Aeolianites and marine deposits of the Wahiba Sands: character and palaeoenvironments. *The Journal of Oman Studies Special Report* 3, 75–95.

Gardner, R.A.M. & Hendry, D.A. (1995) Early silica diagenetic fabrics in Late Quaternary sediments, south India. *Journal of the Geological Society, London* 152, 183–192.

Gardner, R.A.M. & McLaren, S.J. (1993) Progressive vadose diagenesis in late Quaternary aeolianite deposits? In: Pye, K. (Ed.) *The Dynamics and Environmental Context of Aeolian Sedimentary Systems*. Special Publication 72. Bath: Geological Society Publishing House, pp. 219–234.

Gardner, R.A.M. & McLaren, S.J. (1994) Variability in early vadose carbonate diagenesis. *Earth-Science Reviews* 36, 27–47.

Gardner, R.A.M. & McLaren, S.J. (1999) Hydraulic behaviour of coastal dune sands: a pilot study. *Journal of Coastal Research* 15, 936–949.

Gavish, E. & Friedman, G.M. (1969) Progressive diagenesis in Quaternary to late Tertiary carbonate sediments: sequence and time scale. *Journal of Sedimentary Petrology* 39, 980–1006.

Given, R.K. & Wilkinson, B.H. (1985) Kinetic control of morphology, composition and mineralogy of abiotic sedimentary carbonates. *Journal of Sedimentary Petrology* 55, 109–119.

Glennie, K.W. & Singhvi, A.K. (2002) Event stratigraphy, palaeoenvironment and chronology of SE Arabian deserts. *Quaternary Science Reviews* 21, 853–869.

Glennie, K.W., Mudd, G. & Nagtegaal, P.J.C. (1978) Depositional environment and diagenesis of Permian Rotliegendes sandstones in Leman Bank and Sole Pit areas of the UK southern North Sea. *Journal of the Geological Society, London* 135, 25–34.

Godfrey-Smith, D.I., Vaughan, K.B., Gopher, A. & Barkai, R. (2003) Direct luminescence chronology of the Epipalaeoloithic Kebaran site of Naha Hadera V, Israel. *Geoarchaeology* 18, 461–475.

Goudie, A.S. (1991) *The Encyclopaedic Dictionary of Physical Geography*. Oxford: Blackwell.

Goudie, A.S. & Sperling, C.H.B. (1977) Long-distance transport of foraminiferal tests by wind in the Thar Desert. *Journal of Sedimentary Petrology* 47, 630–633.

Gvirtzman, G., Shachnai, E., Bakler, N. & Iiani, S. (1984) Stratigraphy of the Kukar Group (Quaternary) of the coastal plain of Israel. *G.S.I. Current Research*, 70–82.

Hadley, D.G., Brouwers, E.M. & Brown, T.M. (1998) Quaternary palaeodunes, Arabian Gulf coast: age and palaeoenvironmental evolution. In: Alsharhan, A.S., Glennie, K.K., Whittle, G.L. & Kendall, C.G. (Eds) *Quaternary Deserts and Climatic Change*, Rotterdam: Balkema, pp. 123–141.

Halley, R.B. & Harris, P.M. (1979) Freshwater cementation of a 1,000 year-old oolite. *Journal of Sedimentary Petrology* 49, 969–988.

Harris, W.H. & Matthews, R.K. (1968) Subaerial diagenesis of carbonate sediments: efficiency of the solution-reprecipitation process. *Science* **160**, 77–79.

Hearty, P.J. (2002) Revision of the late Pleistocene stratigraphy of Bermuda. *Sedimentary Geology* **153**, 1–21.

Hearty, P.J. (2003) Stratigraphy and timing of eolianite deposition of Rottnest Island, Western Australia. *Quaternary Research* **60**, 211–222.

Hearty, P.J. & Kaufman, D.S. (2000) Whole-rock aminostratigraphy and Quaternary sea-level history of the Bahamas. *Quaternary Research* **54**, 163–173.

Hearty, P.J. & Kindler, P. (1993) New perspectives on Bahamian Geology: San Salvador Island, Bahamas. *Journal of Coastal Research* **9**, 577–594.

Hearty, P.J. & Neumann, A.C. (2001) Rapid sea level and climate change at the close of the last interglaciation (MIS 5e): evidence from the Bahama Islands. *Quaternary Science Reviews* **20**, 1881–1895.

Hearty, P.J., Vacher, H.L. & Mitterer, R.M. (1992) Aminostratigraphy and ages of Pleistocene limestones, Bermuda. *Geological Society of America Bulletin* **104**, 471–480.

Hunter, R.E. (1977) Basic types of stratification in small aeolian dunes. *Sedimentology* **24**, 361–387.

Huntley, D.J., Hutton, J.T. & Prescott, J.R. (1993) The stranded beach-dune sequence of south-east South Australia: a test of thermoluminescence dating, 0–800 ka. *Quaternary Science Reviews* **12**, 1–20.

Huntley, D.J., Hutton, J.T. & Prescott, J.R. (1994) Further thermoluminescence dates from the dune sequence in the south-east of South Australia. *Quaternary Science Reviews* **13**, 201–207.

Hussein, M. (2002) A Quaternary eolianite sequence in the Arabian Gulf coastal region, north-eastern Saudi Arabia: a modern analogue for oomouldic porosity development in an arid setting. *Abstract Cairo 2002: Ancient Oil-New Energy Technical program.* http://aapg.confex.com/aapg/cairo2002/techprogram/paper_62090.htm

Illenberger, W.K. (1996) The geomorphologic evolution of the Wilderness dune cordons, South Africa. *Quaternary International* **33**, 11–20.

Illenberger, W.K. & Verhagen, B.Th. (1990) Environmental history and dating of coastal dunefields. *South African Journal of Science* **86**, 311–314.

James, N.P. (1974) Diagenesis of scleractinian corals in the subaerial vadose environment. *Journal of Palaeontology* **48**, 785–799.

James, N.P. & Bone, Y. (1989) Petrogenesis of Cenozoic temperate water calcarenites, South Australia: a model for meteoric/shallow burial diagenesis of shallow water calcite sediments. *Journal of Sedimentary Petrology* **59**, 191–203.

James, N.P. & Choquette, P.W. (1984) Diagenesis 9 – Limestones – the meteoric diagenetic environment. *Geoscience Canada* **11**, 161–194.

Jones, B. & Kwok Choi, Ng. (1988) The structure and diagenesis of rhizoliths from Cayman Brac, British West Indies. *Journal of Sedimentary Petrology* **58**, 457–467.

Kallweit, H. & Hellyer, P. (2003) A flint 'dagger' from Rumaitha, Emirate of Abu Dhabi, UAE. *Arabian Archaeology and Epigraphy* **14**, 1–7.

Khadkikar, A.S. & Basavaiah, N. (2004) Morphology, mineralogy and magnetic susceptibility of epikarst-terra rossa developed in late Quaternary aeolianite deposits of south-eastern Saurashtra, India. *Geomorphology* **58**, 339–355.

Kindler, P. & Mazzolini, D. (2001) Sedimentology and petrography of dredged carbonate sands from Stocking Island (Bahamas); implications for meteoric diagenesis and aeolianite formation. *Palaeogeography, Palaeoclimatology, Palaeoecology* **175**, 369–379.

Klappa, C.F. (1980) Rhizoliths in terrestrial carbonates: classification, recognition, genesis and significance. *Sedimentology* **27**, 613–629.

Klappa, C.F. (1983) A process response model for the formation of pedogenic calcretes. In: Wilson, R.C.L. (Ed.) *Residual Deposits: surface related weathering processes and materials*. London: Geological Society Special Publication 11, pp. 211–221.

Kocurek, G. (1988) First-order and super bounding surfaces in eolian sequences – bounding surfaces revisited. *Sedimentary Geology* **56**, 193–206.

Kocurek, G. (1991) Interpretation of ancient eolian sand dunes. *Annual Review of Earth and Planetary Sciences* **19**, 43–75.

Kocurek, G. (1999) The aeolian rock record (yes Virginia, it exists, but it really is rather special to create one). In: Goudie, A.S., Livingstone, I. & Stokes, S. (Eds) *Aeolian Environments, Sediments and Landforms*. Chichester: John Wiley & Sons, pp. 239–259.

Kocurek, G. & Dott, Jr. R.H. (1981) Distinction and uses of stratification type in the interpretation of eolian sand. *Journal of Sedimentary Petrology* **51**, 579–595.

Kocurek, G. & Havholm, K. (1993) Eolian sequence stratigraphy – a conceptual framework. In: Weimer, P., Posamentier, H. (Eds) *Siliciclastic Sequence Stratigraphy*. American Association of Petroleum Geologists Memoir **58**, 393–409.

Knox, G.J. (1977) Caliche profile formation, Saldanha Bay, South Africa. *Sedimentology* **25**, 657–674.

Lancaster, N. (1995) *Geomorphology of Desert Dunes*. London: Routledge.

Land, L.S., MacKenzie, F.T. & Gould, S.J. (1967) Pleistocene history of Bermuda. *Bulletin of the Geological Society of America* **78**, 993–1006.

Livingstone, I. & Warren, A. (1996) *Aeolian Geomorphology: an introduction*. London: Longman.

Loope, D.B. (1984) Eolian origin of upper Palaeozoic sandstones, south-east Utah. *Journal of Sedimentary Petrology* **54**, 563–580.

Magaritz, M., Gavish, E., Bakler, N. & Kafri, U. (1979) Carbon and oxygen isotope composition indicators of cementation environments in Recent, Holocene and Pleistocene sediments along the coast of Israel. *Journal of Sedimentary Petrology* **49**, 401–412.

Malan, J.A. (1987) The Bredasdorp Group in the area between Gans Bay and Mossel Bay. *South African Journal of Science* **83**, 506–507.

Martin, A.K. & Flemming, B.W. (1987) Aeolianites of the South African coastal zone and continental shelf as sea-level indicators. *South African Journal of Science* **83**, 507–508.

Martin, G.D., Wilkinson, B.H. & Lohman, K.C. (1986) The role of skeletal porosity in aragonite neomorphism: *Strombus* and *Montastrea* from the Pleistocene Key Largo Limestone, Florida. *Journal of Sedimentary Petrology* **56**, 194–203.

Mattingly, D.J., al-Mashai, M., Aburgheba, H., Balcombe, P., Eastaugh, E., Gillings, M., Leone, A., McLaren, S., Owen, P., Pelling, R., Reynolds, T., Stirling, L., Thomas, D., Watson, D., Wilson, A. & White, K. (1998) The Fezzan Project

1998: preliminary report on the second season of work. *Libyan Studies* **29**, 115–145.

Matthews, R.K. (1974) A process approach to diagenesis of reefs and reef associated limestones. In: Laporte, L.F. (Ed.) *Reefs in Time and Space*. Society of Economic Paleontologists and Mineralogists Special Publication 18, pp. 234–256.

McKee, E.D. (1966) Structures of dunes at White Sands National Monument, New Mexico. *Sedimentology* **7**, 1–69.

McLaren, S.J. (1993) Use of cement types in the palaeoenvironmental interpretation of coastal aeolian sedimentary sequences. In: Pye, K. (Ed.) *The Dynamics and Environmental Context of Aeolian Sedimentary Systems*. Special Publication 72. Bath: Geological Society Publishing House, pp. 235–244.

McLaren, S.J. (1995a) Early diagenetic fabrics in the rhizosphere of Late Pleistocene aeolian sediments. *Journal of the Geological Society, London* **152**, 173–181.

McLaren, S.J. (1995b) The role of sea spray in vadose diagenesis in Late Pleistocene coastal deposits. *Journal of Coastal Research* **11**, 1075–1088.

McLaren, S.J. (2001) Effects of sea spray on vadose diagenesis of late Quaternary aeolianites, Bermuda. *Journal of Coastal Research* **17**, 228–240.

McLaren, S.J. (2003) Aeolianite. In: Goudie, A.S. & Panizza, M. *Encyclopedia of Geomorphology*. London: Routledge.

McLaren, S.J. & Gardner, R.A.M. (2000) New Radiocarbon Dates from a Holocene Aeolianite, Isla Cancun, Quintana Roo, Mexico. *The Holocene* **10**, 757–761.

McLaren, S.J. & Gardner, R.A.M. (2004) Late Quaternary carbonate diagenesis in coastal and desert sands – a sound palaeoclimatic indicator? *Earth Surface Processes and Landforms* **29**, 1441–1459.

McLaren, S.J. & Rowe, P.J. (1996) The reliability of uranium-series mollusc dates from the western Mediterranean basin. *Quaternary Science Reviews (Quaternary Geochronology)* **15**, 709–717.

Moore, Jr, C.H. (1989) *Carbonate Diagenesis and Porosity*. Developments in Sedimentology 46. Amsterdam: Elsevier.

Morrow, N.R. (1971) Small scale packing heterogeneities in porous sedimentary rocks. *American Association of Petroleum Geologists Bulletin* **55**, 514–522.

Morse, J.W. & Mackenzie, F.T. (1990) *Geochemistry of Sedimentary Carbonates*. Developments in Sedimentology 48. Amsterdam: Elsevier.

Muller, G. (1971) Petrology of the Cliff Limestone (Holocene) North Bimini, The Bahamas. *Neues Jarhbuch für Mineralogie Monatschefte* **11**, 507–523.

Muller, G. & Tietz, G. (1975) Regressive diagenesis in Pleistocene aeolianites from Fuerteventura, Canary Islands. *Sedimentology* **22**, 485–496.

Mylroie, J.E. & Carew, J.L. (1995) Karst development on carbonate islands. In: Budd, D.A. Saller, A.H. & Harris, P.M. (Eds) *Unconformities and Porosity in Carbonate Strata*. Memoir 63. Tulsa, OK: American Association of Petroleum Geologists, pp. 55–76.

Nielsen, K.A., Clemmensen, L.B. & Fornós, J.J. (2004) Middle Pleistocene magnetostratigraphy and susceptibility stratigraphy: data from a carbonate aeolian system, Mallorca, western Mediterranean. *Quaternary Science Reviews* **23**, 1733–1756.

Phillips, S.E., Milnes, A.R. & Foster, R.C. (1987) Calcified filaments: an example of biological influences in the formation of calcrete in South Australia. *Australian Journal of Soil Research* **25**, 405–428.

Pierson, B.J. & Shinn, E.A. (1985) Cement distribution and carbonate mineral stabilization in Pleistocene limestones of Hogsty reef, Bahamas. In: Schneidermann, N. & Harris, P.M. (Eds) *Carbonate Cements*. Special Publication 36. Tulsa, OK: Society of Economic Paleontologists and Mineralogists, pp. 153–168.

Pittman, E.D. (1974) Porosity and permeability changes during diagenesis of Pleistocene corals, Barbados, West Indies. *Geological Society of America Bulletin* **85**, 1811–1820.

Playford, P.E. (1988) *Guidebook to the Geology of Rottnest Island*. Guidebook 2. Perth: Geological Survey of Australia.

Playford, P.E., Cockbain, A.E. & Lowe, G.H. (1976) Geology of the Perth Basin, Western *Australia Geological Society Bulletin* **124**, 298.

Prescott, J.R. & Robinson, G.B. (1997) Sediment dating by luminescence: a review. *Radiation Measurement* **27**, 893–922.

Preusser, F., Radies, D. & Matter, A. (2002) A 160,000-year record of dune development and atmospheric circulation in southern Arabia. *Science* **296**, 2018–2020.

Price, D.M., Brooke, B.P. & Woodroffe, C.D. (2001) Thermoluminescence dating of eolianites from Lord Howe Island and south-west Western Australia. *Quaternary Science Reviews* **20**, 841–846.

Roberts, H.H., Ritchie, W. & Mather, A. (1973) Cementation in high latitude dunes. *Coastal Studies Bulletin* **7**, 93–112.

Rose, J., Meng, X. & Watson, C. (1999) Palaeoclimate and palaeoenvironmental responses in the western Mediterranean over the last 140 ka: evidence from Mallorca, Spain. *Journal of the Geological Society, London* **156**, 435–448.

Rust, I.C. (1990) Coastal dunes as indicators of environmental change. *South African Journal of Science* **86**, 299–301.

Saqqa, W. & Atallah, M. (2004) Characterization of the aeolian terrain facies in Wadi Araba Desert, southwestern Jordan. *Geomorphology* **62**, 63–87.

Sayles, R.W. (1931) Bermuda during the ice age. *Proceedings of the American Academy of Arts and Science* **66**, 381–467.

Schmalz, R.F. (1967) Kinetics and diagenesis of carbonate sediments. *Journal of Sedimentary Petrology* **37**, 60–67.

Schroeder, J.H. (1988) Spatial variations in the porosity development of carbonate sediments and rocks. *Facies* **18**, 181–204.

Scoffin, T.P. (1987) *An Introduction to Carbonate Sediments and Rocks*. London: Blackie.

Semeniuk, V. & Meagher, T.D. (1981) Calcrete in Quaternary coastal dunes in south-western Australia: a capillary rise phenomenon associated with plants. *Journal of Sedimentary Petrology* **51**, 47–68.

Singhvi, A.K. & Wintle, A.G. (1999) Luminescence dating of aeolian and coastal sand and silt deposits: applications and implications. In: Goudie, A.S., Livingstone, I. & Stokes, S. (Eds) *Aeolian Environments, Sediments and Landforms*. Chichester: John Wiley & Sons, pp. 293–317.

Sippel, R.F. & Glover, E.D. (1964) The solution alteration of carbonate rocks, the effects of temperature and pressure. *Geochimica et Cosmochimica Acta* **28**, 1401–1417.

Smuts, W.J. & Rust, I.C. (1988) The significance of soil profiles and bounding surfaces in Woody Cape aeolianite. *Palaeoecology of Africa* **19**, 269–276.

Stokes, S. (1997) Dating of desert sequences. In: Thomas, D.S.G. (Ed.) *Arid Zone Geomorphology: Process, Form and Change in Drylands*. Chichester: John Wiley & Sons, pp. 607–637.

Stokes, W.L. (1968) Multiple parallel-truncation bedding planes – a feature of wind-deposited sandstone formations. *Journal of Sedimentary Petrology* **38**, 510–515.

Strasser, A. & Davaud, E. (1986) Formation of Holocene limestone sequences by progradation, cementation and erosion: two examples from the Bahamas. *Journal of Sedimentary Petrology* **56**, 422–429.

Thomas, D.S.G. (1997) Reconstructing ancient arid environments. In: Thomas, D.S.G. (Ed.) *Arid Zone Geomorphology: Process, Form and Change in Drylands*. Chichester: John Wiley & Sons, pp. 577–607.

Thomas, D.S.G., Bateman, M.D., Merhshahi, D. & O'Hara, S. (1997) Development and environmental significance of an eolian sand ramp of last glacial age, Central Iran. *Quaternary Research* **48**, 155–161.

Tsatskin, A. & Ronen, A. (1999) Micromorphology of a Mousterian palaeosol in aeolianites at the site of Habonim, Israel. *Catena* **34**, 365–384.

Tucker, M.E. & Vacher, H.L. (1980) Effectiveness of discriminating beach, dune and river sands by moments and cumulative weight percentages. *Journal of Sedimentary Petrology* **50**, 165–172.

Vacher, H.L. & Rowe, M.P. (1997) Geology and hydrogeology of Bermuda. In: Vacher, H.L. & Quinn, T. (Eds) *Geology and Hydrogeology of Carbonate Islands*. Developments in Sedimentology 54. Amsterdam: Elsevier, pp. 35–90.

Vacher, H.L., Hearty, P.J. & Rowe, M.P. (1995) Stratigraphy of Bermuda: nomenclature, concepts, and status of multiple systems of classification. In: Curran, H.A. & White, B. (Eds) *Terrestrial and Shallow Marine Geology of the Bahamas and Bermuda*. Special Publication 300. Boulder, CO: Geological Society of America, pp. 271–294.

Ward, W.C. (1973) Influence of climate on the early diagenesis of carbonate eolianites. *Geology* **1**, 171–174.

Ward, W.C. (1975) Petrology and diagenesis of carbonate eolianites of north eastern Yucatan Peninsula. *American Association of Petroleum Geologists Studies in Geology* **2**, 500–571.

Williams, A.H. & Walkden, G.M. (2001) Carbonate eolianites from a eustatically influenced ramp-like setting: the Quaternary of the southern Arabian Gulf. In: Abegg, F.E., Harris, P.M. & Loope, D.B. (Eds) *Modern and Ancient Carbonate Eolianites: Sedimentology, Sequence Stratigraphy, and Diagenesis*. Special Publication 71. Tulsa, OK: Society of Economic Paleontologists and Mineralogists, pp. 77–92.

Wright, V.P., Platt, N.H. & Wimbledon, W.A. (1988) Biogenic laminar calcretes: evidence of calcified root-mat horizons in palaeosols. *Sedimentology* **35**, 603–620.

Yaalon, D.H. (1967) Factors affecting the lithification of eolianite and interpretation of its environmental significance in the coastal plain of Israel. *Journal of Sedimentary Petrology* **37**, 1189–1199.

Chapter Six

Tufa and Travertine

Heather A. Viles and Allan Pentecost

6.1 Introduction: Nature and General Characteristics

Tufa and travertine are calcareous sediments precipitated in freshwater environments and found in a wide range of locations around the world. They are of great interest to geomorphologists, ecologists and palaeoenvironmental scientists because of their contributions to landscape and habitat, and their role as important environmental archives (Pentecost, 2005). Usually, such deposits are found where water rich in CO_2 and dissolved calcium carbonate is subjected to degassing of CO_2, upsetting the geochemical balance and leading to the precipitation of calcium carbonate. Tufas and travertines have been recorded from all continents except Antarctica, and are commonly found within areas dominated by limestone geology. The terms 'tufa' and 'travertine' have often been used interchangeably, but may also be given more specific meanings. Thus, travertine (which originates from *lapis tiburtinus*, a Roman building stone) has often been used to denote hard, compact material with low porosity. In this restricted sense, travertine is often used as a building material. On the other hand, the term 'tufa' has often been used to refer to a softer, more friable type of deposit that would not be suited to building purposes. To avoid confusion with tuff (a volcanic ash) it is often called 'calcareous tufa'. In reality, many deposits contain both hard and soft material so such terminology is not very helpful. Other authors use tufa to refer to ambient temperature freshwater carbonates and travertine to refer to carbonates deposited from thermal waters. In this chapter we adopt Pentecost's (1993) scheme and use the phrase 'meteogene tufa and travertine' to refer to deposits formed in waters in which the CO_2 comes primarily from soil and groundwater, and 'thermogene travertine' to refer to deposits formed from waters in which the CO_2 load comes predominantly from the interaction between hot rock and CO_2-rich fluids at depth. Where we use the general phrase 'tufa and travertine' we are referring to both meteogene and thermogene types.

Tufa and travertine sediments are formed both at the Earth's surface and also underground. In caves they occur alongside other examples of carbonate speleothem. The surface forms, which may be distinguished as epigean tufas and travertines, are usually colonised by plants, resulting in a wide range of complex morphologies. The biological component often contributes to their formation, with plants and biofilms providing nuclei for deposition and sometimes aiding the necessary degassing of CO_2 or its removal via photosynthesis. In places, animals can also contribute to tufa deposition, for example through larval structures as described by Durrenfeldt (1978) and Drysdale (1999). The role of plants is highlighted by the use of the term fluvial phytoherm (i.e. a reef-like accumulation influenced by plant activity) to describe tufas and travertines in which the architecture of the deposit at many scales is controlled by both living and dead organisms (Pedley, 1990, 1992, 2000).

Tufa and travertine may be classified using geochemical, fabric and morphological criteria (Pentecost and Viles, 1994). Several classification schemes have been proposed, but none is ideal given the wide range of variability in form and nature both within and between deposits. Pentecost (1993) proposed that tufas and travertines might be classified in the first instance according to the origin of the CO_2 within the water, as this gives a clear isotopic signature. He distinguishes between thermogene deposits, formed from waters in which the CO_2 load comes primarily from the interaction between hot rock and CO_2 rich fluids at depth, and meteogene deposits, formed from water enriched with CO_2 coming from soil and groundwater. Stable carbon isotope compositions of the latter range between about 0 and −11‰. The waters forming thermogene deposits, which are dominantly referred to in the literature as travertines, contain high concentrations of CO_2. Rates of degassing of CO_2 and thus consequent deposition are often very rapid, forming distinctive fabrics with a stable carbon isotope composition of between −4 and +8‰.

The nature and arrangement of sediment components, or fabric, provides another way of classifying tufa and travertine deposits and may be viewed at a range of scales. At the micromorphological scale, as viewed in petrological thin-section, tufas and travertines possess a wide range of fabrics, with differences in porosity, cementation and basic components such as allochems (Figure 6.1). A broad distinction can be made between autochthonous (i.e. formed *in situ*) and allochthonous (i.e. washed in) components, with some tufas (such as pool deposits downstream of barrages) composed largely of autochthonous material washed in from upstream. There is a wide range of fabrics. For example, some meteogene tufas are dominated by fine-grained micrite deposited around the filaments of cyanobacteria, with large pores produced as the cyanobacteria die, which

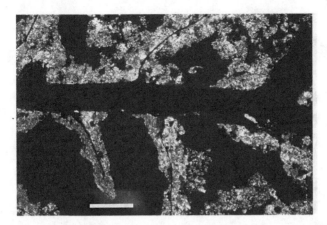

Figure 6.1 Thin section of a sample of tufa from a Holocene paludal deposit at Wateringbury, Kent (c. 6000 years old) viewed under crossed nicols. The mould of a moss stem with attached leaves is encrusted by micrite and microspar. The high porosity of this sample results from the original open framework of the structure (shelter porosity) and the decay of bryophytes. There has been little diagenesis. Scale bar 250 μm.

may become infilled with secondary cementation during diagenesis. Other tufas or travertines may be dominated by larger, sparitic crystals, with little pore space, perhaps as a result of diagenesis.

At the scale of hand specimens, well-circumscribed fabrics can be observed, often owing their distinctive nature to the influence of organisms. For example, some moss, algal and cyanobacterial species are associated with particular meteogene tufa and travertine morphologies. Some cyanobacterial colonies, such as *Phormidium incrustatum*, grow in association with laminated tufas with millimetre-scale layers of alternating light and dark coloured material (Figure 6.2). Another common fabric type is oncoids which consist of pebble-shaped accumulations of calcium carbonate around a central nucleus. More unusual deposit types have been recorded from some tufas and travertines, such as calcite rafts or 'floe' (Carthew et al., 2003a).

Often, it is the gross morphology of tufa and travertine deposits that provides the most obvious basis for classification. The position that the deposit occupies within the hydrological and geomorphological system, sets the basic context for the shape and size of the deposit. For example, barrages are a common meteogene tufa morphology within river channels, as fallen logs and other debris provide a nucleation point for tufa deposition, which itself then modifies the water flow regime. Other locations for tufa and travertine deposition include springs debouching on valley-side slopes, lake margins and marshes. Specific locations of

Figure 6.2 Vertical section through a stream crust colonised by the cyanobacterium *Phormidium incrustatum*. The slightly undulose nature of the upper surface is caused by the coalescence of adjacent colonies. The alternating light and dark layers are probably seasonal and suggest a deposition rate of about 1 mm yr^{-1}. Harrietsham, Kent. Scale bar 5 mm.

formation are found for thermogene travertines, such as fissure ridges. Chafetz and Folk (1984) identify five key types of thermogene travertine, largely based on their studies of travertines at Tivoli, Italy, i.e. terraced-mounds, fissure-ridge, range-front, eroded-sheet and self-built channel forms. Terraced-mounds have also been called rimstone pools by Ekmekci et al. (1995), and are found in great profusion at Pamukkale, western Turkey. Figure 6.3 illustrates some of the many morphological types of tufas and travertines.

Ford and Pedley (1996) recognised four distinctive end-members of a continuum of meteogene tufa and travertine types, i.e. fluvial barrages, lacustrine, paludal and perched springline deposits. Of these, fluvial barrages are perhaps the most easily identified, as they form dam-like structures within rivers over which water flow is accelerated, with still-water pools above and below. Once buried under later sediments, such barrage structures can be located and their architecture identified using ground-penetrating radar, as illustrated by Pedley et al. (2000) and Pedley and Hill (2002). Fluvial barrages have been observed in many parts of the world, with huge impressive examples found in Plitvice National Park, Croatia and Huanglong Scenic District, Sichuan, China, as well as small examples in many streams worldwide, including Cwm Nash, Glamorgan (Viles and Pentecost, 1999). Cascade forms are similar to barrages in their waterfall-like nature, but without the pools above. Examples include the massive Hangguoshu Falls, Guizhou, China and Janet's Foss, Yorkshire, England

Figure 6.3 Some tufa morphologies. (A) Large perched springline deposit on the Napier Range, north-western Australia. Note that there are two phases of deposition apparent, an older one on the right-hand portion and a partially active one on the left-hand side. (B) Suite of small barrages within Cwm Nash, Glamorgan, Wales. Each barrage is around 50 cm high. (C) Back-barrage pool deposits dating from the Holocene period preserved as a coastal cliff face at the seaward end of Cwm Nash, Wales. (D) Small aussen stalactite form from the Napier Range, northwestern Australia. (E) Thin veneers of salt-rich tufas forming from springs emerging on the canyon walls of the Swakop Valley, Namibia.

Figure 6.4 Silver Falls, Tianxing Bridge Park, Guizhou Province, China. The Dabong River descends via a series of travertine cascades and barrages through a thick unit of limestone on the Guizhou Plateau. A locally famous tourist attraction, this cascade is colonised mainly by algae, the strong water flow inhibiting the growth of bryophytes and higher plants.

(Ford and Pedley, 1996). Lacustrine tufas comprise both stromatolites which form in the shallow marginal zone, and deeper water deposits of detrital tufas. Examples of marginal lake tufas include the large tufa mounds found around Pyramid Lake and Big Soda Lake, Nevada (Arp et al., 1999; Rosen et al., 2004). Paludal tufas and travertines form within marshy or spring-fed environments along valley bottoms, where low water flow rates produce widespread, low relief deposits. Calcite within paludal deposits often accumulates around tussocks of marsh vegetation. Glover and Robertson (2003) identify much of the vast Antalya travertine deposits in Turkey as being formed under paludal conditions.

Perched springline deposits cover a range of deposit types which form around springs on valley-side slopes. Such deposits are often characterised by being fan-shaped in plan, with flat tops and steeply sloping fronts. Good examples of perched springline deposits are described from the Rio Tajuña valley in central Spain by Pedley et al. (2003). Meteogene tufas and travertines forming around cascades are among the most common and obvious forms, sometimes providing famous tourist attractions (Figure 6.4). Unusual locations of meteogene tufas and travertines include coastal cliffs (e.g. associated with springs breaking out of Liassic limestone cliffs on the Glamorgan Coast, Wales near Llantwit Major) and around submarine springs, as well as the Yerköprü (travertine bridges) of Turkey (Bayari, 2002), formed by growth of perched springline tufas across a valley. On rare occasions, deposits build from both sides of a river gorge from separate karst springs, meeting above the middle of the river. Rivers are also occasionally diverted by strong growths of tufa, as recorded at Tivoli and

Figure 6.5 Pearl Shoal, Juizhaigou, Sichuan, China. Water from the Beishui River forms an extensive riffle over a growing travertine surface covered in a biofilm of algae and bryophytes before plunging over a large prograding cascade (Pearl Falls). Upstream is a large travertine-dammed lake.

Matlock in Derbyshire (Pentecost, 1999). Another remarkable form, documented only from the mountainous regions of China are the 'river shoals' (Figure 6.5). These consist of large shallow sheets of turbulent water between travertine-dammed lakes, beneath which the tufa builds evenly around a bryophyte and algal biofilm.

6.2 Distribution, Field Occurrence and Geomorphological Relations

Tufas and travertines occur in many different climatic regimes, but may be especially favoured where there is a large supply of water and a high mean annual temperature. A range of papers has provided information about the regional and global distribution of notable tufa and travertine deposits (e.g. Ford, 1989; Pentecost, 1995; Ford and Pedley, 1996; Pentecost and Zhang, 2001). Vast deposits of meteogene tufas occur on the Antalya coastal plain in southwest Turkey, covering an area of $600\,km^2$ with deposit thicknesses of over 250 m in some places (Burger, 1990; Glover and Robertson, 2003), and impressive thermogene terraces are found at Pamukkale, in western Turkey (Dilsiz, 2002). The Dinaric karst area of Croatia, Slovenia, Bosnia and Herzegovina also contains a large range of tufa deposits (Horvatincic et al., 2003), including the classic barrage and pool sequences making up the Plitvice Lakes in the Korana Canyon (Emeis et al., 1987). Famous travertine pools are also found in rivers along the north coast of Jamaica (Porter, 1991), and at Huanglong

Figure 6.6 Large tufa deposit (c. 80 m high) on the edge of the Naukluft Mountains, Namibia.

in China, where the travertine barrages and pools are a UNESCO World Heritage Site (Lu et al., 2000). A total of 88 Quaternary tufa and travertine sites have been identified in China by Pentecost and Zhang (2001), many of which are still actively forming. Most of these occur in the provinces of Guangxi, Guizhou, Sichuan, Yunnan and Xizhang.

Tufa and travertine deposits are found in a wide range of present-day climatic conditions, with examples from both cold and hot extremes, but those occurring at high altitudes/latitudes are mostly thermogene deposits resulting from deep groundwater circulation. Within the high latitudes, Pollard et al. (1999) report on two groups of mineralised springs producing travertine deposits in the Expedition Fiord area of Axel Heiberg Island in the Canadian Arctic. A further, unusual, example comes from Ikka Fjord, Greenland where submarine tufa towers form over submarine springs at the base of the fjord (Buchardt et al., 2001). At the hotter extreme, both warm desert environments and the humid tropics have excellent examples of tufas and travertines. In the Western Desert of Egypt, for example, Crombie et al. (1997) and Nicoll et al. (1999) recognise a range of Pleistocene tufa deposits assumed to date from wetter periods, as are those found in many other parts of North Africa and the Naukluft Mountains of Namibia (Figure 6.6; Brook et al., 1999, Viles et al., 2007). Several recent papers have also indicated the rich variety of deposits to be found in the seasonally-dry tropics within northern Australia in the Barkly karst area (Carthew et al., 2003a,b) and Kimberley area (Viles and Goudie, 1990a), as well as semi-arid parts of northeastern Brazil (Auler and Smart, 2001).

Most tufas and travertines observed today date from the Pleistocene or Holocene epochs, although some deposits are of much greater antiquity.

One well-known example is the Lower Permian Karniowice travertine of Poland, where a rich fossil flora and fauna has been uncovered (Szulc and Cwizewicz, 1989). There is evidence that tufas date back to at least the Tertiary within the Barkly karst area of Queensland, Australia (Carthew et al., 2003b), and they have also been recorded in the Eocene Chadron Formation in South Dakota (Evans, 1999), the middle Miocene Barstow Formation in the Mojave Desert of California, USA (Becker et al., 2001), and in the Miocene Ries crater basin in Germany (Pache et al., 2001).

6.3 Macro- and Micromorphological Characteristics

In the field, active meteogene tufas and travertines possess distinctive morphologies which become translated into a range of facies within fossil deposits. It can often be difficult to recognise both morphologies and facies in the field, as erosional episodes and diagenesis often complicate the picture. Each of the four major meteogene tufa and travertine types recognised by Ford and Pedley (1996) contain a wide range of morphological styles (Table 6.1). Looking first at barrages, the individual dams range greatly in both plan and profile form. In plan, many are curvilinear with a downstream convexity. In profile, barrages may have overhangs on the downstream side, or have vertical faces, or either downstream- or upstream-dipping ramps (as noted by Carthew et al. (2003b) for the barrages on the Gregory River in Queensland, Australia). Many barrages, especially small ones, are exceptionally complicated in plan and profile form, often because of the role of trees and other debris in forming the basic nuclei of the deposit. Barrages often form in suites occupying part of a river's profile, and there have been some studies of the spacing and nature of such barrage systems. For example, Viles and Pentecost (1999) measured the size, shape and spacing of small barrages along Cwm Nash in Glamorgan, Wales. Along one sequence of 28 barrages, they found that the distances between successive barrages averaged 2.67 m and a chi-square test suggested a random distribution. Taken as a whole, it appeared that low stream gradients were associated with a shorter spacing between barrages in this stream, but this appears to be unusual, as reduced gradients normally result in longer spacings.

Lacustrine tufas and travertines also possess a range of distinctive morphologies. Lake stromatolites are often flat-topped and circular in plan view, with some lake margins possessing distinctive tufa towers such as Mono Lake, California (Rieger, 1992) or pinnacles (e.g. Pyramid Lake, Nevada; Arp et al., 1999). The towers form as the CO_2-rich underwater thermal springs react with the high pH lake waters. Many towers are now inactive and exposed owing to the lowering of lake levels in recent times.

Table 6.1 Tufa morphologies and facies characteristics for the four major tufa types

	Fluvial barrage	Lacustrine	Paludal	Perched springline
Morphology	Arcuate dams, often in suites interspersed with pools	Irregular lake margin phytoherms, lime muds in deeper waters	Elongate, following valley-bottom profile, thinning downstream	Fan-shaped in plan, wedge shaped in profile on steep valley sides
Fabric	Undulose to vertical barrage fabrics	Mammilate lobes and vertical fabrics	Sheet-like internal bedforms	Stacked mammillate lobes
Vegetation	Trees, bryophytes, algae (including diatoms) and cyanobacteria	Marginal rushes, grasses, algae and cyanobacteria, charophytes	Grasses, rushes, cyanobacteria and mosses	Mosses, cyanobacteria and algae, some grasses and trees
Animals	Caddis fly larvae locally important			
Vertical curtain fabrics	Common	Absent	Absent	Common
Primary cavities	Common in barrages	Absent	Rare	Abundant
Tufa breccias	Rare	Rare	Rare	Common
Diagenetic features	Compaction in pool sediments. Speleothem in dam cavities	Compaction in lime muds	Compaction of lime muds	Speleothems in cavities

Table 6.1 summarises the basic characteristics of the four major meteogene tufa and travertine types. As Pedley (2000) points out, most meteogene tufas and travertines have fabrics influenced heavily by microbial biofilms. Calcite precipitates in and around such biofilms, to produce a complex, but frequently laminated tufa framework. In many cases, the deposition of calcite onto plant surfaces produces irregular fabrics, with no distinctive bedding characteristics. Most initial deposits of this type contain a high intergranular porosity (see Figure 6.7A). Subsequent decay of the framework plant material produces a secondary mouldic porosity as shown in Figure 6.7B. Movement of water rich in calcium carbonate through this primary deposit results in the development of secondary cements, and compaction under the weight of overlying deposits will also reduce the porosity. The petrofabric of a deposit may therefore change substantially after burial by more recent deposits.

Pedley (2000) proposes a simple model in which abiotic precipitation of calcite in meteogene tufas and travertines produces coarse, bladed spar crystals as a fringe on suitable nucleation surfaces (see Figure 6.7D). Such sparite can develop very quickly in fast-flowing waters. In contrast, under microbiological mediation, micritic calcite is deposited within biofilm layers in both slow- and fast-flowing waters. Chafetz et al. (1994) propose a rather more complicated model from observations of the development of meteogene tufas and travertines at Plitvice Lakes, Croatia. In the Plitvice Lakes, the majority of the carbonate in the barrages and waterfalls was deposited as sparite, with micrite being formed by subsequent neomorphic sparmicritisation engendered by endolithic microbes which are often abundant in the surface layers. Freytet and Verrecchia (1999) identify a rather different situation, where palisadic or prismatic sparite crusts form as a result of early recrystallisation of biologically precipitated micrite, leading to secondary sparites. In many fast-flowing parts of travertine and tufa-depositing streams, such as on the vertical, downstream wall of barrages and waterfalls, moss curtains develop. Several workers have found such mosses to be calcified by large (0.5–2 mm) sparite crystals formed under abiotic processes (Arp et al., 2001). It is unlikely that any one model of petrofabric development will fit all circumstances, because of the wide range of meteogene tufa- and travertine-depositing environments.

Within thermogene travertines, some unusual morphologies and fabrics may be present. Chafetz and Folk (1984) and Chafetz and Guidry (1999) report on the centimetre to metre scale dendritic structures, such as bacterial shrubs, crystal shrubs and ray-crystal crusts, found in such travertines and suggest that bacteria are involved in the production of all types. Similar black, manganese-rich shrubs have been found in thermogene travertines in Morocco, also thought to be of bacterial origin (Chafetz et al., 1998). The precise way in which the microbes produce these structures is unknown, and some dendritic forms, such as those reported by Jones and Renaut (1995), appear to be entirely abiogenic.

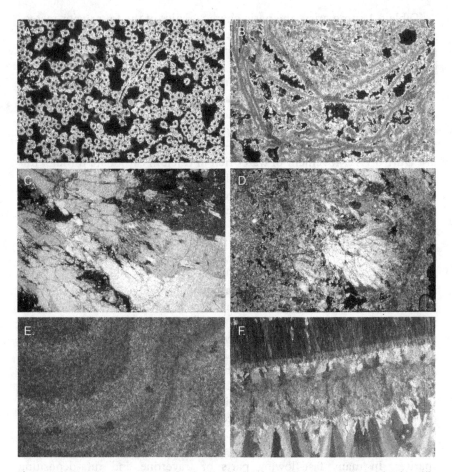

Figure 6.7 (A) Highly porous, actively forming Vaucheria tufa from Fleinsbrunnen Bach, Schwabian Alb, Germany. Radial micrite cement around Vaucheria filaments. Photographed under crossed nicols, width of image 2.5 mm. (B) Leaf relics from ancient tufa barrage at Caerwys, North Wales. Leaves have been partly infilled with micrite, with patchy sparite cements elsewhere leaving mouldic porosity. Photographed under crossed nicols, width of image 2.5 mm. (C) Large bladed spar crystals developed under high water flow rates at Goredale waterfall, Yorkshire. Note patches of cyanobacterial filaments. Photographed under crossed nicols, width of image 2.5 mm. (D) Laminated cyanobacterial tufa from Fleinsbrunnenbach, Schwabian Alb, Germany. Brown micrite layers with filament remains interspersed with large sparite. Photographed under crossed nicols, width of image 2.5 mm. (E) Fine-grained, thinly laminated tufa from Whit Beck, North Yorkshire. In streamwater of pH 11 there appears to be no cyanobacterial or other organic influence on tufa precipitation. Photographed under crossed nicols, width of image 2.5 mm. (F) Laminated sparite from dense, recrystallised laminated tufa in the Naukluft Mountains, Namibia. Photographed under crossed nicols, width of image 2.5 mm.

In some tufas and travertines (both thermogene and meteogene), lami-nated fabrics develop, often where biofilm colonies develop and show sea-sonal growth patterns. Such laminated deposits have been observed from many places, including sites in Europe and Japan, but there are dif-ferences of opinion about the periodicity of the laminations and their environmental significance. A detailed study by Kano et al. (2003) of lami-nated deposits within a tufa-depositing stream in southwest Japan, found the laminations to be clearly annual. The laminations develop within colo-nies of cyanobacteria in fast-flowing waters forming mound-type tufas. Doublets of dark and light coloured laminae were observed, with individual laminae ranging from 0.5 to 3 mm in thickness. The dark laminae are highly porous (62–80%) and comprised dominantly of lightly calcified cyanobac-terial filaments, encrusted with micrite. Conversely, the light laminae are less porous (43–73%) and composed of tubes highly encrusted with micrite. Large oval-shaped pores are often found within the upper parts of the porous laminae, which may be Trichoptera larval retreats (Drysdale, 1999).

Kano et al. (2003) hypothesise that the observed layering results from seasonal changes in precipitation rate, with dense light-coloured layers forming in summer–autumn, and porous dark-coloured layers in winter–spring. Reversed seasonal patterns have been observed by other authors in Europe, where water flow and growth characteristics may be different (Pen-tecost and Spiro, 1990; Freytet and Plet, 1996; Janssen et al., 1999; Arp et al., 2001). In European examples, a porous layer of micrite is precipitated around the growing filaments during spring and summer (see Figure 6.7D). The production of copious algal mucilage during the warmer months appears to prevent the formation of dense crystalline layers, at least in some algal colonies such as *Phormidium*. In winter, when growth slows or even stops, little mucilage is formed and the deposits become more dense and less porous. Cyanobacterially mediated laminations have also been observed in tufas in tropical India (Das and Mohanti, 1997). Similar layers also occur in thermogene travertines as well as in some unusual barrage travertines within highly alkaline small streams. In Italian thermogene travertines, for example, Chafetz and Folk (1984) found seasonal laminations, with bacteri-ally precipitated calcite 'shrubs' forming the summer layer. In contrast, Pentecost et al. (1997) found that laminations within the thermogene trav-ertines at Pamukkale were probably produced by short-term fluctuations in flow regime. In the UK, two streams influenced by lime waste (and thus having pH values of around 12.0) have been found to contain small barrage systems with laminated growth but no evidence of microbial biofilms (Figure 6.7E; Pedley, 2000; Packer, 2004). Laminations are also found within ancient tufa deposits, such as those within the Arbuckle Mountains, Okla-homa (Love and Chafetz, 1988), but such laminations are dominated by sparite, perhaps as a result of diagenetic processes which have obliterated any clear microorganic signatures (see Figure 6.7F). Andrews and Brasier (2005) provide a recent discussion of laminated tufas, their formation and role in understanding climate change.

Other unusual fabrics and components have been found in some meteogene tufa and travertine deposits. Calcite rafts, for example, have been found in inter-barrage pools in the Barkly karst, northern Australia (Carthew et al., 2003b) and they are also known from cave travertines. Within tropical tufa deposits, caddis fly larval nets and retreats form distinctive fabrics, well-preserved in many fossil deposits, which record palaeoflow direction and are indicative of fast-flowing waters (Drysdale et al., 2003). Tufas may also contain other organic components, such as diatom frustules (Figure 6.8a) as well as peloids and other grains (Figure 6.8B).

6.4 Biology, Chemistry, Mineralogy and Petrology

As most tufas and travertines are at least partially deposited as a result of organic processes, there is a strong biological component to many active and recent deposits. After diagenesis, there is often still strong indirect evidence of organisms in the form of a range of trace fossils. Some organisms within tufas can be vital for dating, and also for palaeoenvironmental reconstructions (such as the mollusca studied by Meyrick and Preece (2001) and Meyrick (2003), amongst many others). Pedley (2000) points out that most meteogene tufas and travertines contain a wide range of biota including bacteria, algae, mosses, higher plants and a range of small invertebrates, whereas thermogene travertines usually contain only bacteria and algae. A study by Freytet and Verrecchia (1998) has identified many prokaryotic and eukaryotic algae which contribute to meteogene tufa and travertine deposition, i.e. 44 species of Coccogonophyceae, 122 Hormogonophyceae, 2 Chrysophyceae, 35 Chlorophyceae, 3 Xanthophyceae, 2 diatoms and 3 Rhodophyceae. In the thermogene aragonite travertines of Mammoth Hot Springs, Yellowstone National Park, Fouke et al. (2003) obtained evidence for the occurrence of at least 221 different bacterial taxa using the polymerase chain reaction (PCR) amplification and sequencing of 16S rRNA genes. Different calcification styles are associated with many of these organisms. For example, the cyanobacterium *Phormidium incrustatum* is initially associated with micrite grains attached to the outer layers of the sheath, although these may later neomorphose to form fan-like calcite crystals, whereas the green alga *Vaucheria* (Xanthophyta) sometimes becomes calcified by sparite platelets according to Freytet and Verrecchia (1998). A study of the flora of the thermogene travertines at Pamukkale revealed a diverse assemblage of phototrophs, some of which were calcifying (Pentecost et al., 1997). Seventeen taxa of cyanobacteria were recognised, along with 16 diatoms and 5 Chlorophyceae. Some cyanobacteria

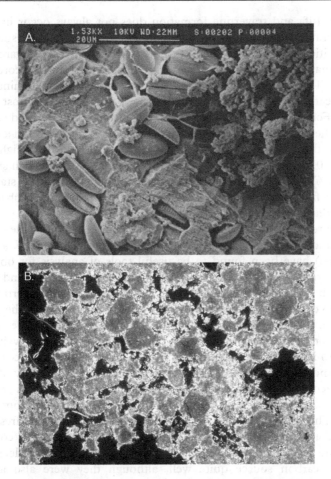

Figure 6.8 (A) Scanning electron microscopy image of diatom frustules within actively forming barrage tufas at Cwm Nash, Glamorgan. Scale bar 20 μm. (B) Detrital tufas from ancient barrage at Caerwys, North Wales. Micritic peloids within sparitic cement with much remaining pore space. Photographed under crossed nicols, width of image 2.5 mm.

(e.g. *Phormidium laminosum*) were producing mineralised layers, whereas others were endolithic forms, etching holes into the travertine.

Precipitation of tufas and travertines can usually be summarised by the following simple chemical equation:

$$Ca + 2HCO_3 \leftrightarrow CaCO_3 + CO_2 + H_2O \qquad (1)$$

In essence, waters rich in dissolved calcium carbonate become chemically disequilibriated as a result of degassing or biochemical removal of CO_2, reducing the amount of calcium carbonate that can be held in solution.

However, tufa and travertine formation does not always occur in waters supersaturated with calcium carbonate unless suitable nuclei or substrates for deposition are present. Such nuclei may be provided by organisms or organic matter such as dead leaves, branches and stems. According to equation (1), the chemistry of most tufas and travertines is dominated by calcium carbonate, although a range of minor minerals is almost always present. For example, Ihlenfeld et al. (2003) found, in a detailed study of tufa geochemistry on samples from Gregory River, northern Queensland, Australia, that calcite formed around 99.2% of the tufa by weight, with quartz (c. 0.25 wt%), kaolinite (c. 0.35 wt%) and nontronite-rich smectite (c. 0.2 wt%). Silica can sometimes be present in the amorphous state (e.g. as diatom frustules) or as doubly terminated quartz prisms, probably the result of diagenesis.

Within recent years aspects of tufa geochemistry, such as stable isotopes of carbon and oxygen, as well as trace element ratios (e.g. Mg/Ca, Sr/Ca and Ca/Ba) have been found to be useful tools of analysis for obtaining palaeoenvironmental information from the calcite within tufas and travertines. Both $\delta^{18}O$ and $\delta^{13}C$ values extracted from tufas and travertines can be used to determine environmental and climatic conditions at the time of deposition. The $\delta^{18}O$ values reflect water temperatures and the isotopic composition of the water at the time of deposition, and the $\delta^{13}C$ values can be used to interpret the source of carbon dioxide within the precipitating waters. Andrews et al. (1997) provide an overview of the utility of stable isotope records from different microbial carbonates such as tufas and travertines to palaeoclimatic and palaeoenvironmental reconstructions within Europe. Their study found, from a detailed analysis of over 80 freshwater carbonate accumulations, that $\delta^{18}O$ values recorded the isotopic composition of the regional precipitation well, and the $\delta^{13}C$ values reflected the dominant carbon source quite well, although they were also affected by altitude and microenvironmental factors. Tufas from stable back-barrage pool environments with high accumulation rates and low microenvironmental variability seem to provide the best record. More recently, Andrews (2006) provides a key review of the processes that influence stable isotopes in tufas and travertines, which will aid future attempts to untangle stable isotope records.

Some comparative studies of tufa and speleothem isotope geochemistry have found that tufas and travertines record changes in the local palaeoenvironment, whereas speleothems record larger climatic changes (e.g. Horvatincic et al., 2003). Speleothems usually form in more stable environmental conditions, with only minor variations in water temperatures, whereas tufas are affected by wide diurnal and seasonal environmental changes as well as strong biotic components. Tufas also tend to be precipitated at more rapid rates than speleothems and deposited in a less regular manner. Trace element ratios within tufas and travertines, such as Mg/Ca and Sr/Ba, can also be used to reconstruct palaeoclimatic and palaeoenvironmental conditions (Ihlenfeld et al., 2003), although their interpretation is often complex and requires an understanding of whether the tufas and travertines were precipitated in equilibrium with the surrounding waters.

Tufa and travertine mineralogy tends to be dominated by low-Mg calcite, although there are some examples of aragonite and disordered dolomite in lacustrine tufas (e.g. Great Salt Lake, Utah; Pedone and Dickson, 2000). The rare calcium carbonate hydrate known as ikaite has also been found in tufa (Buchardt et al., 2000). Vaterite, a rare anhydrous polymorph of calcium carbonate, has also been found to be co-precipitated with calcite in pools within a travertine barrage system at Huanglong, Sichuan, China (Lu et al., 2000).

6.5 Mechanisms of Formation and Accumulation

There has been a long debate about whether the deposition of tufa and travertine is largely an inorganic or an organically mediated process. For tufa and travertine deposition to occur, as we have seen above, it is necessary for carbon dioxide to be removed from the water as well as for there to be suitable nuclei or substrates on which the tufa can develop. In certain circumstances both of these requirements can be met by purely physico-chemical processes, whereas in other places organisms play a role in one or both. It is also clear that the nature of many thermogene deposits discourages the growth of all but the hardiest organisms, and thus the biotic role in such deposits is often quite low in comparison with meteogene types.

Purely physico-chemical deposition of travertine and tufa can occur where water is fast-flowing, as turbulence encourages degassing of CO_2. Such conditions are often found in the upper reaches of many streams, on the crest of waterfalls, and indeed anywhere where gradients and discharges are high. Zhang et al. (2001) and Chen et al. (2004) have investigated the nature of physico-chemical precipitation of tufas and travertines within fast flowing waters at waterfalls, and conclude that aeration, jet-flow and low-pressure effects all contribute to deposition. The effects are themselves induced by accelerated flow velocities and enlargement of the air-water interface at these waterfall sites. Algae and mosses accelerate tufa deposition here but are not the prime causes of the accumulation. A similar combination of inorganic and biotic influences was found by Arp et al. (2001) in a stream in Bavaria, Germany. Hydrochemical data collected here showed that physico-chemical deposition driven by carbon dioxide degassing was the major process, with only minor contributions from organic photosynthesis. Where physico-chemical precipitation dominates, sparitic calcite is usually formed, often with large crystals, particularly when the waters are only slightly supersaturated with calcite.

Some authors have found a much greater organic role in meteogene tufa and travertine deposition, especially in lower reaches of streams, where lower gradients prevail and there is less turbulent degassing of CO_2 (Pedley, 2000). In such situations, carbonate precipitation is frequently a by-product of microbial metabolism. Calcite becomes deposited on microorganisms, and within their associated extracellular polymeric substances (EPS). Micritic calcite is the dominant form of such precipitates. In some cases, microorganic biofilms and their EPS act simply to trap microdetrital calcite already within the water. However, organisms can also contribute to the erosion and breakdown of tufas and travertines. Within Pyramid Lake, Nevada, where thermal spring waters mix with alkaline lake water, Arp et al. (1999) found that mucus within biofilms can modify and even inhibit calcite precipitation. Similar results of microbial inhibition of travertine deposition have been found from Huanglong, Sichuan, China, where algal-covered substrates experienced less calcium carbonate precipitation in back-barrage pools than those without algae. Furthermore, electron microscope evidence of etched calcite under a layer of diatoms and filamentous algae indicates that diatoms may be aiding dissolution of the precipitated calcium carbonate (Lu et al., 2000).

Rates of tufa and travertine deposition have been quantified using a range of techniques including test substrates (e.g. Merz-Preiss and Riding, 1999; Viles and Pentecost, 1999), hydrochemical calculations and mass balance models (Herman and Lorah, 1988), the microerosion meter (Drysdale and Gillieson, 1997) and repeated physical measurements of calcifying organic communities (Pentecost, 1987). Within meteogene deposits, calculated and measured rates from less than 1 mm yr^{-1} to around 0.5 mm yr^{-1} have been recorded, with most values towards the lower end of this scale (Viles and Goudie, 1990b). For example, Drysdale and Gillieson (1997) found a mean deposition rate of 4.15 mm yr^{-1} along Louie Creek, Queensland, Australia, and Merz-Preiss and Riding (1999) found rates of up to 2.2 mm yr^{-1} on tufa barrages in small freshwater streams in southern Germany. Longer-term estimates of deposition rates, from Pleistocene and Holocene barrage tufas in the Mijares River canyon, Spain, give similar rates of 1–5 mm yr^{-1} (Pena et al., 2000). At the famous Plitvice Lake barrages, accumulation rates of 1–3 cm yr^{-1} have been observed, with downstream barrages showing the fastest rates (Chafetz et al., 1994). Higher rates, up to c. 1 m yr^{-1}, have been recorded from thermogene deposits.

Several models of tufa and travertine accumulation have been proposed, including those already discussed for meteogene deposits by Pedley (1992) and Ford and Pedley (1996), and for thermogene deposits by Chafetz and Folk (1984). Recent investigations have focused on trying to establish whether climate has an overriding control on the nature of tufa and travertine depositional styles (Pedley et al., 1996; Carthew et al., 2003b). Pedley et al. (1996) propose two contrasting models applying to barrages in cool temperate and warm semi-arid environments, respectively, derived from comparative studies in Britain and Spain. Both barrage and pool architectures vary between the two models. Barrages are proposed to be vertical, narrow topped arcuate structures in semi-arid examples, instead of thicker, arcuate barrages comprised of downstream dipping sheets in cool temperate settings. Pools within semi-arid systems are deep with stromatolitic forms along their margins, whereas within cool temperate environments pools tend to be shallower with tufa deposits dominated by marginal vegetation stands.

Carthew et al. (2003b) identify different morphological styles within seasonal tropical environments, i.e. upstream-dipping ramps on barrages, and tufa domes forming within pools downstream of barrages. Furthermore, Carthew et al. (2003b) note that calcite rafts and extensive aquatic insect larval features seem to be restricted to seasonal tropical environments. Identifying a clear climatic signal in tufa and travertine deposit architecture can be problematic, however, as studies are based on generally few examples from each climatic location and confounding factors, such as drainage basin characteristics, may exert an equally strong control on fluvial tufas and travertines. Further careful observations from a wider range of sites are required to test the general validity of such climatic models of tufa and travertine deposits.

Several authors have noted the concentration of tufa and travertine deposits within certain, restricted portions of the geological record, notably the Quaternary and early Holocene. Goudie et al. (1993) hypothesised that a 'Late Holocene tufa decline' could be identified within Europe, i.e. that since around 2500 yr BP there has been a marked decline in the deposition of meteogene tufa and travertine. Looking at tufa deposits in several European countries it is clear that there are few large, active deposits in comparison with a wide range of such deposits dating from the early to mid-Holocene. For example, the large deposits dated by Soligo et al. (2002) from central Italy cease around 5 ka. A range of factors has been proposed to explain such a decline in tufa deposition, notably climatic factors influencing discharge or human factors influencing water chemistry and/or catchment conditions (Goudie et al., 1993). Several more recent hypotheses have also been proposed. Controversially, Griffiths and Pedley (1995) propose that climate change over the Holocene produced a 'window' in terms of CO_2 concentrations in the atmosphere during the early Holocene which facilitated extensive tufa deposition. As an alternative, Dramis et al. (1999) propose that changing groundwater temperatures, and particularly the difference between groundwater and air temperatures, played a key role. Where groundwater was particularly cold, caused by deep penetration of water under Late-glacial cold conditions, there would be a significant temperature difference at surface springs, causing considerable deposition owing to the increased solubility of carbon dioxide in colder waters. The groundwater–surface temperature gradient would have gradually lessened over the Holocene as the Late-glacial groundwater passed through the system. As Goudie et al. (1993) point out, it is difficult to test these different hypotheses based on the datasets available. Furthermore, some authors even doubt the reality of the Late Holocene tufa decline, as more active sites of tufa deposition are being found in Europe (e.g. Baker and Simms, 1998), although these are generally small, even if widespread.

6.6 Palaeoenvironmental Significance

Tufa and travertine deposits provide a valuable palaeoenvironmental archive, with datable material and evidence such as organic remains, isotopes and trace elements, which can be used to infer climatic and environmental conditions at the time of formation. Pollen, ostracods, molluscs, leaf impressions, charcoal, Sr/Ba, Mg/Ca and U/Ca ratios, $\delta^{18}O$ and $\delta^{13}C$ can provide insights into the depositional environment and local fauna and flora. Such information is useful not only for reconstructing the conditions when tufa was deposited, but also for testing some of the relationships between climate and tufa formation proposed in the preceding section. An essential prerequisite of such palaeoenvironmental studies is that there should be accurate and precise dating of the tufa and travertines. For many years, accurate age determination (Garnett et al., 2004) has proved problematic, with ^{14}C dating being affected by contamination from 'dead' carbon recycled from limestone. Some workers have attempted to get around this problem by dating molluscan shells and other biogenic calcite within tufa deposits, but here too diagenesis may influence the accuracy of the dates obtained, and often there is no suitable material to date radiometrically within the stratigraphic units of interest. Uranium-series dating has proved to be a powerful dating tool for many epigean tufas and travertines (e.g. Auler and Smart, 2001; Soligo et al., 2002) stretching back over much of the Late Quaternary. However, like ^{14}C, uranium-series dating is also complicated for tufas and travertines, because of open-system behaviour and contamination from detrital and organic particles, although recent techniques have been developed to cope with such complexity (Mallick and Frank, 2002). More recently, Rich et al. (2003) have examined the use of optically stimulated luminescence as a tool for dating tufa and travertine, through dating sand grains contained within spring tufas around saline lakes in Texas, USA.

Meteogene tufas and travertines provide a more useful set of palaeoenvironmental information than do thermogene examples because of the complexity induced in the latter by geothermal warming and endogenic CO_2 production (Liu et al., 2003). Meteogene tufas and travertines provide complex palaeoenvironmental indicators when compared with most speleo-them deposits, because they form under more variable ambient tempera-tures and water flow regimes, rather than the more muted underground regimes within cave systems. As Andrews et al. (2000) note, considerable care must be taken in choosing suitable tufa and travertine deposits for palaeoenvironmental studies as microbial effects may complicate isotope behaviour. Many tufa and travertine deposits are inferred to have been produced in warm, wet conditions (e.g. marine oxygen isotope stages (MIS) 9 and 7 in Croatia (Horvatincic et al., 2000) and MIS 5, 3 and 1 in central Italy (Soligo et al., 2002)). An interesting use of tufas and trav-ertines as palaeoenvironmental indicators was made by Hancock et al. (1999) who coined the term 'travitonics' to describe their use as indicators of tectonic movements. Looking at the relative position and/or deformation of travertines, as well as dating travertines by U-series, helps constrain the ages of fault motions along fissure ridges.

6.7 Relationships to other Terrestrial Geochemical Sediments and Geomorphological Roles

Tufas and travertines often form in conjunction with peats and marls, and can sometimes grade into calcretes (see Chapter 2) or themselves become calcretised or case-hardened. Meteogene tufas and travertines are often recorded from cave entrances, with speleothem (Chapter 7) replacing them out of the photic zone. The large size and fluvial environmental setting of many tufa and travertine deposits means that they are often highly signifi-cant to local geomorphology. Where suites of barrages form along rivers, for example, they can provide a major control on river flow regimes and fluvial profiles. Furthermore, spring tufas may influence slope processes by providing a hard, less permeable cap overlying soils, and may even influ-ence slope stability. Travertines may also be associated with lacustrine or palustrine geochemical sediments (see discussion in Chapter 9).

6.8 Directions for Future Research

Tufas and travertines remain a rich source of scientific data. There are many directions worth pursuing in future work, we list five of the key ones below. First, in order to fully utilise the palaeoenvironmental record within tufas, there need to be improvements in the dating of these deposits. New techniques, refinements of old techniques, or careful selection of material are required to provide more accurate and precise dating over a wide range of timescales. Second, there are many tufa and travertine locations around the world which remain unstudied or the subject of only preliminary investigations. Many such deposits may provide key information about their local and regional environment and climatic setting, as well as adding to current understanding of the nature and development of tufas and travertines. Third, there is a need for further studies of the biology of tufas and the links between biota and the deposition and erosion of tufas and travertines. Fourth, there needs to be more work developing the potential of stable isotope and trace element records, especially on laminated deposits, as records of changing environmental conditions. Finally, it would be good to see more development of general models of both meteogene tufa and travertine and thermogene travertine formation which could account for a range of processes, acting at different scales, in a range of climatic settings.

Acknowledgements

Kathleen Nicoll and Martyn Pedley provided very helpful comments on earlier drafts of this chapter.

References

Andrews, J.E. (2006) Paleoclimatic records from stable isotopes in riverine tufas: synthesis and review. *Earth-Science Reviews* 75, 85–104.

Andrews, J.E. & Brasier, A.T. (2005) Seasonal records of climate change in annually laminated tufas: short review and future prospects. *Journal of Quaternary Science* 20, 341–421.

Andrews, J.E., Riding, R. & Dennis, P.F. (1997) The stable isotope record of environmental and climatic signals in modern terrestrial microbial carbonates from Europe. *Palaeogeography, Palaeoclimatology, Palaeoecology* 129, 171–189.

Andrews, J.E., Pedley, H.M. & Dennis, P.F. (2000) Palaeoenvironmental records in Holocene Spanish tufas: a stable isotope approach in search of reliable climatic archives. *Sedimentology* 47, 961–978.

Arp, G., Thiel, V., Reimer, A., Michaelis, W. & Teitner, J. (1999) Biofilm expolymers control microbialite formation at thermal springs discharging into the alkaline Pyramid Lake, Nevada, USA. *Sedimentary Geology* 126, 159–176.

Arp, G., Wedemeyer, N. & Reitner, J. (2001) Fluvial tufa formation in a hard-water creek (Deinschwanger Bach, Franconian Alb, Germany). *Facies* 44, 1–22.

Auler, A.S. & Smart, P.L. (2001) Late Quaternary paleoclimate in semiarid Northeastern Brazil from U-series dating of travertines and water-table speleothems. *Quaternary Research* 55, 159–167.

Baker, A. & Simms, M.J. (1998) Active deposition of calcareous tufa in Wessex, UK, and its implications for the 'late-Holocene tufa decline'. *The Holocene* 8, 359–365.

Bayari, C.S. (2002) A rare landform: Yerköprü travertine bridges in the Taurids karst range, Turkey. *Earth Surface Processes and Landforms* 27, 577–590.

Becker, M.L., Cole, J.M., Rasbury, E.T., Pedone, V.A., Montanez, I.P. & Hanson, G.N. (2001) Cyclic variations of uranium concentration and oxygen isotopes in tufa from the middle Miocene Barstow formation, Mojave Desert, California. *Geology* **29**, 139–142.

Brook, G.A., Marais, E. & Cowart, J.B. (1999) Evidence of wetter and drier conditions in Namibia from tufas and submerged speleothems. *Cimbebasia* **15**, 29–39.

Buchardt, B., Israelson, C., Seaman, P. & Stockmann, G. (2001) Ikaite towers in Ikka Fjord, southwest Greenland: their formation by mixing of seawater and alkaline spring water. *Journal of Sedimentary Research* **A71**, 176–189.

Burger, D. (1990) The travertine complex of Antalya/Southwest Turkey. *Zeitschrift für Geomorphologie, Supplement Band* **77**, 25–46.

Carthew, K.D., Drysdale, R.N. & Taylor, M.P. (2003a) Tufa deposits and biological activity, Riversleigh, Northwestern Queensland. In: Roach, I.C. (Ed.) *Advances in Regolith*. Boca Raton: CRC Press, pp. 55–59.

Carthew, K.D., Taylor, M.P. & Drysdale, R.N. (2003b) Are current models of tufa sedimentary environments applicable to tropical systems? A case study from the Gregory River. *Sedimentary Geology* **162**, 199–218.

Chafetz, H.C. & Folk, R.L. (1984) Travertines: depositional morphology and the bacterially constructed constituents. *Journal of Sedimentary Petrology* **54**, 289–316.

Chafetz, H.S. & Guidry, S.A. (1999) Bacterial shrubs, crystal shrubs, and ray-crystal shrubs: bacterial vs. abiotic precipitation. *Sedimentary Geology* **126**, 57–74.

Chafetz, H.S., Srdoc, D. & Horvatincic, N. (1994) Early diagenesis of Plitvice Lakes waterfall and barrier travertine deposits. *Geographie Physique et Quaternaire*. **48**, 247–255.

Chafetz, H.S., Akdim, B., Julia, R. & Reid, A. (1998) Mn- and Fe-rich black travertine shrubs: bacterially (and nanobacterially) induced precipitates. *Journal of Sedimentary Research* **A68**, 404–412.

Chen, J.A., Zhang, D.D., Wang, S.J., Xiao, T.F. & Huang, R.E. (2004) Factors controlling tufa deposition in natural waters at waterfall sites. *Sedimentary Geology* **166**, 353–366.

Crombie, M.K., Arvidson, R.E., Sturchio, N.C. El Alfy, Z. & Abu Zeid, K. (1997) Age and isotopic constrains on Pleistocene pluvial episodes in the Western Desert, Egypt. *Palaeogeography, Palaeoclimatology, Palaeoecology* **130**, 337–355.

Das, S. & Mohanti, M. (1997) Holocene microbial tufas: Orissa State, India. *Carbonates and Evaporites* **12**, 204–219.

Dilsiz, C. (2002) Environmental issues concerning natural resources at Pamukkale protected site, southwest Turkey. *Environmental Geology* **41**, 776–784.

Dramis, F., Materazzi, M. & Cilla, G. (1999) Influence of climatic changes on freshwater travertine deposition: a new hypothesis. *Physics and Chemistry of the Earth Part A: Solid Earth and Geodesy* **24**, 893–897.

Drysdale, R.N. (1999) The sedimentological significance of hydropsychid caddisfly larvae (Order: Trichoptera) in a travertine-depositing stream: Louie Creek, Northwest Queensland, Australia. *Journal of Sedimentary Research* **A69**, 145–150.

Drysdale, R. & Gillieson, D. (1997) Micro-erosion meter measurements of travertine deposition rates: A case study from Louie Creek, Northwest Queensland, Australia. *Earth Surface Processes and Landforms* 22, 1037–1051.

Drysdale, R.N., Carthew, K.D. & Taylor, M.P. (2003) Larval caddis-fly nets and retreats: A unique biosedimentary paleocurrent indicator for fossil tufa deposits. *Sedimentary Geology* 161, 207–215.

Durrenfeldt, A. (1978) Untersuchungen zur Besiedlungsbiologie von Kalktuff faunistische, ökologische und elektronenmikroskopische Befunde. *Archive für Hydrobiologie Supplemente* 54, 1–79.

Ekmekci, M., Gunay, G. & Simsek, S. (1995) Morphology of rimstone pools at Pamukkale, Western Turkey. *Cave and Karst Science* 22, 103–106.

Emeis, K.-C., Reichnow, H.-H. & Kempe, S. (1987) Travertine formation in Plitvice National park, Yugoslavia: chemical versus biological control. *Sedimentology* 34, 595–609

Evans, J.E. (1999) Recognition and implications of Eocene tufas and travertines in the Chadron formation, White River group, Badlands of South Dakota. *Sedimentology* 46, 771–789.

Ford, T.D. (1989) Tufa – the whole dam story. *Cave Science* 16, 39–49.

Ford, T.D. & Pedley, H.M. (1996) A review of tufa and travertine deposits of the world. *Earth-Science Reviews* 41, 117–175.

Fouke, B.W., Bonheyo, G.T., Sanzenbacher, B. & Frias-Lopez, J. (2003) Partitioning of bacterial communities between travertine depositional facies at Mammoth Hot Springs, Yellowstone National Park, U.S.A. *Canadian Journal of Earth Science* 40, 1531–1548.

Freytet, P. & Plet, A. (1996) Modern freshwater microbial carbonates: the Phormidium stromatolites (tufa-travertine) of southeastern Burgundy (Paris basin, France). *Facies* 34, 219–238.

Freytet, P. & Verrecchia, E.P. (1998) Freshwater organisms that build stromatolites: a synopsis of biocrystallisation by prokaryotic and eukaryotic algae. *Sedimentology* 45, 535–563.

Freytet, P. & Verrecchia, E.P. (1999) Calcitic radial palisadic fabric in freshwater stromatolites: Diagenetic and recrystallized feature or physicochemical sinter crust? *Sedimentary Geology* 126, 97–102.

Garnett, E.R., Gilmour, M.A., Rowe, P.J., Andrews, J.E. & Preece, R.C. (2004) ^{230}Th/^{234}U dating of Holocene tufas: possibilities and problems. *Quaternary Science Reviews* 23, 947–958.

Glover, C. & Roberston, A.H.F. (2003) Origin of tufa (cool-water carbonate) and related terraces in the Antalya area, SW Turkey. *Geological Journal* 38, 329–358.

Goudie, A.S., Viles, H.A. & Pentecost, A. (1993) The late-Holocene tufa decline in Europe. *The Holocene* 3, 181–186.

Griffiths, H.I. & Pedley, H.M. (1995) Did changes in late Last Glacial and early Holocene atmospheric CO_2 concentrations control rates of tufa precipitation? *The Holocene* 5, 238–242.

Hancock, P.L., Chalmers, R.M.L., Altunel, E. & Cakir, Z. (1999) Travitonics: Using travertines in active fault studies. *Journal of Structural Geology* 21, 903–916.

Herman, J.S. & Lorah, M.M. (1988) Calcite precipitation rates in the field: measurement and prediction for a travertine-depositing stream. *Geochimica et Cosmochimica Acta* **52**, 2347–2355.

Horvatincic, N., Calic, R. & Geyh, M.A. (2000) Interglacial growth of tufa in Croatia. *Quaternary Research* **53**, 185–195.

Horvatincic, N., Krajcar, B.I., Bogomil, O. (2003) Differences in the ^{14}C age, delta ^{13}C and delta [18]O of Holocene tufa and speleothem in the Dinaric karst. *Palaeogeography, Palaeoclimatology, Palaeoecology* **193**, 139–157.

Ihlenfeld, C., Norman, M.D., Gagan, M.K., Drysdale, R.N., Maas, R. & Webb, J. (2003) Climatic significance of seasonal trace element and stable isotope variations in a modern freshwater tufa. *Geochimica et Cosmochimica Acta* **67**, 2341–2357.

Janssen, A., Swennen, R., Podor, N. & Keppens, E. (1999) Biological and diagenetic influences in Recent and fossil tufa deposits from Belgium. *Sedimentary Geology* **126**, 75–95.

Jones, B. & Renaut, R. W. (1995) Noncrystallographic calcite dendrites from hot-spring deposits at Lake Bogoria, Kenya. *Journal of Sedimentary Research* **A65**, 154–169.

Kano, A., Matsuoka, J., Kojo, T. & Fujii, H. (2003) Origin of annual laminations in tufa deposits, southwest Japan. *Palaeogeography, Palaeoclimatology, Palaeoecology* **191**, 243–262.

Liu, Z., Zhang, M. , Li, Q. & You, S. (2003) Hydrochemical and isotope characteristics of spring water and travertine in the Baishuitai area (SW China) and their meaning for palaeoenvironmental reconstruction. *Environmental Geology* **44**, 698–704.

Love, K.M. & Chafetz, H.S. (1988) Diagenesis of laminated travertine crusts, Arbuckle Mountains, Oklahoma. *Journal of Sedimentary Petrology* **58**, 441–445.

Lu, G., Zheng, C., Donahoe, R.J. & Berry, L.W. (2000) Controlling processes in a $CaCO_3$ precipitating stream in Huanglong Natural Scenic District, Sichuan, China. *Journal of Hydrology* **230**, 34–54.

Mallick, R. & Frank, N. (2002) A new technique for precise Uranium-series dating of travertine micro-samples. *Geochimica et Cosmochimica Acta* **66**, 4261–4272.

Merz-Preiss, M. & Riding, R. (1999) Cyanobacterial tufa calcification in two freshwater streams: ambient environment, chemical thresholds and biological processes. *Sedimentary Geology* **126**, 103–124.

Meyrick, R.A. (2003) Holocene molluscan faunal history and environmental change at Kloster Muhle, Rheinland-Pfalz, western Germany. *Journal of Quaternary Science* **18**, 121–132.

Meyrick, R.A. & Preece, R.C. (2001) Molluscan successions from two Holocene tufas near Northampton, English Midlands. *Journal of Biogeography* **28**, 77–93.

Nicoll, K., Giegengack, R. & Leindienst, M. (1999) Petrogenesis of artefact-bearing fossil-spring tufa deposits from Karga Oasis, Egypt. *Geoarchaeology* **14**, 849–863.

Pache, M., Reitner, J. & Arp, G. (2001) Geochemical evidence for the formation of a large Miocene 'travertine' mound at a sublacustrine spring in a soda lake (Wallerstein castle rock, Nordlinger Ries, Germany). *Facies* **45**, 211–230.

Packer, L. (2004) *Spatial and temporal patterns of tufa deposition.* Unpublished BA Dissertation, School of Geography and the Environment, University of Oxford.

Pedley, H.M. (1990) Classification and environmental models of cool freshwater tufas. *Sedimentary Geology* **68**, 143–154.

Pedley, H.M. (1992) Freshwater (Phytoherm) reefs: the role of biofilms and their bearing on marine reef cementation. *Sedimentary Geology* **79**, 255–274.

Pedley, H.M. (2000) Ambient temperature freshwater microbial tufas. In: Riding, R.E. & Awramik, S.M. (Eds) *Microbial Sediments.* Berlin: Springer-Verlag, pp. 179–186.

Pedley, H.M. & Hill, I. (2002) The recognition of barrage and paludal tufa systems by GPR: case studies in the geometry and correlation of hidden Quaternary freshwater carbonate facies. In: Bristow, C.S. & Jol, H.M. (Eds) *Ground Penetrating Radar in Sediments.* Special Publication 211. Bath: Geological Society Publishing House, pp. 207–223.

Pedley, H.M., Andrews, J., Ordonez, S., Garcia del Cura, M.A., Gonzales Martin, J.-A. & Taylor, D. (1996) Does climate control the morphological fabric of freshwater carbonates? A comparative study of Holocene barrage tufas from Spain and Britain. *Palaeogeography, Palaeoclimatology, Palaeoecology* **121**, 239–257.

Pedley, H.M., Hill, I., Denton, P. & Brasington, J. (2000) Three-dimensional modelling of a Holocene tufa system in the Lathkill valley, N. Derbyshire, using ground penetrating radar. *Sedimentology* **47**, 721–735.

Pedley, H.M., Gonzalez Martin, J.A., Ordonez Delgada, S. & Garcia del Cura, M.A. (2003) Sedimentology of Quaternary perched springline and paludal tufas: criteria for recognition, with examples from Gudalajara Province, Spain. *Sedimentology* **50**, 23–44.

Pedone, V.A. & Dickson, J.A.D. (2000) Replacement of aragonite by quasi-rhombohedral dolomite in a late Pleistocene tufa mound, Great Salt Lake, Utah, USA. *Journal of Sedimentary Research* **A70**, 1152–1159.

Pena, J.L., Sancho, C. & Lozano, M.V. (2000) Climatic and tectonic significance of late Pleistocene and Holocene tufa deposits in the Mijares river canyon, eastern Iberian Range, Northeast Spain. *Earth Surface Processes and Landforms* **25**, 1403–1417.

Pentecost, A. (1987) Some observations on the growth rates of mosses associated with tufa and interpretation of some post-glacial bryoliths. *Journal of Bryology* **14**, 543–550.

Pentecost, A. (1993) British travertines: A review. *Proceedings of the Geologists' Association* **104**, 23–39.

Pentecost, A. (1995) The Quaternary travertine deposits of Europe and Asia Minor. *Quaternary Science Reviews* **14**, 1005–1028.

Pentecost, A. (1999) The origin and development of the travertines and associated thermal waters at Matlock Bath, Derbyshire. *Proceedings of the Geologists' Association* **110**, 217–232.

Pentecost, A. (2005) *Travertine.* Berlin: Springer.

Pentecost, A. & Spiro, B. (1990) Stable carbon and oxygen isotope composition of calcites associated with modern freshwater cyanobacteria and algae. *Geomicrobiological Journal* **8**, 17–26.

Pentecost, A. & Viles, H.A. (1994) A review and reassessment of travertine classification. *Geographie Physique et Quaternaire*, **48**, 305–314.

Pentecost, A. & Zhang, Z. (2001) A review of Chinese travertines. *Cave and Karst Science* **28**, 15–28.

Pentecost, A., Bayari, S. & Yesertener, C. (1997) Phototrophic microorganisms of the Pamukkale travertine, Turkey: their distribution and influence on travertine deposition. *Geomicrobiology Journal* **14**, 269–283.

Pollard, W., Omelon, C., Andersen, D. & McKay, C. (1999) Perennial spring occurrence in the Expedition Fiord area of western Axel Heiberg Island, Canadian High Arctic. *Canadian Journal of Earth Sciences* **36**, 105–120.

Porter, A.R.O. (1991) *Jamaica, a Geological Portrait*. Kingston: Institute of Jamaica Publications Ltd.

Rich, J., Stokes, S., Wood, W. & Bailey, R. (2003) Optical dating of tufa via in situ aeolian sand grains: a case example from the Southern High Plains, USA. *Quaternary Science Reviews* **22**, 1145–1152.

Rieger, T. (1992) Calcareous tufa formations. Searles Lake and Mono Lake. *California Geologist* (for 1992), **45**, 99–109.

Rosen, M.R., Arehart, G.B. & Lico, M.S. (2004) Exceptionally fast growth rate of a <100 year old tufa, Big Soda Lake, Nevada: Implications for using tufa as a paleoclimatic proxy. *Geology* **32**, 409–412.

Soligo, M., Tuccimei, P., Barberi, R., Delitala, M.C., Miccadei, E. & Taddeucci, A. (2002) U/Th dating of freshwater travertine from Middle Velino Valley (Central Italy): paleoclimatic and geological implications. *Palaeogeography, Palaeoclimatology, Palaeoecology* **184**, 147–161.

Szulc, J. & Cwizewicz, M. (1989) The Lower Permian freshwater carbonates of the Slawkow Graben, Southern Poland: sedimentary facies and stable isotope studies. *Palaeogeography, Palaeoclimatology, Palaeoecology* **70**, 107–120.

Viles, H.A. & Goudie, A.S. (1990a) Reconnaissance studies of the tufa deposits of the Napier Range, NW Australia. *Earth Surface Processes and Landforms* **15**, 425–443.

Viles, H.A. & Goudie, A.S. (1990b) Tufas, travertines and allied carbonate deposits. *Progress in Physical Geography* **14**, 19–41.

Viles, H.A. & Pentecost, A. (1999) Geomorphological controls on tufa deposition at Nash Brook, South Wales, United Kingdom. *Cave and Karst Science* **26**, 61–68.

Viles, H.A., Taylor, M.P., Nicoll, K. & Neumann, S. (2007) Facies evidence of hydroclimatic regime shifts in tufa deposition sequences from the arid Naukluft Mountains, Namibia. *Sedimentary Geology* **195**, 39–53.

Zhang, D.D., Zhang, Y., Zhu, A. & Cheng, X. (2001) Physical mechanisms of river waterfall tufa (travertines) formation. *Journal of Sedimentary Research* **A71**, 205–216.

Chapter Seven

Speleothems

Ian J. Fairchild, Silvia Frisia, Andrea Borsato
and Anna F. Tooth

7.1 Introduction to Speleothem-Forming Cave Environments

7.1.1 The scope and purpose of this chapter

Speleothems are mineral deposits formed in caves, typically in karstified host rocks (Gunn, 2004). The cave environment is arguably an extension of the surface landscape, because caves are defined as cavities large enough for humans to enter (Hill and Forti, 1997). Speleothem deposits are controlled not only by the distribution, quantity and chemistry of water percolating through the karstic aquifer (a property strongly influenced by the surface geomorphology and macroclimate), but also by the cave's peculiar microclimate, which in turn is controlled by cave geometry, aquifer properties and external microclimates. Despite the abundance of process research, there has been a relative lack of emphasis on the dynamic behaviour of aquifer and cave environments, which has hindered the production of integrated models.

Research on karst terrains has focused mainly on surface geomorphology, the geometry of cave systems, the hydrology and hydrogeochemistry of major springs, and the dating of cave development and the major phases of speleothem formation (Jennings, 1971, 1985; Sweeting, 1972, 1981; White, 1988; Ford and Williams, 1989). Space does not permit us to cover the use of speleothem dating in constraining rates of landscape evolution, but a selection of exemplars from this field are Lauritzen (1990), Atkinson and Rowe (1992), Farrant et al. (1995), Hebdon et al. (1997), Losson and Quinif (2001) and F. Wang et al. (2004).

Since 1990, rapid progress in the development of proxies for palaeoclimate from calcareous speleothems has occurred (Gascoyne, 1992; McDermott, 2004; White, 2004). Pioneering work in this area, from Hendy and Wilson (1968) onwards, was published in leading journals and the same is true of much recent work (e.g. X. Wang et al., 2004; Yuan et al., 2004). An unfortunate side-effect of this success is that, through pressure of space or interest, authors typically focus on their time series results without describing the geomorphological context of the material. Even in the longer accounts in international journals, this context is often not systematically described. Conversely, much high-quality work on karst geomorphology and hydrology has been published in national speleological journals or conference publications, poorly accessible to general readers. Hence, there is a danger that the spectacular contribution that speleothem studies are currently making to palaeoclimate research may be undermined by a lack of understanding of the complexities of cave environments by other climate researchers. Even speleothem workers themselves could be tempted to forget the original context of the speleothem on the laboratory bench. This accounts for the focus of this chapter, which is on calcareous speleothem formation in their geomorphological (including hydrogeological) context, with particular emphasis on issues related to palaeoclimate determination. Hence we focus on particular speleothem forms (Figure 7.1): stalactites (ceiling-growths from cave drips), stalagmites (floor-growths from cave drips) and flowstones (speleothem sheets from thin water flows on walls or floors). For examples of the palaeoclimate time-series themselves see Richards and Dorale (2003), Harmon et al. (2004), White (2004), McDermott (2004), Fairchild et al. (2006a) and references therein.

7.1.2 The CO_2-degassing paradigm for calcareous speleothem formation

Just as one can refer to a planetary physiology (Lovelock, 1988), so the key processes of CO_2 creation, transport and exhalation responsible for calcareous speleothem formation (Figure 7.2A) exemplify cave physiology. Shaw (1997) gave a meticulous review of the development of understanding of this issue from the 18th to the early 20th century, including the realisation that solutes were derived principally by reaction with soil CO_2. A key focus of 20th century geomorphological work on karst became the climatic controls on the development of particular landforms and assemblages (Jennings, 1985), but there were typically few and generalised statements regarding implications for speleothems (e.g. Corbel, 1952). Bögli (1960) correctly emphasised the key control of organically derived CO_2 in soils in driving more carbonate dissolution and stimulating more speleothem formation in tropical climates, but Trudgill (1985) cautioned that too much interpretation had been placed on CO_2 measurements of soil gas, and too little on transport processes.

An important starting point for modern geochemical studies of the carbonate system in cave waters is that of Holland et al. (1964), who related

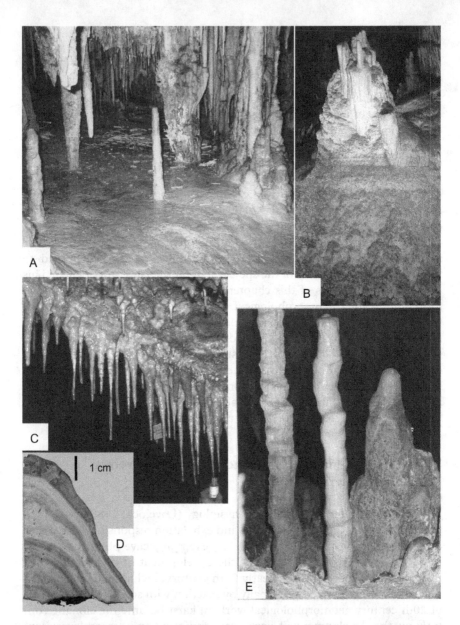

Figure 7.1 Speleothem characteristics. (A) Nerja cave, southeast Spain. Highly decorated cave chamber (view around 1.5 m across) with various stalactites and stalagmites (some joined into columns) with an intervening flowstone sheet. Debris of fallen, fast-growing soda straw stalactites on flowstone in background. (B) Nerja Cave. Phreatic pool deposits, with former water levels shown by horizontal lines, capped by more recent stalagmites. (C) Crag cave, southwest Ireland. Line of stalactites (some are soda straws in their lower parts) representing water seepage from a fracture. Scale shown by collecting bottle, lower right and by the 5 mm minimum diameter of stalactites. (D) Ernesto cave, northeast Italy. Section through conical stalagmite ER78, Ernesto cave (Frisia et al., 2003); the darker top portion post-dates the year 1860. (E) Nerja Cave. Group of stalagmites; height of view approximately 50 cm. The two cylindrical or 'candle-shaped' stalagmites to left show evidence of slight shifts in the lateral position of dripping water over time; the right-hand stalagmite has a more conical morphology, which could reflect either a higher drip rate or a larger fall height. The central stalagmite has a moist surface and may be currently active.

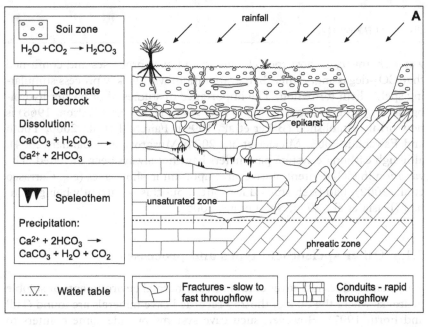

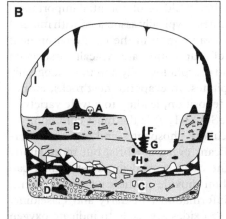

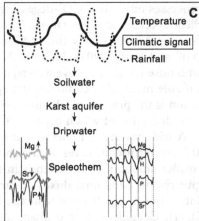

Figure 7.2 (A) Conceptual model of the karst system with its physiology of water flow and CO_2 transport and release (Tooth, 2000). The upper part of the unsaturated zone (the epikarst), having both higher porosity and permeability, is an important source of stored water. (B) Cartoon of speleothem occurrence (black) in relation to cave sedimentational history (Smart and Francis, 1990). Remnants of an old wall-fill (I) probably pre-date sediment layer D as do the speleothem clasts within D. Unit C contains mammal material and flowstones spreading out from the walls and incorporating breccia fragments. Unit B follows deposition of another flowstone and is then cut by a cave stream in which speleothems (F) have subsequently grown including a rimstone pool deposit (G) when the stream course was dammed. Speleothem unit A incorporates human remains and may continue to the present day. Percolating waters allowed local cementation (H), but the exact age is unclear. (C) The concept of karstic capture of high-resolution climatic signals (Fairchild, 2002). Annual temperature and rainfall variations are shown as the input, but longer-term changes are just as relevant. Likewise the examples of captured signals are in terms of trace element variations, but other suitable parameters include speleothem morphology, lamination or isotopic properties. In the trace element diagrams vertical bars are spaced at annual intervals. Left: ultra-high resolution soda straw stalactite, Crag Cave, Ireland. Right: more typical stalagmite record, Grotta di Ernesto, Italy. (Both modified from Fairchild et al., 2001.)

water chemistry to host rock chemistry and cave processes and confirmed that CO_2-degassing, rather than evaporation, was the key process stimulating speleothem formation. The importance of studies of dripwater hydrology and hydrogeochemistry (e.g. Pitty, 1966; Hendy and Wilson, 1968) is now firmly established (Baker et al., 1999a; Fairchild et al., 2000; Tooth and Fairchild, 2003; Mickler et al., 2004; Musgrove and Banner, 2004), although carbonate system parameters have not always been included. The variation in CO_2 content of cave air, important in showcaves (e.g. Carrasco et al., 2002), had been studied less often, but this is currently being rectified (Frisia et al., 2000; Spötl et al., 2005).

7.1.3 Other speleothem types and processes

This chapter does not cover caves in hydrothermal environments or soluble bedrocks where the bulk of the 255 recorded cave minerals are found (Hill and Forti, 1997). However, such cave systems provide some pointers to processes other than CO_2-degassing that could be of potential importance. Slender eccentric twisted forms (helictites), typically associated with mineral coatings sealing cave walls, indicate evaporation in the cave environment (Hill and Forti, 1997). High levels of evaporation are typically associated with raised Mg/Ca in cave waters, and morphologically and mineralogically variable metastable carbonate precipitates. In evaporite host rocks, evaporation is the primary mechanism of deposition, leading to a wide variety of recorded salts, of which gypsum ($CaSO_4.2H_2O$) is the most common.

A wide variety of minerals (particularly phosphates and organic minerals) are formed in caves with significant organic debris, but more subtly smaller amounts of such debris in the cave or overlying karst aquifer can give rise to additional fluxes of carbon dioxide to the cave environment. Intact pollen yields palaeoclimatic information (McGarry and Caseldine, 2004). Coatings of iron or manganese oxides are likely to indicate oxygen consumption in feeding waters, which may have impacts on speleothem chemistry.

Standing bodies of water produce characteristic phreatic growths of calcite crystals with well-developed crystal terminations, typically associated with rimstone dams (gour pools). Dating of such deposits points to varying water levels in the past (Figure 7.1B), although there could be climatic (Bar-Matthews et al., 2003) or hydrological causes for such variations. Cave rafts are precipitates forming at the surface of pools and can occur seasonally in response to higher rates of CO_2-degassing (Andreo et al., 2002; Spötl et al., 2005). Finally, significant air currents may be indicated by orientated speleothems, or by deposits identified as of aerosol origin (Hill and Forti, 1997). Speleothems can accumulate during prolonged periods of the history of caves and can be interrupted by episodes of clastic sedimentation (Sasowsky and Mylroie, 2004), by breakdown of deposits, sometimes induced by seismicity, and by archaeological disturbance. Figure 7.2B (after Smart and Francis, 1990) graphically illustrates the potentially complex stratigraphy.

7.1.4 The capture of climatic signals in speleothems

Figure 7.2C presents the concept of a climatic signal preserved in a speleothem (Fairchild, 2002), a signal that is mediated by the geomorphological and hydrological environment, as discussed in detail in Fairchild et al. (2006a). Current research involves the forward modelling of processes that lead to the transformation of the input signal, and inverse modelling (Kaufmann and Dreybrodt, 2004) by appropriate transfer functions to recover aspects of the original climatic signal. However, progress is limited by the appropriate definition of system attributes, as is illustrated in the next section.

7.2 Distribution, Field Occurrence and Geomorphological/ Hydrological Relations

7.2.1 Ingredients for speleothem formation

Climate determines the timing and quantity of H_2O input, the degree to which it lies as snow, and to which it can potentially be recycled by evaporation. Climate (and the linked variables of altitude and topography) influences colonisation by vegetation. The development of vegetation strongly affects the chemistry of the carbonate system and soil properties (facilitating speleogenesis), increases evapotranspiration, and can also be modified by human activity. The recharge of the aquifer with water (and the transport of CO_2, colloids and solutes in general) is strongly influenced by surface topography and the nature of soils and other surficial deposits, as well as the degree and style of karstification of the aquifer. Karstic aquifers have a complex distribution of pore spaces, including one-dimensional conduits, fractures and 'matrix', which facilitates mixing and gives rise to a wide range of transmission times in the unsaturated zone. Although the conventional definition of speleothems only includes deposits in caves rather than smaller spaces (Hill and Forti, 1997), we need to be aware of processes outwith the caves (including deposition). Speleothems are typically fed by dripwaters with a large component of water derived from matrix storage, although faster flowpaths are relatively more important for flowstones. A crucial, somewhat neglected, property is the nature of cave ventilation, which removes CO_2 and allows speleothem formation to continue. In the following three sections we group the controlling factors to address the key issues of inflowing water quantity, water quality and in-cave processes.

7.2.2 Controls on quantity of speleothem-forming water inflow to caves

Climatic aridity clearly limits speleothem formation. However, speleothems are able to form even in semi-arid areas given sufficient infiltration from heavy rains and adequate storage capacity of the karstic aquifer. The timing of episodes of past speleothem growth in currently dry caves has proved extremely important in elucidating changes in atmospheric circulation (e.g. Burns et al., 2001; Vaks et al., 2003; X. Wang et al., 2004).

Vegetation and animal activity in soils play important roles in generating macropores, which locally focus recharge into the karst aquifer (e.g. Beven and Germann, 1982; Tooth and Fairchild, 2003). A reduction in evapotranspiration by local deforestation could lead to an increase in seepage water in caves, which might be recognisable in palaeostudies by carbon isotope evidence.

The aquifer structure causes a complex response to a water infiltration event. Recharge is focused into surface depressions (dolines), which are fed by lateral flow through a zone of enhanced permeability (the epikarst or subcutaneous zone) constituting the upper part of the karst aquifer. Dolines typically feed major conduits that are the primary channels of karst drainage. Speleothems are fed by a combination of seepage and fracture water, consistent with observations in quarried karstic aquifers indicating that the system consists of a matrix with a network of discontinuities (fractures or conduits). Models of complex aquifer behaviour (Smart and Friederich, 1986; Vaute et al., 1997; Perrin et al., 2003) are consistent with test results using artificial tracers (Bottrell and Atkinson, 1992), demonstrating dominantly vertical penetration with a range of flow-through times to drips in shallow caves ranging from days to over a year. Studies of frequently measured or automatically logged discharge (Smart and Friederich, 1986; Baker et al., 1997a, Genty and Deflandre, 1998; Baker and Brunsdon, 2003; Tooth and Fairchild, 2003; Fairchild et al., 2006b) illustrate the range of hydrological responses that can occur to infiltration events. Although fracture-fed flow normally results in flow increases, there can be local flow decrease (underflow) or non-linear behaviour, which relate to phenomena such as pockets of gas phase in the flow path, and hydrological thresholds in general. Seasonal-scale variations in drip rate are a feature of many climates (e.g. Figure 7.3A; Genty and Deflandre, 1998). If a physically continuous water fill of cavities exists, pressure increases will be transmitted virtually instantaneously and so drip rates are typically in phase with the seasonal variation in infiltration. Cases where they are out of phase (e.g. Nerja cave, Andreo et al., 2002; Liñan et al., 2002) perhaps reflect a significant gas phase in the system.

7.2.3 Controls on composition of inflowing waters

The greater abundance of speleothems in tropical climates is mirrored both by higher mean soil pCO_2 at higher mean annual temperatures (Harmon et al., 1975) and the positive relationship between growing season soil pCO_2 and mean annual actual evapotranspiration (Brook et al., 1983). These relationships arise because higher pCO_2 waters can carry higher dissolved carbonate loads and hence can precipitate more when degassed (Figure 7.4A). However, the soil data exhibit a large scatter and local controls can also be important in determining soil pCO_2 (Miotke, 1974; Tooth, 2000; Davis et al., 2001). Nevertheless, speleothem growth is inhibited in cooler climates, and distinct warmer periods of speleothem growth can be recognised in cool temperate regions during the late Quaternary (Baker et al., 1993a).

Drake (1983), following Drake and Wigley (1975), attempted to explain data scatter with the important conceptual distinction of a 'coincident system' (one in which carbonate is dissolved to saturation at the pCO_2 of the soil environment) versus a 'sequential system' in which solutions discharging into the epikarst are undersaturated for $CaCO_3$ (carbonate-poor soils) and reach saturation under conditions closed to resupply of CO_2. Drake (1983) also cautioned against the notion of fixed seasonal values for soil pCO_2 and water hardness; water dripping during high discharge events could be undersaturated, for example. Also, growth rate could be higher than expected for the climatic regime where, for example, high pCO_2 and carbonate hardness arise from organic sources of CO_2 within the epikarst, or where there is a source of strong acid, particularly sulphuric acid from pyrite oxidation (Atkinson, 1977, 1983; Spötl et al., 2004).

Earlier literature suggested that the seasonal pattern in pCO_2 that is commonly found in karst soils is transmitted to the cave environment, leading to seasonal changes in water hardness and growth rate of speleothems (e.g. Moore, 1962; Gams, 1965; Pitty, 1966, 1968). This has been sufficiently influential as to form the starting point for explaining seasonal variations in the Ca content of cave waters (Genty et al., 2001b). However, the original work did not demonstrate this phenomenon using conservative tracers, and most systems seem to behave differently: some, for example, are buffered by seasonally invariant CO_2 values in the epikarst (Atkinson, 1977; Fairchild et al., 2000). Two other effects are more likely to lead to these changes. First, a reduction in Ca content with only small changes in other cations is likely to be due to precipitation of $CaCO_3$ up-flow of the drip site (Holland et al., 1964). This was termed *prior precipitation* by

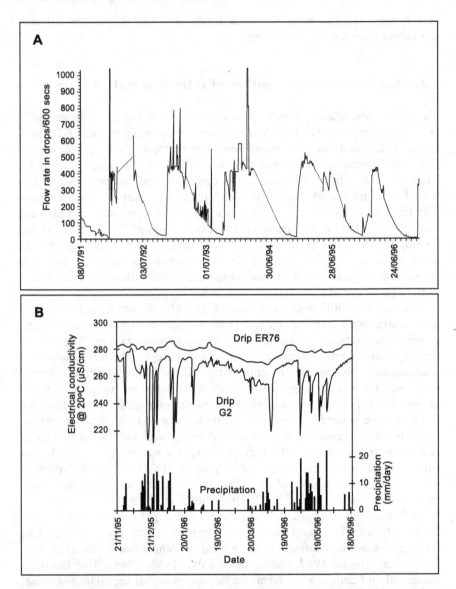

Figure 7.3 (A) Seasonal variations in drip rate with superimposed short-term hydrological events from a stalactite in Pere-Nöel cave in Belgium; there is a close relationship between flow rate and periods of high water excess (Genty and Deflandre, 1998). (B) Variations in cation loads, as monitored by electroconductivity, of drip waters in response to seasonal patterns and individual infiltration events for two drips at Ernesto cave (Frisia et al., 2000).

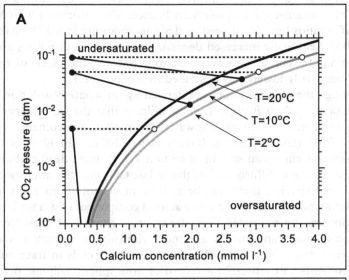

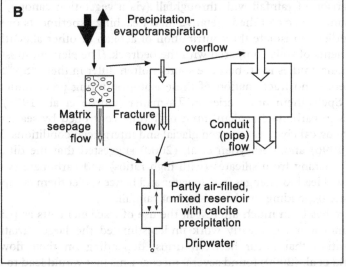

Figure 7.4 (A) Dissolved Ca loads resulting from dissolution of pure limestone to saturation and their relationship with (soil or epikarst) pCO_2 (figure 3 in Kaufmann, 2003). The calcite saturation lines are the curved lines sloping from upper right to bottom left and are distinguished by temperature (2, 10 and 20°C). The horizontal dashed lines refer to 'open' or coincident system dissolution whereby equilibrium with calcite is attained whilst still in contact with excess CO_2. The 'closed' or sequential system (lines sloping down to right) arises where water with a fixed CO_2 content subsequently reacts with $CaCO_3$ in the subsoil or karstic aquifer to reach equilibrium. (B) Plumbing model illustrating processes affecting dripwater hydrology and hydrochemistry as used in Fairchild et al. (2006b).

Fairchild et al. (2000), who showed that it was enhanced during the dry season at Clamouse cave in southern France. The associated high Mg/Ca and Sr/Ca ratios have been used to develop an aridity index (McMillan et al., 2005). Seasonally increased degassing occurs when drip rates are slow and the aquifer contains more gas spaces. Second, the pCO_2 of cave air can be seasonally lowered as described in section 7.2.4.

Although the bulk of cations are derived by carbonate dissolution in soil and epikarst, carbon isotope studies indicate that the bulk of carbon is derived from organic sources: dripwaters and modern speleothems typically have 80–95% modern carbon (Genty et al., 2001a). In Drake's (1983) coincident system (open system of section 7.4.2), most dead carbon from aquifer dissolution diffuses out of the soil as CO_2 into the atmosphere.

For other species, there can be significant contributions from marine aerosols and pollutants. The marine aerosol component of wet atmospheric deposition can be estimated from the chloride content of dripwaters, but pollution-related components are more difficult to determine directly. However, Frisia et al. (2005) demonstrated that trends in trace sulphate in speleothems can represent a record of atmospheric sulphur pollution. Comparison of rainfall with throughfall (via a vegetation canopy) illuminates some aspects of the vegetative flux. Leaching experiments or separation studies can isolate the contribution of aeolian or other allochthonous components of soils compared with the bedrock. The element most useful in sourcing studies is Sr, because sources often differ in their [87]Sr/[86]Sr ratio and there is no fractionation of these isotopes during precipitation from water. Speleothem time series in Tasmania (Goede et al., 1998) reveal changes in carbonate aeolian input over time controlled by sea level and meteorological changes between glacial and interglacial conditions. Banner et al. (1996) and Verheyden et al. (2000) suggested that the differential rate of leaching from silicates (with high ratios) and carbonate (with low ratios) will lead to changes in dripwater and hence speleothem composition over time depending on the amount of rainfall.

There has been much interest in the use of trace elements as palaeoclimatic indicators, but early work underestimated the large variations in composition that occur between drips, depending on their flow path. Fairchild et al. (2000) found several mechanisms that would lead to enrichment in trace elements in or immediately following dry conditions: incongruent dissolution related to preferential retention of Ca in soils due to freezing or evaporation; prior calcite precipitation; and enhanced dissolution of dolomite over long time periods (Roberts et al., 1998). Since then, new data and modelling approaches (Baker et al., 2000; Tooth and Fairchild, 2003; Fairchild et al., 2006b) emphasise the importance, at low flow, of low-permeability seepage aquifer compartments, which are often enriched in particular trace elements (Figure 7.4B).

The effects of infiltration events are quite variable spatially and range from an increase in flow with no change in chemistry in an entirely seepage-fed drip, to a rapid change in flow rate and composition in shallow fracture-fed sites (Tooth and Fairchild, 2003). Very rapid infiltration can be reflected in dilution of the water (Figure 7.3B, Frisia et al., 2000), although the opposite can be found in cases where the waters normally are saturated and a reduction in prior calcite precipitation (Figure 7.4B) is the main consequence of the infiltration event (Fairchild et al., 2006b). The introduction of soil-derived tracers (i.e. fluorescent humic and fulvic acids) is characteristic (Baker et al., 1997a), and such colloids may even be associated with suspended sediment in extreme cases. Speleothem evidence at several European sites (Baker et al., 1993b; Genty et al., 1997; Frisia et al., 2000; Proctor et al., 2000; Huang et al., 2001) suggests that the introduction of impurities can be largely confined to a brief interval in the autumn each year, perhaps reflecting a critical level of aquifer recharge at this phase of the hydrological year. This provides one of the key mechanisms for the development of annual properties of speleothems (as discussed in section 7.2.5).

7.2.4 Cave factors that control the distribution and rate of growth of speleothems

The shape of cave passages influences the distribution of speleothems, but the rate of ventilation of caves in relation to the rate of input of fluids is also a critical control on speleothem growth. The development of cave passages is the subject of speleogenesis (see Klimchouk et al., 2000). Caves can form below, at or above the water table, but the water table model is most relevant for tubular sub-horizontal cave passages that have the optimal shape for speleothem accumulation. Davis (1930) independently rediscovered the insight of Grund (1910; in Sweeting, 1981) that base-level fall causes caves to evolve from a stage of active enlargement to a stage of filling with sediment and speleothems. Hence, the southern Chinese caves of the Guilin district (Sweeting, 1995) show progressively older, and better-decorated caves with increasing altitude. There are many examples in more complex environments where cave formation is multistage, and in any given case, there will be a strong control by rock structure, including both primary bedding and secondary features such as faults and joints.

Ventilation is a key parameter since a better-ventilated cave will have lowered pCO_2 and hence enhanced speleothem growth rates. A special case occurs when a cave stream dominates cave ventilation. Here, relative humidity will always be 100% and the cave air pCO_2 will be controlled by that of the stream, which can vary seasonally (Troester and White, 1984). More generally, the climate and geometry of the cave passages control ventilation. Although models of ventilation of caves with a large opening can be derived from classic physics (Cigna, 1967; Wigley, 1967; Wigley and Brown, 1971, 1976), it is hard to derive the appropriate controlling parameters quantitatively even in systems with very simple geometry (Atkinson et al., 1983; De Freitas and Littlejohn, 1987). Radon, derived from the ^{238}U decay-series, is an effective tracer of circulation, and enhanced winter circulation, associated with lower Rn values, is commonly found (Hakl et al., 1997; Dueñas et al., 1999). Although ventilation is influenced by wind direction, the key control is a pressure difference between the cave interior and the external atmosphere in response to synoptic weather systems and the constant cave temperatures. This is most effective in chimney-type circulation where there are both upper and lower entrance points for air (Cigna, 1967; Wigley and Brown, 1976). Caves that descend from a single entrance will be expected to ventilate much more during the winter (Mavlyudov, 1997). Dramatic changes in $p\mathrm{CO}_2$ and $\delta^{13}\mathrm{C}$ of water and cave air in response to seasonal changes in ventilation have been demonstrated by detailed monitoring at the Austrian Obir cave (Spötl et al., 2005). Our unpublished system calculations at the Ernesto cave suggest that day-to-day cave breathing is effective in regulating CO_2 levels at this near-surface site, and the seasonal fall in CO_2 (Frisia et al., 2000; Huang et al., 2001) could relate to a cut-off of epikarst gas supply by the seasonal filling of the aquifer with water. In both these sites, humidity remains very high, but in others, variations in humidity permitting seasonal evaporation may occur.

7.3 Macro- and Micromorphological Characteristics of Flowstones, Stalactites and Stalagmites

7.3.1 Flowstones

Flowstones (Figure 7.1A) are widespread coverings of cave floor and walls that accrete roughly parallel to the host surface and may occur tens or hundreds of metres downstream of the water source (Ford and Williams, 1989). They have in common a tendency to display undulations in surface morphology and the lamina structure is dominantly parallel and continuous, but in detail there are many sub-types reflecting local slopes, water supply and other factors (Hill and Forti, 1997). Laminae arise primarily because of variations in impurity content and they may also fluoresce under ultraviolet excitation (Shopov, 2004). An advantage for palaeoenvironmental study is that flowstones can be cored with relatively little damage to the cave environment, and they can grow over tens of thousands of years. Conversely, there are issues of representativeness of a small core, and dating can be compromised by impurities. Flowstones often form under intermittent or weakly supersaturated thin flows of water and so typically accrete slowly (10–$100\,\mu\mathrm{m\,yr}^{-1}$).

7.3.2 Soda straws and other stalactites

Stalactites (Figures 7.1A, C and 7.5) and stalagmites (Figure 7.1A, B, D and E) are related to dripping water and hence are called dripstones. Short et al. (2005) have applied a developed form of free boundary dynamics theory to predict the gravitationally influenced conical shape of solid stalactites. The variable geometry of zones where water emerges causes the variety of downward-growing forms (Figure 7.1C). For example, downwards-extending draperies or curtains develop along water trickle courses and can be translucent where crystal growth axes are perpendicular to their width (Ford and Williams, 1989). Some workers have cut solid stalactites to derive sections for successful palaeoclimatic work (Bar-Matthews et al., 1999), but others have avoided them because of concerns over lamina geometry (Smart and Francis, 1990), or perhaps for fear of insufficient control on lateral changes in composition on their surface (Hendy, 1971).

Soda straw stalactites are hollow, with a diameter minimised by the surface tension of the water drop at their tip (Figures 7.1C and 7.5A, D; Curl, 1972). They have a wall only 0.1–0.4 mm thick (Figure 7.5B), and so accrete primarily downwards with preferentially orientated crystals (Moore, 1962). Self and Hill (2003), drawing on the works of V.I. Stepanov and V.A. Maltsev, show that soda straws can continue to develop at the tip of solid stalactites (Figure 7.1C), since water is drawn inside the soda straw only close to the tip. Soda straws drip relatively slowly, although the relationship to drip rates on solid stalactites has not been systematically investigated. Whilst soda straws can occur in dense inverted forests and sometimes break off under their own weight (Figure 7.1A), they can extend up to several metres in length.

Soda straw stalactites commonly display a lateral banding (Figure 7.5A, D) with regularity and spacing that suggests an annual origin. First described by Moore (1962), this property is surprisingly unmentioned in later systematic compilations. The banding sometimes reflects an undulating outer surface (Moore, 1962) and hence indicates external accretion, but far more commonly represents growth steps on the inner wall (Figure 7.5B). A particularly neat way of confirming the annual origin of bands was found by Huang et al. (2001), who found equally spaced internal impurity layers, homologous with layers in stalagmites from the same cave and known to be annual in origin (Frisia et al., 2000). Soda straws consist of a few crystals that extend longitudinally downwards (Figure 7.5A). Some have feathery, dendritic terminations (Self and Hill, 2003), which are said to 'recrystallise', or more likely be overgrown to form a more solid wall. More commonly, the wall tapers at the tip over the last year of growth (Figure 7.5B). Crystal surfaces are covered with myriad rhombic terminations of crystallites that make up the macroscopic crystals (Figure 7.5B, C). Although soda straws can be used to derive short, but unusually high-resolution

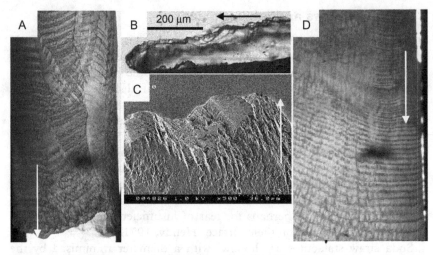

Figure 7.5 Soda straw stalactites from Ernesto cave (A, C, D) and Crag Cave (B) with arrows indicating direction of growth (originally downward). (A) and (D) are transmitted light views of the original straws, and are 5 mm wide; (B) is a transmitted light view of a thick polished section through the stalactite wall and (C) is an scanning electron microscopy image. These stalactites show a thin calcite wall with a smooth exterior (base of image B) and a crystal-rough interior. The wall tapers to around 100 µm wide at its tip (tip of arrow in A; left-hand side of B; top of C). The internal wall displays myriad rhombic crystal terminations (B, C) with larger steps (tip of arrow in B; sub-horizontal lines in A and D) representing an annual rhythm of change of growth rate. The tip of the arrow in (D) represents the onset of faster growth at the end of the Little Ice Age in the 1860s.

proxy environmental records (Fairchild et al., 2001), they are difficult to handle and analyse. A key point is that the banding, which is visible *in situ* in the field when in the range of 0.1 to 1 mm spacing, points to a fundamental seasonality in the saturation state of karst waters, and hence the lateral rate of growth of the straws. This would otherwise take a large monitoring effort to uncover.

Baldini (2001) addressed the neglected issue of the relative volume of stalactites versus their underlying stalagmites, and found a correlation with drip rate that can be explained by the incomplete degassing of fast drips from soda straw stalactites (Moore, 1962). Rocques (1969; in Bögli, 1980) showed that CO_2-degassing of droplets should be largely completed within a few minutes, although the rate of this process diminishes rapidly with time. This can be confirmed by the pH change that occurs following the placement of a drip from a soda straw on a suitable electrode. Conversely, it follows that where water has spent some time percolating on the outside of a solid stalactite, it should already be close to equilibrium with atmospheric CO_2. Studies of pH of cave droplets ought to consider such issues more systematically in order to develop the ideas introduced by Baldini (2001).

7.3.3 Stalagmites: macromorphology

Stalagmite shape has attracted more systematic attention. Franke (1965) developed a theory of the control of stalagmite width by flow rate, with particular reference to cylindrical (candle) types (Figure 7.1A, E). Curl (1973) also gave a theoretical justification for the minimum diameter of cylindrical forms (see also section 7.5.2). However, Gams (1981) showed that the width also increases with drip fall height because of splash effects. These effects tend to cause a change in morphology, with a sunken central splash cup developing (Bögli, 1980). A central depression can also arise by dissolution through periodic undersaturation of the drip water (Frisia, 1996). Franke (1965) mentioned other morphological types including those that have petal-like extensions from a cylindrical core, attributed to variations in growth rate and greater fall height (giving rise to splash effects). Franke (1965) referred also to conical forms in which there is significant growth on the sides of the structures (Figure 7.1D, E): the morphology is attributed by him to a decreasing growth rate and by Gams (1981) to decreasing fall height as the stalagmite grows.

7.3.4 Stalagmites: fabric types

The internal composition of stalagmites may consist of either a single or several different fabrics. The term fabric, or texture, indicates the geometry and the spatial arrangement of single crystals that compose a synchronous layer (Grigor'ev, 1961; Stepanov, 1997; Self and Hill, 2003). Speleothem crystal morphology is related to parent water flow and chemistry, in terms of drip rate and chemistry, capillary or gravitational supply of ions to growth sites, rate of CO_2 outgassing, and the variability of these factors (Genty, 1992; González et al., 1992; Jones and Kahle 1993; Kendall, 1993; Genty and Quinif, 1996). Ultimately, crystal morphology is generated by the dominant atomic growth mechanisms and so Frisia et al. (2000) included crystal microstructure when defining fabrics. A key concept was introduced by Kendall and Broughton (1978) who pointed out that stalagmite crystals are, in fact, composite crystals formed by individual crystallites. Crystallites form separate terminations on the growth surface (e.g. inset in Figure 7.6A and Figure 7.7C), but have a minor space between them that is removed by lateral crystallite coalescence (overgrowth) just behind the growth front, or the space can remain as a fluid inclusion. Where speleothem surfaces are flooded, competitive growth fabrics occur, as in the model of Dickson (1993). In practice, the thin solution film (typically up to 100 μm) on the speleothem surface, limits the size of crystal units that can compete in this way (Broughton, 1983).

Kendall and Broughton (1978) drew on the BCF (i.e. Burton, Cabrera and Frank) crystal growth model (Burton et al., 1951). Consider a crystal made of building 'blocks': its surfaces could be smooth, if only one face of the each block is exposed at the surface, or show steps, when two faces of the blocks are exposed, or kinked, when three faces of the blocks are exposed. Step and kink sites are favourable for growth reactions and the number of such sites exposed to the surface may be increased when the density of crystal defects increases. Crystal defects may originate through, for example, the incorporation of foreign ions, misfits at impinging growth fronts, and condensation of vacancies (Wenk et al., 1983), and commonly indicate either high growth rate or the availability of foreign ions or particles. Similar misfits could be caused by fluctuating flow rates, periodic exposure of the growing crystal faces to the cave air, rapid outgassing, or dissolution. In this light, Frisia et al. (2000) defined several fabric types as described below and shown in Figure 7.6.

1   'Columnar fabric' (Figure 7.6A,B) consists of crystals with a length to width ratio ≥6, with usually straight crystal boundaries, uniform extinction and c-crystallographic axis perpendicular to the substrate. Crystals have a low density of crystal defects, the most common of which are dislocations. Where the length to width ratio is ≤6:1 the fabric is called fibrous (Folk, 1965). Its definition varies from that proposed by Kendall and Broughton (1978) because it includes the low density of defects that explains the ordered stacking of crystallites in each 'composite crystal', all of which have the same orientation with respect to the substrate. The crystallites that compose columnar fabric show flat faces at the solid–fluid interface. They commonly form through the spiral growth mechanism: the advance of monomolecular steps nucleated on screw dislocations. Columnar fabric commonly forms translucent stalagmites, and can show annual laminae, but whatever the cause of the lamination, it does not perturb crystal growth. The fabric commonly forms at relatively low dripwater supersaturation and concentrations of foreign ions, and relatively constant flow conditions. Genty and Quinif (1996) distinguished porous and compact types of 'palisadic' columnar fabric, with a compact palisade fabric being indicative of higher drip rate.

2   'Microcrystalline fabric' (Figure 7.6C), not to be confused with the randomly orientated crystals of carbonate mud-micrite textures, has been discriminated from columnar fabric on the basis of the irregular stacking of crystallites and the high density of crystal defects. It has been observed in annually laminated alpine stalagmites, where it forms milky, opaque and porous layers. The misorientation of some crystallites with respect to their substrate yields composite crystals with serrated to interfingered

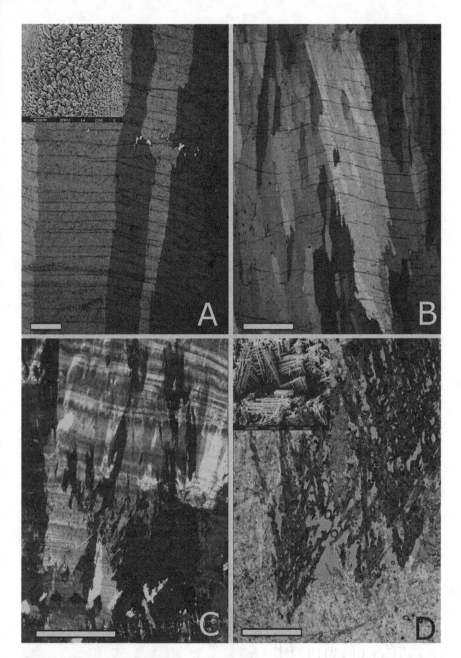

Figure 7.6 Stalagmite calcite fabrics (all thin-section photographs in crossed polars; scale bars = 2 mm). (A) Columnar fabric in CC4 from Crag Cave (Ireland). The straight features within crystals are cleavage planes. The upper box is a scanning electron microscopy micrograph showing the active stalagmite top, where the tips of crystallites composing the larger columnar individuals emerge. (B) Columnar fabric, from AL1, Grotta di Aladino (Italy). Note the irregular crystal boundaries and the cleavage planes. (C) Microcrystalline fabric from ER78, Grotta di Ernesto (Italy). Note the interfingered boundaries, intercrystalline porosity, growth laminae, nucleation sites, large voids (black in the photograph), and the absence of cleavage planes. (D) Dendritic fabric in CC3 from Crag Cave (Ireland). The upper box is a scanning electron microscopy micrograph showing the scaffold-like arrangement of crystallites.

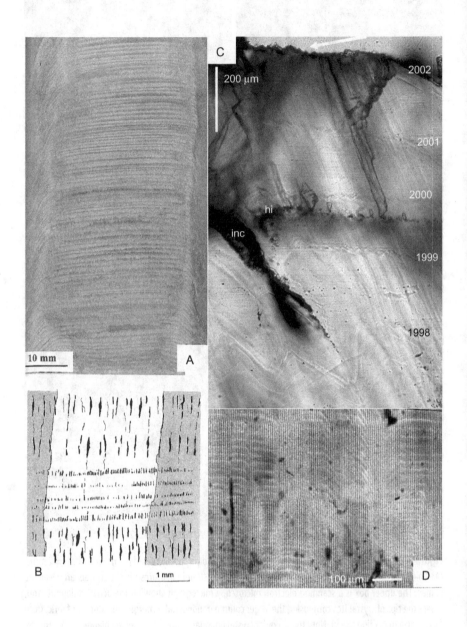

boundaries, and patchy, rather than uniform, extinction. Crystallites are characterised by dislocations, repetitive twinning, and lamellae, indicative of high disturbance of the system related to flow variability and periodic input of growth inhibitors. Supersaturation does not seem to play an important role, because microcrystalline fabric forms at the same supersaturation conditions as columnar fabric.

3 'Dendritic fabric' (Figure 7.6D) displays crystallites in branches similar to a dendrite crystal and its high density of crystal defects: twins, lamellae sub-grain boundaries and dislocations. The branching composite crystals are characterised by many macrosteps and macrokinks, which can become new growth sites, thus explaining the scaffold-like appearance given by repetitive branching. It is a very porous fabric, typical of stalagmite formation under fluctuating discharge, strong outgassing phenomena, and periodic capillary flow.

4 'Fan fabric' is typical of aragonite, and it is formed by acicular (needle-like) or ray (with square termination) crystals radiating from a central nucleus. The single aragonite crystals typically show microtwinning (Frisia et al., 2002).

7.3.5 Stalagmites: annual lamination (couplets and organic-bearing laminae)

Sometimes, there is a repeated stacking of the fabrics (or laminae within fabrics) in an ordered succession, reflecting changes in the composition of the drip waters. Two main types have been shown to occur on an annual scale: (1) seasonal alternations of fabric and/or mineralogy defining

---

Figure 7.7 Stalagmite laminae. (A, B) Stalagmite PNst4, Père-Noël cave (from Genty and Quinif, 1996). (A) Polished hand specimen of stalagmite with flat-lying lamina in core, and steeply-dipping laminae on flanks. Laminae are couplets: alternating white-porous and dark-compact layers. (B) Section of thin-section illustrating that the white-porous laminae represent trains of inclusions; laminae are much thinner, and the inclusions smaller, in the central area; crystals differentiated by shading. (C) Stalagmite sample Obi 84, from Obir Cave, southeast Austria (sample of Professor C. Spötl). The sample was collected in December 2002. Narrow autumnal laminae show zig-zag terminations of crystallites. The crystallite boundaries are aligned NNW–SSE in the image and crystallite terminations at the top surface are arrowed. The 1998 lamina displays high-relief, which appears to presage the incorporation of an air inclusion (labelled 'inc') and this lamina also is a doublet. Within the 2000 lamina there is a growth hiatus (laterally discontinuous outside the field of view), labelled 'hi' corresponding to a line of fluid inclusions and the development of some calcite nuclei with different optical orientations. (D) Part of a photomosaic produced by Dr C. Proctor of UV-fluorescence of sample SU96-7 from Tartair cave, Northwest Scotland (Fairchild et al., 2001; Proctor et al., 2002). Narrow vertical lines are imaging artefacts and black areas are fluid inclusions. The narrow fluorescent laminae reveal parallel growth of calcite crystallites.

couplets, and (2) discrete pulses of impurities during infiltration events (Baker and Genty, 2003).

1 'Couplets' arise where there are seasonal variations in drip water composition and flowrate, leading to different crystal growth fabrics and chemistries. In settings with a pronounced dry season, deposition of two different phases of $CaCO_3$ can occur. In a speleothem from Drotsky's cave in northwest Botswana, precipitation of calcite in the wet season, becoming progressively more Mg-rich over time, is followed by aragonite in the dry season (Railsback et al., 1994). In France and Belgium, Genty and Quinif (1996) and Genty et al. (1997) have recognised the typical occurrence of couplets of white porous calcite and dark compact calcite, resulting from seasonal variations in supersaturation and rate of dripping of supplying water (Figure 7.7A, B), and Yadava et al. (2004) found similar couplets in aragonitic speleothems from India. Calcitic couplets occur in speleothems in the Guadalupe Mountains of New Mexico, where thin laminae represent dry conditions in time periods associated with archaeological evidence of cultural changes in indigenous peoples (Polyak and Asmerom, 2001). Perrette et al. (2005) show that there can be changes in the wavelength of UV fluorescence within presumed annual laminae, reflecting changing proportions of soil-derived humic and fulvic acids.

2 'Impurity pulse laminae' (Figure 7.7C, D) typically have enhanced fluorescence under UV excitation (Baker et al., 1993b; Shopov et al., 1994), reflecting enhanced input from soil. In some cases, the laminae are also clearly visible in thin sections in transmitted light (Fairchild et al., 2001; Frisia et al., 2003), or are primarily recognisable in this way. The existence of such laminae reflects a specific time of year when there is excessive infiltration, but their existence also relies on the hydrological functioning of the aquifer. The filling of the aquifer beyond a critical level may provide the mechanism for flushing of impurities through previously air-filled cavities at a specific stage in the hydrological year. This time in northwest Europe is mid-autumn. However, the annual origin of layering is not universal (Baker and Genty, 2003). For example, Baker et al. (1999b, 2002) describe the occurrence of doublets. Here, in specific years, two infiltration events occur and the second can be associated with snowmelt or an excessively wet period in the winter.

7.4 Chemistry of Speleothems

7.4.1 Dating methods

Whereas interval dating can be carried out by counting annual layers (e.g. Baldini et al., 2002), radiometric techniques are needed to provide the overall time constraints on long-term speleothem deposition. The most widely applicable radiometric technique is $^{230}Th-^{234}U-^{238}U$ disequilibrium dating, which can be used between a few hundred years (limited by determination of ^{230}Th) to around 500 ka (Edwards et al., 1987). Uranium is incorporated into $CaCO_3$ as the uranyl ion UO_2^{2+}, derived from the dominant aqueous species $(UO_2(CO_3)_3^{4-})$, whereas Th is practically insoluble and so will be incorporated into speleothems only with non-carbonate phases. Contamination by Th offers analytical challenges, but these can be overcome with sufficient knowledge of primary $^{230}Th/^{232}Th$ ratios (Richards and Dorale, 2003). Precision is limited by primary U concentration, which is variable over several orders of magnitude. Uranium is enriched in aragonite compared with calcite, due to its favourable coordination environment (Reeder et al., 2000). The rapid technological evolution during the past 20 years from alpha-spectometry to thermal ionisation mass spectrometry (TIMS) (Edwards, et al., 1987) and to multicollector inductively coupled plasma mass spectrometry (MC-ICPMS) dramatically increased the resolution and the precision of the dating system. The sample size decreased from 10–100 g (α-spectrometry) to 10–500 mg (TIMS and MC-ICPMS), and the precision (2σ) in the ^{230}Th analyses ameliorated from 2–10% to 0.1–0.4% (Goldstein and Stirling, 2003). In addition to dating issues, $^{234}U/^{238}U$ ratio variations have also proved useful as palaeohydrological proxies (Ayalon et al., 1999; Frumkin and Stein, 2004).

U-series dating of speleothems has played a significant part in Quaternary science in recent years. A significant geomorphological application is the estimation of rates of denudation of the landscape (Atkinson and Rowe, 1992). The groundwater-fissure deposit at Devil's Hole, Nevada, which grew for 0.5 Myr (Winograd et al., 1992), showed some discrepant dates of inferred glacial terminations compared with Milankovitch theory. Glacial termination II also occurred earlier than predicted from Milankovitch theory in a climatically sensitive Alpine site (Spötl et al., 2002, 2004), although this is not the case in China (Yuan et al., 2004). The ages of the Dansgaard–Oeschger events during the last glacial cycle have been most accurately determined using speleothems (e.g. Spötl and Mangini, 2002; Genty et al., 2003). Finally, data from submerged Bahamian speleothems have been used to extend the calibration of the ^{14}C timescale (Beck et al., 2001).

Other dating techniques include different isotopes from the uranium decay series. In particular, $^{231}Pa-^{235}U$ can be applied within the 0–200 ka time span (Edwards et al., 1997), whereas U–Pb dating can be used for old speleothems with high U content and low Pb (Richards et al., 1998), but applications have been limited to date. For samples older than 500 ka, low-precision ages can be obtained from palaeomagnetic methods, electron-spin resonance (Grün, 1989) and $^{234}U/^{238}U$ disequilibrium (Ludwig et al., 1992). For young samples, ^{14}C dating was used before U-series dating became firmly established, but is limited by uncertainty over the percentage of dead carbon in a given sample incorporated during growth (Genty et al., 2001a).

7.4.2 Stable isotope composition

Stable isotope studies have formed the major type of geochemical investigation of speleothems (Harmon et al., 2004; McDermott, 2004). Such work was stimulated initially by the prospect of palaeotemperature analysis, given that cave interiors usually approximate the mean annual external temperature (variations with depth are discussed by Leutscher and Jeannin, 2004). The oxygen isotopic composition of calcium carbonate ($\delta^{18}O_c$), formed in isotopic equilibrium with a water of fixed composition ($\delta^{18}O_w$), decreases with increasing temperature. However, it is now recognised that changes in the water composition form the main agent for change in $\delta^{18}O_c$ over time, and so one current research focus is the technically challenging extraction of coeval water from fluid inclusions to determine the $\delta^{18}O_w$ composition directly, or via the meteoric water line, from δD (Matthews et al., 2000; Dennis et al., 2001; McGarry et al., 2004).

In most cave systems, equilibrium isotopic fractionation is modified by kinetic effects (Mickler et al., 2004) resulting from factors such as high supersaturation of the water (Kim and O'Neil, 1997), and outgassing and evaporation (González and Lohmann, 1988). Fortunately, cave carbonates often form in quasi-isotopic equilibrium, that is from waters at relatively low supersaturation, constant drip rate, and cave relative humidity of c. 100%. The only calcite fabric normally associated with strong kinetic isotope fractionations is dendritic fabric, in the case of carbon isotopes. Aragonite is slightly heavier in needle crystals compared with fibres (Frisia et al., 2002).

Where only limited evaporation occurs at the surface or in the epikarst, the mean isotopic composition ($\delta^{18}O_w$) of cave drip waters reflects the mean annual isotopic composition of precipitation ($\delta^{18}O_p$, Yonge et al., 1985). In regions with semi-arid or arid climate, evaporative processes lead to heavier $\delta^{18}O_w$ values than precipitation (Bar-Matthews et al., 1996). In principle, different parts of the aquifer could preferentially store isotopically distinct water from different seasons. Although not necessarily significant (McDermott et al., 1999), the site-specific nature of this issue makes modern monitoring studies advisable.

The oxygen isotope composition of atmospheric precipitation ($\delta^{18}O_p$) becomes increasingly negative with decreasing temperature due to progressive removal of vapour in a Rayleigh fractionation process, but changes in water source characteristics render the temperature dependence of $\delta^{18}O_p$ variable and site-dependent. For mid- to high-latitude regions, the $\delta^{18}O_p$ dependence on temperature averages $+0.59 \pm 0.09‰$ °C^{-1} (Dansgaard, 1964; Rozanski et al., 1993). This positive dependence exceeds the calculated calcite-water fractionation at equilibrium (about $-0.24‰$ per°C). As temperature usually has a larger effect on the $\delta^{18}O_p$ than on the calcite-water fractionation, a positive correlation between temperature and $\delta^{18}O_c$

should be expected in many mid- and high-latitude sites. Heavier $\delta^{18}O_c$ has thus been taken to reflect warmer mean annual temperatures (Dorale et al., 1992, 1998; Gascoyne, 1992; McDermott et al., 1999, 2001; Paulsen et al., 2003). The tenability of the approach must, however, be tested for each site; for example the opposite relationship in north Norway was found by Lauritzen and Lundberg (1999).

In many regions, it has been observed that the most intense rainfall events have the lowest $\delta^{18}O_p$. Thus, in rainy periods or seasons the $\delta^{18}O_c$ would reflect this 'amount effect' on $\delta^{18}O_p$. This has been observed in monsoon regions, where the most depleted $\delta^{18}O_c$ values correspond to heavier monsoon rain intensity (Fleitmann et al., 2003). A high correlation between high-resolution (annual) $\delta^{18}O_c$ series from Oman speleothems and the residual ^{14}C ($\Delta^{14}C$), provided well-constrained evidence of solar activity influence on the Indian Ocean monsoon (Neff et al., 2001; Fleitmann et al., 2003). A similar correlation has been demonstrated also in continental humid settings (Niggeman et al., 2003), and can be interpreted as solar forcing effects on European storm tracks (Shindell et al., 2001). Thus, the $\delta^{18}O_c$ of speleothems is a powerful tool to detect solar forcing effects. Caves in Israel have been particularly well calibrated for the influences of rainfall amount and source effects on $\delta^{18}O_c$ (Bar-Matthews et al., 1996, 1999, 2000; Kolodny et al., 2003).

The equilibrium carbon isotope composition of aragonite is around 2‰ greater in aragonite than calcite, but aragonite is typically slighter heavier than expected because of kinetic effects (Frisia et al., 2002). Where calcite later replaces the thermodynamically unstable aragonite, it can preserve evidence of the original heavier $\delta^{13}C$ aragonite composition (Frisia et al., 2002). The carbon isotopic composition of cave carbonates ($\delta^{13}C_c$) depends on the $\delta^{13}C$ composition of the dissolved inorganic carbon (DIC), on growth rate (Turner, 1982), on changes with the gaseous phase and the supersaturation state of the water with respect to calcium carbonate (Richards and Dorale, 2003). As for the case of oxygen, the carbon isotopic composition of the parent water seems to be in many cases the most important factor determining the $\delta^{13}C_c$ of speleothem carbonates. The carbon dissolved in drip waters mainly derives from three sources: atmospheric CO_2, soil CO_2 and dissolution of the karstic host rock. Variations in the isotopic composition of soil pCO_2 have been held to be important (Denniston et al., 2000; Frappier et al., 2002; Genty et al., 2003; Paulsen et al., 2003). For example, the $\delta^{13}C$ composition of soil CO_2 varies according to the photosynthetic pathway of plants. The $\delta^{13}C$ values of respired soil CO_2 of C3 trees and shrubs adapted to a relatively cold and wet climate typically range between −26 and −20‰, and those of C4-type drought-adapted grasses range between −16 and −10‰ (Cerling, 1984). The control on $\delta^{13}C$ by a change in type of photosynthetic pathway is well illustrated by studies in the American mid-west (e.g. Dorale et al., 1998). Under dry conditions, however, the $\delta^{13}C$ of respired soil CO_2 of C3 plants becomes heavier due to restricted stomatal conductance (Paulsen et al., 2003), and under humid conditions it is lighter. Thus, given that all other environmental conditions are the same, the $\delta^{13}C_c$ of cave carbonates should be lighter when the vegetation above the cave mostly consists of C3 plants living under unlimited water availability conditions, and heavier when the vegetation consists of C4 plants, or under arid settings. Genty et al. (2003) also indicate that cool climates inhibit soil CO_2 production and yield heavier $\delta^{13}C_c$.

The $\delta^{13}C_c$ values become heavier also in the case of within-cave phenomena, mostly related to outgassing (Baker et al., 1997b). At equilibrium, $\delta^{13}C$ of DIC is isotopically heavier by around 10‰ compared with gaseous CO_2. Hence, speleothem calcite deposited in ventilated cave passages are likely to be characterised by heavier $\delta^{13}C_c$ than speleothem calcite formed in secluded, non-ventilated passages, and supersaturations are likely to be higher, leading to disequilibrium fabrics. Calcite fabrics developed under disequilibrium conditions, such as the dendritic fabric, show ^{13}C enrichment with respect to the columnar fabric (Frisia et al., 2000). Equilibrium fractionation modelled as a progressive Rayleigh distillation process was able to account well for the variations in $\delta^{13}C$ of drip waters in Soreq cave (Bar-Matthews et al., 1996). Additional kinetic effects due to fast degassing could also occur (Hendy, 1971) and Spötl et al. (2005) showed that kinetic enhancement of carbon isotope fractionation becomes important in strongly ventilated cave passages.

7.4.3 Trace element compositions and their controls

Controls on the water composition were discussed in section 7.2.3, but additional complexities arise by the partitioning of species into $CaCO_3$. In principle, three groups of elements or species can be distinguished.

1   Those that substitute for Ca or CO_3 in the $CaCO_3$ lattice (e.g. Sr, Ba and U in aragonite and calcite, and Mg and SO_4 in calcite). Here a partition coefficient (K) may be defined: $Tr/Cr_{CaCO3} = K \times Tr/Cr_{solution}$, where Tr is the trace element or species and Cr is the carrier Ca or CO_3. Coefficient K varies with mineral species and may vary with temperature, growth kinetics or other factors (Morse and Bender, 1990). It is also dependent on the crystallographic form, which may account for the complex geochemical zoning in aragonites recorded by Finch et al. (2001, 2003). For calcite, partition coefficients apply to rhombohedral crystals (Huang and Fairchild, 2001), and the values represent a mean of the two forms of incorporation at antipathetic growth steps as observed in experimental products (Paquette and Reeder, 1995).

2   Those that are incorporated interstitially on a molecular scale in $CaCO_3$ (e.g. PO_4, Na and F in calcite). Typically these components are preferentially incorporated at growth defects and so tend to be more abundant where crystals have more defects, related to either rate or style of growth. Hydrogen, as measured by ion probe (Fairchild et al., 2001), probably reflects either molecular water or fluid nano-inclusions.

3   Those that are present in fluid or solid inclusions within the $CaCO_3$. In cases where solid inclusions are brought in by high flows, the distribution of such elements could be hydrologically significant.

Finch et al. (2003) and Treble et al. (2003) have been able to derive empirical transfer functions for rainfall in South African and west Australian examples, by comparison of data from selected trace elements (mostly in the first group) with instrumental climatic data. McMillan et al. (2005) showed that annual trace element trends in stalagmites from Clamouse cave follow those expected from seasonal dryness and that they are superimposed on a long period of relative drought 1100–1200 yr BP. Hellstrom and McCulloch (2000) emphasised the utility of combined trace element and isotopic data in palaeoclimatic studies.

Annual variations in trace elements are now known to be a normal feature of caves in shallow karst environments (Roberts et al., 1998; Fairchild et al., 2001; Finch et al., 2001; Huang et al., 2001; Fairchild, 2002) and subannual variations can also be observed where speleothems grow sufficiently quickly. These variations can be related mainly either to changes in the ion ratios in solution in relation to hydrological effects, or to seasonal changes in supersaturation of karst waters, that is the same two controls that are responsible for petrographic laminae. In section 7.6.2, such variations are related to potential controlling factors.

7.4.4 Organic geochemistry

Total organic carbon measurements have also been used (Batiot et al., 2003) to trace karstic hydrology, but the UV-fluorescing fraction (e.g. Baker and Genty, 1999; Baker et al., 1999a) appears to be selectively incorporated in speleothems (McGarry and Baker, 2000). Relatively little molecular organic geochemistry has been carried out. Lauritzen et al. (1994) made a pioneering study of long-term records of amino acid racemisation and there is currently interest in assessing the environmental associations of different types of organic molecule. For example, Xie et al. (2003) demonstrated changes in the relative incorporation of low and high molecular weight lipids that correlated with trends in $\delta^{13}C$. Given also increasing evidence for the involvement of microbial films in $CaCO_3$ precipitation on speleothems (e.g. Cacchio et al. (2004) who show that different bacterial strains generate precipitates of differing $\delta^{13}C$ composition), this will be a rich field for the future.

7.5 Mechanisms of formation

7.5.1 The CO_2-degassing model and its consequence for speleothem growth rate

In general, more highly supersaturated solutions will tend to precipitate $CaCO_3$ more quickly. Supersaturation arises by a set of reactions, which involve a loss of volatile acid (carbon dioxide) from the solution:

$$H_2CO_3 \rightarrow CO_2 \text{ (aqueous)} \rightarrow CO_2 \text{ (gas)} \qquad (1)$$

$$HCO_3^- + H^+ \rightarrow H_2CO_3 \qquad (2)$$

Because the last reaction reduces H^+ proportionally much more than HCO_3^-, it stimulates the following reaction:

$$HCO_3^- \rightarrow H^+ + CO_3^{2-} \qquad (3)$$

The increase in carbonate (CO_3^{2-}) ions raises the saturation state of the solution as defined by the saturation index for calcite (SI_{cc}), where: $SI_{cc} = \log (IAP/K_s)$ where IAP (ionic activity product) = $(Ca^{2+})(CO_3^{2-})$ in a given solution and K_s is the solubility product. Where the IAP and K_s are equal, SI_{cc} is zero and the solution is at equilibrium with calcite; positive values of SI_{cc} correspond to supersaturation. Growth may be inhibited if $SI_{cc} < 0.05$–0.15, or if there are species in solution which block growth sites. Values of SI_{cc} rarely exceed 0.5–0.7 in caves, but commonly approach 1 in surface karst streams where degassing is rapid (Dandurand et al., 1982). Extremely high growth rates also occur where speleothem growth is sourced from hyperalkaline waters flowing through concrete structures or lime kilns (Moore, 1962), but here the controlling reactions are different (Baker et al., 1999b).

The most fundamental advances in our understanding of the physical controls on growth rate of calcite have come from the work of Dreybrodt and co-workers, as summarised in Dreybrodt (1988, 1999), Baker et al. (1998) and Kaufmann (2003). The rate of growth (R) thus depends on both the cationic and anionic behaviour in the system, but in many cases can be simplified to be a function of Ca concentration, as follows:

$$R = \alpha(c - c_{eq})$$

where α is a rate constant (units of cm s^{-1}), c is the Ca concentration (mol L^{-1}) in the solution, and c_{eq} is the Ca concentration at equilibrium with calcite. Kaufmann (2003), following Dreybrodt (1999), substitutes c_{eff} for c_{eq} where c_{eff} is the effective equilibrium value allowing for kinetic inhibition and equates to a SI_{cc} value of around 0.05. Values of α depend on temperature and the pCO_2 of the cave air.

The analysis considers the simultaneous operation of three processes, each of which could limit growth rate under particular circumstances.

1 Reactions at the surface of the calcite crystal. These are given by the PWP model of Plummer et al. (1978), which applies also to calcite dissolution. These reactions are rate-limiting where the solution is only weakly supersaturated.

2 As a result of $CaCO_3$ precipitation, CO_3^{2-} ions are consumed which drives reactions (3), (2) and (1) as above and results in CO_2-degassing, in addition to that which will have been occurring in any case if the solution had not had time to reach equilibrium with the cave atmosphere (solution pCO_2 > cave atmospheric pCO_2). The process of CO_2-formation becomes rate-limiting if the film of water between the calcite surface and the cave atmosphere is very thin, and diminishes up to film thicknesses of 200 μm. Most speleothems have film thicknesses within this range.

3 Mass transport of species away from the growing calcite surface. This becomes important at fast growth rates or when there is a thick stagnant film of water (>400 μm), as in gour pools, or small depressions on flowstone or stalagmite surfaces.

Baker and Smart (1995), Baker et al. (1998) and Genty et al. (2001b) have carried out the most extensive field-tests of the Dreybrodt growth model in cave environments and have found generally good agreement within a factor of two for stalagmites, with somewhat more variability for flowstones. Genty et al. (2001b) have provided the first extensive analysis of seasonal variations in growth rate in relation to Ca concentrations, which in most cases bear a relationship to depositional temperature.

In Figure 7.8C, we combine the algorithms expressing the Dreybrodt model given in Baker et al. (1998) and Kaufmann (2003) with the field data on the variation of Ca^{2+} content of drip waters with cave temperature given by Genty et al. (2001b), to express the maximal growth rates expected with temperature at different sites. The growth rates could be even faster with film thickness of 200–400 μm, but such thicknesses are not typical. Growth rates around the maximal level are found, for example, at the Ernesto cave today (Frisia et al., 2003) at 6.5°C. However, there are several reasons why non-porous speleothems fail to grow this fast.

1 The film thickness may be thinner. At 50 μm thickness and 10°C using the conditions of Figure 7.8C, the growth rate reduces by about one-third.

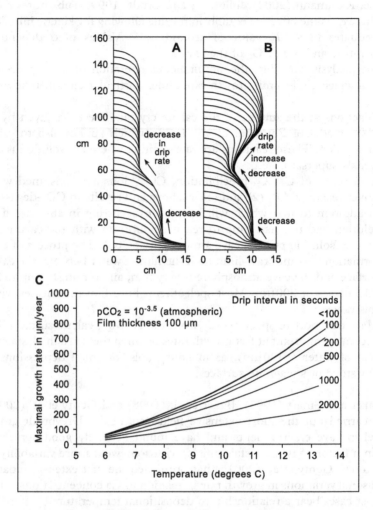

Figure 7.8 (A,B) Examples of modelled speleothem macromorphologies illustrating the concept of equilibrium shapes developed by constant rates of vertical accretion, but modified by changing drip rate, where the diameter is a positive function of drip rate (Dreybrodt, 1999, figure 3). The equilibrium radius in centimetres is given by $(V/\pi\alpha\,\Delta T)^{0.5}$, where ΔT is the time interval between drips, V is the drip volume and α is the growth rate constant as introduced in section 7.5.1. (C) Modelled maximal growth rates of speleothems under a stagnant fluid layer by combining the predictions of the Dreybrodt growth model with the field data on Ca concentration of drip waters in Genty et al. (2001). Factors such as prior calcite precipitation are known to have reduced Ca (Fairchild et al., 2000, 2006b) at certain of the sites of Genty et al. (2001b), so their linear regression line is modified to rise from 50 mg L^{-1} at 6°C to 160 mg L^{-1} at 14°C. Reasons for growth rates to lie below these maximal values are discussed in the text.

2 The initial Ca concentration of cave water might be limited by lower-than-expected soil pCO_2 because of relative aridity (Genty et al., 2001b) or where the solutions follow a sequential evolution (Drake, 1983) and reach saturation at lower pCO_2 values than the soil.

3 The cave may be insufficiently ventilated. If pCO_2 is ten times atmospheric, the growth rate diminishes by around 25% (or rather more at lowered Ca concentrations such as if factors 2 or 4 apply – see figure 3 in Baker et al., 1998). When more seasonal cave pCO_2 data (e.g. Huang et al., 2001; Spötl et al., 2005) become available, the effects of seasonally enhanced ventilation can be added to the approach of Genty et al. (2001b).

4 Prior calcite precipitation may already have occurred. This must be a major factor for drips falling from active solid stalactites and it has tended to be underestimated in the past.

5 The drip rate is slow so that growth is limited by the supply of fresh, supersaturated solution (see lower lines on Figure 7.8C). An example of this effect is the correlation between thickness of calcitic laminae and rainfall in Drotsky's Cave, Botswana (Railsback et al., 1994), where the calcite layers alternate with aragonite layers formed in the dry season. Generally, at very slow drip rates, it becomes more likely that evaporative effects also come into play. Even in the more humid climate of Belgium, Genty and Quinif (1996) found a good correlation between water excess and thickness of annual laminae.

6 The solution composition varies significantly from pure $CaCO_3$ (e.g. pure dolomite aquifer). A contrary case arises when sulphuric acid from pyrite oxidation enhances limestone dissolution (e.g. Atkinson et al., 1983).

Given that the basic geometric aspects of the cave remain relatively constant for long periods, the above data suggest that temperature changes may influence decadal to millennial variations in growth rate of stalagmites. Frisia et al. (2003) have shown that the growth rate of speleothems in the Ernesto cave is particularly sensitive to temperature. Proctor et al. (2000, 2002) describe a different case from the peat-covered shallow karst system at the Tartair cave in Scotland. Here, a quantitative model was produced by comparison of time series of annual growth rates with instrumental climatic data to support the concept that soil CO_2 production is more strongly dependent on rainfall than temperature at this site. Relatively dry conditions promote higher CO_2 production in what would otherwise be a water-logged soil, and enhance $CaCO_3$ dissolution and speleothem growth.

7.5.2 Stalagmite morphology

Curl (1973) extended the insights of Franke (1965) and provided a theoretical analysis of the expected minimum diameter of a stalagmite (around 3 cm), which arises from spreading of a water droplet over a flat surface, in relation to typical volumes of the water drop and thickness of the water film. Dreybrodt (1988, 1999) has further extended the analysis, and the modelled effects of changing water supply are illustrated in Figure 7.8A, B. Genty and Quinif (1996) have interpreted a Belgian stalagmite in this way using laminae thickness as an independent proxy for drip rate.

Kaufmann (2003) has graphically illustrated the effects of pCO_2, temperature and drip rate on speleothem morphology in relation to known long-term climate change. This is an interesting exercise, but too many factors could change over the long intervals being modelled for the results to be considered realistic. A number of stalagmites also have more conical shapes in which significant growth occurs on their flanks.

7.6 Summary of Use of Speleothems in Palaeoclimate Determination

At a coarse level of time resolution, episodes of speleothem growth record relatively wet periods in semi-arid climates, or relatively warm periods in cool temperate climates. The precise dating of speleothems allows them to be used to test Milankovitch theory and to calibrate the timing of palaeoclimate fluctuations. Counting of physical or chemical laminae allows the duration of climatically significant intervals to be determined.

Used qualitatively, a number of parameters exhibit changes that can be interpreted in palaeoclimatic terms, given an understanding of the directional sense of the calibration. Examples are laminae thickness for temperature or rainfall, $\delta^{18}O$ for temperature or rainfall, $\delta^{13}C$ for type or amount of vegetation or degree of cave air circulation, Mg and/or Sr for aridity, $^{87}Sr/^{86}Sr$ for rainfall or changes in aeolian activity or source. Quantitative transfer functions have been derived for the above parameters in specific circumstances, but their degree of applicability over time needs to be viewed cautiously. Finally, it should be remembered that the ground surface

→

Figure 7.9 (A) Cross-section through the Alpine Ernesto cave, Trentino province, Italy. (B) Interpretation of Mg and $\delta^{13}C$ records through stalagmite ER76 (McDermott et al., 1999; Frisia et al., 2003) from the cave. The cave is at 1165 m altitude and has a temperature of 6.5°C. Holocene palaeoclimatic reconstructions have been carried out on speleothems using isotopes and fabrics (McDermott et al., 1999), isotopes and trace elements (Frisia et al., 2001) and lamina thickness (Frisia et al., 2003). In these diagrams we highlight other potential controls on long-term evolution. The cave is situated under a steep forested hillside and variations in the degree of forest cover may well have controlled the $\delta^{13}C$ composition of the speleothems, with the increase in the last millennium probably related to human-induced deforestation (Frisia et al., 2001). The overlying very thin debris fan feeds from a gully in the cliff of Rosso Ammonitico limestones

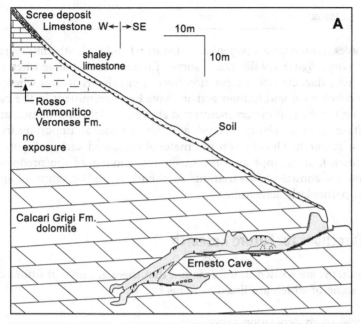

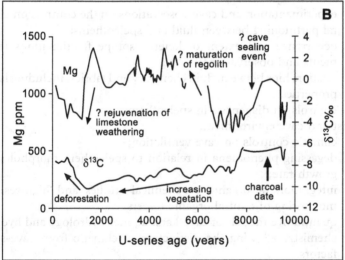

Figure 7.9 *Continued*

that overlie the Lower Jurassic dolomites in which the cave is set. Movement of limestone down the fan supplied abundant calcite in the soil overlying the cave in which cave waters derive their characteristic chemistry and which is captured by the speleothems. Arguably, the long-term evolution of the Mg/Ca content of drip waters and hence the Mg content of speleothems may reflect the geomorphological evolution of the regolith. Intermittent debris movement is required at least until 9 ka, which is the date of an episode of cave occupation, dated by the charcoal in a hearth in the cave. Only micromammal remains are found in the cave after this time and it can be inferred that it was closed by debris movement and only reopened after its contemporary discovery in 1983. The major shifts in Mg content could be interpreted as reflecting changing availability for dissolution of limestone and dolomite fragments in the soil zone, although climatic changes may also be influential.

over caves is often anthropogenically disturbed and speleothems can record this activity. Figure 7.9 illustrates some of the issues that arise when trying to extract palaeoclimate information from speleothems at sites where both geomorphological and human activity have been prominent in the past.

Several of the parameters mentioned above can yield information at high resolution, even to subannual level, and this provides an important frontier area for research. Drawing on the material discussed earlier, Figure 7.10 provides a first attempt at some qualitative models of the predominant controls on annual-scale variations in chemistry and/or fabric of speleothems, particularly stalagmites.

7.7 Directions for Future Research

Speleothems are so rich in information that there are myriad future directions, some of which are listed as items below.

1   Speleothem deposition processes:
    (a)   experimentation and cave observations on the controls on elemental partitioning between fluid and speleothems;
    (b)   occurrence and extent of kinetic isotope fractionations (experiments and observations);
    (c)   relationships between fluid chemistry and fabrics, including laminae properties;
    (d)   controls on diagenesis in speleothems.
2   Karst and cave environment:
    (a)   climatic controls on cave ventilation;
    (b)   degassing phenomena in relation to speleothem morphology and growth rate;
    (c)   integration of organic geochemical studies and high-resolution monitoring of natural organic tracers;
    (d)   quantitative models of cave-karst aquifer hydrology and hydrogeochemistry allowing the distinction of climatic from cave-specific factors.
3   Palaeoclimate:
    (a)   duplication of time series;
    (b)   multiproxy data sets – data-gathering is some way behind advances in technology at present and new proxies are being discovered every year;
    (c)   integration with other palaeoclimate recorders (e.g. lakes, peat bogs, ice cores, marine records);
    (d)   interchange of information with climatic modellers;
    (e)   testing of forcing models of climate change (orbital, solar, ocean versus atmospheric circulation etc.).

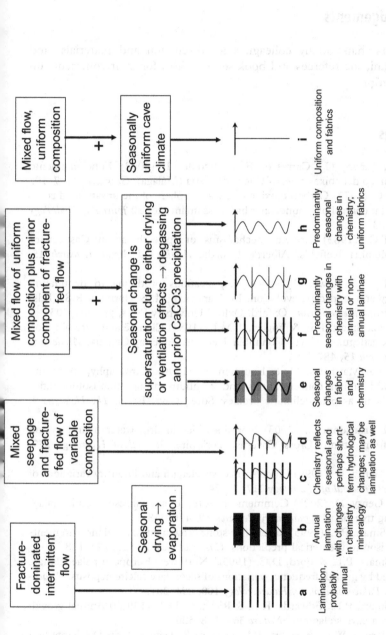

Figure 7.10 Diagrammatic relationships between the flow-related and cave-related geomorphological factors and the high-resolution properties of speleothems (mainly stalagmites). Possible examples are (a) Ballynamintra (Fairchild et al., 2001); (b) Drotsky's Cave, Botswana, Railsback et al. (1994); (c) or (d) Villars (inferred from water chemistry of Baker et al., 2000) (d) Crag cave stalactite (Fairchild et al., 2001), Clamouse (McDermott et al., 1999; Frisia et al., 2002; McMillan et al., 2005); (e) Père-Noël and other Belgian caves, and Villars, France (Genty and Quinif, 1996; Genty et al., 1997), (f) some Obir stalagmites (Spötl et al., 2005 and unpublished); (g) Ernesto (Fairchild et al., 2001; Huang et al., 2001; Obir (Fig. 7.7c); (h) Crag stalagmite 8.2 ka event (Baldini et al., 2002); (i) not recorded for certain: may only occur in some deep caves.

Acknowledgements

The authors thank many colleagues for discussion and materials, and James Baldini, the referees and book series editor, for their comments on the manuscript.

References

Andreo, B., Liñan, C., Carrasco, F. & Vadillo, I. (2002) Funcionamiento hidrodinámico del epikarst de las Cueva de Nerja (Málaga). *Geogaceta* **31**, 7–10.

Atkinson, T.C. (1977) Carbon dioxide in the atmosphere of the unsaturated zone: an important control of groundwater hardness in limestones. *Journal of Hydrology* **35**, 111–123.

Atkinson, T.C. (1983) Growth mechanisms of speleothems in Castleguard Cave, Columbia Icefields, Alberta, Canada. *Arctic and Alpine Research* **15**, 523–536.

Atkinson, T.C. & Rowe, P.J. (1992) Applications of dating to denudation chronology and landscape evolution. In: Ivanovich, M. & Harmon, R.S. (Eds) *Uranium-series disequilibirum*, Oxford: Oxford University Press, pp. 669–703.

Atkinson, T.C., Smart, P.L. & Wigley, T.M.L. (1983) Climate and natural radon levels in Castleguard Cave, Columbia Icefields, Alberta, Canada. *Arctic and Alpine Research* **15**, 487–502.

Ayalon, A., Bar-Matthews, M. & Kaufman, A. (1999) Petrography, strontium, barium and Uranium concentrations, and strontium and uranium isotope ratios in speleothems as palaeoclimatic proxies: Soreq Cave, Israel. *The Holocene* **9**, 715–722.

Baker, A. & Brunsdon, C. (2003) Non-linearities in drip water hydrology: an example from Stump Cross Caverns, Yorkshire. *Journal of Hydrology* **277**, 151–163.

Baker, A. & Genty, D. (1999) Fluorescence wavelength and intensity variations of cave waters. *Journal of Hydrology* **217**, 19–34.

Baker, A. & Genty, D. (2003) Comment on 'A test of annual resolution in stalagmites using tree rings'. *Quaternary Research* **59**, 476–478.

Baker, A. & Smart, P.L. (1995) Recent flowstone growth rates: Field measurements in comparison to theoretical predictions. *Chemical Geology* **122**, 121–128.

Baker, A., Smart, P.L. & Ford, D.C. (1993a) Northwest European palaeoclimate as indicated by growth frequency variations of secondary calcite deposits. *Palaeogeography, Palaeoclimatology, Palaeoecology* **100**, 291–301.

Baker, A., Smart, P.L., Edwards, R.L. & Richards, D.A. (1993b) Annual growth banding in a cave stalagmite. *Nature* **364**, 518–520.

Baker, A., Barnes, W.L. & Smart, P.L. (1997a) Stalagmite Drip Discharge and Organic Matter Fluxes in Lower Cave, Bristol. *Hydrological Processes* **11**, 1541–1555.

Baker, A., Ito, E., Smart, P.L. & McEwan, R.F. (1997b) Elevated and variable values of ^{13}C in speleothems in a British cave system. *Chemical Geology* **136**, 263–270.

Baker, A., Genty, D., Dreybrodt, W., Barnes, W.L., Mockler, N.J. & Grapes, J. (1998) Testing theoretically predicted stalagmite growth rate with Recent annually laminated samples: Implications for past stalagmite deposition. *Geochimica et Cosmochimica Acta* **62**, 393–404.

Baker, A., Mockler, N.J. & Barnes, W.L. (1999a) Fluorescence intensity variations of speleothem-forming groundwaters: Implications for paleoclimate reconstruction. *Water Resources Research* **35**, 407–413.

Baker, A., Proctor, C.J. & Barnes, W.L. (1999b) Variations in stalagmite luminescence laminae structure at Poole's Cavern, England, AD 1910–1996: calibration of a palaeoprecipitation proxy. *The Holocene* **9**, 683–688.

Baker, A., Genty, D. & Fairchild, I.J. (2000) Hydrological characterisation of stalagmite drip waters at Grotte de Villars, Dordogne, by the analysis of inorganic species and luminescent organic matter. *Hydrology and Earth System Sciences* **4**, 439–449.

Baker, A., Proctor, C. & Barnes, W.L. (2002) Stalagmite lamina doublets: a 1000 year proxy record of severe winters in northwest Scotland. *International Journal of Climatology* **22**, 1339–1345.

Baldini, J. (2001) Morphological and dimensional linkage between recently deposited speleothems and drip water from Browns Folly Mine, Wiltshire, England. *Journal of Cave and Karst Studies* **63**, 83–90.

Baldini, J.U.L., McDermott, F. & Fairchild, I.J. (2002) Structure of the '8,200 year' cold event revealed by a high-resolution speleothem record. *Science* **296**, 2203–2206.

Banner, J.L., Musgrove, M., Asmerom, Y., Edwards, R.L. & Hoff, J.A. (1996) High-resolution temporal record of Holocene ground-water chemistry: Tracing links between climate and hydrology. *Geology* **24**, 1049–1053.

Bar-Matthews, M., Ayalon, A., Matthews, A., Sass, E. & Halicz, L. (1996) Carbon and oxygen isotope study of the active water-carbonate system in a karstic Mediterranean cave: Implications for palaeoclimate research in semiarid regions. *Geochemica et Cosmochimica Acta* **60**, 337–347.

Bar-Matthews, M., Ayalon, A., Kaufman, A. & Wasserburg, G.J. (1999) The Eastern Mediterranean palaeoclimate as a reflection of regional events: Soreq cave, Israel. *Earth and Planetary Science Letters* **166**, 85–95.

Bar-Matthews, M., Ayalon, A. & Kaufman, A. (2000) Timing and hydrological conditions of Sapropel events in the Eastern Mediterranean, as evident from speleothems, Soreq Cave, Israel. *Chemical Geology* **169**, 145–156.

Bar-Matthews, M., Ayalon, A., Gilmour, M., Matthews, A. & Hawkesworth, C.J. (2003) Sea-land oxygen isotopic relationships from planktonic foraminifera and speleothems in the Eastern Mediterranean region and their implication for paleorainfall during interglacial intervals. *Geochimica et Cosmochimica Acta* **67**, 3181–3199.

Batiot, C., Liñan, C., Andreo, B., Emblach, C., Carrasco, F. & Blavoux, B. (2003) Use of Total Organic Carbon (TOC) as tracer of diffuse infiltration in a dolomitic karstic system: the Nerja Cave (Andalusia, southern Spain). *Geophysical Research Letters* **30**(22) 2179, doi:10.1029/2003GL018546.

Beck, J.W., Richards, D.A., Edwards, L.A., Silverman, B.W., Smart, P.L., Donahue, D.J., Hererra-Osterheld, S., Burr, G.S., Calsoyas, L., Jull, A.J.T. & Biddulph,

D. (2001) Extremely large variations of atmospheric ^{14}C concentration during the last glacial period. *Science* **292**, 2453–2458.

Beven, K. & Germann, P. (1982) Macropores and water flow in soils. *Water Resources Research* **18**, 1311–1325.

Bögli, A. (1960) Lakosung unde Karrenbilding. *Zeitschrift fur Geomorphologie, Suppl.* **2**, 4–21 (translated by M. Fargher in Sweeting, M.M. (Ed.) 1981 *Karst geomorphology*, Stroudsbrug, PA: Hutchinson Ross, pp. 64–89).

Bögli, A. (1980) *Karst Hydrology and Physical Speleology*. Berlin: Springer-Verlag.

Bottrell, S.H. & Atkinson, T.C. (1992) Tracer study of flow and storage in the unsaturated zone of a karstic limestone aquifer. In: Hötzl, H. & Werner, A. (Eds) *Tracer Hydrology*. Rotterdam: Balkema, pp. 207–211.

Brook, G.A., Folkoff, M.E. & Box, E.O. (1983) A world model of soil carbon dioxide. *Earth Surface Processes and Landforms* **8**, 79–88.

Broughton, P.L. (1983) Environmental implications of competitive growth fabrics in stalactitic carbonate. *International Journal of Speleology* **13**, 31–41.

Burns, S.J., Fleitmann, D., Matter, A., Neff, U. & Mangini, A. (2001) Speleothem evidence from Oman for continental pluvial events during interglacial periods. *Geology* **29**, 623–626.

Burton, W.K., Cabrera, N. & Frank, F.C. (1951) The growth of crystals and the equilibrium structure of their surfaces. *Philosophical Transactions of the Royal Society of London* **A243**, 299–358.

Cacchio, P., Contento, R., Ercole, C., Cappuchio, G., Martinez, M.P. & Lepidi, A. (2004) Involvement of microorganisms in the formation of carbonate speleothems in the Cervo Cave (L'Aquila-Italy). *Geomicrobiology Journal* **21**, 497–509.

Carrasco, F., Vadillo, I., Liñan, C., Andreo, B. & Durán, J.J. (2002) Control of environmental parameters for management and conservation of Nerja Cave (Malaga, Spain). *Acta Carsologica* **31**, 105–122.

Cerling, T.E. (1984) The stable isotopic composition of soil carbonate and its relationship to climate. *Earth and Planetary Science Letters* **71**, 229–240.

Cigna, A.A. (1967) An analytical study of air circulation in caves. *International Journal of Speleology* **1–2**, 41–54.

Corbel, J. (1952) A comparison between the karst of the Mediterranean region and of north western Europe. *Transactions of the Cave Research Group of Great Britain* **2**, 3–25.

Curl, R.L. (1972) Minimum diameter stalactites. *Bulletin of the National Speleological Society* **34**, 129–136.

Curl, R.L. (1973) Minimum diameter stalagmites. *Bulletin of the National Speleological Society* **35**, 1–9.

Dandurand, J.L., Gout, R., Hoefs, J., Menschel, G., Schott, J. & Usdowski, E. (1982) Kinetically controlled variations of major components and carbon and oxygen isotopes in a calcite-precipitating spring. *Chemical Geology* **36**, 299–315.

Dansgaard, W. (1964) Stable isotopes in precipitation. *Tellus* **16**, 436–468.

Davis, W.M. (1930) Origin of limestone caverns. *Geological Society of America Bulletin* **41**, 475–628.

Davis, J., Amato, P. & Kiefer, R. (2001) Soil carbon dioxide in a summer-dry subalpine karst, Marble Moutntains, California, USA. *Zeitschrift für Geomorphologie* **45**, 385–400.

De Freitas, C.R. & Littlejohn, R.N. (1987) Cave climate: assessment of heat and moisture exchange. *Journal of Climatology* **7**, 553–569.

Dennis, P.F., Rowe, P.J. & Atkinson, T.C. (2001) The recovery and isotopic measurement of water from fluid inclusions in speleothems. *Geochimica et Cosmochimica Acta* **65**, 871–884.

Denniston, R.F., González, L.A., Asmerom, Y., Reagan, M.K. & Recelli-Snyder, H. (2000) Speleothem carbon isotopic records of Holocene environments in the Ozark Highlands, USA. *Quaternary International* **67**, 61–67.

Dickson, J.A.D. (1993) Crystal-growth diagrams as an aid to interpreting the fabrics of calcite aggregates. *Journal of Sedimentary Petrology* **63**, 1–17.

Dorale, J.A., González, L.A., Reagan, M.K., Pickett, D.A., Murrell, M.T. & Baker, R.G. (1992) A high-resolution record of Holocene climate change in speleothem calcite from Cold Water Cave, northeast Iowa. *Science* **258**, 1626–1630.

Dorale, J.A., Edwards, R.L., Ito, E. & González, L.A. (1998) Climate and vegetation history of the mid-continent from 75 to 25 ka: a speleothem record from Crevice Cave, Missouri, USA. *Science* **282**, 1871–1874.

Drake, J.J. (1983) The effects of geomorphology and seasonality on the chemistry of carbonate groundwater. *Journal of Hydrology* **61**, 223–236.

Drake, J.J. & Wigley, T.M.L. (1975) The effect of climate on the chemistry of carbonate groundwater. *Water Resources Research* **11**, 958–962.

Dreybrodt, W. (1988) *Processes in Karst Systems*. Springer-Verlag, Berlin.

Dreybrodt, W. (1999) Chemical kinetics, speleothem growth and climate. *Boreas* **28**, 347–356.

Dueñas, C., Frenández, M.C., Cañete, S., Carretero, J. & Liger, E. (1999) ^{222}Rn concentrations, natural flow rate and the radiation exposure levels in the Nerja Cave. *Atmospheric Environment* **33**, 501–510.

Edwards, R.L., Chen, J.H. & Wasserburg G.J. (1987) ^{238}U- ^{234}U- ^{230}Th- ^{232}Th systematics and the precise measurements of time over the past 500,000 years. *Earth and Planetary Science Letters* **81**, 175–192.

Edwards, R.L., Cheng, H., Murrell, M.T. & Goldstein, S.J. (1997) Protactinium-231 dating of carbonates by thermal ionization mass spectrometry: Implications for Quaternary climate change. *Science* **276**, 782–786.

Fairchild, I.J. (2002) High-resolution speleothem trace element records: potential as climate proxies. In: Carrasco, F., Durán, J.J. & Andreo, B. (Eds) *Karst and Environment*. Malaga (Spain): Fundación Cueva de Nerja, pp. 377–380.

Fairchild, I.J., Borsato, A., Tooth, A.F., Frisia, S., Hawkesworth, C.J., Huang, Y., McDermott, F. & Spiro, B. (2000) Controls on trace element (Sr-Mg) compositions of carbonate cave waters: implications for speleothem climatic records. *Chemical Geology* **166**, 255–269.

Fairchild, I.J., Baker, A., Borsato, A., Frisia, S., Hinton, R.W., McDermott, F. & Tooth, A.F. (2001) High-resolution, multiple-trace-element variation in speleothems. *Journal of the Geological Society, London* **158**, 831–841.

Fairchild, I.J., Smith, C.L., Baker, A., Fuller, L., Spötl, C., Mattey, D., McDermott, F. & E.I.M.F. (2006a) Modification and preservation of environmental signals in speleothems. *Earth-Science Reviews* **75**, 105–153.

Fairchild, I.J., Tuckwell, G.W., Baker, A. & Tooth, A.F. (2006b) Modelling of dripwater hydrology and hydrogeochemistry in a weakly karstified aquifer (Bath, UK): implications for climate change studies. *Journal of Hydrology* **321**, 213–231.

Farrant, A.R., Smart, P.L., Whitaker, F.F. & Tarling, D.H. (1995) Long-term Quaternary uplift rates inferred from limestone caves in Sarawak, Malaysia. *Geology* 23, 357–360.

Finch, A.A., Shaw, P.A., Weedon, G.P. & Holmgren, K. (2001) Trace element variation in speleothem aragonite: potential for palaeoenvironmental reconstruction. *Earth and Planetary Science Letters* 186, 255–267.

Finch, A.A., Shaw, P.A., Holmgren, K. & Lee-Thorp, J. (2003) Corroborated rainfall records from aragonitic stalagmites. *Earth and Planetary Science Letters* 215, 265–273.

Fleitmann, D., Burns, S.J. & Mudelsee, M. (2003) Holocene forcing of the Indian monsoon recorded in a stalagmite from Southern Oman. *Science* 300, 1737–1739.

Folk, R.L. (1965) Some aspects of recrystallization in ancient limestones. In: Pray, L.C. & Murray, R.C. (Eds) *Dolomitization and Limestone Diagenesis*: Special Publication 13. Tulsa, OK: Society of Economic Paleontologists and Mineralogists, pp. 14–48.

Ford, D.C. & Williams, P.W. (1989) *Karst Geomorphology and Hydrology*. London: Unwin Hyman.

Franke, H.W. (1965) The theory behind stalagmite shapes. *Studies in Speleology* 1, 89–95.

Frappier, A., Sahagian, D., González, L.A. & Carpenter, S.J. (2002) El Nino events recorded by stalagmite carbon isotopes. *Science* 298, 565.

Frisia, S. (1996) Petrographic evidences of diagenesis in speleothems: some examples. *Speleochronos* 7, 21–30.

Frisia, S., Borsato, A., Fairchild, I.J. & McDermott, F. (2000) Calcite fabrics, growth mechanisms, and environment of formation in speleothems from the Italian Alps and southwestern Ireland. *Journal of Sedimentary Research* 70, 1183–1196.

Frisia, S., Borsato, A., McDermott, F., Spiro, B., Fairchild, I., Longinelli, A., Selmo, E., Pedrotti, A., Dalmeri, G., Lanzinger, M. & van der Borg, K. (2001, for 1998) Holocene climate fluctuations in the Alps as reconstructed from speleothems. *Prehistoria Alpina* (Museo Tridentino di Scienze Naturali) 34 (for 1998), 111–118.

Frisia, S., Borsato, A., Fairchild, I.J., McDermott, F. & Selmo, E.M. (2002) Aragonite-calcite relationships in speleothems (Grotte de Clamouse, France): environment, fabrics, and carbonate geochemistry. *Journal of Sedimentary Research* 72, 687–699.

Frisia, S., Borsato, A., Preto, N. & McDermott, F. (2003) Late Holocene annual growth in three Alpine stalagmites records the influence of solar activity and the North Atlantic Oscillation on winter climate. *Earth and Planetary Science Letters* 216, 411–424.

Frisia, S., Borsato, A., Fairchild, I.J. & Susini, J. (2005). Variations in atmospheric sulphate recorded in stalagmites Variations in atmospheric sulphate recorded in stalagmites by synchrotron micro-XRF and XANES analyses. *Earth and Planetary Science Letters* 235, 729–740.

Frumkin, A. & Stein, M. (2004) The Sahara-East Mediterranean dust and climate connection revealed by strontium and uranium isotopes in a Jerusalem speleothem. *Earth and Planetary Science Letters* 217, 451–464.

Gams, I. (1965) Über die Faktoren die Intensität der Sintersedimentation bestimmen. *Proceedings 4th International Congress of Speleology, Ljubliana* 3, 117–126.

Gams, I. (1981) Contribution to morphometrics of stalagmites. *Proceedings of the 8th International Congress of Speleology, Bowling Green, Kentucky,* pp. 276–278.

Gascoyne, M. (1992) Palaeoclimate determination from cave deposits. *Quaternary Science Reviews* **11**, 609–632.

Genty, D. (1992) Les speleothems du tunnel de Godarville (Belgique) – un example exceptionnel de concrétionnment moderne – intérêt pour l'étude de las cinétiquede precipitation de las calcite et de sa relation avec les variations d'environnements. *Spéléochronos* **4**, 3–29.

Genty, D. & Quinif, Y. (1996) Annually laminated sequences in the internal structure of some Belgian stalagmites – importance for paleoclimatology. *Journal of Sedimentary Research* **66**, 275–288.

Genty, D. & Deflandre, G. (1998) Drip flow variations under a stalactite of the Père Noël cave (Belgium). Evidence of seasonal variations and air pressure constraints. *Journal of Hydrology* **211**, 208–232.

Genty, D., Baker, A. & Barnes, W. (1997) Comparaison entre les lamines luminescentes et les lamines visibles annuelles de stalagmites. *Comptes Rendus Academie Sciences, Paris* **325**, 193–200.

Genty, D., Baker, A., Massault, M., Proctor, C., Gilmour, M., Pons-Branchu, E. & Hamelin, B. (2001a) Dead carbon in stalagmites: Carbonate bedrock palaeodissolution vs. ageing of soil organic matter. Implications for ^{13}C variations in speleothems. *Geochimica et Cosmochimica Acta* **65**, 3443–3457.

Genty, D., Baker, A. & Vokal, B. (2001b) Intra- and inter-annual growth rate of modern stalagmites. *Chemical Geology* **176**, 191–212.

Genty, D., Blamart, D., Ouahdi, R., Gilmour, M., Baker, A., Jouzel, J. & Van-Exer, S. (2003) Precise dating of Dansgaard-Oeschger climate oscillations in western Europe from stalagmite data. *Nature* **421**, 833–838.

Goede, A., McCulloch, M., McDermott, F. & Hawkesworth, C. (1998) Aeolian contribution to strontium and strontium isotope variations in a Tasmanian speleothem. *Chemical Geology* **149**, 37–50.

Goldstein, S.J. & Stirling C.H. (2003) Techniques for measuring Uranium-series nuclides: 1992–2002. In: Bourdon B., Henderson G.M., Lundstrom C.C. & Turner S.P. (Eds) Uranium-series geochemistry. *Reviews in Mineralogy and Geochemistry* **52**, 23–57.

González, L.A. & Lohmann, K.C. (1988) Controls on mineralogy and composition of spelean carbonates: Carlsbad Caverns, New Mexico. In: James, N.P. and Choquette, P.W. (Eds) *Paleokarst,* New York: Springer-Verlag, pp. 81–101.

González, L.A., Carpenter, S.J. & Lohmann, K.C. (1992) Inorganic calcite morphology – roles of fluid chemistry and fluid-flow. *Journal of Sedimentary Petrology* **62**, 382–399.

Grigor'ev, D.P. (1961) *Onthogeny of minerals.* L'vov Izdatel'stvo L'vovskogo Univ., English Translation 1965, Jerusalem: Israel Program for Scientific Translation.

Grün, R. (1989) Electron spin resonance (ESR) dating. *Quaternary International* **1**, 65–109.

Gunn, J. (Ed.) (2004) *Encyclopedia of Caves and Karst Science.* New York: Fitzroy Dearborn.

Hakl, J., Hunyadi, I., Csige, I., Gécky, G., Lénart, K.L. & Várhegyi, A. (1997) Radon transport phenomena studied in karst caves – international experiences on radon levels and exposures. *Radiation Measurements* **28**, 675–684.

Harmon, R.S., White, W.B., Drake, J.J. & Hess, J.W. (1975) Regional hydrochemistry of North American carbonate terrains. *Water Resources Research* **11**, 963–967.

Harmon, R.S., Schwarcz, H.P., Gascoyne, M., Hess, J.W. & Ford, D.C. (2004) Paleoclimate information from speleothems: the present as a guide to the past. In: Sasowsky, I.D. & Mylroie, J. (Eds) *Studies of Cave Sediments. Physical and Chemical Records of Palaeoclimate.* New York: Kluwer Academic, pp. 199–226.

Hebdon, N.J., Atkinson, T.C., Lawson, T.J. & Young, I.R. (1997) Rate of glacial valley deepening during the Late Quaternary in Assynt, Scotland. *Earth Surface Processes and Landforms* **22**, 307–315.

Hellstrom, J.C. & McCulloch, M.T. (2000) Multi-proxy constraints on the climatic significance of trace element records from a New Zealand speleothem. *Earth and Planetary Science Letters* **179**, 287–297.

Hendy, C.H. (1971) The isotopic geochemistry of speleothems – I. The calculation of the effects of different modes of formation on the isotopic composition of speleothems and their applicability as palaeoclimatic indicators. *Geochimica et Cosmochimica Acta* **35**, 801–824.

Hendy, C.H. & Wilson, A.T. (1968) Palaeoclimate data from speleothems. *Nature* **219**, 48–51.

Hill, C. & Forti, P. (1997) *Cave Minerals of the World,* 2nd edn. Huntsville, AL: National Speleological Society.

Holland, H.D., Kirsipu, T.V., Huebner, J.S. & Oxburgh, U.M. (1964) On some aspects of the chemical evolution of cave water. *Journal of Geology* **72**, 36–67.

Huang, Y. & Fairchild, I.J. (2001) Partitioning of Sr^{2+} and Mg^{2+} into calcite under karst-analogue experimental conditions. *Geochimica et Cosmochimica Acta* **65**, 47–62.

Huang, Y., Fairchild, I.J., Borsato, A., Frisia, S., Cassidy, N.J., McDermott, F. & Hawkesworth, C.J. (2001) Seasonal variations in Sr, Mg and P in modern speleothems (Grotta di Ernesto, Italy). *Chemical Geology* **175**, 429–448.

Jennings, J.N. (1971) *Karst.* London: MIT Press.

Jennings, J.N. (1985) *Karst Geomorphology.* Oxford: Basil Blackwell.

Jones, B. & Kahle, C.F. (1993) Morphology relationships, and origin of fiber and dendrite calcite crystals. *Journal of Sedimentary Petrology* **63**, 1018–1031.

Kaufmann, G. (2003) Stalagmite growth and palaeo-climate: the numerical perspective. *Earth and Planetary Science Letters* **214**, 251–266.

Kaufmann, G. & Dreybrodt, W. (2004) Stalagmite growth and palaeo-climate: an inverse approach. *Earth and Planetary Science Letters* **224**, 529–545.

Kendall, A.C. (1993) Columnar calcite in speleothems – discussion. *Journal of Sedimentary Petrology* **63**, 550–552.

Kendall, A.C. & Broughton, P.L. (1978) Origin of fabric in speleothems of columnar calcite crystals. *Journal of Sedimentary Petrology* **48**, 550–552.

Kim, S.T. & O'Neil, J.R. (1997) Equilibrium and nonequilibrium oxygen isotope effects in synthetic carbonates. *Geochimica et Cosmochimica Acta* **61**, 3461–3475.

Klimchouk, A.B., Ford, D.C., Palmer, A.N. & Dreybrodt, W. (Eds) (2000) *Speleogenesis. Evolution of Karst Aquifers.* Huntsville, Alabama: National Speleological Society.

Kolodny, Y., Bar-Matthews, M., Ayalon, A. & McKeegan, K.D. (2003) A high spatial resolution delta O-18 profile of a speleothem using an ion-microprobe. *Chemical Geology* **197**, 21–28.

Lauritzen, S.-E. (1990) Autogenic and allogenic denudation in carbonate karst by the multiple basin method – an example from Svartisen, North Norway. *Earth Surface Processes and Landforms* **15**, 157–167.

Lauritzen, S.E. & Lundberg, J. (1999) Calibration of the speleothem delta function: an absolute temperature record for the Holocene in northern Norway. *The Holocene* **9**, 659–669.

Lauritzen, S.-E., Haugen, J.E., Lovlie, R. & Giljenielsen, H. (1994) Geochronological potential of isoleucine epimerisation in calcite speleothems. *Quaternary Research* **41**, 52–58.

Leutscher, M. & Jeannin, P.-Y. (2004) Temperature distribution in karst systems: the role of air and water fluxes. *Terra Nova* **16**, 344–350.

Liñan, C., Carrasco, F., Andreo, B., Jiménez de Cisneros, C. & Caballero, E. (2002) Caracterización isotópica de las agues do goteo de las Cueva de Nerja y de su entorno hidrogeológicao (Málaga, Sur de España). In: Carrasco, F., Durán, J.J. & Andreo, B. (Eds) *Karst Environments*. Malaga: Fundación Cueva de Nerja, pp. 243–249.

Losson, B. & Qunif, Y. (2001) La capture de la Moselle: nouvelles données chronologiques par datations U/Th sue spéléothèmes. *Karstologia* **37**, 29–40.

Lovelock, J. (1988) *The Ages of Gaia*. New York: W.W. Norton.

Ludwig, K.R., Simmons, K.R., Szabo, B.J., Winograd, I.J., Landwehr, J.M., Riggs, A.C. & Hoffman, R.J. (1992) Mass-spectrometric ^{230}Th–^{234}U–^{238}U dating of the Devils Hole calcite vein. *Science* **258**, 284–287.

McDermott, F. (2004) Palaeo-climate reconstruction from stable isotope variations in speleothems: a review. *Quaternary Science Reviews* **23**, 901–918.

McDermott, F., Frisia, S., Huang, Y., Longinelli, A., Spiro, B., Heaton, T.H.E., Hawkesworth, C.J., Borsato, A., Keppens, E., Fairchild, I.J., van der Borg, K., Verheyden, S. & Selmo, E. (1999) Holocene climate variability in Europe: evidence from δ^{18}O and textural variations in speleothems. *Quaternary Science Reviews* **18**, 1021–1038.

McDermott, F., Mattey, D.P. & Hawkesworth, C. (2001) Centennial-scale Holocene climate variability revealed by a high-resolution speleothem delta O-18 record from SW Ireland. *Science* **294**, 1328–1331.

McGarry, S. & Baker, A. (2000) Organic acid fluorescence: applications to speleothem palaeoenvironmental reconstruction. *Quaternary Science Reviews* **19**, 1087–1101.

McGarry, S.F. & Caseldine, C. (2004) Speleothem palynology: an undervalued tool in Quaternary studies. *Quaternary Science Reviews* **23**, 2389–2404.

McGarry, S.F., Bar-Matthews, M., Matthew, A., Vaks, A., Schilman, B. & Ayalon, A. (2004) Constraints on hydrological and palaeotemperature variations in the Eastern Mediterranean region in the last 140 ka given by the δD values of speleothem fluid inclusions. *Quaternary Science Reviews* **23**, 919–934.

McMillan, E., Fairchild, I.J., Frisia, S. & Borsato, A. (2005) Calcite-aragonite trace element behaviour in annually layered speleothems: evidence of drought in the Western Mediterranean 1200 years ago. *Journal of Quaternary Science* **20**, 423–433.

Matthews, A., Ayalon, A. & Bar-Matthews, M. (2000) D/H ratios of fluid inclusions of Soreq cave (Israel) speleothems as a guide to the Eastern Mediterranean Meteoric Line relationships in the last 120 ky. *Chemical Geology* **166**, 183–191.

Mavlyudov, B.R. (1997) Caves climatic systems. *Proceedings of the 12th International Congress of Speleology, La Chaux-le-Fonds, Switzerland, 10–17 August 1997,* pp. 191–194.

Mickler, P.J., Banner, J.L., Stern, L., Asmerom, Y., Edwards, R.L. & Ito, E. (2004) Stable isotope variations in modern tropical speleothems: evaluating equilibrium vs. kinetic effects. *Geochimica et Cosmochimica Acta* **68**, 4381–4393.

Miotke, F.-D. (1974) Carbon dioxide and the soil atmosphere. *Abhandlungen zur Karst-Und Höhlenkunde, Reihe A, Speläologie,* Heft 9.

Moore, G.W. (1962) The growth of stalactites. *National Speleological Society Bulletin* **24**, 95–105.

Morse, J.W. & Bender, M.L. (1990) Partition coefficients in calcite: Examination of factors influencing the validity of experimental results and their application to natural systems. *Chemical Geology* **82**, 265–277.

Musgrove, M. & Banner, J.L. (2004) Controls on the spatial and temporal variability of vadose dripwater geochemistry: Edwards Aquifer, central Texas. *Geochimica et Cosmochimica Acta* **68**, 1007–1020.

Neff, U., Burns, S.J., Mangini, A., Mudelsee, M., Fleitmann, D. & Matter, A. (2001) Strong coherence between solar variability and the monsoon in Oman between 9 and 6 kyr ago. *Nature* **411**, 290–293.

Niggeman, S.M., Mangini, A., Richter, D.K. & Wurth, G. (2003) A paleoclimate record f the last 17,600 years in stalagmites from the B7 cave, Sauerland, Germany. *Quaternary Science Reviews* **22**, 555–567.

Paulsen, D.E., Li, H.-C. & Ku, T.-L. (2003) Climate variability in central China over the last 1270 years revealed by high-resolution stalagmite records. *Quaternary Science Reviews* **22**, 691–701.

Paquette, J. & Reeder, R.J. (1995) Relationship between surface structure, growth mechanism, and trace element incorporation in calcite. *Geochimica et Cosmochimica Acta* **59**, 735–749.

Perrette, Y., Delannoy, J.-J., Desmet, M., Lignier, V. & Destobmes, J.-L. (2005) Speleothem organic matter content imaging. The use of a fluorescence index to characterize the maximum emission wavelength. *Chemical Geology* **214**, 193–208.

Perrin, J., Jeannin, P.-V. & Zwahlen, F. (2003) Epikarst storage in a karst aquifer: a conceptual model based on isotopic data, Milandre test site, Swizerland. *Journal of Hydrology* **279**, 106–124.

Pitty, A.F. (1966) An approach to the study of karst water. *University of Hull, Occasional Papers in Geography* **5**, 438–445.

Pitty, A.F. (1968) Calcium carbonate content of karst water in relation to flow-through time. *Nature* **217**, 939–940.

Plummer, L.N., Wigley, T.M.L. & Parkhurst, D.L. (1978) The kinetics of calcite dissolution in CO_2-water systems at 5° to 60°C and 0.0 to 1.0 atm CO_2. *American Journal of Science* **278**, 179–216.

Polyak, V.J. & Asmerom, Y. (2001) Late Holocene climate and cultural changes in southwestern United States. *Science* **294**, 148–151.

Proctor, C.J., Baker, A., Barnes, W.L. & Gilmour, M.A. (2000) A thousand year speleothem proxy record of North Atlantic climate from Scotland. *Climate Dynamics* **16**, 815–820.

Proctor, C.J., Baker, A. & Barnes, W.L. (2002) A three thousand year record of North Atlantic climate. *Climate Dynamics* **19**, 449–454.

Railsback, L.B., Brook, G.A., Chen, J., Kalin, R. & Fleischer, C.J. (1994) Environmental controls on the petrology of a late Holocene speleothem from Botswana with annual layers of aragonite and calcite. *Journal of Sedimentary Research* **A64**, 147–155.

Reeder, R.J., Nugent, M., Lamble, G.M., Tait, C.D. & Morris, D.E. (2000) Uranyl incorporation into calcite and aragonite: XAFS and luminescence studies. *Environmental Science and Technology* **34**, 638–644.

Richards, D.A. & Dorale, J.A (2003) Uranium-series chronology and environmental applications of speleothems. *Review in Mineralogy and Geochemistry* **52**, 407–460.

Richards, D.A., Bottrell, S.H., Cliff, R.A., Ströhle, K. & Rowe, P. (1998) U-Pb dating of a speleothem of Quaternary age. *Geochemica Cosmochimica Acta* **62**, 3683–3688.

Roberts, M.S., Smart, P.L. & Baker, A. (1998) Annual trace element variations in a Holocene speleothem. *Earth and Planetary Science Letters* **154**, 237–246.

Rozanski, K., Araguas-Araguas, L. & Gonfiantini, R. (1993) Isotopic patterns in modern global precipitation. In: Swart, P.K., Lohman, K.L., McKenzie, J.A. & Savin. S. (Eds) *Climate Change in Continental Isotopic Record*. Monograph 78. Washington, DC: American Geophysical Union, pp. 1–36.

Sasowsky, I.D. & Mylroie, J. (Eds) (2004) *Studies of Cave Sediments*. Kluwer, New York.

Self, C.A. & Hill, C.A. (2003) How speleothems grow: an introduction to the ontogeny of cave minerals. *Journal of Cave and Karst Studies* **65**, 130–151.

Shaw, T.R. (1997) Historical Introduction. In: Hill, C. & Forti, P. (Eds) *Cave Minerals of the World*, 2nd edn. Huntsville, Alabama: National Speleological Society, pp. 27–43.

Shindell, D.T., Schmidt, G.A., Mann, M.E., Rind, D. & Waple, A. (2001) Solar forcing of Regional Climate Change during the Maunder Minimum. *Science* **294**, 2149–2152.

Shopov, Y.Y. (2004) Speleothems: luminescence. In: Gunn, J. (Ed.) *Encyclopedia of Cave and Karst Science*, New York: Fitzroy Dearborn, pp. 695–697.

Shopov, Y.Y., Ford, D.C. & Schwarcz, H.P. (1994) Luminescent microbanding in speleothems: High-resolution chronology and palaeoclimate. *Geology* **22**, 407–410.

Short, M.B., Baygents, J.C., Beck, J.W., Stone, D.A., Toomey, R.S. & Goldstein, R.E. (2005) Stalactite growth as a free-boundary problem: a geometric law and its platonic ideal. *Physical Review Letters* **94**, 018501(4).

Smart, P.L. & Francis, P.D. (1990) *Quaternary Dating Methods – a User's Guide*. Technical Guide 4. London: Quaternary Research Association.

Smart, P.L. & Friederich, H. (1986) Water movement and storage in the unsaturated zone of a maturely karstified aquifer, Mendip Hills, England. *Proceedings of the Conference on Environmental Problems in Karst Terrains and their Solutions*, 28–30 October, Bowling Green, Kentucky, National Water Wells Association, pp. 57–87.

Spötl, C. & Mangini, A. (2002) Stalagmite from the Austrian Alps reveals Dansegaard-Oeschger events during isotope stage 3: Implications for the absolute chronology of Greenland ice cores. *Earth and Planetary Science Letters* **203**, 507–518.

Spötl, C., Mangini, A., Frank, N., Eichstädter, R. & Burns, S.J. (2002) Start of the last interglacial period at 135 ka: Evidence from a high Alpine speleothem. *Geology* **30**, 815–818.

Spötl, C., Mangini, A., Burns, S.J., Frank, N. & Pavuza, R. (2004) Speleothems from the high-alpine Spannagel cave, Zillertal Alps (Austria). In: Sasowsky, I.D. & Mylroie, J. (Eds) *Studies of Cave Sediments. Physical and Chemical Records of Palaeoclimate.* New York: Kluwer Academic, pp. 243–256.

Spötl, C., Fairchild, I.J. & Tooth, A.F. (2005) Speleothem deposition in a dynamically ventilated cave, Obir Caves (Austrian Alps). Evidence from cave air and drip water monitoring. *Geochimica et Cosmochimica Acta,* **69**, 2451–2468.

Stepanov, V.I. (1997) Notes on mineral growth from the archive of V.I. Stepanov. *Proceedings University Bristol Spaelaeological Society* **21**, 25–42.

Sweeting, M.M. (1972) *Karst Landforms.* London: Macmillan.

Sweeting, M.M. (Ed.) (1981) *Karst Geomorphology.* Benchmark Paper in Geology 59. Stroudsburg, PA: Hutchison Ross.

Sweeting, M.M. (1995) *Karst in China.* Berlin: Springer-Verlag.

Tooth, A.F. (2000) *Controls on the geochemistry of speleothem-forming karstic drip waters.* Unpublished PhD thesis, Keele University.

Tooth, A.F. & Fairchild, I.J. (2003) Soil and karst aquifer hydrological controls on the geochemical evolution of speleothem-forming drip waters, Crag Cave, southwest Ireland. *Journal of Hydrology* **273**, 51–68.

Treble, P., Shelley, J.M.G. & Chappell, J. (2003) Comparison of high resolution sub-annual records of trace elements in a modern (1911–1992) speleothem with instrumental cliate data from southwest Australia. *Earth and Planetary Science Letters* **216**, 141–153.

Troester, J.W. & White, W.B. (1984) Seasonal fluctuations in the carbon dioxide partial pressure in a cave atmosphere. *Water Resources Research* **20**, 153–156.

Trudgill, S. (1985) *Limestone Geomorphology.* London: Longman.

Turner, J.V. (1982) Kinetic fractionation of C-13 during calcium-carbonate precipitation. *Geochimica et Cosmochimica Acta* **46**, 1183–1191.

Vaks, A., Bar-Matthews, M., Ayalon, A., Schilman, B., Gilmour, M., Hawkesworth, C.J., Frumkin, A., Kaufman, A. & Matthews, A. (2003) Paleoclimate reconstruction based on the timing of speleothem growth and oxygen and carbon isotope composition in a cave located in the rain shadow in Israel. *Quaternary Research* **59**, 182–193.

Vaute, L., Drogue, C., Garrelly, L. & Ghelfenstein, M. (1997) Relations between the structure of storage and the transport of chemical compounds in karstic aquifers. *Journal of Hydrology* **199**, 221–236.

Verheyden, S., Keppens, E., Fairchild, I.J., McDermott, F. & Weis, D. (2000) Mg, Sr and Sr isotope geochemistry of a Belgian Holocene speleothem: implications for paleoclimate reconstructions. *Chemical Geology* **169**, 131–144.

Wang, F., Hongchun, L., Rixiang, Z. & Feizhou, Q. (2004) Later Quaternary downcutting rates of the Qianyou River from U/Th speleothem dates, Qinling mountains, China. *Quaternary Research* **62**, 194–200.

Wang, X., Auler, A.S., Edwards, R.L., Cheng, H., Cristall, P.S., Smart, P.L., Richards, D.A. & Shen, C.-C. (2004) Wet periods in northeastern Brazil over the past 210 kyr linked to distant climate anomalies. *Nature* **432**, 740–743.

Wenk, H.-R., Barber, D.J. & Reeder, R.J. (1983) Microstructures in carbonates. In: Reeder R.J. (Ed.) Carbonates: Mineralogy and Geochemistry, *Reviews in Mineralogy*, 11, Washington: Mineralogical Society of America, pp. 301–367.

White, W.B. (1988) *Geomorphology and Hydrology of Karst Terrains*. New York: Oxford University Press.

White, W.B. (2004) Palaeoclimate records from speleothems in limestone caves. In: Sasowsky, I.D. & Mylroie, J. (Eds) *Studies of Cave Sediments. Physical and Chemical Records of Palaeoclimate*, New York: Kluwer Academic, pp. 135–175.

Wigley, T.M.L. (1967) Non-steady flow through a porous medium and cave breathing. *Journal of Geophysical Research* 72, 3199–3205.

Wigley, T.M.L. & Brown, M.C. (1971) Geophysical applications of heat and mass transfer in trubulent pipe flow. *Boundary Layer Meteorology* 1, 300–320.

Wigley, T.M.L. & Brown, M.C. (1976) The Physics of Caves. In: Ford, T.D. & Cullingford, C.H.D. (Eds) *The Science of Speleology*, London: Academic Press, pp. 329–358.

Winograd, I.J., Coplen, T.B., Landwehr, J.M., Riggs, A.C., Ludwig, K.R., Szabo, B.J., Kolesar, P.T. & Revesz, K.M. (1992) Continuous 500,000-year climate records from vein calcite in Devil's Hole, Nevada. *Science 258*, 255–260.

Xie, S., Yi, Y., Huang, J., Hu, C., Cai, Y., Collins, M. & Baker, A. (2003) Lipid distribution in a subtropical southern China stalagmite as a record of soil ecosystem response to paleoclimate change. *Quaternary Research 60*, 340–347.

Yadava, M.G., Ramesh, R. & Pant, G.B. (2004) Past monsoon rainfall variations in peninsular India recorded in a 331-year-old speleothem. *The Holocene 14*, 517–524.

Yonge, C., Ford, D.C., Gray, J. & Schwarcz, H.P. (1985) Stable isotope studies of cave seepage water. *Chemical Geology 58*, 97–105.

Yuan, D., Cheng, H., Edwards, R.L., Dykoski, C.A., Kelly, M.J., Zhang, M., Qing, J., Lin, Y., Wang, Y., Wu, J., Dorale, J.A., An, Z. & Ci, Y. (2004) Timing, duration, and transitions of the last interglacial Asian monsoon. *Science 304*, 575–578.

Chapter Eight

Rock Varnish

Ronald I. Dorn

8.1 Introduction: Nature and General Characteristics

Most earth scientists thinking about geochemical sediments envisage stratigraphic sequences, not natural rock exposures. Yet, rarely do we see the true
colouration and appearance of natural rock faces without some masking
by biogeochemical curtains. Geochemical sediments known as rock coatings (Table 8.1) control the hue and chroma of bare-rock landscapes. Tufa
and travertine (Chapter 6), beachrock (Chapter 11) and nitrate efflorescences (Chapter 12) exemplify circumstances where geochemical sediments
can cover rocks. Perhaps because of its ability to alter a landscape's appearance dramatically (Figure 8.1), the literature on rock varnish remains one
of the largest in the general arena of rock coatings (Chapter 10 in Dorn,
1998).

Rock varnish (often called 'desert varnish' when seen in drylands) is a
paper-thin mixture of about two-thirds clay minerals cemented to the host
rock by typically one-fifth manganese and iron oxyhydroxides. Upon examination with secondary and backscattered electron microscopy, the accre-
tionary nature of rock varnish becomes obvious, as does its basic layered
texture imposed by clay minerals (Dorn and Oberlander, 1982). Manganese enhancement, two orders of magnitude above crustal values, remains
the geochemical anomaly of rock varnish and a key to understanding its
genesis.

Field observations have resulted in a number of informal classifications.
Early field geochemists recognised that varnish on stones in deserts differs
from varnish on intermittently flooded rock surfaces (Lucas, 1905). Another
example of differentiating varnish involves position on a desert pavement
clast (Mabbutt, 1979): black varnish rests on the upper parts of a pavement
clast, a shiny ground-line band of varnish occurs at the soil–rock–atmosphere interface, and an orange coating is found on the underside of

Table 8.1 Different types of rock coatings (adapted from Dorn, 1998)

Coating	Description
Carbonate skin	Coating composed primarily of carbonate, usually calcium carbonate, but sometimes combined with magnesium
Case hardening agents	Addition of cementing agent to rock matrix material; the agent may be manganese, sulphate, carbonate, silica, iron, oxalate, organisms, or anthropogenic
Dust film	Light powder of clay- and silt-sized particles attached to rough surfaces and in rock fractures
Heavy metal skins	Coatings of iron, manganese, copper, zinc, nickel, mercury, lead and other heavy metals on rocks in natural and human-altered settings
Iron film	Composed primarily of iron oxides or oxyhydroxides
Lithobiontic coatings	Organic remains form the rock coating, for example lichens, moss, fungi, cyanobacteria, algae
Nitrate crust	Potassium and calcium nitrate coatings on rocks, often in caves and rock shelters in limestone areas
Oxalate crust	Mostly calcium oxalate and silica with variable concentrations of magnesium, aluminum, potassium, phosphorus, sulphur, barium, and manganese. Often found forming near or with lichens. Usually dark in colour, but can be as light as ivory
Phosphate skin	Various phosphate minerals (e.g. iron phosphates or apatite) that are mixed with clays and sometimes manganese; can be derived from decomposition of bird excrement
Pigment	Human-manufactured material placed on rock surfaces by people
Rock varnish	Clay minerals, Mn and Fe oxides, and minor and trace elements; colour ranges from orange to black due to variable concentrations of different manganese and iron oxides
Salt crust	The precipitation of chlorides on rock surfaces
Silica glaze	Usually clear white to orange shiny luster, but can be darker in appearance, composed primarily of amorphous silica and aluminum, but often with iron
Sulphate crust	Composed of the superposition of sulphates (e.g., barite, gypsum) on rocks; not gypsum crusts that are sedimentary deposits

Figure 8.1 This road cut between Death Valley, California and Las Vegas, Nevada illustrates the ability of 10–100 μm-thick rock varnish (darkens debris slope surface on upper half of image) to mask the appearance of the host igneous and metamorphic rocks (light-appearing road cut on lower half of image).

cobbles in contact with the soil (Engel and Sharp, 1958). Detailed discussions also distinguish different relationships among varnish and the underlying rock (Haberland, 1975).

An issue in classification that the introductory reader should be aware of involves the fairly frequent use of the term 'desert varnish' or 'rock varnish' to describe completely different types of rock-surface alterations (e.g. Howard, 2002). Even experienced desert geomorphologists (e.g. Bull, 1984; McFadden et al., 1989) have mistaken surfaces such as basalt ventifacts and geothermal deposits for varnish. Consider just Antarctic research, which has been consistently unable to distinguish between iron films and weathering rinds usually termed either desert varnish (Glasby et al., 1981) or rock varnish (Ishimaru and Yoshikawa, 2000) from manganese-rich coatings (Glazovskaya, 1958; Dorn et al., 1992).

The problem for any researcher new to the topic of manganese-rich varnish is that the vast majority of papers are written by investigators who collect only a few samples from a few locales and simply assume that their samples, somehow, are equivalent to varnishes collected in completely different biogeochemical settings. The only formal classification of rock varnish thus far presented (Dorn, 1998, pp. 214–224) maintains the goal of forcing varnish researchers to ensure that they have sampled the same variant in any comparisons. For the sake of clarity and brevity, this chapter presents illustrations only for varnishes formed in subaerial settings.

8.2 Distribution, Field Occurrence and Geomorphological Relations

Rock varnish grows on mineral surfaces exposed by erosional processes, either in weathering-limited landscapes such as deserts (e.g. White, 1993), or in settings dominated by rapid erosion such as glaciated terrain (e.g. Whalley et al., 1990). Thus, one first-order control on distribution rests with geomorphological processes exposing rock faces. Another first-order control on distribution rests on whether the varnish formed in a geochemical landscape (Perel'man, 1961) favourable for stability; exposure to minimal acidity helps maintain the manganese and iron in their oxidised and immobile condition in varnish. A second-order issue turns on the nature of the erosional process. Spalling often exposes planar joint faces that already have rock coatings developed in subsurface settings (Coudé-Gaussen et al., 1984; Villa et al., 1995; Dorn, 1998).

Rock varnish exists in geomorphological catenas (Haberland, 1975; Palmer, 2002), much in the way that soils vary along slopes (Jenny, 1941). Taking a simple basalt hill in the western Mojave Desert of California, Palmer (2002) mapped the abundance of black rock varnish, orange iron films, and surfaces experiencing rapid-enough erosion that they lacked field-observable rock coatings (Figure 8.2). Rock varnish occurs more frequently on metre-sized colluvium at the hill crest. In contrast, iron films, that initially formed on clasts formerly buried in soil, increasingly dominate on subaerial slope positions experiencing greater soil erosion from grazing (Palmer, 2002) or Holocene erosion (Hunt and Wu, 2004).

Palmer's (2002) general observation that different geomorphological positions influence rock varnish carries over to a single boulder. Just considering varnish thickness, millimetre-scale topographic highs host thinner varnishes than 'microbasins' or broad depressions a few millimetres across (Figure 8.2). Depressions more than a few millimetres deep favour retaining enough moisture to host varnish-destroying microcolonial fungi (Figure 8.3), cyanobacteria, or even lichens. Granodiorite boulders on a moraine in the Karakoram Mountains, for example, illustrate the importance of aspect where varnish grows on south-facing surfaces, while lichens and salt efflorescences occupy north-facing surfaces (Waragai, 1998).

The varnish literature is filled with misunderstandings surrounding its distribution, field occurrence and geomorphological relations. Consider one of the most common errors, equating darkness in a field sample with age, a generalisation derived from observing the darkening of alluvial-fan units over time. If geomorphological units darken over time, would not individual hand samples? This is not necessarily the case. The darkest varnishes are usually those that grow first in rock crevices (Coudé-Gaussen et al., 1984; Villa et al., 1995) and are then 'born dark' as spalling exposes

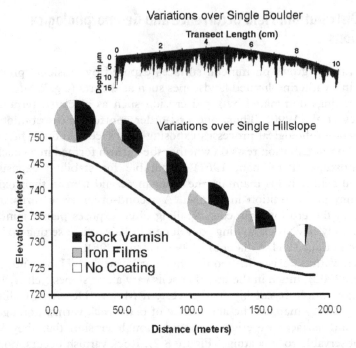

Figure 8.2 Rock varnish varies considerably over short distances, over a single boulder and over a single hillslope. The upper part of this figure illustrates that varnish thickness varies considerably over a single granodiorite boulder on the Tioga-3 moraine at Bishop Creek, about 19 ± 1 ka ^{36}Cl years old (Phillips et al., 1996). The thickest varnishes grow in shallow 'dish shaped' microbasins. Deeper rock depressions typically host varnishes experiencing erosion from acid-generating fungi and cyanobacteria. The lower portion of this figure presents a rock coating catena (adapted from Palmer, 2002) for three rock coating types, and shows variability on a basaltic hillslope in the Mojave Desert, California. Note that rock varnish dominates the more stable hill crest; iron films increase in abundance towards the slope bottom, and rock surfaces mostly lacking either coating occurs in a similar abundance on all slope positions.

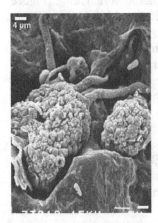

Figure 8.3 Originally named black globular units (Borns et al., 1980) for their appearance under a hand lens, microcolonial fungi are common inhabitants on desert rocks (Staley et al., 1982) that experience warm season convective precipitation. Originally thought to be agents of varnish (Staley et al., 1983) formation, the lack of manganese-enhancement by these and other fungi (Dorn and Oberlander, 1982) and their ability to dissolve varnish (Dorn, 1994) makes these organisms either adventitious or destructive agents, but not formative (Dragovich, 1993).

fracture varnish. Rather than detail each misunderstanding here, Table 8.2 summarises these peculiar components of the literature.

8.3 Macro- and Micromorphological Characteristics

Regions with abundant vegetative cover do have rock varnish on bare-rock surfaces, but these are too small in extent to alter signatures in remotely-sensed imagery. That is why arid regions host a large number of remote-sensing investigations about the macromorphological characteristics of varnish (Kenea, 2001). Alluvial fans are favourite study sites (White, 1993; Farr and Chadwick, 1996; Milana, 2000), in part because of the dramatic colour changes produced by progressive increases in the aerial extent of varnishing (Figure 8.4) and increasing darkness as fracture varnishes are exposed by boulder spalling (Villa et al., 1995).

Regional climatic variations alter the micromorphological character of rock varnishes, where abundance of aeolian dust appears to be a key factor. The vast majority of rock varnishes have micromorphologies that range between lamellate and botryoidal (Figure 8.5). A paucity of dust allows a clustering of varnish accretion, thus producing botryoidal forms. An abundance of dust smothers the tendency of Mn accumulation around nucleation centres, producing lamellate micromorphologies. A variety of other factors, however, can influence micromorphology, including: clay mineralogy, manganese abundance, epilithic organisms such as microcolonial fungi, morphology of the underlying rock, varnish thickness, aeolian abrasion, proximity to soil, and other microenvironmental factors (Dorn, 1986).

The dominant cross-sectional texture of rock varnish is layering (Figure 8.4C, E). Even the highest resolution transmission electron microscopy (HRTEM) views of varnish consistently show layering textures (Krinsley et al., 1995) imposed by its clay mineralogy (Figure 8.4E). Even when the surface morphology is botryoidal, cross-sectional texture remains layered internally within each botryoidal mound (Figure 8.5B). All meaningful theories of varnish genesis must explain varnish textures as well as varnish chemistry.

8.4 Chemistry and Mineralogy

Analytical chemist Celeste Engel completed the first Masters thesis research on rock varnish and teamed up with geomorphologist Robert Sharp to write a seminal paper on rock varnish in the Mojave Desert (Engel and Sharp, 1958). The major elements are O, H, Si, Al, and Fe in approximately equal abundance with Mn. The key mystery of varnish formation (von Humboldt, 1812; Lucas, 1905; Engel and Sharp, 1958; Jones, 1991) is

Table 8.2 A few of the misunderstandings in the literature surrounding rock varnish and its environmental relations. See chapter 10 in Dorn (1998) for a more detailed account of the history of thought on varnish

Topic	Misunderstanding	Resolution
Accretion or rock weathering	A century of debate exists on whether varnish accretes on (Potter and Rossman, 1977) or derives from the underlying rock (Hobbs, 1918)	Low- (Dorn and Oberlander, 1982) and high-resolution (Krinsley, 1998) electron microscopy and trace element geochemistry (Thiagaran and Lee, 2004) fails to show evidence of anything other than an accretionary process
Adventitious organisms	Varnishes host a wide array of organisms (Kuhlman et al., 2005, 2006; Schelble et al., 2005) such as lichens (Laudermilk, 1931), microcolonial fungi (Borns et al., 1980; Staley et al., 1983; Gorbushina et al., 2002) gram-positive bacteria (Perry et al., 2003), cyanobacteria and other organisms (Krumbein, 1969)	The simple presence of organics in association with varnish bears little importance to varnish genesis, unless biotic remains explicitly link to varnish-formation processes. Spatially precise geochemical analyses reveal the difference between adventitious organisms (Dragovich, 1993) and those bacteria that can be seen to enhance Mn and Fe through in situ analyses (Dorn and Oberlander, 1982; Krinsley, 1998)
Case hardening	Rock varnish is too thin to be an effective case hardening agent	Cation leaching (Dorn and Krinsley, 1991) from varnish provides materials for redeposition within the weathering rind – thus case hardening the outer shell of rock (Haberland, 1975)

Darkness shows time	Darker appearing surfaces are older than younger appearing surfaces (McFadden et al., 1989; Reneau, 1993)	This misunderstanding causes a number of errors in geomorphological analyses. Eleven factors other than time influence darkness (table 10.7 in Dorn, 1998)
Desert restriction	Desert varnish occurs in deserts and not elsewhere, and this distribution somehow relates to mode of formation	Rock varnish occurs in every terrestrial weathering environment (Dorn and Oberlander, 1982; Douglas et al., 1994; Lucas, 1905; Whalley et al., 1990). More frequent occurrence in deserts relates to greater landscape geochemical stability (Perel'man, 1961) for Mn and Fe cements
Lithology preference	Rock varnish grows best on certain lithologies such as fine-grained extrusive rocks	A more important issue is host rock stability. Rock varnish forms on all lithologies whose surface stability exceeds rates of varnish formation
Lustre	The shiny appearance of varnish relates to age or wind abrasion	Neither is true. Some sheen derives from very smooth and Mn-rich varnishes. The shiniest varnishes, found at the ground line on desert pavement cobbles, are only a few microns thick and intercalate with silica glaze
Manganese and iron abundance	When averaged over an area, the total accumulation of manganese and iron increases over time	Major problems confound this naive belief, including exposure of manganese-rich fracture coatings (Douglas et al., 1994), erosion of faster-forming varnishes (Dorn and Oberlander, 1982), and role of microenvironments (Whalley et al., 1990)

Table 8.2 *Continued*

Topic	Misunderstanding	Resolution
Mars analog	Assuming that varnish is a biogenic rock coating, the occurrence of Mn-rich varnishes on Mars provides evidence of life or former life	The use of varnish as a bioindicator of Martian life is untenable until abiotic origins of Mn-enhancement are falsified for Martian conditions
Surface stability	Since cosmogenic nuclide analyses show erosion of rock surfaces (Nishiizumi et al., 1993), and geomorphological surfaces hosting varnish such as desert pavements experience instability (Peterson, 1981), how can varnish exist for 10^5 years (Liu and Broecker, 2000; Liu, 2003)?	Rock surface erosion rarely occurs in a linear fashion. Larger clasts can host zero erosion rates, while adjacent clasts experience much more rapid erosion. Linear rates of boulder erosion remain unusual. Similarly, landforms can exhibit signs of instability such as erosion of fines, even though particular rock faces can remain stable
Thickness	Thickness increases over time	The thickest varnishes grow in wetter micropositions that foster erosion of the underlying weathering rind or on spalled fracture faces. Thus, although varnish does thicken over time if all other formation factors are held constant, the thinnest varnishes can be the oldest at any given site

Figure 8.4 Rock varnish on Hanaupah Canyon alluvial fan, Death Valley, viewed at different scales. (A) SPOT satellite image where darkness first increases and then decreases with age. As varnish coverage increases, older sections of the Bar and Channel segment darken, reaching maximum darkness on the smooth pavement (B) segment. Channel incision starts to lighten appearance as varnished cobbles erode into gullies, with lightening increasing as channel incision progresses to the point where rounded ballenas expose a mix of calcrete and varnished cobbles. (C–E) Electron microscope imagery all revealing layering patterns in rock varnish, collected from Hanaupah Canyon alluvial fan, Death Valley. (C) Back-scattered electron image reveals average atomic number, where the bright layers are rich in manganese and iron and the darker layers are richer in clay minerals. (D) Scanning electron microscopy image shows shape, with the arrows identifying the contact between the rock in the lower right and the layered varnish. (E) High resolution transmission electron microscopy image is at a much greater magnification, showing that there are layers within layers, where the layered nature of clay minerals (Potter and Rossman, 1977) imposes the basic texture.

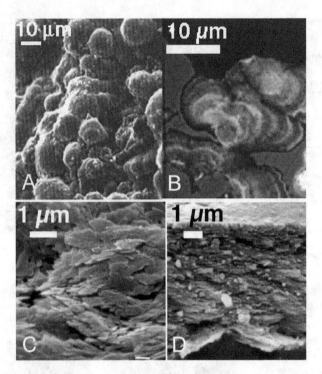

Figure 8.5 Varnish micromorphology form ranges from botryoidal to lamellate (A and C). Two types of imagery show botryoidal varnishes from Kitt Peak, Arizona: (A) the topography by secondary electrons; (B) the same structures from the bottom upwards with back-scattered electrons – showing the layering structures inside each nucleation centre. (C and D) Scanning electron microscopy images of lamellate clay minerals accreting on rock varnish in Death Valley, California. (C) Individual clay platelets overlap as they cement onto the surface. (D) The clays impose a lamellate structure in cross-section, as first noticed by Potter and Rossman (1977).

how to explain the great enrichment in Mn, normally a trace element in soils and rocks. Varnish Mn : Fe ratios vary from less than 1 : 1 to over 50 : 1, in contrast with Mn : Fe ratios of about 1 : 60 in the Earth's crust. Early microprobe analyses showed measurements higher than 15% Mn (Hooke et al., 1969). Concentrations of >80% Mn in focused spots occur on budding bacteria forms (Dorn, 1998).

Mn (II) is the most important soluble form, generally stable in the pH range 6–9 in natural waters (Morgan and Stumm, 1965). Since Mn (III) is thermodynamically unstable in a desert environment lacking abundant humic acids, Mn (IV) is the primary insoluble oxyhydroxide found in varnish, with average valencies of about +3.8 to +3.9 (Potter and Rossman, 1979a; McKeown and Post, 2001). The general Mn (II) oxidation reaction is:

$$Mn^{2+} + 1/2 \, O_2 + H_2O \rightarrow MnO_2 + 2H^+ \tag{1}$$

for an oxidation state of (IV).

The inherent heterogeneity of varnish becomes abundantly clear in analyses of minor elements. In order of decreasing concentration, these tend to be: Ca, Mg, Na, K, Ti and P. Like Mn, minor elements vary considerably (0.5–1.5%), with Ba and Sr as the most abundant trace elements, along with Cu, Ni, Zr, Pb, V, Co, La, Y, B, Cr, Sc and Yb found in all varnishes in order of decreasing concentration (Engel and Sharp, 1958). Elements Cd, W, Ag, Nb, Sn, Ga, Mo and Zn were found in some, but not all varnishes (Lakin et al., 1963). With the acquisition of data from newer techniques using electron microprobes (Liu, 2003), neutron activation (Bard, 1979), particle induced X-ray emission (PIXE; Dorn et al., 1990), inductively coupled plasma mass spectrometry (Thiagarajan and Lee, 2004), and ion microprobes (O'Hara et al., 1989), heterogeneity remains the rule of chemical composition at all spatial scales: from micron to micron within a single depth profile or chemical transect, between locations on the same rock, between different rocks from the same area and between different areas (Table 8.3). The palaeoenvironmental significance of variations in varnish chemistry with depth are discussed in section 8.6.

The following generalisations remain unchanged from the original observations of Engel and Sharp (1958).

1   Silica and aluminum, taken together, comprise the bulk of rock varnish – consistent with clays being the dominant mineralogy.
2   Manganese and iron oxides typically comprise one-quarter to one-third of rock varnish with high point-to-point variability at scales from nanometres to kilometres.
3   Minor elements display variable patterns. Elements Mg, K and Ca are correlated with clays in cation-exchange positions. Barium often correlates with S in barium sulphates (O'Hara et al., 1989; Dorn et al., 1990; Cremaschi, 1996). In other cases, Ba correlates with Mn (Liu, 2003). Ti can correlate with Fe in titanomagnetite detrital grains (Mancinelli et al., 2002), but more often Ti is not well correlated with any major element (Dorn, 1998).
4   Trace heavy metals generally correlate with Mn abundance, and sometimes with Fe, due to the scavenging properties of these oxyhydroxides (Jenne, 1968; Tebo et al., 2004; Thiagarajan and Lee, 2004).

Varnish minerals were originally reported to be amorphous (Engel and Sharp, 1958), with goethite (Scheffer et al., 1963) and ferric chamosite (Washburn, 1969) as important components. Seminal research conducted with infrared spectroscopy, X-ray diffraction and electron microscopy at the California Institute of Technology revealed that the bulk of rock varnish

Table 8.3 Examples of elemental variation exhibited in bulk chemical analyses of rock varnishes found in desert regions. Results are normalised to 100% with measurements by particle induced X-ray emission (Cahill, 1986), with NA indicating not available and BLD below the limit of detection

Site	Position	Element																		
		Na	Mg	Al	Si	P	S	K	Ca	Ti	Mn	Fe	Ni	Cu	Zn	Rb	Sr	Zr	Ba	Pb
Trail Fan, Death Valley	Former rock fracture	BLD	0.14	23.74	39.09	0.49	0.7	3.45	4.87	1.52	10.87	13.47	0.13	0.12	0.27	BLD	BLD	0.29	0.85	BLD
Manix Lake, Mojave Desert	>1 m above soil	1.1	3.44	25.77	32.35	1.15	0.3	2.11	1.35	0.84	12.47	18.09	BLD	0.22	0.3	0.25	0.21	0.22	0.19	0.74
Makanaka Till, Hawaii	With silica skin	0.62	1.98	21.13	29.77	0.69	0.2	3.3	4.89	0.73	13.6	21.13	BLD	0.33	0.49	BLD	BLD	BLD	0.16	0.98
Sinai Peninsula, Egypt	>1 m above soil	0.28	1.5	22.94	32.81	BLD	BLD	2.42	2.91	0.68	11.97	22.94	BLD	0.25	0.42	BLD	0.42	BLD	0.18	0.27
Petroglyph, South Australia	>1 m above soil	0.17	1.21	22.81	33.34	0.53	BLD	2.79	2.18	0.65	21.7	13.26	BLD	0.44	0.44	BLD	BLD	BLD	0.14	0.34
Ingenio, Peru Desert	At soil surface	NA	2.11	20.45	45.88	0.53	1.13	2.91	6.22	0.85	4.94	12.03	BLD	0.04	0.16	BLD	0.11	BLD	2.42	0.22
Ayers Rock, Australia	From rock fracture	NA	1.58	28.77	35.69	BLD	BLD	2.11	1.45	1.19	11.91	16.57	BLD	BLD	BLD	BLD	BLD	BLD	0.73	BLD

is composed of clay minerals (Potter and Rossman, 1977), dominantly illite, montmorillonite, and mixed-layer illite–montmorillonite. As previously noted, the layering seen in varnish at all scales reflects this clay mineralogy (Figures 8.4 and 8.6). Subsequent research has confirmed the dominance of clay minerals (Potter and Rossman, 1977, 1979c; Dorn and Oberlander, 1982; Krinsley et al., 1990, 1995; Israel et al., 1997; Dorn, 1998; Krinsley, 1998; Diaz et al., 2002; Probst et al., 2002), although there is some disagreement (Perry et al., 2006).

Clay minerals are cemented to the underlying rock by oxides of manganese and iron. Birnessite ($[Na,Ca]Mn_7O_{14} \cdot 2.8H_2O$) and birnessite-family minerals are the dominant manganese minerals in black varnish, and haematite is a major iron oxide in both black varnish and orange iron films (Potter and Rossman, 1977, 1978, 1979a,b,c; McKeown and Post, 2001). Birnessite-family minerals are found frequently in microbial deposits (Tebo and He, 1998; Tebo et al., 2005). McKeown and Post's (2001, p. 712) summary perspective suggests that:

> [e]ven if analysis methods are improved, the situation will remain complicated by the flexibility and great variety of Mn oxide structures. The common elements of these structures enable them to easily intergrow with and transform with one another. Furthermore, many of the phases, particularly the layered structures, readily exchange interlayer cations in response to even slight changes in chemistry on a microscale.

Imagery by HRTEM (Krinsley et al., 1995; Krinsley, 1998) reveals that the iron and manganese oxides that cement clay minerals together exist in this layered structure at the nanometre scale, derived from what appears to be sheaths of bacteria (Figure 8.7). The granular texture of the bacteria sheaths can be seen releasing nanometre-scale Mn–Fe that could be unit cells (Figure 8.6) into the weathered edges of the clay minerals (Fig 8.8) (Dorn, 1998). The ability to undergo incomplete transformation between phases at the nanometre scale would certainly be consistent with a disordered todorokite/birnessite-like phase (cf. Potter and Rossman, 1979a; McKeown and Post, 2001).

Detritus, both organic and inorganic, often settles in morphological depressions on varnish. These pieces are sometimes trapped by accreting rock varnish, leading to a wide variety of other minerals that are sometimes found in varnish samples such as quartz, feldspars and magnetite (Potter and Rossman, 1979a; Dorn, 1998, p. 198; Mancinelli et al., 2002).

8.5 Mechanisms of Formation or Accumulation

The genesis of most varnish constituents is explained easily. The clay mineral building blocks of varnish are ubiquitous in terrestrial weathering

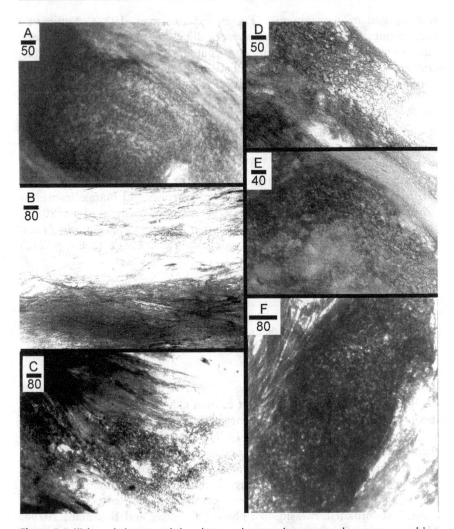

Figure 8.6 High resolution transmission electron microscopy imagery reveals manganese and iron minerals that appear to be moving from the granular remnants of bacterial sheaths into adjacent clay minerals. The granular textures found between layered varnish consist of Mn–Fe precipitates on bacterial remains (Krinsley, 1998). Images are of varnishes from: (A) Nasca, Peru; (B, E) Death Valley; (C, D) Kaho'olawe, Hawai'i; (F) Antarctica. Scale bars in nanometres.

environments where rock coatings are found (Dorn, 1998) and hence require simple dust deposition by aeolian processes. Minor and trace elements come adsorbed to clays, with subsequent enhancement by scavenging properties of Mn–Fe oxyhydroxides (Jenne, 1968; Forbes et al., 1976; Thiagarajan and Lee, 2004).

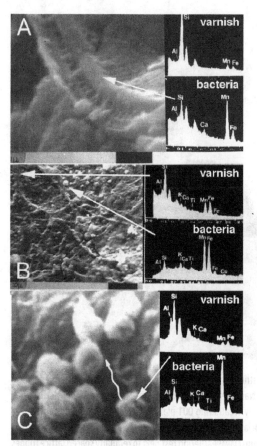

Figure 8.7 Budding bacteria morphologies actively concentrate manganese from (A) Crooked Creek, Kentucky, (B) Negev Desert and (C) Antarctica, as seen by energy dispersive 'spots' focused on the bacteria, compared with defocused analyses on adjacent uncolonised varnish. The Negev Desert hyphae resemble *Metallogenium* in morphology (Perfil'ev et al., 1965). The squiggly arrow in the Antarctica frame shows hyphae extending from bacterial-sized cells, exhibiting a *Pedomicrobium* budding bacteria morphology.

Explaining the great manganese enhancement in varnish (von Humboldt, 1812; Lucas, 1905; Engel and Sharp, 1958) has long been recognised as a key to understanding varnish formation. How clay minerals and manganese selectively interact, however, is just as important a key in discriminating relevant formative processes as concentrating Mn (Potter and Rossman, 1977). The mutual dependency of clay minerals and manganese can be realised by understanding what happens when one is lacking. Dust films, silica glaze and other rock coatings may result from clay mineral accretion without Mn (Dorn, 1998). Conversely, a different type of rock coating, heavy-metal skins, accrete when manganese precipitation occurs without clay minerals (see Chapter 8 in Dorn, 1998). Desert varnish accumulates only when and where nanometre-scale fragments of manganese, as seen in HRTEM imagery, cement broken and decayed fragments of clay minerals to rock surfaces.

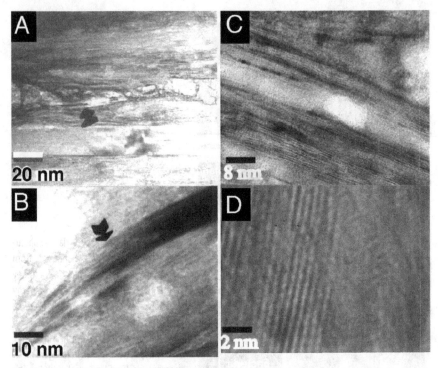

Figure 8.8 Clay minerals appear to be weathering by the insertion of Mn–Fe in samples from (A and B) Antarctica, (C) Peru and (D) Death Valley. Nanometre-scale cells of Mn–Fe appear within detrital clay mineral grains (arrows in A and B), essentially weathering 001 planes of exfoliation (Robert and Tessier, 1992). In many cases, the feathering is associated with granular textures (cf. Figure 8.6). At higher magnifications (C, D), disorganised wavy layers from clay weathering rest next to regularly spaced lattice fringes (with spacing that is consistent with illite, smectite, chlorite and interstratified clay – textural interstratification) (Robert et al., 1990). Scale bars in nanometres.

Four general conceptual models have been proposed to explain varnish formation (Figure 8.9). This section first presents the four proposed hypotheses of varnish genesis and then evaluates strengths and weaknesses.

8.5.1 Abiotic hypothesis of varnish genesis

Proponents of the abiotic hypothesis purport that varnish forms without the aid of biological enhancement of Mn (Linck, 1901; Engel and Sharp, 1958; Hooke et al., 1969; Moore and Elvidge, 1982; Smith and Whalley, 1988). To increase Mn:Fe ratios two to three orders of

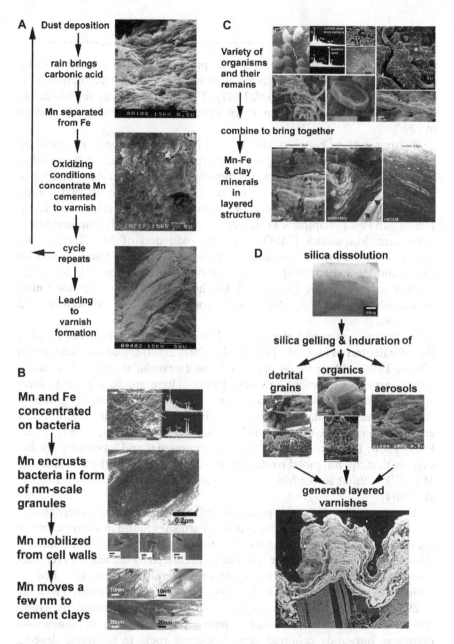

A

Dust deposition

↓

rain brings
carbonic acid

↓

Mn separated
from Fe

↓

Oxidizing
conditions
concentrate Mn
cemented
to varnish

↓

cycle
repeats

↓

Leading
to
varnish
formation

B

Mn and Fe
concentrated
on bacteria

↓

Mn encrusts
bacteria in form
of nm-scale
granules

↓

Mn mobilized
from cell walls

↓

Mn moves a
few nm to
cement clays

C

Variety of
organisms
and their
remains

↓

combine to bring together

Mn-Fe
& clay
minerals
in
layered
structure

D   silica dissolution

↓

silica gelling & induration of

detrital   organics   aerosols
grains

↓

generate layered
varnishes

Figure 8.9 Conceptual models of rock varnish formation. (A) Abiotic enhancement by iterations of acid solutions separating Mn^{2+} followed by oxidising conditions concentrates Mn in varnish (Hooke et al., 1969; Smith and Whalley, 1988). (B) Polygenetic clay–bacteria interactions where varnish formation starts with the oxidation and concentration of Mn (and Fe) by bacteria. Wetting events dissolve nanometre-scale fragments of Mn. Ubiquitous desert dust supplies interstratified clay minerals. The Mn-bacterial fragments fit into the weathered edges of clays, cementing clays much like mortar cements brick. (C) Lithobionts or their organic remains (e.g. spores, polysaccharides and other humic substances) play a role in binding varnish and concentrating Mn and Fe oxides and oxyhydroxides. (D) Silica binding of detrital grains, organics and aerosols (Perry and Kolb, 2003; Perry et al., 2006).

magnitude above crustal values, abiotic processes rely on greater mobility of divalent manganese over ferrous iron where small pH fluctuations dissolve Mn but not Fe (Krauskopf, 1957). The Mn released by slightly acidic precipitation is then fixed in clays after water evaporation or change in pH.

Redox reactions between Mn (II) and Mn (IV) are largely governed by pH. In natural waters, Mn (II) oxidation requires pH values >8.5 to oxidise homogeneously within weeks to months without microbial assistance (Morgan and Stumm, 1965). Laboratory experiments reveal that abiotic oxidation from Mn (II) to Mn (IV) requires two steps (Hem and Lind, 1983); Mn first precipitates as an oxyhydroxide (MnOOH) that then forms tetravalent Mn oxides (MnO_2). Abiotic oxidation of Mn (II) may be increased by the presence of abundant Fe oxyhydroxides found in varnish (Junta and Hochella, 1994), but Fe-catalysed oxidation results in Mn (III)-bearing oxyhydroxides (Junta and Hochella, 1994) that are not found in varnish samples (cf. Potter and Rossman, 1979c; McKeown and Post, 2001).

In contrast, microbial oxidation of Mn (II) to Mn (IV) is quite rapid at the near neutral pH values (Tebo et al., 1997, 2004) found on rock varnish (Dorn, 1998). The process appears to be extracellular where Mn oxides create casts on cell walls (Krinsley, 1998). There are both gram-positive and gram-negative bacteria that form a diverse phylogenetic group of sheathed and appendaged bacteria capable of oxidising Mn (e.g. Perfil'ev et al., 1965; Khak-mun, 1966; Dorn and Oberlander, 1981, 1982; Peck, 1986; Hungate et al., 1987). Environmental and laboratory studies suggest that the microbial oxidation of Mn (II) results in the formation of Mn (IV) without lower Mn (III) valences (Tebo and He, 1998; Tebo et al., 2004).

The geography of varnish poses critical obstacles, making purely abiotic formation of varnish difficult to explain. Varnishes found in wetter climates are typically higher in manganese than those in arid, more alkaline settings (Dorn, 1990) – a result that would not be predicted by the high pH requirements of abiotic oxidation. Black, Mn-rich varnishes occur in a host of moist environments such as within joint faces in glacial Norway (Whalley et al., 1990) and Iceland (Douglas et al., 1994). In contrast, hyperarid deserts are dominated by varnishes with a relative paucity of manganese; orange to burgundy coloured varnishes cover rocks in hyperarid deserts, such as the Atacama, and locally alkaline settings such as those adjacent to deflating salt playas in western North America (Dorn, 1998). Repeated oscillations in pH are the key to abiotic concentration of manganese over iron (cf. Engel and Sharp, 1958; Hooke et al., 1969; Smith and Whalley, 1988). However, the very environments where one would expect the least number of these pH fluctuations are the humid environments where

Mn-rich varnishes are most common (Dorn, 1990, 1998) – providing a geographical conundrum for the abiotic enhancement mechanism.

Other geographical problems with the abiotic hypothesis (Dorn, 1998, p. 242–243) include: rapid pH fluctuations in the range to mobilise and oxidise Mn where varnish does not commonly occur (in such places as the rainshadows of Mauna Loa and Mauna Kea on Hawai'i); purely inorganic Mn fixation experiments yielding too low concentrations of Mn; millimetre-scale distribution of varnish mimicking the pattern of microbial colonisation; an inability to explain botryoidal varnish textures by abiotic processes; and an inability to produce true varnish experimentally without microorganisms.

One of the most serious problems with the abiotic explanation rests in a simple analysis of rates of varnish formation. Based on detailed analyses of over 10,000 microbasins (Liu and Broecker, 2000, 2007; Liu, 2003, 2006), consistent with independent observations (Dorn, 1998), rates of varnish formation are on the order of a few microns per 1000 years. If Mn is enhanced abiotically, there would be no reason for such a slow rate of varnish formation. Multiple dust deposition and carbonic acid wetting iterations take place annually, even in drought years. Under this model, some Mn would be leached from dust with each event and in turn mix with other constituents to accrete in varnish layers. Such progressive leaching would be relentless without a rate-limiting step, and varnish accretion would be from 10^2 to 10^4 times faster than real varnishes, that is if abiotic enhancement led to varnish formation.

Although arguments potentially could be made against each of these problems with an abiotic explanation of Mn-enhancement, such accommodations have not yet been forthcoming in the literature. Abiotic processes are certainly involved in varnish formation (cf. Bao et al., 2001), including trace element enhancement with wetting (Thiagarajan and Lee, 2004) and the process of clay cementation (Potter and Rossman, 1977; Potter, 1979, pp. 174–175; Dorn, 1998). However, these processes alone would not generate a rock varnish.

8.5.2 Lithobionts or their organic remains produce and bind varnish

The general biological argument includes of a number of researchers who report observing microbial remains growing on varnish (e.g. Allen et al., 2004), or report organic compounds such as bacterial DNA (e.g. Perry et al., 2002) within varnish. Lichens (Laudermilk, 1931; Krumbein, 1971), cyanobacteria (Scheffer et al., 1963; Krumbein, 1969), microcolonial fungi (see Figure 8.3; cf. Staley et al., 1982; Allen et al., 2004, pp. 25), pollen

(White, 1924), peptides (Linck, 1928), refractory organic fragments (Staley et al., 1991), gram-negative non-sporulating cocci bacteria (Sterflinger et al., 1999), and amino acids from gram-positive chemo-organotrophic bacteria (Warsheid, 1990; Nagy et al., 1991; Perry et al., 2003) illustrate only spatial correlation with varnish (see Dorn (1998, pp. 238–9, 243) for a more complete list of references). Fatty acid methyl esters analysed from Mojave Desert varnish do not provide any evidence that adventitious gram-positive bacteria responsible for these fatty acids play a role in varnish formation (Schelble et al., 2005). A great leap forward in the study of microbes potentially involved in varnish genesis, however, rests in understanding phylogenetic affiliations and in estimating the total number of microbes actually present – in this case in fracture varnish collected from desert pavement cobbles in the Whipple Mountains of eastern California after winter rains (Kuhlman et al., 2005, 2006). About 10^8 cells of bacteria per gram dry weight of varnish summarises the story: varnish represents a habitat occupied by a plethora of bacteria, with a significant presence of gram-negative Proteobacteria groups and Actinobacteria (Kuhlman et al., 2005, 2006). Spatial contiguity of organisms and organic remains with rock varnish, however, is simply insufficient to explain great Mn-enhancement or clay cementation.

The aforementioned growing number of observations of apparently adventitious organisms and organic matter suffers from a core problem: there is no process by which these observed organisms or their remains produce rock varnish. The most cited of these purported agents – micro-colonial fungi (see Figure 8.3 and Borns et al., 1980; Staley et al., 1983; Gorbushina et al., 2002) – either dissolve or co-exist with varnish (Dragovich, 1993; Dorn, 1998). Even research on cultured Mn-oxidising bacteria (e.g. Krumbein, 1969; Krumbein and Jens, 1981; Palmer et al., 1985; Hungate et al., 1987) miss a key criteria. The bottom line is that seeing is believing, in this case morphologically similar shapes actually seen enhancing Mn *in situ* (Figure 8.7) (cf. Perfil'ev et al., 1965; Khak-mun, 1966; Dorn and Oberlander, 1981, 1982; Peck, 1986). Research into the microbial habitat of rock varnish is just starting, and the future rests in following in the footsteps of Kuhlman et al. (2005, 2006), as long as this research is matched with *in situ* studies.

Rates of varnish formation in desert environments should make readers very sceptical of the growing plethora of biological agents purported to be involved in varnishing. In places where varnishing rates have been measured in warm desert environments, typically one to twenty bacterial diameters accumulate each thousand years (Liu, 1994; Liu and Broecker, 2000). Rates would be several orders of magnitude higher if all of the observed rock-surface organisms, extracted organic matter associated with varnish and rock surfaces, and cultured biological agents actually play a role in making desert varnish. Consider the types of amino acids extracted from varnishes; the apparent lack of D-alloisoleucine and concentrations of serine suggest that the extracted amino acids are only centuries old – not the thousands of years that would indicate a formative role. If all of the 10^8 microbes per gram are involved in varnish formation, one should be able to see desert rock surfaces encrusted at rates such as seen in travertine contexts (cf. Chafetz et al., 1998). Furthermore, if these commonly seen organisms made varnish, cross-sections would look like biotically generated coatings where fossilised remains are common (Francis, 1921; Ferris et al., 1986; Konhauser et al., 1994; Chafetz et al., 1998; Kennedy et al., 2003).

Fossilised casts of Mn-enhanced bacterial remains are rare in desert varnish; the intra-varnish remains seen in HRTEM and back-scatter electron (BSE) imagery often resemble gram-negative appendaged bacteria in their morphology (Dorn and Meek, 1995; Krinsley, 1998; Dorn, 1998). Simply finding organic remains such as amino acids, DNA or even culturing Mn-oxidisers obfuscates the key question: what is the *in situ* evidence that certain microbes supply the Mn (and Fe) that cement clays to rock surfaces? Researchers wishing to advance any single agent to the role of formative agents must present clear evidence of *in situ* textural connections to formative processes involving Mn enhancement.

8.5.3 Silica binding model

The most recent conceptual model proposed to explain varnish formation focuses on silica as the key agent (Perry and Kolb, 2003; Perry et al., 2006). The hypothesis starts with dissolution of silica from anhydrous and hydrous minerals. The next step has the silica gelling, condensing and hardening. A romantic article title 'baking black opal in the desert sun' (Perry et al., 2006) emphasises the importance of the hot desert environment to this model. In the process of silica binding, detrital minerals, organics and dust from the local environment are bound into varnish. There is no ambiguity about the key role of silica in the envisioned process: '[s]ilicic acid and (di)silicic acid and polymers form complexes with ions and organic molecules, especially those enriched in hydroxy amino acids . . . The centrality of silica to this mechanism for cementing and forming coatings means that silica glazes and desert varnish should be considered part of the same class' (Perry et al., 2006, p. 520).

There is no question that the importance of amorphous hydrated silica (opal) and silica minerals (e.g. moganite) have not been emphasised in varnish research. Furthermore, the process of silica binding as envisioned (Perry and Kolb, 2003; Perry et al., 2006) would be compatible with other models of varnish formation. In other words, the movement of water that remobilises varnish constituents (Dorn and Krinsley, 1991; Krinsley, 1998) would be compatible with and could explain observations of silica (Perry and Kolb, 2003; Perry et al., 2006).

There are, however, multiple fatal flaws with silica binding as the key process in varnish formation. First, the silica binding process does not explain Mn-enhancement seen in varnish or the Mn-minerals seen in varnish. Second, although silica dissolution is slow on rock surfaces, it is not a rate-limiting step capable of slowing varnish accretion to the scale of a few microns per millennia. The formation of silica glaze within decades (Curtiss et al., 1985; Gordon and Dorn, 2005), perhaps from silica binding (Perry and Kolb, 2003; Perry et al., 2006) or perhaps from another process (chapter 13 in Dorn, 1998), indicates that this model cannot explain varnish accretion. Third, the silica binding model rides on the opinion that clay minerals are not a dominant part of rock varnish (Perry et al., 2006), a view incompatible with data sets several orders of magnitude more extensive (Potter and Rossman, 1977, 1979c; Dorn and Oberlander, 1982; Krinsley et al., 1990, 1995; Israel et al., 1997; Dorn, 1998; Krinsley, 1998; Diaz et al., 2002; Probst et al., 2002; Liu, 2003; Liu and Broecker, 2007). Fourth, this model does not accommodate the geography of rock coatings. For example, why would silica glazes dominate on Hawaiian basalt flows (Curtiss et al., 1985; Gordon and Dorn, 2005) and not manganiferous rock varnish? Abundant examples of this juxtaposition exist (e.g. Dorn, 1998), and one of the closest Mars analogue exists on the Tibetan Plateau, where rock varnish rests adjacent to silica glaze and other rock coatings (Dorn, 1998, pp. 367–371). The silica binding model cannot explain this geographical pattern. Fifth, 'baking' is a requirement of the silica binding model, but varnish occurs in subsurface and cold-climate settings where the requisite heat and light do not exist (Anderson and Sollid, 1971; Douglas, 1987; Dorn and Dragovich, 1990; Whalley et al., 1990; Dorn et al., 1992; Douglas et al., 1994; Villa et al., 1995; Dorn, 1998). Sixth, the silica binding process does not explain the characteristics of the varnish microlaminations data set discussed in a later section.

8.5.4 Polygenetic model

The polygenetic model of rock varnish formation combines biotic enhancement of Mn with abiotic processes (Dorn, 1998). It builds on the mineralogy research of Potter and Rossman (1979a,b,c) and studies of bacteria fossil remains within varnish layers (Dorn and Meek, 1995; Krinsley et al., 1995; Dorn, 1998; Krinsley, 1998). In brief, weathered remains of Mn-rich bacterial casts (Figure 8.6) cement weathered clay minerals to rock surfaces.

Varnish formation probably starts with ubiquitous dust deposition on rock surfaces; budding bacteria (Dorn and Oberlander, 1982) actively concentrate Mn and Fe (Figure 8.7). Some of these bacterial 'sheaths' become microfossils (Dorn and Meek, 1995; Dorn, 1998; Krinsley, 1998) that then decay (Figure 8.6). Enhancement of Mn by other microbes, such as gram-positive bacteria, could certainly contribute; but *in situ* evidence of Mn-enhancement and preservation of Mn–Fe 'bacterial casts' has heretofore been lacking for morphologies other than budding or append-aged bacterial forms. The exact mechanism for oxidation of Mn^{2+} is uncertain, perhaps involving heteropolysaccharides containing the protein that catalyses the oxidation reaction with molecular oxygen. Mechanisms for Mn^{2+} binding and oxidation, however, have yet to be demonstrated for varnish.

Decay of Mn–Fe-encrusted bacterial casts mobilises nanometre-scale Mn and Fe and then cements clay minerals (Figure 8.8). At the highest HRTEM magnifications, irregular layers of Mn–Fe-cemented clays rest next to regularly spaced lattice fringes with spacing that is consistent with illite, smectite, chlorite and interstratified clay – textural interstratification. The reader should note that this discussion is at the nanometre-scale and that the interlayering of different types of rock coatings discussed elsewhere is at the much coarser micrometre scale. Potter (1979, pp. 174–175) argued for this step without the benefit of supporting HRTEM imagery:

> Deposition of the manganese and iron oxides within the clay matrix might then cement the clay layer . . . the hexagonal arrangement of the oxygens in either the tetrahedral or octahedral layers of the clay minerals could form a suitable template for crystallisation of the layered structures of birnessite. The average 0–0 distance of the tetrahedral layer is 3.00 Å in illite-montmorillonite mixed-layered clays, which differs only 3.4 percent from the 2.90 Å distance of the hexagonally closed-packed oxygens in birnessite . . .

Varnish formation, then, is a nanometre-scale marriage between clay minerals and the millennial accumulation and decay of Mn–Fe-encrusted bacterial casts. Nanometre-sized Mn–Fe remains of bacteria both feather and cement clay minerals to the rock and to prior varnish.

Any successful model of varnish formation, however, must also explain additional issues associated with varnish chemistry. First, iron is enhanced slightly in varnish over crustal and adjacent soil concentrations. Although most of the iron derives from the clay minerals, some is biotically enhanced (see Figure 8.7 and Dorn and Oberlander, 1982; Adams et al., 1992; Sterflinger et al., 1999) as the nanometre-scale Fe seen in the bacterial casts (Krinsley, 1998) is also found reworked into the weathered remnants of clay minerals seen in HRTEM imagery. Considerable HRTEM work, however, is needed to better document processes of minor iron enhancement.

Second, a number of trace elements are enhanced in rock varnish (Engel and Sharp, 1958; Lakin et al., 1963; Bard, 1979). This enrichment is explained abiotically by Thiagarajan and Lee (2004); trace element enrichment starts when acidic pH values in rainwater *c.* 5.7 dissolve rare earths, Co, Ni, Pb, Sr, Rb, Cs and other trace elements from dust particles on rock surfaces. Coprecipitation of these trace elements with Mn–Fe oxyhydroxides leaves behind the aeolian dust particles that then blow away. The net process is the sort of concept proposed for the abiotic enhancement of manganese – analogous to a water filter that slowly traps water impurities; in the case of varnish, the dust particles slowly leave behind trace elements trapped by the Mn–Fe minerals. In addition, the opal observed by Perry et al. (2006) could be precipitated as water moves throughout and redistributes varnish constituents (Dorn and Krinsley, 1991; Krinsley, 1998).

8.5.5 Evaluating model strengths and weaknesses

In addition to the great enhancement of Mn, other criteria have been proposed to adjudicate these different hypotheses (Table 8.4). No single model explains all observed aspects of Table 8.4, but three of the proposed models run into severe difficulties when they are tasked to explain key varnish characteristics: abiotic enhancement of manganese; lithobionts or their organic remains produce varnish; and silica binding. These three problematic models, for example, have no severe rate-limiting step. The processes suggested would all produce rock varnish very rapidly, from 100 to 10,000 times faster than the $1-10 \mu m \, kyr^{-1}$ rates observed in nature (Table 8.4).

The viable polygenetic model, in contrast, turns on the very slow processes by which manganese enhancement takes place and very slow processes by which manganese and iron interact with clay minerals. A significant advantage of the polygenetic model is that it does not suffer from the fatal flaws extant in the other three models (Table 8.5). A minor advantage of the polygenetic model is that it is fully compatible with processes in the silica binding model and the presence of adventitious lithobionts and their organic remains in varnish. An issue for the polygenetic model, however, is that the abiotic enhancement hypothesis has not been falsified – a very difficult task given the rate of varnish formation.

8.6 Palaeoenvironmental Significance

Varnish sometimes serves as an indicator of palaeoenvironments even before the Quaternary (Dorn and Dickinson, 1989; Marchant et al., 1996). A buried Miocene hillslope in southern Arizona, for example, hosts a fossilised debris slope where many of the small boulders are covered by fossilised varnish (Dorn and Dickinson, 1989). Simply finding an isolated clast of rock varnish in buried contexts cannot be used to infer any

Table 8.4 Criteria that have been used to adjudicate competing models of rock varnish formation

Criteria	A successful model must be able to explain . . .
Accretion rate	. . . typical rates of accretion on the order of 1–10 µm per millennia (Dorn, 1998; Liu and Broecker, 2000). Although faster-growing varnishes occur (Dorn and Meek, 1995), such varnish accretion rates (Liu, 2003, 2006; Liu and Broecker, 2007) demand an extreme rate-limiting step for whatever process makes varnish
Clay minerals	. . . the dominance of clay minerals in rock varnish (Potter and Rossman, 1977, 1979c; Dorn and Oberlander, 1982; Krinsley, 1998; Krinsley et al., 1990; Krinsley et al., 1995; Israel et al., 1997; Dorn, 1998; Broecker and Liu, 2001; Diaz et al., 2002; Probst et al., 2002)
Fe behaviour	. . . the differential enhancement of iron in different varnish microlamination (VML) (Liu et al., 2000; Broecker and Liu, 2001; Liu and Broecker, 2007) and different places (Adams et al., 1992; Dorn, 1998; Allen et al., 2004)
Laboratory creation	. . . the creation of artificial varnish coatings (Krumbein and Jens, 1981; Dorn and Oberlander, 1981, 1982; Jones, 1991) may be considered by some to be a vital criteria. However, given the extraordinary time-scale jump between any laboratory experiment and natural varnish formation, and the extreme rate-limiting step involved in natural varnish formation, rigid application of this criteria may be problematic
Lithobionts and organic remains	. . . the occurrence of different types of lithobionts and the nature of organic remains. The plethora of *in situ* and cultured organisms, as well as spores, polysaccharides and other humic substances associated with varnish (see review in Dorn, 1998; Gonzalez et al., 1999; Netoff, 2001; Gorbushina et al., 2002; Kurtz and Netoff, 2001; Perry et al., 2002, 2003; Perry and Kolb, 2003; Allen et al., 2004; Viles and Goudie, 2004; Kuhlman et al., 2005, 2006; Schelble et al., 2005) must be explained

Table 8.4 *Continued*

Criteria	A successful model must be able to explain . . .
Mn enhancement	. . . the enhancement of Mn typically more than a factor of 50 above potential source materials (von Humboldt, 1812; Lucas, 1905; Engel and Sharp, 1958)
Mn-mineralogy	. . . Mn mineralogy characteristic of birnessite-family minerals (Potter, 1979; Potter and Rossman, 1979a,b; McKeown and Post, 2001; Probst et al., 2001). Very difficult to study because of its small size, Mn oxides range from large tunnel phases to layer phases such as the birnessite family notable in varnish
Not just a few samples	. . . observations at sites around the world. The cost of microanalytical investigations demands analyses of only a few samples at a time. Nonetheless, Oberlander (1994, p 118) emphasises: 'researchers should be warned against generalising too confidently from studies of single localities'
Paucity of microfossils	. . . the extremely infrequent occurrence of preserved microfossils. Examination of numerous sedimentary microbasins (Liu, 2003, 2006; Liu and Broecker, 2007), and decades of research has generated only a few observations of microfossils (Dorn and Meek, 1995; Dorn, 1998; Krinsley, 1998; Flood et al., 2003)
Rock coating geography	. . . why does rock varnish grow in one place and other rock coatings elsewhere. Over a dozen major types of coatings form on terrestrial rock surfaces (Dorn, 1998). A successful model must be able to why, for example, rock varnish grows with iron films, silica glaze, phosphate skins and oxalate crusts in the Khumbu of Nepal (Dorn, 1998, pp. 360–361) and with dust films, carbonate crust, phosphate film, silica glaze and oxalate crusts in Tibet (Dorn, 1998, pp. 367–369).

Table 8.4 *Continued*

Criteria	A successful model must be able to explain . . .
Rock varnish variety and geography	. . . why different types of rock varnishes occur where they occur. Rock varnish classification must be dealt with, a very complex topic summarised in Dorn (1998: 214–224). Consider just the geography of light. Although light can be a key issue for some rock coatings (McKnight et al., 1988; Nienow et al., 1988), formation of varnish in locales with little or no light (Douglas, 1987; Dorn and Dragovich, 1990; Whalley et al., 1990; Douglas et al., 1994; Villa et al., 1995; Dorn, 1998) places strong constraints on light-related processes.
Varnish microlamination	. . . the revolution in VML understanding. Over ten thousand sedimentary microbasins analysed by Liu (Liu, 1994, 2003, 2006; Liu and Broecker, 2000, 2007; Liu et al., 2000; Broecker and Liu, 2001), a method subject to blind testing (Marston, 2003), reveals clear late Pleistocene and Holocene patterns in abundance of major varnish constituents connected to climate change. The characteristics of this single largest varnish data set must be explained.

particular past climate (Dorn, 1990). However, entire slopes or pavements (Marchant et al., 1996) covered with varnished boulders permits palaeo-climatic inference. In southern Arizona, a reasonable inference is that the occurrence of rock varnish in this buried position (Dorn and Dickinson, 1989) indicates an arid to semi-arid climate.

Microstratigraphic layers in actively forming varnish also provide pal-aeoenvironmental insights. A wide variety of varnish properties have been used to infer Quaternary palaeoenvironmental change. Micromorphologi-cal change (Dorn, 1986), trace element geochemistry such as lead profiling (Dorn, 1998; Fleisher et al., 1999), ^{17}O in sulphates (Bao et al., 2001), organic carbon ratios (Dorn et al., 2000), and interlayering with other rock coatings (Dragovich, 1986; Dorn, 1998) exemplify the potential of varnish microstratigraphy to inform on environmental change.

One key to extracting palaeoenvironmental data rests in sampling the same type of varnishes at sites used to calibrate the varnish proxy record

Table 8.5 Performance of alternative rock varnish conceptual models with respect to adjudicating criteria detailed in Table 8.4

Criteria	Model			
	Abiotic Mn enhancement	Lithobionts or their remains	Polygenetic clay–bacteria–inorganic interactions	Silica binding
Accretion rate	There is no rate-limiting step that would slow the accretion rate to fit observations. Even given the lowest dust deposition rates and fewest wetting–acidification events, this process would be 10^2 faster than observed rates	If only a fraction of the reported organic remains or agencies precipitated Mn and Fe, formation would be more than 10^3 faster than observed rates	The rare Mn enhancement by a few budding bacteria, observed in situ, would generate observed rates	The silica binding process has no rate-limiting step to slow growth. Under this model, there is no reason why decadal growth rates for silica glazes (Curtiss et al., 1985; Gordon and Dorn, 2005) would not also apply to rock varnish
Clay minerals	Clay minerals are not important, and hence there is no reason why they should dominate in varnish	Clay minerals are not important, and hence there is no reason why they should dominate in varnish	Clay minerals are critical to this model, and the model explains their dominance. Mn mobilised from bacteria fit into octahedral positions, cementing clays	Proponents assert that clay minerals are not important, and hence there is no reason why they should dominate in varnish

Fe behaviour	Highly oxidising conditions in dry climates or in microsites would lead to the observed Fe behaviour	Different organisms might favour the observed iron behaviour in different conditions and settings	Highly oxidising conditions in dry climates or in microsites would lead to the observed behaviour, since these bacteria do not favour highly oxidising settings	The observed iron behaviour is not important to this process.
Laboratory creation	Replication has occurred with similarities and substantial differences	Replication has occurred with similarities and substantial differences	Replication has occurred with similarities and substantial differences	Replication has occurred with similarities and substantial differences
Lithobionts and organic remains	The occurrence of typically <1% of organic matter is readily explained by purely adventitious incorporation as Mn enhancement occurs	The occurrence is a prerequisite of this model	The occurrence of <1% of organic matter is readily explained by adventitious organics incorporated into varnish	The occurrence of organic matter is predicted by silica binding processes

Table 8.5 *Continued*

Criteria	Model			
	Abiotic Mn enhancement	Lithobionts or their remains	Polygenetic clay-bacteria-inorganic interactions	Silica binding
Mn enhancement	This model was developed to explain Mn enhancement	Abundant culturing research documents the presence of Mn-oxidising organisms	This model explicitly explains Mn enhancement, followed by clay cementation	Mn enhancement is not explained by silica dissolution or reprecipitation
Mn mineralogy	Mn mineralogy is not explained.	Mn mineralogy is consistent with biotic enhancement of Mn	This model explains the fit of nanometer-scale Mn oxides into clay minerals, with detailed HRTEM observations (Dorn, 1998; Krinsley, 1998) supporting earlier hypotheses (Potter, 1979, pp. 174–175; Potter and Rossman, 1977).	Mn mineralogy is not explained.

Not just a few samples	The model's discussion in the literature has been fit to only a few site contexts	Individual studies involve only a very few sites, due to the expenses involved, but cumulative observations have a wide distribution	This model has been applied globally and is consistent with Liu's varnish microlamination (VML) observations of over 10^4 microsedimentary basins	The model is too new to have been assessed at sites globally or articulated to the VML data set.
Paucity of microfossils	This model would predict the paucity of microfossils	Several studies are notable for presence of microfossils, while others are notable for analysis of organic residues only	This model explains the paucity of microfossils, because Mn-encrusted cell walls are redistributed at the nanometer scale	Model proponents hope to use microfossils to explore extraterrestrial life, but this makes little sense given poor preservation
Rock coating geography	The presence of rock varnish in a plethora of settings that lack repeated oxidation cycles challenges this model	This model has not yet explained the geography of rock coatings	This model explains the geography of rock coatings (Dorn, 1998)	This model has not yet explained why rock varnish or silica glazes exists at one locale and not in other geographic settings

Table 8.5 *Continued*

	Model			
Criteria	Abiotic Mn enhancement	Lithobionts or their remains	Polygenetic clay–bacteria–inorganic interactions	Silica binding
Rock varnish variety and geography	This model has not yet explained the variety and geography of rock varnish	This model has not yet explained the variety and geography of rock varnish	This model explains the variety and geography of rock varnish (chapter 10 in Dorn, 1998)	This model requires light and heat. Yet, both are known to be unnecessary for varnish formation (Anderson and Sollid, 1971; Douglas, 1987; Dorn and Dragovich, 1990; Whalley et al., 1990; Dorn et al., 1992; Douglas et al., 1994; Villa et al., 1995; Dorn, 1998)
Varnish microlamination	This model could explain VML observations, whereby conditions of greater alkalinity and fewer wetting events would generate drier VML events and more wetting events would generate wetter VML events	Model proponents have not yet explored the implications of the VML record	This model explains VML observations. Specific HRTEM observations in different VML reveal more nanometre-scale bacterial fragments in wet periods	Wording accommodates alternating climatic changes producing VML. However, the silica binding process does not explain VML

and unknown sites (Dorn, 1998, pp. 214–224). Another key rests in avoiding post-depositional modification of varnish strata, such as erosion by microcolonial fungi (Figure 8.3) or cation leaching and redepositional processes (Dorn and Krinsley, 1991; Krinsley, 1998).

Of the variety of potential palaeoenvironmental and dating tools yet proposed for varnish (Table 8.6), the most potent method is varnish micro-laminations (VML). Black, yellow and orange layers in varnish reflect regional climatic changes (Dorn, 1990; Cremaschi, 1996; Liu and Dorn, 1996; Zhou et al., 2000; Broecker and Liu, 2001; Diaz et al., 2002; Lee and Bland, 2003; Liu, 2003, 2006), where black layers, rich in manganese, record wet intervals. Orange and yellow layers with less manganese record drier periods (Jones, 1991). A blind test of the VML method (Liu, 2003; Marston, 2003), communicated and refereed by the editor of *Geomorphology*, resulted in this conclusion:

> This issue contains two articles that together constitute a blind test of the utility of rock varnish microstratigraphy as an indicator of the age of a Quaternary basalt flow in the Mohave Desert. This test should be of special interest to those who have followed the debate over whether varnish microstratigraphy provides a reliable dating tool, a debate that has reached disturbing levels of acrimony in the literature. Fred Phillips (New Mexico Tech) utilised cosmogenic ^{36}Cl dating, and Liu (Lamont-Doherty Earth Observatory, Columbia University) utilised rock varnish microstratigraphy to obtain the ages of five different flows, two of which had been dated in previous work and three of which had never been dated. The manuscripts were submitted and reviewed with neither author aware of the results of the other. Once the manuscripts were revised and accepted, the results were shared so each author could compare and contrast results obtained by the two methods. In four of the five cases, dates obtained by the two methods were in close agreement. Independent dates obtained by Phillips and Liu on the Cima 'I' flow did not agree as well, but this may be attributed to the two authors having sampled at slightly different sites, which may have in fact been from flows of contrasting age. Results of the blind test provide convincing evidence that varnish microstratigraphy is a valid dating tool to estimate surface exposure ages (Marston, 2003).

The method works in the western USA (Liu et al., 2000), western China (Zhou et al., 2000), North Africa (Cremaschi, 1996) and Patagonia (T. Liu, personal communication, 2001), with the notation that these different regions require different calibrations. Liu's (2003, 2006) calibration for the Great Basin of the western USA extends back to before 140,000 ka. Greyscale illustrations do not do the method justice, but Figure 8.10 shows that progressively older varnishes display progressively more detailed VML sequences.

The great power of the VML method rests in obtaining two types of information at once: palaeoenvironmental sequence and calibrated age. For

Table 8.6 Different methods that have been used to assess rock varnish chronometry. The refinement of age resolution is either relative, calibrated by independent age control, correlated to discrete events, or numerical based on radiometric measurements

Method and age resolution	Synopsis of method
Accumulation of Mn and Fe: calibrated age	As more varnish accumulates, the mass of manganese and iron gradually increases. Occasionally this old idea is resurrected, but it has long ago been demonstrated to yield inaccurate results in tests against independent control (Bard, 1979; Dorn, 2000)
Appearance: relative age	The appearance of a surface darkens over time as varnish thickens and increases in coverage. However, there are too many exceptions to permit accurate or precise assignment of ages based on visual appearance. For example, varnish can form in under 100 years in selected microenvironments (Krumbein, 1969; Dorn and Meek, 1995). There is no known method that yields reliable results (Dorn, 2000), although individuals have tried to make visual measurements using different techniques
Cation-ratio dating: calibrated age	Rock varnish contains elements that are leached (washed out) at fast rates and elements that are not leached rapidly (Dorn and Krinsley, 1991; Krinsley, 1998). Over time, a ratio of leached to immobile declines over time (Dorn, 2000). While differences in field and laboratory sampling procedures have yielded problematic results in multiple investigations attempting replication (e.g. Bamforth, 1997; Watchman, 2000), if the correct type of varnish is used, the method performs well in blind tests (Loendorf, 1991). It has been replicated in such places as China (Zhang et al., 1990), Israel (Patyk-Kara et al., 1997), South Africa (Whitley and Annegarn, 1994) southern Nevada (Whitney and Harrington, 1993), and elsewhere

Table 8.6 *Continued*

Method and age resolution	Synopsis of method
Foreign material analysis: correlated age	Rock carvings made historically may have used steel. The presence of steel remains embedded in a carving would invalidate claims of antiquity, whereas presence of such material as quartz would be consistent with prehistoric age (Whitley et al., 1999). Meticulous sampling and examination of cross-section with electron microprobe is required to determine if steel remains exist in a petroglyph groove
Lead profiles: correlated age	Twentieth century lead and other metal pollution is recorded in rock varnish, because the iron and manganese in varnish scavenges lead and other metals. This leads to a 'spike' in the very surface layer from 20th century pollution. Confidence is reasonably high, because the method (Dorn, 1998) has been replicated (Fleisher et al., 1999; Thiagarajan and Lee, 2004; Hodge et al., 2005) with no publications yet critical of the technique. The method uses sampling and measurement of chemical changes with depth on the scale of microns or finer, analysed with an instrument such as an electron microprobe, ion microprobe or alpha spectrometer
Organic carbon ratio: calibrated age	Organic carbon exists in an open system in the rock varnish that covers petroglyphs. This method compares the more mobile carbon and the more stable carbon. The method is best used in soil settings (Harrison and Frink, 2000), but it has been applied experimentally to rock varnish in desert pavements (Dorn et al., 2000). The method requires careful mechanical extraction of rock varnish, that is then subjected to basic soil wet chemistry procedures

Table 8.6 *Continued*

Method and age resolution	Synopsis of method
Radiocarbon dating of carbonate: numerical age	Calcium carbonate sometimes forms over varnish, and can be radiocarbon dated, providing a minimum age for such features as rock art. The method has been used in Australia (Dragovich, 1986) and eastern California (Cerveny et al., 2006)
Radiocarbon dating of organic material: numerical age	The hope is that carbon trapped by coating provides a minimum age for the petroglyph. First developed in 1986, Watchman (1997) and Dorn (1997) both found the presence of organic carbon that pre-dates and post-dates the exposure of the rock surface — in this case a blind test of petroglyph dating in Portugal. The only person who still uses organic carbon of unknown residues in radiocarbon dating (Huyge et al., 2001; Watchman, 2000), Watchman now admits that he has not tested results against independent controls (Watchman, 2002; Whitley and Simon, 2002a,b).
Radiocarbon dating of oxalate: numerical age	The inorganic mineral oxalate (e.g. whewellite: $CaC_2O_4 \cdot H_2O$) sometimes deposits on top of or underneath rock varnish. Because this mineral contains datable carbon, the radiocarbon age can provide a minimum age for the underlying or overlying varnish. The most reliable research on radiocarbon dating of oxalates in rock surface contexts has been conducted in west Texas (Rowe, 2000; Russ et al., 2000)
Varnish microlaminations: correlated age	Climate fluctuations change the patterns of layers deposited in varnish. The confidence level is high, because the method (Liu, 2006) has been replicated in a rigorous blind test discussed in the text (Marston, 2003), and the method is based on analyses of over 10,000 rock microbasins

example, research on the alluvial fans of Death Valley (Figure 8.10) permit the first-ever high-resolution assessment of the linkage between fan chronology and palaeoenvironmental fluctuations (Liu and Dorn, 1996). In another example, microlaminations permit explorations of the relationship between archaeology and palaeoclimate (Dorn, 2000, 2006; Dorn et al., 2000).

For those interested in applying varnish to Quaternary research, VMLs offer the ability to link geomorphological and archaeological chronology with regional palaeoenvironmental fluctuations (Liu, 2006). Before they embark on this methodology, researchers should internalise two issues. First, the making of ultra-thin sections is extraordinarily difficult; sections a bit too thick are opaque, whereas sections too thin erode away easily – requiring great patience and a light enough laboratory touch. Second, Marston's (2003) careful handling of this blind test of microlaminations, as well as other blind tests (Loendorf, 1991; Arrowsmith et al., 1998), underscores the importance of understanding classification (Dorn, 1998, pp. 214–224) prior to the sampling phase.

8.7 Relationships to other Terrestrial Geochemical Sediments

Rock varnish rarely exists as a discrete geochemical sedimentary deposit. Aspect (Waragai, 1998) and slope position (Palmer, 2002), for example, influence metre-scale facies changes. Vertical interlayering takes place on the scale of microns with every other type of rock coating (Table 8.1). Examples of interlayering include: with oxalates and lithobiontic coatings in waterflow locales; with nitrate crusts in caves; with silica glaze in desert pavements; with phosphates near locales of decomposition of bird excrement; with carbonate skins in archaeological ground figures; with case hardening agents; with anthropogenic pigment in rock art; or with heavy metal skins in periglacial meltwater streams. At least 102 publications have thus far examined varnish interlayered with other types of rock coatings – thus unknowingly mixing 'apples and oranges' in possible misinterpretations.

There have been a number of suggestions in the literature (cf. chapters 10 and 13 in Dorn, 1998) equating silica glazes, coatings of largely amorphous silica, with abiotic formation of desert varnish. In silica glaze formation, soluble Al–Si complexes $[Al(OSi(OH)_3)^{2+}]$ are first released from the weathering of phyllosilicate minerals. These Al–Si complexes are then stabilised through very gentle wetting such as dew deposition or movement of capillary water on rock surfaces. Once stabilised, silicic acid and these Al–Si complexes more readily bond the silica glaze coating to the rock (Dorn, 1998, p. 319). In HRTEM imagery, nanometre-scale bands of silica glaze do experience facies changes with varnish, and in BSE imagery,

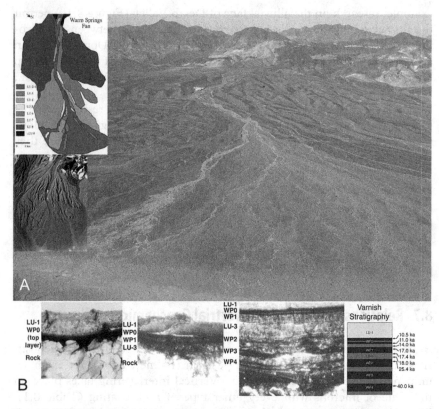

Figure 8.10 Varnish microlaminations (VML) are the most powerful 'tool' yet developed in rock varnish research. This figure illustrates the potential of this tool to yield unique insights in geomorphological research. (A) Looking at this oblique photograph of the Warm Springs alluvial fan in Death Valley, there would be a tendency for most of today's desert geomorphologists to believe in the importance of climatic change in fan development (Farr and Chadwick, 1996; Harvey et al., 1999). Such thinking, however, is not supported by the VML method. Although VML provides the highest resolution palaeoclimatic information yet available for alluvial fans, this detailed mapping fails to reveal a clear relationship between climatic changes and alluvial-fan development (Dorn, 1996). Climatic change may play a role, but this hypothesis is not testable – even with the highest resolution method of VML. High-magnitude and high-frequency palaeoclimatic fluctuations occur with greater frequency than can be related to aggradational events. The inset map is modified from Liu and Dorn (1996), with the inset satellite image from NASA. (B) Varnish microlaminations from Death Valley, California, illustrate progressively more complex stratigraphic sequences in late Pleistocene varnishes. The uppermost LU-1 layer appears yellow in ultrathin section. The other light grey layers appear orange. Dark layers appear black. This figure was modified from Liu and Dorn (1996), with the calibration based on Liu (2003).

micrometre-scale bands of silica glaze interlayer with varnish. Thus, there could also be a cementation role for silica glaze in varnish accretion.

Interlayering of different types of rock coatings, while creating the strong potential for gross misinterpretation, creates excellent opportunities to apply rock coatings to preservation of stone. Natural settings illustrate the importance of mixing different types of cementing agents (Figure 8.11). Iron films play the natural analogue of artificial bonding agents where sandstone grains are held tightly. Iron films create a harder rock face than a more porous cementing mixture of varnish and silica glaze. However, the intercalation of varnish and silica glaze permits much more interstitial water to exit such lithologies as porous sandstone. The stronger cementing agent, iron film, enhances decay in the underlying weathering rind (Tratebas et al., 2004). Whereas spalling under the iron film-hardened sandstone results in rapid erosion of the decayed sandstone, the 'varnish + silica glaze' mixture displays slower and less dramatic erosion (Tratebas et al., 2004).

Yet another example of using the interlayering of varnish with other rock coatings is seen in the prehistoric modification of landforms. Consider geoglyphs created by the disturbance of a desert pavement. Boulders are often piled, reorienting the natural varnish sequence. Finding subaerial black varnish shoved into the soil results in the 'freezing' of the varnish microlaminations sequence (Liu, 2003) underneath the superposition of a soil-contact iron film. At the same time, subaerial microlaminations form on top of the orange varnish. Boulders thrust far enough into the ground may form calcrete, material with potential for understanding the palaeoen-

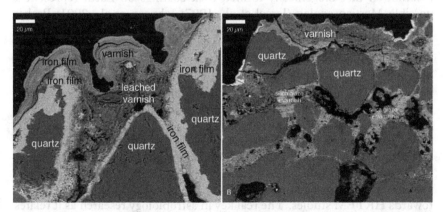

Figure 8.11 Rock varnish interlayers with iron film and silica glaze at Whoopup Canyon, Wyoming. (A) Iron film (back-scatter electron image) acts as a case-hardening agent, and rock varnish accretes on top of the iron film exposed by petroglyph manufacturing. (B) Varnish actively assists in case hardening (back-scatter electron image) when the leached cations reprecipitate with silica glaze in sandstone pores.

vironment (Deutz et al., 2001) and timing (Chen and Polach, 1986) of the disturbance. Thus, understanding all variations of rock varnish and its interdigitation with other rock coatings creates immense possibilities in analysing ancient modifications to stone (N. Cerveny et al., 2006).

8.8 Summary and Directions for Future Research

Desert varnish is a rock coating that is brown to black colour and is characterised by clay minerals cemented to rock surfaces by oxyhydroxides of manganese and iron. The individual Mn–Fe minerals are nanometre in size, a result of the decay of bacterial casts. These remnants of bacteria dissolve and reprecipitate in clay minerals – cementing the mix of clays and oxyhydroxides to rock surfaces. Since varnish forms in all terrestrial environments, 'rock varnish' is often used as a synonym, but the term 'desert varnish' is common because varnishes are most geochemically stable in arid regions. The constituents in varnish accrete on the host rock with thicknesses typically less than 0.1 mm. Usually dull in lustre, its occasional sheen comes from a smooth surface micromorphology in combination with manganese enrichment at the very surface of the varnish. The most useful application of rock varnish rests in the layering pattern of black (manganese-rich) layers and orange (manganese-poor) layers, which reveal past environmental changes experienced by rock surfaces.

Rock varnish is beginning to see an infusion of research funding related to the search for life on Mars by NASA (DiGregorio, 2002), using the assumption that biogenically formed rock coatings on Earth represent a possible Martian analog (Israel et al., 1997; Johnson et al., 2001; Probst et al., 2001; Gorbushina et al., 2002; Wierzchos et al., 2003; Allen et al., 2004; Kuhlman et al., 2005, 2006; Perry et al., 2006). One need in this research arena rests in understanding the transition between Mn–Fe casts of Mn-oxidising organisms on varnish surfaces to their preservation within varnish, and onto their remobilisation into a clay matrix. Cold-dry contexts such as the Tibetan Plateau (Dorn, 1998) and Antarctica (Dorn et al., 1992; Krinsley, 1998) reveal evidence of this process. Simply studying the remains of organisms, most of which are purely adventitious, in warm deserts will inevitably lead to misinterpretation and research dead-ends. Exploring how biogenic remains in the coldest and driest terrestrial settings translate into biogenic coatings, again, requires a penchant towards HRTEM studies. The real key in astrobiology research as it relates to varnish, however, does not rest in understanding terrestrial varnish; the largest gap is the need to falsify abiotic explanations of Mn enhancement on Mars. Imagine when the astrobiology community discovers Mn-rich rock coatings on Mars, claiming evidence of life – only to encounter the reality that simple mechanisms of abiotic enhancement of Mn have not been falsified for Mars.

The greatest need for future research in terrestrial varnish research rests with individuals who are able to carry out extraordinary care in the preparation of laboratory samples such as microlaminations (Liu, 2003; Marston, 2003). This powerful varnish technique involves Pleistocene (Liu, 2003) and Holocene microlamination sequences (Liu and Broecker, 2007) that exist in rapidly forming varnishes and can be found in microenvironments of greater moisture in a largely arid setting (Dorn et al., 2000; Broecker and Liu, 2001). Robust calibrations of these sequences now provide archaeologists and geomorphologists with a powerful tool to understand the interface of chronology and palaeoenvironmental changes over the past few millennia (Liu, 2006).

Another need rests in linking varnishes seen at the satellite remote sensing scale (White, 1993; Farr and Chadwick, 1996; Milana, 2000; Kenea, 2001) to centimetre-scale remote-sensing sensors placed in such platforms as balloons or helicopters. Bridging scales between ground and satellite (Figure 8.4) remains a giant void. Palmer (2002) illustrates the potential in quantitative landform-scale catenas of spatial variability of rock coatings. Certainly, such field work must be calibrated by textural and chemical analyses using such equipment as an electron microprobe. However, a core research question that remains unresolved is a detailed exploration of spatial heterogeneity of varnish on the scale of single landforms – a topic brought in focus by the realisation that 99% of varnish publications are based on only a few 'grab bag' samples. Field-based (with laboratory calibration) studies of varnish spatial heterogeneity will be needed to create the solid backbone of future advances in varnish research.

Acknowledgements

Special thanks to the late James Clark for assistance in electron microscopy. Research was supported in part by sabbatical support from Arizona State University.

References

Adams, J.B., Palmer, F. & Staley, J.T. (1992) Rock weathering in deserts: mobilization and concentration of ferric iron by microorganisms. *Geomicrobiology Journal* **10**, 99–114.

Allen, C., Probst, L.W., Flood, B.E., Longazo, T.G., Scheble, R.T. & Westall, F. (2004) Meridiani Planum hematite deposit and the search for evidence of life on Mars – iron mineralization of microorganisms in rock varnish. *Icarus* **171**, 20–30.

Anderson, J.L. & Sollid, J.L. (1971) Glacial chronology and glacial geomorphology in the marginal zones of the glaciers, Midtadlsbreen and Nigardsbreen, South Norway. *Norsk Geografisk Tidsskrift* **25**, 1–35.

Arrowsmith, J.R., Rice, G.E. & Hower, J.C. (1998) Documentation of carbon-rich fragments in varnish-covered rocks from central Arizona using electron microscopy and coal petrography. *Science*.

Bamforth, D. (1997) Cation-ratio dating and archaeological research design: A response to Harry. *American Antiquity* **62**, 121–129.

Bao, H., Michalski, G.M. & Thiemens, M.H. (2001) Sulfate oxygen-17 anomalies in desert varnishes. *Geochimica et Cosmochimica Acta* **65**, 2029–2036.

Bard, J.C. (1979) *The development of a patination dating technique for Great Basin petroglyphs utilizing neutron activation and X-ray fluorescence analyses.* PhD dissertation, Department of Anthropology, University of California, Berkeley.

Borns, D.J., Adams, J.B., Curtiss, B., Farr, B., Palmer, T., Staley, J. & Taylor-George, S. (1980) The role of microorganisms in the formation of desert varnish and other rock coatings: SEM study. *Geological Society of America Abstracts with Program* **12**, 390.

Broecker, W.S. & Liu, T. (2001) Rock varnish: recorder of desert wetness? *GSA Today* **11** (8), 4–10.

Bull, W.B. (1984) Alluvial fans and pediments of southern Arizona. In: Smiley, T.L., Nations, J.D., Péwé, T.L. & Schafer, J.P. (Eds) *Landscapes of Arizona. The Geological Story.* New York: University Press of America, pp. 229–252.

Cahill, T.A. (1986) Particle-Induced X-ray Emission. In: Whan, R.E. (Eds) *Metals Handbook*, 9th edn, Vol. 10, *Materials Characterization.* Metals Park: American Society for Metals, pp. 102–108.

Cerveny, N., Kaldenberg, R., Reed, J., Whitley, D.S., Simon, J. & Dorn, R.I. (2006). A new strategy for analyzing the chronometry of constructed rock features in deserts. *Geoarchaeology* **21**: 181–203.

Chafetz, H.S., Akdim, B., Julia, R. & Reid, A. (1998) Mn- and Fe-rich black travertine shrubs: Bacterially (and nanobacterially) induced precipitates. *Journal of Sedimentary Research* **68**, 404–412.

Chen, Y. & Polach, J. (1986) Validity of C-14 ages of carbonates in sediments. *Radiocarbon* **28**(2A), 464–472.

Coudé-Gaussen, G., Rognon, P. & Federoff, N. (1984) Piegeage de poussières éoliennes dans des fissures de granitoides due Sinai oriental. *Compte Rendus de l'Academie des Sciences de Paris* **II**(298), 369–374.

Cremaschi, M. (1996) The desert varnish in the Messak Sattafet (Fezzan, Libryan Sahara), age, archaeological context and paleo-environmental implication. *Geoarchaeology* **11**, 393–421.

Curtiss, B., Adams, J.B. & Ghiorso, M.S. (1985) Origin, development and chemistry of silica-alumina rock coatings from the semiarid regions of the island of Hawaii. *Geochemica et Cosmochimica Acta* **49**, 49–56.

Deutz, P., Montanez, I.P., Monger, H.C. & Morrison, J. (2001) Morphology and isotope heterogeneity of Late Quaternary pedogenic carbonates: Implications for paleosol carbonates as paleoenvironmental proxies. *Palaeogeography, Palaeoclimatology, Palaeoecology* **166**, 293–317.

Diaz, T.A., Bailley, T.L. & Orndorff, R.L. (2002) SEM analysis of vertical and lateral variations in desert varnish chemistry from the Lahontan Mountains, Nevada. *Geological Society of America Abstracts with Programs*, 7–9 May Meeting, http://gsa.confex.com/gsa/2002RM/finalprogram/abstract_33974.htm.

DiGregorio, B.E. (2002) Rock varnish as a habitat for extant life on Mars. In: Hoover, R.B., Levin, G.V., Paepe, R.R. & Rozanov, A.Y. (Eds) *Instruments, Methods, and Missions for Astrobiology IV.* Bellingham, WA: Society of Photo-Optical Instrumentation Engineers, pp. 120–130.

Dorn, R.I. (1986) Rock varnish as an indicator of aeolian environmental change. In: Nickling, W.G. (Eds) *Aeolian Geomorphology*, London: Allen & Unwin, pp. 291–307.

Dorn, R.I. (1990) Quaternary alkalinity fluctuations recorded in rock varnish microlaminations on western U.S.A. volcanics. *Palaeogeography, Palaeoclimatology, Palaeoecology* **76**, 291–310.

Dorn, R.I. (1994) Dating petroglyphs with a 3-tier rock varnish approach. In: Whitley, D.S. & Loendorf, L. (Eds) *New Light on Old Art: Advances in Hunter–Gatherer Rock Art Research.* Monograph Series 36. Los Angeles: Institute of Archaeology, California University, pp. 12–36.

Dorn, R.I. (1996) Climatic hypotheses of alluvial-fan evolution in Death Valley are not testable. In: Rhoads, B.L. & Thorn, C.E. (Eds) *The Scientific Nature of Geomorphology.* New York: Wiley, pp. 191–220.

Dorn, R.I. (1997) Constraining the age of the Côa valley (Portugal) engravings with [14]C. *Antiquity* **71**, 105–115.

Dorn, R.I. (1998) *Rock Coatings.* Amsterdam: Elsevier.

Dorn, R.I. (2000) Chronometric techniques: Engravings. In: Whitley, D.S. (Eds) *Handbook of Rock Art Research.* Walnut Creek: Altamira Press, pp. 167–189.

Dorn, R.I. (2006) Petroglyphs in Petrified Forest National Park: Role of rock coatings as agents of sustainability and as indicators of antiquity. *Bulletin of Museum of Northern Arizona* **63**, 52–63.

Dorn, R.I. & Dickinson, W.R. (1989) First paleoenvironmental interpretation of a pre-Quaternary rock varnish site, Davidson Canyon, south Arizona. *Geology* **17**, 1029–1031.

Dorn, R.I. & Dragovich, D. (1990) Interpretation of rock varnish in Australia: Case studies from the Arid Zone. *Australian Geographer* **21**, 18–32.

Dorn, R.I. & Krinsley, D.H. (1991) Cation-leaching sites in rock varnish. *Geology* **19**, 1077–1080.

Dorn, R.I. & Meek, N. (1995) Rapid formation of rock varnish and other rock coatings on slag deposits near Fontana. *Earth Surface Processes and Landforms* **20**, 547–560.

Dorn, R.I. & Oberlander, T.M. (1981) Microbial origin of desert varnish. *Science* **213**, 1245–1247.

Dorn, R.I. & Oberlander, T.M. (1982) Rock varnish. *Progress in Physical Geography* **6**, 317–367.

Dorn, R.I., Cahill, T.A., Eldred, R.A., Gill, T.E., Kusko, B., Bach, A. & Elliott-Fisk, D.L. (1990) Dating rock varnishes by the cation ratio method with PIXE, ICP, and the electron microprobe. *International Journal of PIXE* **1**, 157–195.

Dorn, R.I., Krinsley, D.H., Liu, T., Anderson, S., Clark, J., Cahill, T.A. & Gill, T.E. (1992) Manganese-rich rock varnish does occur in Antarctica. *Chemical Geology* **99**, 289–298.

Dorn, R.I., Stasack, E., Stasack, D. & Clarkson, P. (2000) Through the looking glass: Analyzing petroglyphs and geoglyphs with different perspectives. *American Indian Rock Art* **27**, 77–96.

Dragovich, D. (1986) Minimum age of some desert varnish near Broken Hill, New South Wales. *Search* **17**, 149–151.

Dragovich, D. (1993) Distribution and chemical composition of microcolonial fungi and rock coatings from arid Australia. *Physical Geography* **14**, 323–341.

Douglas, G.R. (1987) Manganese-rich rock coatings from Iceland. *Earth Surface Processes and Landforms* **12**, 301–310.

Douglas, G.R., McGreevy, J.P. & Whalley, W.B. (1994) Mineralogical aspects of crack development and freeface activity in some basalt cliffs, County Antrim, Northern Ireland. In: Robinson, D.A. & Williams, R.B.G. (Eds) *Rock Weathering and Landform Evolution*. Chichester: Wiley, pp. 71–88.

Engel, C.G. & Sharp, R.S. (1958) Chemical data on desert varnish. *Geological Society of America Bulletin* **69**, 487–518.

Farr, T.G. & Chadwick, O.A. (1996) Geomorphic processes and remote sensing signatures of alluvial fans in the Kun Lun mountains, China. *Journal of Geophysical Research – Planets* **101**(E10), 23091–23100.

Ferris, F.G., Beveridge, T.J. & Fyfe, W.S. (1986) Iron-silica crystallite nucleation by bacteria in a geothermal sediment. *Nature* **320**, 609–611.

Fleisher, M., Liu, T., Broecker, W. & Moore, W. (1999) A clue regarding the origin of rock varnish. *Geophysical Research Letters* **26**(1), 103–106.

Flood, B.E., Allen, C. & Longazo, T. (2003) Microbial fossils detected in desert varnish. *Astrobiology* **2**, 608.

Forbes, E.A., Posner, A.M. & Quirk, J.P. (1976) The specific adsorption of divalent Cd, Co, Cu, Pb, and Zn on goethite. *Journal of Soil Science* **27**, 154–166.

Francis, W.D. (1921) The origin of black coatings of iron and manganese oxides on rocks. *Proceedings of the Royal Society of Queensland* **32**, 110–116.

Glasby, G.P., McPherson, J.G., Kohn, B.P., Johnston, J.H., Keys, J.R., Freeman, A.G. & Tricker, M.J. (1981) Desert varnish in southern Victoria Land, Antarctica. *New Zealand Journal of Geology and Geophysics* **24**, 389–397.

Gonzalez, I., Laiz, L., Hermosin, B., Caballero, B., Incerti, C. & Saiz-Jimenez, C. (1999) Bacteria isolated from rock art paintings: the case of Atlanterra shelter (south Spain). *Journal of Microbiological Methods* **36**, 123–127.

Glazovskaya, M.A. (1958) Weathering and primary soil formations in Antarctica. *Scientific Paper of the Institute, Moscow University, Faculty of Geography* **1**, 63–76.

Gorbushina, A.A., Krumbein, W.E. & Volkmann, M. (2002) Rock surfaces as life indicators: New ways to demonstrate life and traces of former life. *Astrobiology* **2**, 203–213.

Gordon, S.J. & Dorn, R.I. (2005) Localized weathering: Implications for theoretical and applied studies. *Professional Geographer* **57**, 28–43.

Haberland, W. (1975) Untersuchungen an Krusten, Wustenlacken und Polituren auf Gesteinsoberflachen der nordlichen und mittlerent Saharan (Libyen und Tchad). *Berliner Geographische Abhandlungen* **21**, 1–77.

Harrison, R. & Frink, D.S. (2000) The OCR carbon dating procedure in Australia: New dates from Wilinyjibari Rockshelter, southeastern Kimberley, Western Australia. *Australian Archaeology* **51**, 6–15.

Harvey, A.M., Wigand, P.E. & Wells, S.G. (1999) Response of alluvial fan systems to the late Pleistocene to Holocene climatic transition: contrasts between the margins of pluvial Lakes Lahontan and Mojave, Nevada and California, USA. *Catena* **36**, 255–281.

Hem, J.D. & Lind, C.J. (1983) Nonequilibrium models for predicting forms of precipitated manganese oxides. *Geochimica et Cosmochimica Acta* **47**, 2037–2046.

Hobbs, W.H. (1918) The peculiar weathering processes of desert regions with illustrations from Egypt and the Soudan. *Michigan Academy of Sciences Annual Report* **20**, 93–98.

Hodge, V.F., Farmer, D.E., Diaz, T. & Orndorff, R.L. (2005) Prompt detection of alpha particles from ^{210}Po: another clue to the origin of rock varnish? *Journal of Environmental Radioactivity* **78**, 331–342.

Hooke, R.L., Yang, H. & Weiblen, P.W. (1969) Desert varnish: an electron probe study. *Journal of Geology* **77**, 275–288.

Howard, C.D. (2002) The gloss patination of flint artifacts. *Plains Anthropologist* **47**, 283–287.

Hungate, B., Danin, A., Pellerin, N.B., Stemmler, J., Kjellander, P., Adams, J.B. & Staley, J.T. (1987) Characterization of manganese-oxidizing (MnII → MnIV) bacteria from Negev Desert rock varnish: implications in desert varnish formation. *Canadian Journal of Microbiology* **33**, 939–943.

Hunt, A.G. & Wu, J.Q. (2004) Climatic influences on Holocene variations in soil erosion rates on a small hill in the Mojave Desert. *Geomorphology* **58**, 263–289.

Huyge, D., Watchman, A., De Dapper, M. & Marchi, E. (2001) Dating Egypt's oldest 'art': AMS ^{14}C age determinations of rock varnishes covering petroglyphs at El-Hosh (Upper Egypt). *Antiquity* **75**, 68–72.

Ishimaru, S. & Yoshikawa, K. (2000) The weathering of granodiorite porphyry in the Thiel Mountains, Inland Antarctica. *Geografiska Annaler* **A82**, 45–57.

Israel, E.J., Arvidson, R.E., Wang, A., Pasteris, J.D. & Jolliff, B.L. (1997) Laser Raman spectroscopy of varnished basalt and implications for in situ measurements of Martian rocks. *Journal of Geophysical Research – Planets* **102**(E12), 28705–28716.

Jenne, E.A. (1968) Controls on Mn, Fe, Co, Ni, Cu and Zn concentrations in soils and water: the significant role of hydrous Mn and Fe oxides. In: Gould, R.F. (Eds) *Trace Inorganics in Water*. Washington, DC: American Chemical Society, pp. 337–387.

Jenny, H. (1941) *Factors of Soil Formation*. New York: McGraw-Hill.

Johnson, J.R., Ruff, S.W., Moersch, J., Roush, T., Horton, K., Bishop, J., Cabrol, N.A., Cockell, C., Gazis, P., Newsom, H.E. & Stoker, C. (2001) Geological characterization of remote field sites using visible and infrared spectroscopy: Results from the 1999 Marsokhod field test. *Journal of Geophysical Research – Planets* **106**(E4), 7683–7711.

Jones, C.E. (1991) Characteristics and origin of rock varnish from the hyperarid coastal deserts of northern Peru. *Quaternary Research* **35**, 116–129.

Junta, J.L. & Hochella Jr. M.F. (1994) Manganese (II) oxidation at mineral surfaces: a microscopic and spectroscopic study. *Geochimica et Cosmochimica Acta* **48**, 897–902.

Kenea, N.H. (2001) Influence of desert varnish on the reflectance of gossans in the context of Landsat TM data, southern Red Sea Hills, Sudan. *International Journal of Remote Sensing* **22**, 1879–1894.

Kennedy, C.B., Scott, S.D. & Ferris, F.G. (2003) Characterization of bacteriogenic iron oxide deposits from Axial Volcano, Juan de Fuca Ridge, Northeast Pacific Ocean. *Geomicrobiology Journal* **20**, 199–214.

Khak-mun, T. (1966) Iron- and manganese-oxidizing microorganisms in soils of South Sakhalin. *Microbiology* **36**(2), 276–281.

Konhauser, K.O., Fyfe, W.S., Schultze-Lam, S., Ferris, F.G. & Beveridge, T.J. (1994) Iron phosphate precipitation by epilithic microbial biofilms in Arctic Canada. *Canadian Journal of Earth Science* **31**, 1320–1324.

Krauskopf, K.B. (1957) Separation of manganese from iron in sedimentary processes. *Geochimica et Cosmochimica Acta* **12**, 61–84.

Krinsley, D. (1998) Models of rock varnish formation constrained by high resolution transmission electron microscopy. *Sedimentology* **45**, 711–725.

Krinsley, D., Dorn, R. I. & Anderson, S. (1990) Factors that may interfere with the dating of rock varnish. *Physical Geography* **11**, 97–119.

Krinsley, D.H., Dorn, R.I. & Tovey, N.K. (1995) Nanometer-scale layering in rock varnish: implications for genesis and paleoenvironmental interpretation. *Journal of Geology* **103**, 106–113.

Krumbein, W.E. (1969) Über den Einfluss der Mikroflora auf die Exogene Dynamik (Verwitterung und Krustenbildung). *Geologische Rundschau* **58**, 333–363.

Krumbein, W.E. (1971) Biologische Entstehung von wüstenlack. *Umschau* **71**, 210–211.

Krumbein, W.E. & Jens, K. (1981) Biogenic rock varnishes of the Negev Desert (Israel): An ecological study of iron and manganese transformation by cyanobacteria and fungi. *Oecologia* **50**, 25–38.

Kuhlman, K.R., Allenbach, L.B., Ball, C.L., Fusco, W.G., La Duc, M.T., Kuhlman, G.M., Anderson, R.C., Stuecker, T., Erickson, I.K., Benardini, J. & Crawford, R.L. (2005). Enumeration, isolation, and characterization of ultraviolet (UV-C) resistant bacteria from rock varnish in the Whipple Mountains, California. *Icarus* **174**, 585–595.

Kuhlman, K.R., Fusco, W.G., Duc, M.T.L., Allenbach, L.B., Ball, C.L., Kuhlman, G. M., Anderson, R.C., Erickson, K., Stuecker, T., Benardini, J., Strap, J.L. & Crawford, R.L. (2006) Diversity of microorganisms within rock varnish in the Whipple Mountains, California. *Applied and Environmental Microbiology* **72**, 1708–1715.

Kurtz, H.D. & Netoff, D.I. (2001) Stabilization of friable sandstone surfaces in a desiccating, wind-abraded environment of south-central Utah by rock surface microorganisms. *Journal of Arid Environments* **48**(1), 89–100.

Lakin, H.W., Hunt, C.B., Davidson, D.F. & Odea, U. (1963) Variation in minor-element content of desert varnish. *U.S. Geological Survey Professional Paper* **475-B**, B28–B31.

Laudermilk, J.D. (1931) On the origin of desert varnish. *American Journal of Science* **21**, 51–66.

Lee, M.R. & Bland, P.A. (2003) Dating climatic change in hot deserts using desert varnish on meteorite finds. *Earth and Planetary Science Letters* **206**, 187–198.

Linck, G. (1901) Über die dunkelen Rinden der Gesteine der Wüste. *Jenaische Zeitschrift für Naturwissenschaft* **35**, 329–336.

Linck, G. (1928) Über Schutzrinden. *Chemie die Erde* **4**, 67–79.

Liu, T. (1994) *Visual microlaminations in rock varnish: a new paleoenvironmental and geomorphic tool in drylands.* PhD dissertation, Department of Geography, Arizona State University, Tempe.

Liu, T. (2003) Blind testing of rock varnish microstratigraphy as a chronometric indicator: results on late Quaternary lava flows in the Mojave Desert, California. *Geomorphology* **53**, 209–234.

Liu, T. (2006). VML Dating Lab. http://www.vmldatinglab.com/

Liu, T. & Broecker, W.S. (2000) How fast does rock varnish grow? *Geology* **28**, 183–186.

Liu, T. & Broecker, W. (2007) Holocene rock varnish microstratigraphy and its chronometric application in drylands of western USA. *Geomorphology* **84**, 1–21.

Liu, T. & Dorn, R.I. (1996) Understanding spatial variability in environmental changes in drylands with rock varnish microlaminations. *Annals of the Association of American Geographers* **86**, 187–212.

Liu, T., Broecker, W.S., Bell, J.W. & Mandeville, C. (2000) Terminal Pleistocene wet event recorded in rock varnish from the Las Vegas Valley, southern Nevada. *Palaeogeography, Palaeoclimatology, Palaeoecology* **161**, 423–433.

Loendorf, L.L. (1991) Cation-ratio varnish dating and petroglyph chronology in southeastern Colorado. *Antiquity* **65**, 246–255.

Lucas, A. (1905) *The blackened rocks of the Nile cataracts and of the Egyptian deserts,* Cairo: National Printing Department.

Mabbutt, J.A. (1979) Pavements and patterned ground in the Australian stony deserts. *Stuttgarter Geographische Studien* **93**, 107–123.

Mancinelli, R.L., Bishop, J.L. & De, S. (2002) Magnetite in desert varnish and applications to rock varnish on Mars. *Lunar and Planetary Science* **33**, 1046. pdf.

Marchant, D.R., Denton, G.H., Swisher, C.C. & Potter, N. (1996) Late Cenozoic Antarctic paleoclimate reconstructed from volcanic ashes in the Dry Valleys region of southern Victoria Land. *Geological Society of America Bulletin* **108**, 181–194.

Marston, R.A. (2003) Editorial note. *Geomorphology* **53**, 197.

McFadden, L.D., Ritter, J.B. & Wells, S.G. (1989) Use of multiparameter relative-age methods for age estimation and correlation of alluvial fan surfaces on a desert piedmont, eastern Mojave Desert. *Quaternary Research* **32**, 276–290.

McKeown, D.A. & Post, J.E. (2001) Characterization of manganese oxide mineralogy in rock varnish and dendrites using X-ray absorption spectroscopy. *American Mineralogist* **86**, 701–713.

McKnight, D.M., Kimball, B.A. & Bencala, K.E. (1988) Iron photoreduction and oxidation in an acidic mountain stream. *Science* **240**, 637–640.

Milana, J.P. (2000) Characterization of alluvial bajada facies distribution using TM imagery. *Sedimentology* **47**, 741–760.

Morgan, J.J. & Stumm, W. (1965) The role of mulvaleng metal oxides in limnological transofrmations as exemplified by iron and manganese. In: Jaag, O. (Ed.)

Second Water Pollution Research Conference, Vol. 1. New York: Pergamon Press, pp. 103–131.

Moore, C.B. & Elvidge, C.D. (1982) Desert varnish. In: Bender, G.L. (Ed.) *Reference handbook on the deserts of North America*. Westport: Greenwood Press, pp. 527–536.

Nagy, B., Nagy, L.A., Rigali, M.J., Jones, W.D., Krinsley, D.H. & Sinclair, N. (1991) Rock varnish in the Sonoran Desert: microbiologically mediated accumulation of manganiferous sediments. *Sedimentology* **38**, 1153–1171.

Nienow, J.A., McKay, C.P. & Friedmann, E.I. (1988) The cryptoendolithic microbial environment in the Ross Desert of Antarctica: light in the photosynthetically active region. *Microbial Ecology* **16**, 271–289.

Nishiizumi, K., Kohl, C., Arnold, J., Dorn, R., Klein, J., Fink, D., Middleton, R. & Lal, D. (1993) Role of in situ cosmogenic nuclides ^{10}Be and ^{26}Al in the study of diverse geomorphic processes. *Earth Surface Processes and Landforms* **18**, 407–425.

Oberlander, T.M. (1994) Rock varnish in deserts. In: Abrahams, A. & Parsons, A. (Eds) *Geomorphology of Desert Environments*. London: Chapman and Hall, pp. 106–119.

O'Hara, P., Krinsley, D.H. & Anderson, S.W. (1989) Elemental analysis of rock varnish using the ion microprobe. *Geological Society America Abstracts with Programs* **21**, A165.

Palmer, E. (2002) Feasibility and implications of a rock coating catena: analysis of a desert hillslope. M.A. Thesis. Department of Geography, Arizona State University, Tempe.

Palmer, F.E., Staley, J.T., Murray, R.G.E., Counsell, T. & Adams, J.B. (1985) Identification of manganese-oxidizing bacteria from desert varnish. *Geomicrobiology Journal* **4**, 343–360.

Patyk-Kara, N.G., Gorelikova, N.V., Plakht, J., Nechelyustov, G.N. & Chizhova, I.A. (1997) Desert varnish as an indicator of the age of Quaternary formations (Makhtesh Ramon Depression, Central Negev). *Transactions (Doklady) of the Russian Academy of Sciences/Earth Science Sections* **353A**, 348–351.

Peck, S.B. (1986) Bacterial deposition of iron and manganese oxides in North American caves. *National Speleological Society Bulletin* **48**, 26–30.

Perel'man, A.I. (1961) Geochemical principles of landscape classification. *Soviet Geography Review and Translation* **11**(3), 63–73.

Perfil'ev, B.V., Gabe, D.R., Gal'perina, A.M., Rabinovich, V.A., Sapotnitskii, A.A., Sherman, É.É. & Troshanov, É.P. (1965) *Applied Capillary Microscopy. The Role of Microorganisms in the Formation of Iron–Manganese Deposits*. New York: Consultants Bureau.

Perry, R.S. & Kolb, V.M. (2003) Biological and organic constituents of desert varnish: Review and new hypotheses. In: Hoover, R.B. & Rozanov, A.Y. (Eds) *Instruments, Methods, and Missions for Astrobiology VII*. Bellingham, WA: Society of Photo-Optical Instrumentation Engineers, pp. 202–217.

Perry, R.S., Dodsworth, J., Staley, J.T. & Gillespie, A. (2002) Molecular analyses of microbial communities in rock coatings and soils from Death Valley, California. *Astrobiology* **2**, 539.

Perry, R.S., Engel, M., Botta, O. & Staley, J.T. (2003) Amino acid analyses of desert varnish from the Sonoran and Mojave deserts. *Geomicrobiology Journal* **20**, 427–438.

Perry, R.S., Lynne, B.Y., Sephton, M.A., Kolb, V.M., Perry, C.C. & Staley, J.T. (2006) Baking black opal in the desert sun: The importance of silica in desert varnish. *Geology* **34**, 737–540.

Peterson, F. (1981) *Landforms of the Basin and Range Province, Defined for Soil Survey.* Technical Bulletin 28. Nevada Agricultural Experiment Station.

Phillips, F.M., Zreda, M.G., Plummer, M.A., Benson, L.V., Elmore, D. & Sharma, P. (1996) Chronology for fluctuations in Late Pleistocene Sierra Nevada glaciers and lakes. *Science* **274**, 749–751.

Potter, R.M. (1979) The tetravalent manganese oxides: clarification of their structural variations and relationships and characterization of their occurrence in the terrestrial weathering environment as desert varnish and other manganese oxides. PhD dissertation, Earth and Planetary Science, California Institute of Technology, Pasadena.

Potter, R.M. & Rossman, G.R. (1977) Desert varnish: the importance of clay minerals. *Science* **196**, 1446–1448.

Potter, R.M. & Rossman, G.R. (1978) Manganese oxides in the terrestrial weathering environment. *Geological Society of America Abstracts with Programs* **10**, 473.

Potter, R.M. & Rossman, G.R. (1979a) The manganese- and iron-oxide mineralogy of desert varnish. *Chemical Geology* **25**, 79–94.

Potter, R.M. & Rossman, G.R. (1979b) Mineralogy of manganese dendrites and coatings. *American Mineralogist* **64**, 1219–1226.

Potter, R.M. & Rossman, G.R. (1979c) The tetravalent manganese oxides: identification, hydration, and structural relationships by infrared spectroscopy. *American Mineralogist* **64**, 1199–1218.

Probst, L., Thomas-Keprta, K. & Allen, C. (2001) Desert varnish: preservation of microfossils and biofabric (Earth and Mars?). *GSA Abstracts With Programs* http://gsa.confex.com/gsa/2001AM/finalprogram/abstract_27826.htm.

Probst, L.W., Allen, C.C., Thomas-Keprta, K.L., Clemett, S.J., Longazo, T.G., Nelman-Gonzalez, M.A. & Sams, C. (2002) Desert varnish – preservation of biofabrics and implications for Mars. *Lunar and Planetary Science* **33**, 1764.pdf.

Reneau, S.L. (1993) Manganese accumulation in rock varnish on a desert piedmont, Mojave Desert, California, and application to evaluating varnish development. *Quaternary Research* **40**, 309–317.

Robert, M. & Tessier, D. (1992) Incipient weathering: some new concepts on weathering, clay formation and organization. In: Martini, I.P. & Chesworth, W. (Eds) *Weathering, Soils & Paleosols.* Amsterdam: Elsevier, pp. 71–105.

Robert, M., Hardy, M., Elsass, F. & Righi, D. (1990) Genesis, crytallochemistry and organization of soil clays derived from different parent materials in temperate regions. *International Conference of Soil Science, Fourteenth* **7**, 42–47.

Rowe, M.W. (2000) Dating by AMS radiocarbon analysis. In: Whitley, D.S. (Eds) *Handbook of Rock Art Research.* Walnut Creek: Altamira Press, pp. 139–166.

Russ, J., Loyd, D.H. & Boutton, T.W. (2000) A paleoclimate reconstruction for southwestern Texas using oxalate residue from lichen as a paleoclimate proxy. *Quaternary International* **67**, 29–36.

Scheffer, F., Meyer, B. & Kalk, E. (1963) Biologische ursachen der wüstenlackbildung. *Zeitschrift für Geomorphologie* **7**, 112–119.

Schelble, R., McDonald, G., Hall, J. & Nealson, K. (2005). Community structure comparison using FAME analysis of desert varnish and soil, Mojave Desert, California. *Geomicrobiology Journal* **22**, 353–360.

Smith, B.J. & Whalley, W.B. (1988) A note on the characteristics and possible origins of desert varnishes from southeast Morocco. *Earth Surface Processes and Landforms* 13, 251–258.

Staley, J.T., Palmer, F. & Adams, J.B. (1982) Microcolonial fungi: common inhabitants on desert rocks? *Science* 215, 1093–1095.

Staley, J.T., Jackson, M.J., Palmer, F.E., Adams, J.C., Borns, D.J., Curtiss, B. & Taylor-George, S. (1983) Desert varnish coatings and microcolonial fungi on rocks of the Gibson and Great Victoria Desert, Australia. *BMR Journal of Australian Geology and Geophysics* 8, 83–87.

Staley, J.T., Adams, J.B., Palmer, F., Long, A., Donahue, D.J. & Jull, A.J.T. (1991) *Young ¹⁴Carbon Ages of Rock Varnish Coatings from the Sonoran Desert.* Unpublished Manuscript.

Sterflinger, K., Krumbein, W.E., Lallau, T. & Rullkötter, J. (1999) Microbially mediated orange patination of rock surfaces. *Ancient Biomolecules* 3, 51–65.

Tebo, B.M. & He, L.M. (1998) Microbially mediated oxidative precipitation reactions. In: Sparks, D.L. & Grundl, T.J. (Eds) *Mineral-Water Interfacial Reactions*, Washington, DC: American Chemical Society, pp. 393–414.

Tebo, B.M., Ghiorse, W.C., van Waasbergen, L.G., Siering, P.L. & Caspi, R. (1997) Bacterially mediated mineral formation: Insights into manganese (II) oxidation from molecular genetic and biochemical studies. *Reviews in Mineralogy* 35, 225–266.

Tebo, B.M., Bargar, J.R., Clement, B.G., Dick, G.J., Murray, K.J., Parker, D., Verity, R. & Webb, S.M. (2004) Biogenic manganese oxides: Properties and mechanisms of formation. *Annual Review Earth and Planetary Science* 32, 287–328.

Tebo, B.M., Johnson, H.A., McCarthy, J.K. & Templeton. A.S. (2005). Geomicrobiology of manganese (II) oxidation. *TRENDS in Microbiology* 13, 421–428.

Thiagarajan, N. & Lee, C.A. (2004) Trace-element evidence for the origin of desert varnish by direct aqueous atmospheric deposition *Earth and Planetary Science Letters* 224, 131–141.

Tratebas, A., Cerveny, N. & Dorn, R.I. (2004). The effects of fire on rock art: Microscopic evidence reveals the importance of weathering rinds. *Physical Geography* 25, 313–333.

Viles, H.A. & Goudie, A.S. (2004) Biofilms and case hardening on sandstones from Al-Quawayra, Jordan. *Earth Surface Processes and Landforms* 29, 1473–1485.

Villa, N., Dorn, R.I. & Clark, J. (1995) Fine material in rock fractures: aeolian dust or weathering? In: Tchakerian, V.P. (Eds) *Desert Aeolian Processes*. London: Chapman and Hall, pp. 219–231.

Von Humboldt, A. (1812) *Personal Narrative of Travels to the Equinoctial Regions of America During the Years 1799–1804 V. II* (Translated and Edited by T. Ross in 1907). London: George Bell & Sons.

Waragai, T. (1998) Effects of rock surface temperature on exfoliation, rock varnish, and lichens an a boulder in the Hunza Valley, Karakaram Mountains, Pakistan. *Arctic and Alpine Research* 30, 184–192.

Warsheid, T. (1990) Untersuchungen zur Biodeterioration von Sandsteinen unter besonderer Berücksichtigung der chemoorganotrophen Bakterien. PhD dissertation, Geomicrobiology, Universität Oldenburg, Oldenburg.

Washburn, A.L. (1969) Desert varnish. In: Washburn, A.L. (Ed.) *Weathering, Frost Action and Patterned Ground in the Mesters District, Northeast Greenland.* Copenhagen: Reitzels, pp. 14–15.

Watchman, A. (1997) Differences of Interpretation for Foz Côa Dating Results. *National Pictographic Society Newsletter* **8**, 7.

Watchman, A. (2000) A review of the history of dating rock varnishes. *Earth-Science Reviews* **49**, 261–277.

Watchman, A. (2002) A reply to Whitley and Simon. *International Newsletter on Rock Art* **34**, 11–12.

Whalley, W.B., Gellatly, A.F., Gordon, J.E. & Hansom, J.D. (1990) Ferromanganese rock varnish in North Norway: a subglacial origin. *Earth Surface Processes and Landforms* **15**, 265–275.

White, C.H. (1924) Desert varnish. *American Journal of Science* **7**, 413–420.

White, K. (1993) Image processing of Thematic Mapper data for discriminating piedmont surficial materials in the Tunisian Southern Atlas. *International Journal of Remote Sensing* **14**, 961–977.

Whitley, D.S. & Annegarn, H.J. (1994) Cation-ratio dating of rock engravings from Klipfontein, Northern Cape Province, South Africa. In: Dowson, T.A. & Lewis-Williams, J.D. (Eds) *Contested Images: Diversity in Southern African Rock Art Research.* Johannesburg: University of the Witwatersrand Press, pp. 189–197.

Whitley, D.S. & Simon, H.M. (2002a) Recent AMS radiocarbon rock engraving dates. *International Newsletter on Rock Art* **32**, 11–16.

Whitley, D.S. & Simon, H.M. (2002b) Reply to Huyge and Watchman. *International Newsletter on Rock Art* **34**, 12–21.

Whitley, D.S., Dorn, R.I., Simon, H.M., Rechtman, R. & Whitley, T.K. (1999) Sally's rockshelter and the archaeology of the vision quest. *Cambridge Archaeological Journal* **9**, 221–247.

Whitney, J.W. & Harrington, C.D. (1993) Relict colluvial boulder deposits as paleoclimatic indicators in the Yucca Mountain region, southern Nevada. *Geological Society of America Bulletin* **105**, 1008–1018.

Wierzchos, J., Ascaso, C., García-Sancho, L. & Green, A. (2003) Iron-rich diagenetic ninerals are biomarkers of microbial activity in Antarctic rocks. *Geomicrobiology Journal* **20**, 15–24.

Zhang, Y., Liu, T. & Li, S. (1990) Establishment of a cation-leaching curve of rock varnish and its application to the boundary region of Gansu and Xinjiang, western China. *Seismology and Geology (Beijing)* **12**, 251–261.

Zhou, B.G., Liu, T. & Zhang, Y.M. (2000) Rock varnish microlaminations from northern Tianshan, Xinjiang and their paleoclimatic implications. *Chinese Science Bulletin* **45**, 372–376.

Chapter Nine

Lacustrine and Palustrine Geochemical Sediments

Eric P Verrecchia

9.1 Introduction

Lakes can be described as any natural terrestrial depression filled up with free and quiet water without a connection to the ocean. The surface of lakes varies from several tens of thousands of kilometres^2 to a minimum of $10,000 \, m^2$ (or $10,000 \, m^3$). When the water body has a mean depth of <1 m, the environment is considered as palustrine. The fact that some lakes can be temporary makes the transition from lacustrine to palustrine environments relatively complex. In addition, in palustrine environments, soils can develop and contribute to the genesis and/or diagenesis of hardened terrestrial geochemical accumulations.

The way in which lakes function is described by a science called limnology (Wetzel, 2001), which is not the subject of this chapter. Nevertheless, some basic concepts have to be recalled in order to explain the origin of biogeochemical precipitation of some minerals. Ninety per cent of lakes originate in either (a) tectonic basins, e.g. East African Rift lakes (Tiercelin and Vincens, 1987) or the Dead Sea (Begin et al., 1985), or (b) inherited or active glacial landscapes, e.g. Alpine lakes (Ariztegui and Wildi, 2003) or North American lakes (Hutchinson, 1957). Less than 10% of the world's lakes have another origin such as crater lakes, natural dams, karstic lakes, large ponds in wet areas, ephemeral lakes and oxbows in floodplains, or pluvial lakes in the arid zone (Wetzel, 2001). The diversity of lake settings as well as their latitudinal and altitudinal positions provide multiple conditions, potentially leading to a wide variety of geochemical deposits.

The biogeochemistry of lakes is directly influenced by the thermal regime of the water body. The thermal regime is at the origin of the way waters are thermally stratified during the year. A typical thermal stratification shows three different water bodies. A warmer and lighter mass of water is found near the surface, called the epilimnion (Figure 9.1). Colder and heavier water stays at the bottom of the lake, forming the hypolimnion. Between these two water bodies, the water forming the thermal transition is called the metalimnion or thermocline. A lake's thermal regime is defined depending on the way these various water bodies are mixed during the year. In addition to exogenic factors such as the wind or the fluvial inflow, it is the change in water density which determines the water body mixing. The maximum density of water is reached at 4°C. In lakes, the water body stratifies in response to this water density, defining the various thermal lake types (Håkanson and Jansson, 1983).

Conventionally, lakes are considered monomictic when they undergo only one period of water mixing and circulation, during summer. They can be either cold monomictic (the water body never reaches temperatures higher than 4°C) or warm monomictic (the temperature never drops below 4°C). For example, Lake Neuchâtel (Switzerland) is a warm monomictic lake although it is situated in a temperate inland environment. During summer in Swiss temperate latitudes, the surficial water is warmed up by the sun and the temperature reaches 20–25°C in the epilimnion, whereas the hypolimnion stays at temperatures slightly higher than 4°C. Neverthe-less, if in this type of lake the watershed includes cold tributary rivers or streams (resulting from cold precipitation at altitude or snow melt), sudden cold influxes dive to the bottom of the lake because of differences in water densities (a process called underflow) and lead to a mixing of waters, often accompanied by bottom sediment reworking.

Lakes are described as dimictic when the water circulates twice a year, generally in spring and autumn. From summer to the end of autumn, the epilimnion water cools down. When it reaches 4°C, it dives into the hypo-limnion, pushing the hypolimnion water to the surface, resulting in an autumn turnover. From winter to the end of spring, the epilimnion water warms up from a temperature below 4°C. When it reaches 4°C, it dives again into the hypolimnion resulting in the spring turnover. Finally, some lakes have a polymictic regime, i.e. with many water body mixings during the year. These thermic properties are fundamental because they determine water body circulations and directly influence sedimentation and chemical properties of the water body as well as the conditions at the lake-bottom–sediment interface (allowing the input of organic matter, oxygen, sediment reworking, etc.). When a lake is lacking complete circulation of the water body, it is called meromictic (Figure 9.2). In meromictic lakes, the deepest water contains no dissolved oxygen. Sediments can stay undisturbed and provide a detailed record of the geological lake history. The monimolim-nion is the deepest part of this type of lake. It is the most dense and is not involved in annual circulation.

The chemocline defines the boundary that separates the mixolimnion from the monimolimnion, i.e. where the salinity gradient is the steepest.

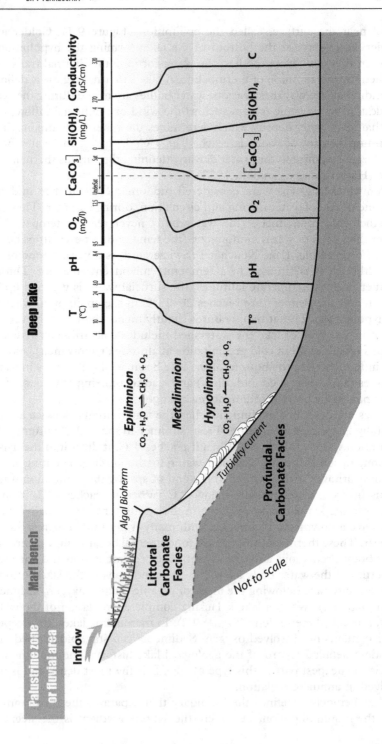

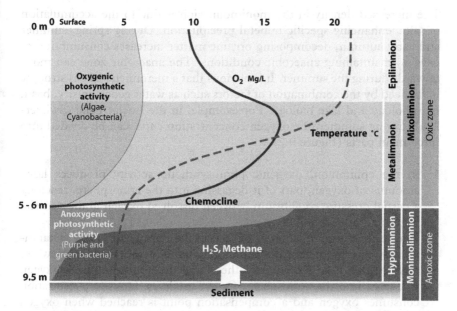

Figure 9.2 Sketch showing the relationships between oxygen, temperature and biogenic activity in a meromictic lake, at noon in summer (Loclat Lake, Switzerland; data from Aragno, 1981, and personal communication). The mixolimnion constitutes the oxic zone. The mixolimnion includes the metalimnion and the epilimnion, both defined by the temperature curve. The anoxic zone is called the monimolimnion.

◄───

Figure 9.1 Diagrammatic cross-section of a typical hard–water temperate lake during summer (adapted from Dean, 1981). The various profiles refer to Lake Neuchâtel (Lambert, 1999). Marl bench and deep lake views are given in Figure 9.3B, C. Profiles describe physical and chemical properties. Temperature (T in °C) varies from 4°C (hypolimnion) to 24°C (surficial epilimnion). This profile defines the water-body stratification of the lake. The pH curve shows that the pH does not drop below 8.0. The lake is slightly alkaline. Calcite ($[CaCO_3]$ curve) is partly stable in the water body: the epilimnion and the metalimnion are supersaturated in calcite (concentration curve is to the right of the dotted line), whereas calcite dissolves in the hypolimnion (see also conductivity curve). The oxygen (O_2) curve shows degassing in the upper part of the water body and a relative accumulation of oxygen in the metalimnion due to oxygenic photosynthesis. The low values in the hypolimnion are explained by consumption of oxygen during organic matter decay and respiration (see also Figure 9.2). The silica curve is related to a large decrease in silica in the epilimnion due to silica exhaustion during spring diatom blooms. The increase in silica concentration with depth is caused by partial dissolution of diatom frustules and silica release from bottom sediments. The conductivity curve provides a good picture of the concentration of ions in solution. In the epilimnion, the conductivity is low because most of the ions are trapped in calcite crystals. In hard-water lakes, the most abundant ions are Ca^{2+}, HCO_3^- and CO_3^{2-}. Their precipitation as $CaCO_3$ decreases the conductivity. However, as soon as the conditions are no longer stable for carbonate minerals, endogenic calcite dissolves, increasing the conductivity in the hypolimnion.

The increased density in the monimolimnion is due to the accumulation of salts, enhancing specific mineral precipitation. During spring, summer and early autumn, decomposing organic matter increases consumption of oxygen, maintaining anaerobic conditions. The anaerobic zone can move upwards during the summer. It is obvious that a meromictic lake is strongly influenced by the combination of factors such as water geochemistry, basin morphology and lake biology. For example, in the Loclat Lake (Switzerland) during summer, the oxygen concentration curve can be divided into three main parts (Figure 9.2).

1   In the epilimnion, oxygenic photosynthetic activity produces large amounts of oxygen, part of it degassing into the atmosphere, reaching a partial pressure equilibrium between dissolved oxygen in the surficial water and atmospheric oxygen.
2   The relative accumulation of oxygen in the upper part of the metalimnion is due to the stratification of water and the absence of water mixing. Due to active photosynthesis and the impossibility for oxygen to escape into the atmosphere, O_2 accumulates. However, respiration consumes oxygen and a compensation point is reached when oxygen consumption is balanced by oxygen production.
3   In the lower part of the metalimnion, this compensation point is attained and surpassed. Respiration is still present but oxygen production decreases due to the decreasing amount of oxygenic photosynthetic microorganisms. Finally, oxygen is totally consumed and the anoxic zone starts.

In the anoxic monimolimnion, hydrogen sulphide and methane can be produced by sulphur-reducing and methanogenic bacteria. In the deep lake anoxic zone, below the photic boundary, fermentative microorganisms are mainly responsible for the carbon cycle. In the anoxic zone of shallow-water lakes, purple and green bacteria are able to accomplish anoxygenic photosynthesis using pigments catching light wavelengths different from those of oxygenic microorganisms from the surface, and are therefore able to penetrate to the lake bottom. Anoxygenic photosynthetic microorganisms can use H_2S produced by sulphur-reducing bacteria as the electron donor. The chemical sequence during purple and green bacteria photosynthetic activity leads to sulphate ion production: $H_2S \rightarrow S^0 \rightarrow SO_4^{2-}$. Although the sulphate ion is in solution, it can reach supersaturation at the water–sediment interface and/or in the sediment, enhancing precipitation of gypsum in the presence of Ca^{2+} ions. Methane is consumed by methanotroph microorganisms and/or can be oxidised once the chemocline is reached (presence of oxygen) by methane in solution or as gas bubbles. The organic matter produced in the upper part of the lake and deposited on the lake bottom is not decayed in an anoxic environment: a black organic layer can accumulate over many years, leading to lacustrine black shale deposition and preservation.

A lake that is not meromictic is called holomictic. Holomictic lakes have one of the mixing types described previously, i.e. warm or cold monomictic, dimictic, etc. In conclusion, holomictic and meromictic properties refer to chemical characteristics of the water bodies, whereas epi-, meta- and hypolimnion refer to the thermic regime of lakes. Obviously, both concepts are linked.

9.2 Nature and General Characteristics

The nature of geochemical sediments associated with lakes and palustrine environments depends on the water chemistry as well as the lake and/or marsh climatic and geomorphological settings. This high dependency on the nature of the lake watershed and the hydrochemical composition of the phreatic waters make it difficult to propose a clear nomenclature of lacustrine geochemical sediments. Nevertheless, Håkanson and Jansson (1983) suggest that minerals in lake sediments be separated according to their source (allogenic, endogenic or authigenic).

Allogenic minerals are derived from outside the lake. They are supplied from streams and surface runoff, shore erosion, floating ice and atmospheric fallout. Endogenic minerals are related to the biogeochemical processes involved in the lake water body. These minerals precipitate or flocculate in the water and settle to the bottom of the lake. These minerals can record some biogeochemical periodical events, such as diatom blooms during spring, which concentrate the silica dissolved in water into the algal test. Calcium carbonate is also a good indicator of the variation of seasonal biogeochemical conditions during the year (Dean, 1981) as well as over longer time periods. Temporal variations in pH and Eh (redox) conditions can also be recorded by iron precipitates. In palustrine carbonate soils, the successive reduced and oxidised states of iron (Fe^{2+} and Fe^{3+} respectively) lead to the formation of yellowish patches called marmorisation (Freytet and Plaziat, 1982; see below). Authigenic minerals form inside the lake sediments and/or during early diagenesis, due to emersion (or emergence), for example. Endogenic and authigenic minerals constitute lacustropalustrine geochemical sediments. Therefore, a distinction has to be made between detrital and geochemical sediments. Detrital lacustrine deposits are not the subject of this chapter, although they can be found associated with geochemical sediments (e.g. Figure 9.9H). Detrital sediments are constituted by reworked rock fragments and minerals, whatever their grain size, trapped and settled in the lacustrine basin during floods (from rivers or watershed runoff) or atmospheric inputs (for dust size particles). Geochemical sediments result from direct or mediated precipitation of minerals from the lake water body, i.e. mainly from ions in solution, or are formed inside lake sediments during early and/or late diagenesis.

Lacustrine mineral nomenclature can also be devised using climatic settings and concentration of water solutions and by including the evaporitic or non-evaporitic nature of the minerals (Table 9.1). Magny (1991) proposes another type of sediment classification for lakes from the Jura Mountains (Figure 9.3A). This classification takes into account three different poles: organic, detrital and carbonate. These poles are related to the geomorphological–climatic settings described as biostasic (stable environment dominated by vegetation and soil processes leading to a chemical and organic precipitation inside lakes) and rhexistasic (unstable environment dominated by erosive processes leading to detrital accumulations in lacustrine basins). These two terms, rhexistasy and biostasy, have been coined by Erhart (1967) who proposes a general theory linking soil genesis, terrestrial and marine sedimentation through geological times.

9.3 Geomorphological Relations

9.3.1 Present-day environments

The geomorphology of a lacustrine basin is directly influenced by the lake's general setting (rift lakes, Alpine lakes, inherited Last Glaciation lakes, etc.). Information on the various sizes and morphological characteristics of world lakes can be found in Wetzel (2001). At the field scale, lakes can have steep borders, characterised by cliffs. Very steep borders can represent a geomorphological hazard when landslides occur along unstable cliffs. A large amount of rocks collapsing into the lake may cause dangerous water movements and waves, causing lake turnovers, resuspension of sediments, and sudden and violent shore erosion. Lakes can also be surrounded by palustrine plains, in contact with benches along the lakeshore. Benches are accumulations of sediments (mainly endogenic) and various amounts of organic matter, depending on the bench sedimentation rate. Benches gently and regularly slope downward to the upper talus, which steeply dives towards the deep lake (Figure 9.3B–D). In hard-water lakes, benches are formed by relatively pure lacustrine carbonates (chalk and marls), composed of biologically induced microcrystalline calcium carbonate (or micrite, mainly from photosynthesis), as well as gastropod and ostracod micrite (Murphy and Wilkinson, 1980; Freytet and Verrecchia, 2002; see Figure 9.3D). Murphy and Wilkinson (1980) noted that vertical bench sedimentary sequences are nearly identical to lateral progradational facies presently being deposited on the talus and lake bottom (Figure 9.3D).

An idealised stratigraphic column for temperate-region lacustrine carbonate deposits is made of the following beds (Murphy and Wilkinson, 1980).

Table 9.1 Main minerals and mineral groups associated with lacustrine geochemical sediments and their possible origins (modified, augmented and compiled from Stoffers and Fischbeck, 1974; Krumbein, 1975; Håkanson and Jansson, 1983; Campy and Meybeck, 1995; Sebag et al., 1999; Giralt et al., 2001; Braissant and Verrecchia, 2002; Dupraz et al., 2004). The '+' sign indicates the frequency of occurrence. Absence of sign means minerals have only been observed

Allogenic source		Endogenic source		Authigenic source
Eolian origin	Watershed origin	Non-evaporitic	Evaporitic	Early diagenetic
Quartz (+++)	Illite (+++)	Opaline silica (diatoms +++)	Gypsum (+++)	Fe-Mn oxides (+++)
Kaolinite (+++)	Smectite (+++)	Calcite (+++)	Halite (+++)	Mg-Mn carbonates (++)
Sulphate (Gypsum +)	Quartz (+++)	Dolomite (+)	Calcite (+)	Phosphates
Opaline silica (+)	Calcite (+++)	Vaterite	High magnesian calcite	Sodium silicate
Carbonate (Calcite +)	Kaolinite (++)	Monohydrocalcite	Aragonite	Zeolites
Mixed clays	Feldspar (++)	Low magnesian calcite	Protodolomite	Sulphides
Fe-Mn oxides	Mixed clays (+)	Protodolomite	Dolomite	Fluorides
	Dolomite		Magnesian silicate	Palygorskite
	Oxalates		Sodium silicate	Sepiolite
	Nontronite		Fluorides	Nontronite
	Palygorskite		Bromides	Bentonite
	Fe-Mn oxides		Natron	Dolomite
	Amphiboles		Trona	
	Pyroxenes			
	Pyrite			
	Apatite			

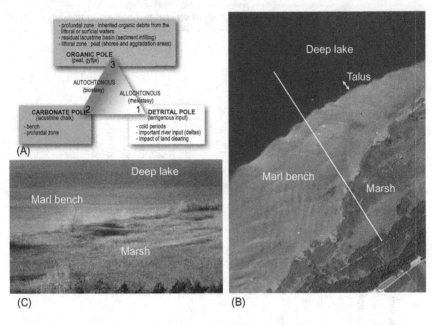

(A)

(C)

(B)

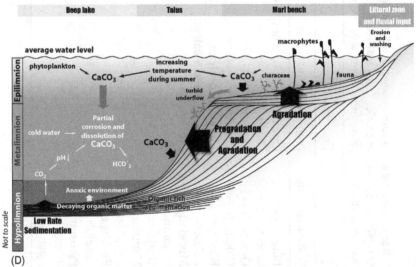

(D)

Figure 9.3 (A) Classification of lacustrine sedimentation in Jura Mountain lakes according to Magny (1991). The geochemical sediments are mainly concentrated around the carbonate pole. (B) Aerial and (C) field view from the palustrine (marsh) zone towards the deep lake (Lake Neuchâtel, Switzerland) in a hard-water lake environment. The water level in the marsh zone can fluctuate, sometimes flooding the entire area. The marl bench is mainly constituted by biogeochemical carbonate deposits (biogenic calcite of bacterial and algal origin, shells, charophytes). In this area, detrital sediments are reduced to dust and atmospheric inputs. The transition to the deep lake is steep and sudden. The white line on (B) shows the shot direction of view (C) (aerial view from 'Système d'Information du Territoire Neuchâtelois Swisstopo ©'). (D) Morphological relationships between sedimentation, environments and processes in a hard-water lake associated with a local high sedimentation rate.

1   The lake basement is composed of various outwash pebbles, moraines and/or sandy deposits.

2   If the lake is deep enough, organic matter can be preserved in its lower part, forming the bottomset sediments, often enriched in organic matter with periodic layers of clays in the anoxic zone of the monimolimnion. Respiration, methanotrophic activity and organic decay increase the CO_2 partial pressure, decreasing the pH, thereby enhancing calcite dissolution.

3   If the whole water body is rich enough in oxygen, organic matter remains only as traces and can be mixed with clays due to potential bioturbation.

4   As soon as the calcite saturation level is reached, biogeochemical carbonate crystals accumulate, mixed with remains of ostracods; sediments prograding towards the lake centre form foresets. Foresets can include turbidite beds a few centimetres thick due to turbid underflow. Calcium carbonate precipitates only when *Characeae* are present. Along the talus, and with depth, the water cools down increasing calcite dissolution.

5   On the upper part of the talus, micrite can be mixed with gastropod shells and fragments.

6   Sediment aggradation forms the lacustrine topset on the marl bench and micrite is mainly composed of crushed *Chara*, algal and cyanobacterial micrite, biologically induced micrite, sandy aggregates, and freshwater fauna and shell fragments (such as some ostracods, gastropods such as *Planorbis* and *Limnea*, and bivalves such as *Unionidae*); in some lakes, where waves are strong and regular enough, pisolitic and oncolitic gravels may form; at the contact with the shore, sediments can be washed. Shore erosion contributes to increased detrital sediment input. During summer, calcium carbonate directly precipitates as micrite (carbonate mud) and/or microsparite from supersaturated waters (due to increasing temperature and photosynthetic activity).

7   Sediments are enriched in organic matter forming calcareous peat close to the shore, e.g. to the palustrine environment.

8   Above the mean water level, the palustrine environment begins; it can include peats, calcareous marsh and ponds (Figure 9.3B, C), and hydromorphic (pseudogley) soils depending on the nature of the morphoclimatic setting.

9.3.2 Traces of palaeolakes in landscapes

When lakes form in large topographic depressions, such as rift valleys or wide glacial U-shaped valleys, their biogeochemical sediments are differentially eroded, forming hills with flat tops (Figure 9.4A, D). Erosion of lake sediments can also lead to relief inversions. For example, ephemeral lakes can form between sand ridges even in desert areas (e.g. as in the Simpson Desert, Australia, in December 1974; Dewolf and Mainguet, 1976). Sedimentation of limestone and early diagenetic silicification transform lake sediments into hard stones. With time, the palaeolake sediments are less eroded than the surrounding sediments, leading to the formation of longitudinal hills (e.g. Oligocene from the Paris Basin) and relief inversion (Figure 9.4B, D). Another example of such interdunal sedimentation is provided by Holocene lacustro-palustrine sediments inside the Saharian Great Western Erg (Fontes et al., 1985; Gasse et al., 1987). Ponds, leading to palustrine deposits, formed in wadi valleys blocked by sand dunes. Limestones and diatomites are today eroded, forming small hills between the sand dunes and deflation corridors, and sometimes isolating carbonate towers formed by sub-lacustrine spring travertines (Callot, 1991) (see Chapter 6). Finally, palaeolake sediments can be perched above present-day alluvial floodplains (Figure 9.4C, D). They formed in ancient oxbows and should not be confused with cemented palaeosols developed on terraces.

9.4 Macro- and Micromorphological Characteristics

9.4.1 Field and macroscale characteristics

Geochemical sediments related to lakes are easily observed and recognised in recent environments as well as in the geological record. Usually

───────────────────────────►

Figure 9.4 Simplified sketch of the geomorphological evolution of some lacustro-palustrine landscapes. (A) Rift and natural dam lakes are mainly infilled with fine geochemical material such as calcite, aragonite and/or gypsum, with a part being detrital (e.g. proglacial Pleistocene and/or Alpine lakes, Lake Lisan, Dead Sea Rift, Israel; see Figure 9.8F, G). After a climate change or tectonic event, the lake basin is eroded by streams or wadis, creating pseudoterraces and small hills with flat tops. (B) Biogeochemical sediments (diatomites, limestones, marls) deposited between ridges of aeolian formations. Lacustro-palustrine deposits can be laterally associated with palaeosols (presence of vegetation). After a climate change, rocks from palaeolakes resist wadi erosion and wind deflation and corrasion to form hills and mesas (hammada in North Africa). The relief is inverted. (C) In large river floodplains, abandoned meanders form oxbow lakes. After river erosion, they constitute remains of an old floodplain surface. (D) Left: bottom aragonitic and silty sediments of Pleistocene Lake Lisan eroded by wadis, aerial view, Dead Sea Rift, Israel. See details in Figure 9.8F, G. Centre: relief inversion in Matmata loessic deposits (Pleistocene, Tunisia). The hill top is capped by palustrine limestones, interpreted as calcretes (with a thick laminar crust), which were deposited in carbonate swamps in a lower topographic position between loess dunes and ridges. Right: lacustro-palustrine oxbow deposits interpreted as calcrete associated with a Rio Grande terrace (New Mexico, USA). This palaeolacustro-palustrine carbonate deposit is above the present-day river level.

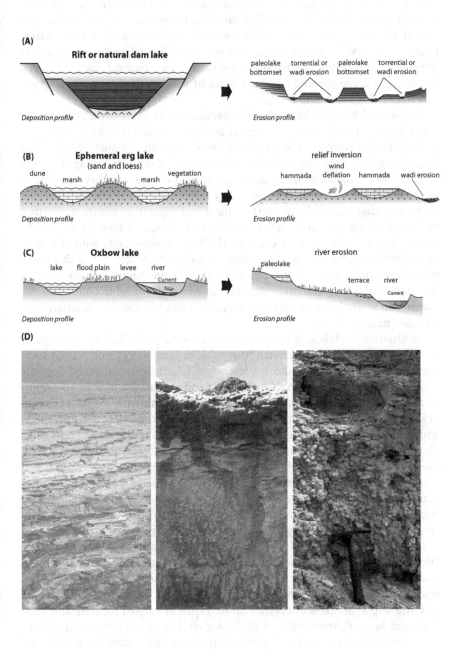

(A)
Rift or natural dam lake

Deposition profile

paleolake   torrential or   paleolake   torrential or
bottomset   wadi erosion   bottomset   wadi erosion

Erosion profile

(B)
Ephemeral erg lake
(sand and loess)

dune   marsh   vegetation
            marsh

Deposition profile

relief inversion
wind
hammada   deflation   hammada   wadi erosion

Erosion profile

(C)
Oxbow lake

lake   flood plain   levee   river
                                    Current

Deposition profile

river erosion
paleolake

            terrace   river
                        Current

Erosion profile

(D)

identified as horizontal, stratified, and/or laminated beds, they show a fairly well preserved homogeneity, emphasised by a characteristic sediment chemistry and fauna. In general, lacustrine limestones appear in the geological record as thick beds (from tens of centimetres to a few metres; Figure 9.5A), with bioturbations, sometimes with algal bioherms, or traces of roots and emersion (transition to palustrine limestones; Figure 9.5C, E). These beds are separated by marly beds of various thickness, from <1 cm to >10 cm (Figure 9.5A, B, E).

Laminated beds in lacustrine limestones result from variations in seasonal organic productivity, precipitation and redissolution of minerals, as well as detrital inputs, including accidental turbidites (Figure 9.5D). The lack of lamination preservation in some lacustrine deposits is explained by bioturbation (vegetation, animal burrows) i.e. sedimentary structures occurring in the oxic zone can be totally disturbed. In the anoxic zone, the absence of digging and burrowing animals and vegetation allows lamination preservation. In the geological record, it is common to observe laminated anoxic and dark layers between two thick calcareous beds with traces of bioturbation, sometimes including emersion features (palaeosols–palustrine limestones). There can be various explanations for this, such as lake eutrophication, rare bioturbation (due to unfavourable conditions for plants and animals), a meromictic regime or fast successions of emergence and submergence.

In Quaternary lakes, it is common to observe periodical successions of layers, called 'varves'. The term 'varves' (De Geer, 1912) or 'varval sediments' (Figure 9.6A) should be restricted to seasonal sediment deposits related to glaciers although they are also used to describe sediments organised as couplets deposited in non-glacial environments. Kelts and Hsü (1978) described such sediments in Lake Zurich as couplets 2–5 mm thick, which can be partially interpreted as geochemical sediments. These couplets are composed by a first layer divided into an organic substrate, with blue-green algae, iron sulphides and fine mineral detritus (late autumn and winter deposit) and a lighter-coloured layer formed by diatom frustules (early spring blooms). The second layer of the couplet is also paired and is composed of large calcite crystals (up to 30 μm) with rare diatoms (late spring) and a micritic layer with aggregates of plankton (summer and autumn). In meromictic lakes, such as Green Lake in New York State (Brunskill, 1969; Thompson et al., 1990), couplets form in the monimolimnion. From November to May, a dark organic layer forms (0.2 mm thick). From June to October, organic matter is associated with calcite (micrite and aggregates of crystals up to 64 μm in length) forming a lighter-coloured layer 0.5 mm thick. Calcite forms during 'whitings' (the open-water precipitation of calcium carbonate) due to the coccoid cyanobacteria *Synechoccocus sp.* when the lake waters are stratified.

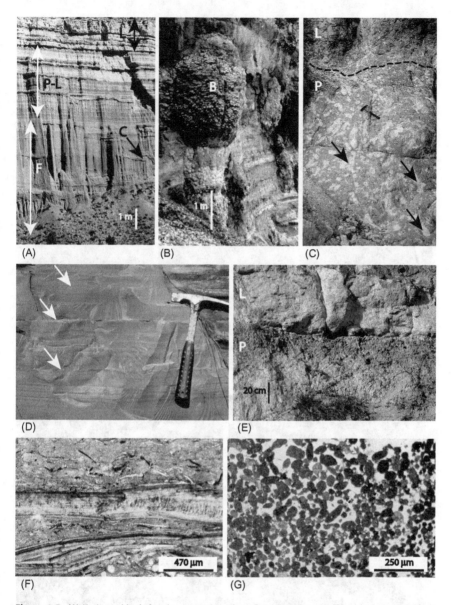

Figure 9.5 (A) Horizontal beds forming a transition from floodplain deposits [F] with channels [C] and palaeosols to a palustro-lacustrine environment [P-L], and lacustrine limestones [L]. Miocene, Teruel Basin, Spain. (B) Lacustrine deposits with stromatolitic bioherms [B]. Limagne Plain, central France, Oligocene. (C) Palustrine limestone [P] with abundant root traces (arrows). The transition with the upper thick lacustrine [L] dolomitic limestone is sharp. Root traces are the only remains of life. These limestones are azoic. Anzal Basin, southern Morocco, Upper Eocene to Lower Oligocene. (D) Lacustrine bottomset sediments enriched in organic matter and showing thin turbiditic layers (arrows). Manosque Basin, southeastern France, Oligocene. (E) Palustrine limestone [P] with a well developed palaeosol at the top in a sharp contact with a thick bed of lacustrine limestone [L]. Numerous traces of marmorisation. Aquitaine Basin, France, Oligocene. (F) Various types of crushed shell fragments in a lacustrine mud. Minervois, southern France, Lutetian. (G) Lacustrine bioclastic and oolitic sand deposited near a shore. Limagne Plain, central France, Oligocene.

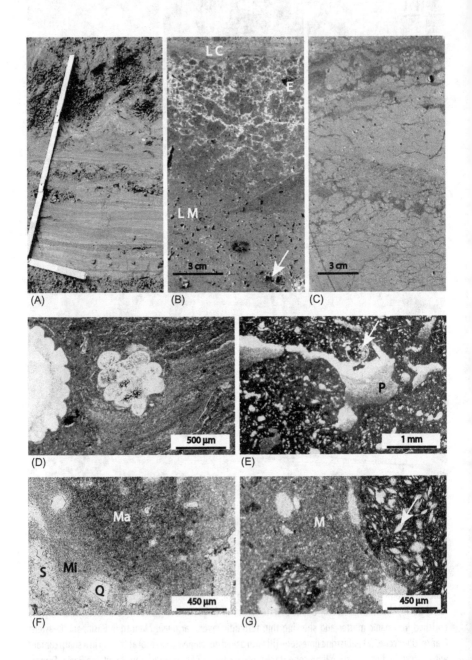

(A)

(B)

(C)

(D)

(E)

(F)

(G)

9.4.2 Petrography of lacustrine and palustrine deposits

The study by Freytet and Plaziat (1982) and the recent reviews by Freytet and Verrecchia (2002), Verrecchia (2002) and Alonso-Zarza (2003) provide excellent overviews of the petrography of lacustrine and palustrine deposits, and contain abundant photographs, information and references. In this section, only the main characteristic features associated with biogeochemical lacustro-palustrine carbonate sediments will be described. Evaporitic sediments are the subject of another chapter (Chapter 10), and detrital sediments are not the topic of this book. As such, carbonates and organic sediments are the only facies that remain.

In lacustrine carbonate biogeochemical deposits, various morphotypes are used to describe the amount of biogenic remains, detrital input and intensity of sediment reworking. Muds form the main part of lacustrine biogeochemical limestones. This is the reason why they appear so homogeneous in outcrops. Nevertheless, biogeochemical lacustrine deposits can also form marls, i.e. a mixture of carbonate, varying between 20 and 80 wt.%, and other compounds such as fine siliciclastic particles and/or organic matter. The carbonate phase in marl is constituted by an accumulation of biochemically precipitated minerals by cyanobacteria, *Characeae*, diatoms (stalk), planktonic organisms, epiphytic organisms on leaves of aquatic plants and

◀ ───

Figure 9.6 (A) 'Glacial' varves from a Last Glacial Maximum lake, Val de Travers, Switzerland. The regular annual sedimentation is occasionally disturbed by turbidites. Compare with Figure 9.5D. (Photograph courtesy of Professor M. Burkhard.) (B) Slab of a transition from lacustrine (LM) to palustrine environment with emergence (E) and desiccation cracks infilled with secondary calcite, Oligocene, Paris Basin, France. The lacustrine mud includes casts of *Planorbis*, a freshwater Gastropod (arrow). The sequence ends with a laminar crust, a terrestrial stromatolite (LC). This type of sequence is often confused with pedogenic calcretes in the literature. (C) Succession of lacustrine mud deposits undergoing short emergence. Desiccation cracks and nodulisation separate clusters which are reworked when the water level rises. Oligocene, Paris Basin, France. (D) Lamina of dark micrite and microsparite with ostracod test fragments and *Chara* encrustations. A gyrogonite has fallen to the bottom of the lake disturbing non-compacted laminae, different from varves shown in (A). Cesseras, Languedoc, southern France, Middle Eocene. (E) Pedogenic pseudomicrokarst in emerged lacustrine mud (see gastropod shell, arrow). The porosity results from roots and is progressively infilled with grain-size sorted material (P). Palustrine limestone from St-Jean de Minervois, southern France, Eocene. (F) Traces of pedogenesis in emerged micrite (Mi). Iron segregation (Ma: marmorisation) is related to water table fluctuations and changes in redox conditions. Detrital quartz (Q) is surrounded by phreatic sparite that also infilled pores (S). Maastrichtian, southern France. (G) Palustrine micritic limestone (M) infilled by a dark secondary micrite associated with gypsum crystals (arrow). These features demonstrate a change in the evaporative and ecological conditions. Plantorel, southern France, Lower Paleocene.

small animals. Another type of white carbonate-rich lacustrine deposit is lacustrine chalk. It has a similar origin to marl, but is mainly deposited in the pelagic to hemi-pelagic zone of the lakes and often remains undisturbed. All of these biogeochemical sediments are characterised by a homogeneous and mainly micritic fabric due to partial to intense bioturbation.

Although frequently laminated and homogeneous, lacustrine chalks include juxtaposed irregular or rounded clusters of pellets, peloids and clots in thin-sections. In marls, the micritic matrix may contain some microsparitic crystals, bioclasts and eventually quartz grains associated with rare thin turbidite beds. Laminations can be emphasised by thin algal mats and/or turbidites, sometimes including wood, organic or shell debris. The presence of microsparitic crystals is difficult to interpret: they can result from early phreatic cementation, whitings or partial biomineralisation of microbial mats. The succession of laminations, related to slow and quiet sedimentation, can be disturbed by organisms. An increase in sedimentation rate leads to the formation of massive limestones, which can be characterised by the abundance of marls, traces of burrows, presence of gastropod and ostracod shell fragments (Figure 9.5F) or *Chara* gyrogonites (Figure 9.6D). These marls and chalks constitute the parent material for further pedogenic transformation when emerged and result in the formation of palustrine limestones.

Algal-rich and stromatolitic limestones are characterised by the direct influence of algae and microorganisms (Figure 9.5B). Although the macroscopic morphology of these biochemical deposits is extremely varied, thin sections often allow algal remains to be identified, either as species (Freytet and Verrecchia, 1998) or as microsymbiotic algal mats (e.g. Gasse et al., 1987; Dupraz et al., 2004). In the deep anoxic zone of meromictic lakes, the organic matter can be preserved, forming shales and sapropels. In contrast, algal deposits close to the lake shore can be reworked by waves and form bioclastic and oolitic sands (Figure 9.5G).

Emersion of lacustrine muds leads to the formation of palustrine facies (Figure 9.6B, C). Diagnostic features of emersion are desiccation cracks, nodulisation and root traces (Figures 9.5C and 9.6B, C, E). Emersion obviously results in the development of soils, which may or may not be under the influence of the water table. Pedogenic processes depend on the climate and the role of biota. In areas influenced by water table fluctuations, pedogenesis is frequently identified by traces of marmorisation, i.e. mobilisation of Fe^{2+} during floods and formation of Fe^{3+}-rich aggregates during oxidation, when the water level is low (Figures 9.5E and 9.6F). The result is a sediment with mottled pink, purple, red and yellow patches in the accumulation area of ferric iron, and grey or white in the area depleted in iron. Manganese and calcium can also migrate with iron, resulting in a complex fabric of mottled patches and ferruginous nodules. Such palustrine soils can be identified as pseudogley palaeosols when found in the fossil record. Burrows constitute another feature and may be observed as striotubules (a micromorphologic term for infillings formed by stacked watch-glass shapes with alternating colour and grain size) infilled with a secondary sediment and sometimes traces of evaporitic minerals, such as gypsum, indicating evaporation and desiccation (Figure 9.6G). In thin-section, burrows in palustrine limestones often exhibit circles and concentric ellipses emphasised by differences in grain size and by the orientation of the bioclasts.

Burrows should not be confused with root voids. Klappa (1980) distinguished five basic types of root traces in sediments, collectively called rhizoliths: root moulds (voids), root casts (infilling of voids), root tubules (cemented cylinders around roots), rhizocretions (pedodiagenetic mineral accumulation around roots) and root petrification (mineral impregnation of plant tissue). Such root traces are extremely common in palustrine limestones. Nevertheless, root traces can be complex. Roots may impregnate the soil matrix (i.e. a subcutanic feature) and generate various types of recrystallisation using proton force. After death and decomposition of roots, their voids can be enlarged and infilled by internal sediment, vadose silt and various cements (coatings, calcitans, vadose or phreatic cements). These multiple generations of sediments and cements due to the fluctuation of the water level (Figure 9.6E) are described as the palustrine 'pseudomicrokarst' facies (Freytet and Plaziat, 1982).

To conclude, a typical palustrine microfacies results from the succession of events related to water-level fluctuations. After emergence, the first features to appear are desiccation cracks, vertical and horizontal joint planes, the latter being often related to dry algal mats. Initial pedogenesis then allows the soil to be structured. Clots and matrix nodules form, leading to craze and curve planes. The nodulisation process can be so intense that the resulting sediment has a brecciated appearance (Figure 9.6B, C). As the soil develops, vegetation and animals create root and burrow traces, forming voids which will be infilled with various secondary sediments and cements, related to the fluctuation of the water table and the soil-water regime. At this stage, it is somewhat difficult to distinguish a palustrine limestone from a conventional 'hardpan calcrete' (see Chapter 2) in thin-section.

9.5 Chemistry and Mineralogy

9.5.1 General composition of waters

The chemistry of lake waters is extremely variable in composition because of the diversity of the watershed lithology, the climatic water balance and the residence time of the water in the lacustrine basin. Therefore, the lake water chemistry is primarily determined by the watershed geology and atmospheric inputs (Campy and Macaire, 2003). Due to this origin, the major ions in lake waters are: Ca^{2+}, Mg^{2+}, Na^+, K^+, Cl^-, SO_4^{2-}, HCO_3^- and CO_3^{2-}. Various types of water bodies can be distinguished according to the concentration of waters as well as their main composition. If the concentration of total dissolved ions is <$0.1\,gL^{-1}$, the lake is called oligohaline. Very low ionic concentrations occur when the water input is only rainfall or when the watershed has a low ionic production, such as metamorphic basins in temperate climates. In mesohaline lakes (total dissolved ion concentration is <$1\,gL^{-1}$), the watershed normally comprises soluble bedrocks. Peri-alpine, North American, and some of the East African Rift lakes belong to this category. When the concentration of dissolved ions reaches the ocean's mean value ($\approx 35\,gL^{-1}$), the water is brackish and the lake regime is euryhaline. These lakes are either situated close to the sea and hydrogeologically in contact with it (paralic environment) or they have inflow from watersheds enriched in carbonate and evaporitic rocks, with the lake basin concentrating the solutions. This case is particularly true for lakes with a high evaporation rate, leading to temporary concentration of the solutions and precipitation of specific minerals (see Table 9.1). Some East African Rift lakes as well as the Aral Sea belong to this category. Total concentration of dissolved ions higher than sea water also exists: these lakes are described as hyperhaline. The lake water is a true brine, sometimes reaching concentrations >$400\,gL^{-1}$ (e.g. the Dead Sea, North African chotts, playas, or South American salinas).

9.5.2 Chemistry and mineralogy of euryhaline to hyperhaline lakes

The chemistry and mineralogy of euryhaline to hyperhaline lakes are related to the composition of the undersaturated inflowing waters, and during evaporation and concentration of the solutes, to the composition of the brines. Therefore, the general chemical composition is determined initially by inflowing waters, which include runoff, rain and groundwater, and evolves with the ratio of evaporation to water supply through time, i.e. brine chemistry changes as evaporative concentration progresses. This change in chemistry results in mineralogical sequences depending on the climatic settings and the composition of the inflowing waters. Three different pathways (Figure 9.7) have been described in the literature (Hardie et al., 1978; Risacher, 1992; Arakel and Hongjun, 1994; Campy and Meybeck, 1995). One is alkaline and two are neutral. In a first step, low magnesian calcite precipitates whatever the chosen pathway, with possible authigenesis of clay minerals. In a second step, alkaline and neutral pathways have to be distinguished. The difference is due to the ratio between alkalinity (i.e., in a first approximation, the concentration of HCO_3^- anions) and the total concentration of Ca^{2+} and Mg^{2+} cations (Figure 9.7). In the alkaline pathway, the concentration of hydrogeno-carbonate anions is more than twice the total concentration of calcium and magnesium cations. The high alkalinity enhances dissolution of silica, leading to precipitation of various magnesium and sodium silicates and zeolites (Reeves Jr., 1978; Sebag et al., 2001; Figure 9.8A–C). Silica is provided by fossil diatoms (diatomites; Figure 9.8A) that grew during favourable conditions, i.e. in more diluted lacustrine waters. However, when the biogeochemical environment changes to drier conditions, silica is dissolved, carbonate precipitates and evaporitic sedimentation occurs with precipitation of secondary amorphous silica and sodium silicates (Sebag et al., 1999).

In the neutral pathways, the total concentration of calcium and magnesium cations is higher than the concentration in hydrogeno-carbonate anion. Minerals precipitated along one of these pathways are identical at the start of the process to what is observed in temperate hard-water lakes (when they evaporate, the sequence is almost complete to the step of gypsum precipitation). Waters can concentrate much more in arid environments than under temperate climates, due to high and rapid evaporation, leading to a sequential precipitation of various minerals (Figure 9.8D–G). The nature of the minerals also depends on the amount of sulphate anions available in the residual waters (Figure 9.7). In conclusion, five different brine compositions occur, depending on the initial composition of the undersaturated inflowing waters, as well as the rate of evaporation leading to brine concentration and supersaturation (Figures 9.7 and 9.8A–G). More information about terrestrial evaporites and efflorescence crusts is given in Chapters 10 and 12.

9.5.3 Chemistry and mineralogy of temperate hard-water lakes

Most temperate hard-water lakes follow the neutral pathway involving $HCO_3^- < (Ca^{2+} + Mg^{2+})$ (Figure 9.7). The presence of water rich in calcium and hydrogeno-carbonate ions enhances calcium carbonate precipitation through inorganic and organic processes. Lake pH is an important characteristic of carbonate kinetics (Figure 9.1). Lake pH is mainly regulated by photosynthesis, respiration and mineralisation of the organic matter due to the fact that most of the anions in solution are constituted by the two species of carbonates, CO_3^{2-} and HCO_3^-. Photosynthetic activity uptakes dissolved carbon dioxide and displaces equilibria, resulting in a pH increase. In contrast, respiration, as well as mineralisation of organic matter, lead to CO_2 production and favour the opposite reaction. Consequently, the annual evolution of lake pH is characterised by a progressive increase in the epilimnion during the stratified period, i.e. from May to November. Concomitantly, pH decreases in the metalimnic and hypolimnic zones, mainly because of the mineralisation of the organic matter at the bottom of the lake.

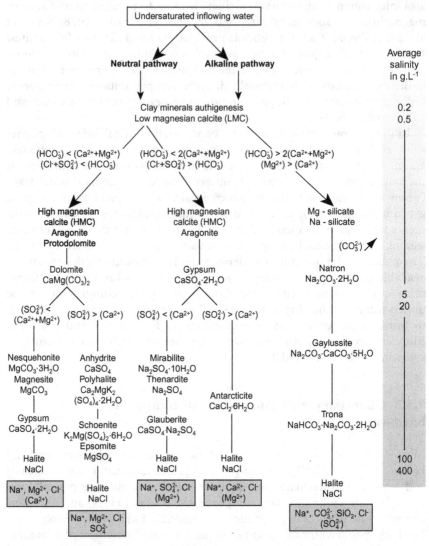

Figure 9.7 Simplified chart showing the evaporite precipitation sequence from waters of various compositions. Alkaline and neutral pathways are shown. The alkaline pathway is common in salinas, playas and apolyhaline lakes. The neutral pathway can be divided into two different sequences according to the ratio of the concentrations of HCO_3^- to $(Ca^{2+} + Mg^{2+})$. The main ions still in solution in the residual brines are given at the end of the sequence (grey frames). On the right-hand side, the average salinity of the water is provided as an indicator of total ion concentration ($g\,L^{-1}$).

Figure 9.8 (A) Clayey and calcareous diatomite from northern Lake Chad. Various sedimentary layers can be seen: they correspond to transition from a lacustrine environment (bottom) to a palustrine environment (top), from sub-arid to arid conditions. (Photograph courtesy of Professor A. Durand.) (B) Spherule-like crystals of kenyaite (hydrous sodium silicate) precipitated in apolyhaline interdunal ponds, Lake Chad.

Figure 9.8 *Continued*

(C) Zeolite crystals (Z) inside a crack between a mass of magadiite (hydrous sodium silicate, S). Same environment as (C). (D) Dead Sea brine showing regular salt deposits related to the fluctuation of the lake water level, Israel. (E) Close-up of salt deposits, mainly constituted by halite and sylvite. (F) Lake Lisan regular varval deposits composed of detritic marl (white arrow) and endogenous aragonite (black arrow), Israel. Lake Lisan was the Pleistocene precursor of the Dead Sea. (G) Same sediments as in (F). The bottom part (1) is constituted by flat and regular beds, whereas the top part of the outcrop (2) is very disturbed, probably an earthquake that shook sediments still partly saturated with water, forming seismites.

Therefore, in temperate hard-water lakes with a monomictic or dimictic regime, the lake chemistry evolves seasonally. During the summer, the epilimnion is supersaturated in calcite (mainly low-magnesian calcite). Molluscs, ostracods, charophytes and cyanobacteria strongly contribute to the precipitation of $CaCO_3$. This point is particularly true along the marl bench (Figures 9.1–9.3). Plants, such as charophytes (Figure 9.9A, B), algae (e.g. diatoms; Figure 9.9C–E) and cyanobacteria (Figure 9.9F, G) extract CO_2 and H_2O from lake water during photosynthesis yielding precipitation of $CaCO_3$ at the surface of leaves, stems, as well as in cyanobacterial mucilagenous sheaths (i.e. inside microorganic extra cellular polymeric substances called EPS; Figure 9.9E, F) or directly on the sheath surface (Figure 9.9G). In addition, because of the slight supersaturation of low-magnesian calcite in the epilimnion due to an increase in temperature (facilitating CO_2 degassing) and CO_2 consumption during phytoplankton activity, low-magnesian calcite can precipitate as micro- to nanocrystals that fall down slowly to the lake bottom as a sedimentary rain (Figure 9.9H, I). This calcite can accumulate on the marl benches, whereas it is more likely to dissolve in the deeper hypolimnion because of a lower pH resulting from decaying and mineralisation of organic matter (Dean, 1981). Nevertheless, in some lakes (e.g. Lake Neuchâtel, Switzerland), the summer

--➤

Figure 9.9 Scanning electron micrographs (SEM). (A) *Chara* sp. stem encased in externally precipitated calcite, Lake Neuchâtel (Switzerland). *Chara sp.* can contain up to 50% $CaCO_3$ by dry weight. (B) Calcareous outer cover of the female reproductive oogonium of a *Chara* sp., Lake Neuchâtel (Switzerland). The external cortical tubes are spirally arranged giving the grain a ribbed appearance. At the end of autumn, a new sediment layer on the marl bench is formed by a carbonate mud of low magnesian calcite derived from crushed and decaying *Chara*. (C) Diatom frustule (D) with its organic coating (arrow). Lake Neuchâtel (Switzerland). Diatoms constitute the main part of biogenic silica precipitated in hard-water lakes as well as pluvial lakes from the arid zone (see Figure 9.8A). (D) Another species of diatom (D) with a stalk, Lake Neuchâtel (Switzerland). This type of stalk is often a site for calcite precipitation (Freytet & Verrecchia, 1998). (E) A third type of diatom (D) with its organic coating associated with calcite crystals (C). Calcite nucleates inside an extra cellular polymeric substance (EPS) mat (arrow). Lake Neuchâtel (Switzerland). Low temperature SEM (LTSEM). (F) Accumulation of calcite (C) inside and outside an *Oscillatoria* sp. (a cyanobacteria, arrow). Numerous calcite crystals are associated with EPS. Lake Neuchâtel (Switzerland). LTSEM. (G) Cyanobacterial sheath (S) cast encrusted by calcite (C). Cyanobacteria greatly contribute to $CaCO_3$ accumulation and O_2 production in the photic zone. (H) Lacustrine chalk (low magnesian calcite mud) mainly composed of micrite (M). Some detrital calcitic grains (C) are visible. Some platelets (arrows) are agglomerated inside the micrite: they are composed of inherited clay minerals from the lake watershed. Lake Neuchâtel (Switzerland). (I) Example of bio-induced precipitation of calcite. This type of crystal (C), showing slow growth, precipitates at the surface of *Potamogeton* sp. leaves. During wave agitation of this aquatic plant and/or after plant death, calcite crystals fall to the lake bottom, contributing to lacustrine marl deposition.

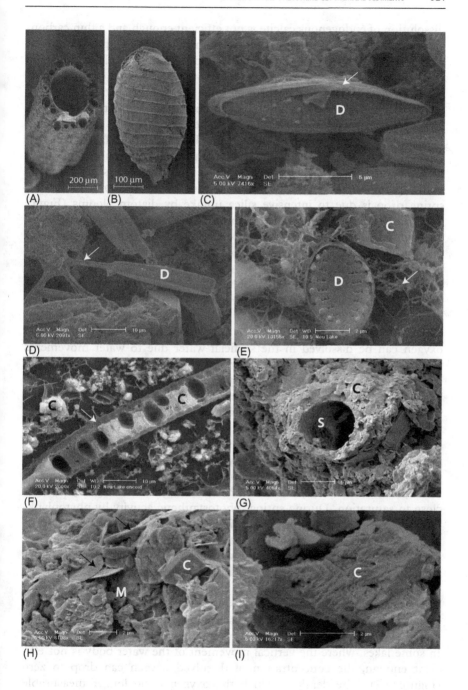

dissolution of endogenous calcite is not efficient enough and a thin carbonate layer can form. As observed by Kelts and Hsü (1978), preservation of a carbonate layer can generate periodic deposits: summer calcite is covered by autumn and winter detrital material (clays) and settled organic debris when the lake is back to its winter thermic stratification regime.

Regarding silica, various amounts can be dissolved in hard-water lakes. For example, in Lake Neuchâtel (Lambert, 1999), the concentration of dissolved silica reaches values $>3\,mg\,L^{-1}$ and stays fairly constant ($>2\,mg\,L^{-1}$) with depth during the period of the homothermal regime (from December to March). Nevertheless, during spring, dissolved silica decreases in the epilimnion ($0.5\,mg\,L^{-1}$) to values at the limit for diatom growth. This spring drop is due to intensive silica uptake by diatom blooms (Figure 9.9C–E). But at the same time, dissolved silica concentration increases in the hypolimnion (to 2.5–$3\,mg\,L^{-1}$), probably due to partial dissolution of diatoms, enhanced by zooplankton consumption, as well as to silica release from bottom sediments.

Another important component of hard-water lakes is the amount of dissolved oxygen within the water column. When the partial pressure of oxygen in the epilimnion pO_2 is lower than in the atmosphere (i.e. the water body is undersaturated with respect to oxygen), parts of the atmospheric oxygen can be dissolved in the surficial water due to water movements (waves). During photosynthetic activity in the warm season, dissociation of water leads to the production of oxygen, increasing the pO_2, with oxygen being used by heterotroph organisms for respiration. When the surficial water body pO_2 is higher than the atmospheric pO_2, oxygen degasses into the atmosphere. In Lake Neuchâtel (Lambert, 1999), variations in dissolved oxygen during the year are strictly correlated to photosynthetic activity as well as to the water-body stratification. In March, when the regime is homothermic, the amount of dissolved oxygen is regular throughout the lake's depth and reaches values around $12\,mg\,L^{-1}$, which is close to saturation. During spring, photosynthetic activity by phytoplankton increases the amount of dissolved oxygen in the epilimnion to concentrations $>14\,mg\,L^{-1}$. In August, the dissolved oxygen concentration decreases close to the surface (re-equilibration with atmospheric pO_2), and on a larger scale, in the bottom part of the metalimnion and close to the bottom part of the lake, at the interface between the hypolimnion and the sediment. The first shift is due to the intense respiration activity, which exhausts part of the oxygen produced. The second shift, at the contact with the sediment, is related to the decaying (i.e. oxidation) of the organic matter (Figure 9.1). In some lakes, where the vertical movement of the water body is not efficient enough, the concentration of dissolved oxygen can drop to zero (Figure 9.2). The depth at which the oxygen is no longer measurable defines the limit between the oxic and anoxic environment. In this part of the hypolimnion, oxygenic photosynthesis is replaced by anoxygenic photosynthesis, and cyanobacteria and algae by purple and green bacteria. In the anoxygenic photosynthesis, H_2O is replaced by H_2S and methanotrophs consume CH_4, both H_2S and CH_4 being produced inside the bottom sediment and released in the water body. In these lakes, organic matter can be more easily preserved in the sediments: there is no oxidation and bioturbation is extremely limited to almost non-existent. In these cases, black shales containing pyrite can accumulate and be used as palaeoenvironmental proxies.

9.5.4 The unusual case of carbon-dioxide-rich lakes

In some rare cases, the chemistry of specific crater lakes is driven by the CO_2 concentration in the hypolimnion. For example, Lake Nyos, situated in the Oku volcanic field (Cameroon), is inside a crater of a maar formed about 400 years ago (Lockwood and Rubin, 1989). The lake is about 1800 m wide and 208 m deep. Lake Nyos is a meromictic lake with a chemocline. The source of CO_2 is due to the presence of a low-temperature reservoir of free carbon dioxide below the lake bottom (Evans et al., 1993). Convective cycling of lake water through the sediments allows CO_2 transportation into the lake from an underlying breccia-filled volcanic pipe (diatreme). The hypolimnion (monimolimnion) also contains biologically produced methane. The highest concentrations of CO_2 and CH_4 have been measured in 1990 as $0.30\,\text{mol}\,\text{kg}^{-1}$ and $1.7\,\text{mol}\,\text{kg}^{-1}$, respectively (Evans et al., 1993). Total dissolved-gas pressure near the lake bottom is 1.06 MPa, i.e. 10.5 atm. This figure corresponds to only 50% of the hydrostatic pressure (21 atm): if the CO_2 pressure increases abnormally or if some water movement (lake overturn) occurs, the gas can be rapidly released into the atmosphere. In 1986, rainwater from heavy summer showers may have been blown to one side of the lake by strong August winds. Being colder, and therefore denser than the warmer lake water, the rainwater mass dived down one side of the lake, displacing the hypolimnion waters. This convective and accidental overturn resulted in the ascent and decompression of the bottom water, causing the dissolved gas to exsolve and bubble upward at dramatic speeds. The bubbles themselves may have lowered the overall density of the gas–water mixture, resulting in even greater rates of ascent, decompression and exsolution (Kling et al., 1989). The result was a rapid and violent expulsion of CO_2, with dramatic consequences. On 21 August 1986, a cloud of CO_2 was released from the lake bottom. Because CO_2 is more dense than air, it stayed at ground level and flowed down the surrounding valleys. The cloud travelled as far as 25 km from the lake. One thousand seven hundred deaths were caused by suffocation and 845 people were hospitalised. So much gas escaped from this single event, that the surface level of Lake Nyos dropped by approximately 1 m (Evans et al., 1993).

9.6 Relationships to other Terrestrial Geochemical Sediments

Lacustrine geochemical sediments can be associated with travertines (see Chapter 6). In some geomorphological settings, streams and waterfalls flow into a lake, producing oncolites and travertines in association with lacustrine deposits: the procession of waterfalls and lakes in Plitvice (Croatia) is an excellent example. Another interesting case is provided by travertine peaks at the border between Ethiopia and Djibouti Republic. In this northern part of the East African rift, a wide depression of $6000\,km^2$ and $170\,m$ deep has been occupied by Lake Abhé. Situated in a tectonic graben, water was provided to this palaeolake by artesian wells along fault lines, between 10 and 4 ka (Gasse, 2000). Large masses of calcareous travertines have been deposited around these wells at the bottom of the palaeolake (Rognon, 1985). The present-day landscape is dominated by the travertine peaks, which are much more resistant to erosion than the surrounding lacustrine diatomites and muds.

Although purely lacustrine geochemical sediments are often clearly identifiable, palustrine facies frequently lead to confusion, as is often the case for formations called 'calcretes' (see Chapter 2). The succession of layers composed of an emerged lacustrine mud, traces of roots, nodulisation cracks and nodules, and finally, a top laminar crust (Figure 9.5B; Freytet and Verrecchia, 2002) could easily be interpreted as a monogenic calcrete palaeosol. It is obvious that catena relationships exist between palustrine environments and interfluve carbonate-rich soils. Nevertheless, palustrine carbonates should not be confused with 'calcretes', the definition of which is sometimes (imprecisely) applied to embrace deposits formed in palustrine settings. Attention should be paid to microscopic features, successions of cements, and general geomorphological relationships between forms and formations before calling any terrestrial accumulation of $CaCO_3$ a 'calcrete', especially when potentially associated with a lacustro-palustrine environment.

9.7 Palaeoenvironmental Significance and Directions for Future Research

Lake and palustrine biogeochemical sediments are remarkable proxies for palaeonvironmental reconstruction, in terms of space and time (Figure 9.10). Because lakes are related to specific watershed, water balance and morphoclimatic settings, their sediments record variations in salinity, evaporation rate, organic activity and production rate, as well as biodiversity. Lacustro-palustrine sediments have been used for a considerable time in palaeoclimatic reconstruction (e.g. Freytet and Plaziat, 1982; Gasse et al.,

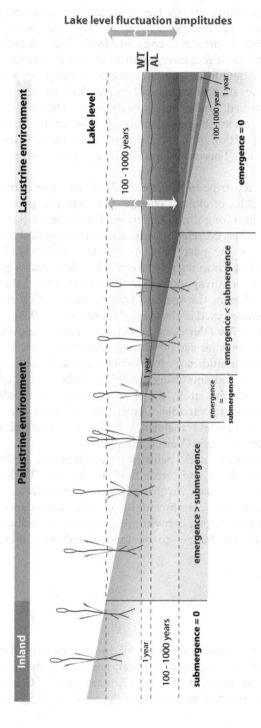

Figure 9.10 Sketch showing the relationship between space and time in a palustro-lacustrine environment. The lake level varies during the season around a water table average level (WTAL) with a limited amplitude. Nevertheless, lakes can undergo extremely important secular or millenar level rises or falls due to changes in the geomorphological settings and/or climate. Regarding the influence of the water level (relationships between emergence and submergence), various environments can be defined. The sedimentary record and the facies obviously change in relation to the position along the submergence gradient. In addition, it is difficult to relate a sediment thickness to a time period: only true glacial varves can be interpreted in terms of periodic annual deposits.

1987; Brauer and Negendank, 2004). Recently, emphasis has been put on isotopic composition of authigenic and biogenic carbonates and diatom silica as palaeoclimate proxies (Leng and Marshall, 2004, and references therein). Nevertheless, direct interpretation of data is fallacious: 'the interpretation of isotopic data from lacustrine succession requires a knowledge of the local processes that might control and modify the signal. Their effects need to be quantified, and a robust calibration using modern lake systems is necessary to establish the relationship between the measured signal, the isotopic composition of the host waters, and climate' (Leng and Marshall, 2004, p. 811). This is undoubtedly a future challenge for researchers.

Other research uses lacustrine biogeochemical sediments as chronometers and/or holistic variables of climate change. For example, present-day varves have been used for a long time (Anderson, 1961) for timescale and climate calibration. However, correlation with actual astronomical data has only been documented in the past 10 years (Anderson, 1993). Recent research on Last Glacial Maximum varves from a Jura Mountains palaeolake demonstrates that solar sun spots were in effect at this time: the three solar harmonics deduced from the spectral analysis of the Wolf number (Berger, 1992) have been detected in the varve thicknesses, i.e. 9.31 ± 0.17, 11.38 ± 0.25 and $22.30 \pm 97 \, yr$ (Verrecchia and Buoncristiani, unpublished data). Correlations between solar activity during geological times and lake records is a hot topic. Other studies try to track climatic and periodic fluctuations in shell growth increments using analytical (Schöne et al., 2004) or holistic approaches. For the latter, the shell periostracum topology is considered as an environmental variable. Detection of growth cycles using spectral methods, such as wavelet transform, constitutes a powerful tool to investigate environmental changes (Verrecchia, 2004).

In conclusion, lacustrine and palustrine geochemical sediments remain an extremely promising and fertile field of research. Their study needs a pluridisciplinary approach involving sedimentary petrology, biogeochemistry, isotopic chemistry, freshwater biology, limnology and signal processing, allowing terrestrial palaeoenvironments and palaeoclimate, as well as present-day climate change, to be investigated with increasing accuracy.

Acknowledgements

A part of this work has been inspired by fruitful discussions with Professor Pierre Freytet. Professor M. Aragno (Neuchâtel) kindly provided information on Loclat Lake, Switzerland. Dr C. Dupraz and L. Chalumeau (Neuchâtel University) provided help with the figures and LTSEM.

References

Alonso-Zarza, A.M. (2003) Paleoenvironmental significance of palustrine carbonates and calcretes in the geological record. *Earth-Science Reviews* **60**, 261–298.

Anderson, R.Y. (1961) Solar-terrestrial climatic patterns in varved sediments. *Annals of the New York Academy of Science* **95**, 424–439.

Anderson, R.Y. (1993) The varve chronometer in Elk Lake: record of climatic variability and evidence for solar-geomagnetic ^{14}C climate connection. In: Bradbury, J.P. & Dean, W. E. (Eds) *Elk Lake, Minnesota; Evidence for Rapid Climate Change in the North-central United States*. Special Paper 276. Boulder, CO: Geological Society of America, pp. 45–67.

Aragno, M. (1981) Responses of microorganisms to temperature. In: Lange, O.L., Nobel, P.S., Osmond, C.B. & Ziegler, H. (Eds) *Encyclopedia of Plant Physiology*, Vol. I-12A, *Physiological Plant Ecology*. Springer-Verlag, Berlin, pp. 339–369.

Arakel, A.V. & Hongjun, T. (1994) Seasonal evaporite sedimentation in desert playa lakes of the Karinga Creek drainage system, Central Australia. In: Renaut, R.W. & Last W.M. (Eds) *Sedimentology and Geochemistry of Modern and Ancient Saline Lakes*. Special Publication 50. Tulsa, OK: Society of Economic Paleontologists and Mineralogists, pp. 91–100.

Ariztegui, D. & Wildi, W. (Ed.) (2003) Lake systems from Ice Age to Industrial Time. *Eclogae Geologicae Helveticae Special Issue* **96**, S1–S133.

Begin, Z.B., Broecker, W., Buchbinder, B., Druckman, Y., Kaufman, A., Magaritz, M. and Neev, D. (1985). *Dead Sea and Lake Lisan Levels in the Last 30,000 Years*. Report 29/85. Jerusalem: Geological Survey of Israel.

Berger, A. (1992) *Le Climat de la Terre*. Bruxelles: De Boeck Université.

Braissant, O. & Verrecchia, E.P. (2002) Microbial biscuits of vaterite in Lake Issyl-Kul (Republic of Kyrgyzstan) – discussion. *Journal of Sedimentary Research* **72**, 944–946

Brauer, A. & Negendank, J. (Eds) (2004) High resolution lake sediment records in climate and environment variability studies: European lake drilling program. *Quaternary International* **122**, 1–133.

Brunskill, G.T. (1969) Fayetteville Green Lake, New-York. II, Precipitation and sedimentation of calcite in a meromicric lake with laminated sediments. *Limnology Oceanography* **14**, 830–847.

Callot, Y. (1991) Histoire d'un massif de dunes, le Grand Erg Occidental (Algérie). *Sécheresse* **2**, 26–39.

Campy, M. & Macaire J.-J. (2003) *Géologie de la Surface*. Paris: Dunod.

Campy, M. & Meybeck, M. (1995) Les sédiments lacustres. In: Pourriot, R. & Meybeck, M. (Eds) *Limnologie Générale*. Paris: Masson, pp. 185–226.

De Geer, G. (1912) A geochronology of the last 12,000 years. *11th International Geological Congress* **1910**, 241–253.

Dean, W.E. (1981) Carbonate minerals and organic matter in sediments of Modern North temperate hard-water lakes. Special Publication 31. Tulsa, OK: Society of Economic Paleontologists and Mineralogists, pp. 213–231.

Dewolf, Y. & Mainguet, M. (1976) Une hypothèse éolienne et téctonique sur l'alignement et l'orientation des buttes tertiaires du bassin de Paris. *Revue de Géographie physique et de Géologie dynamique* **18**, 415–426.

Dupraz, C., Visscher, P., Baumgartner L.K. & Reid, P.R. (2004) Microbe-mineral interactions: early carbonate precipitation in a hypersaline lake (Eleuthera Island, Bahamas). *Sedimentology* 41, 745–765.

Erhart, H. (1967) *La genèse des sols en tant que phénomène géologique – Esquisse d'une théorie géologique et géochimique, biostasie et rhexistasie*. Paris: Masson et Cie Editeurs.

Evans, W.C., Kling, G.W., Tuttle, M.L., Tanyileke, G. & White, L.D. (1993) Gas buildup in Lake Nyos, Cameroon: the recharge process and its consequences. *Applied Geochemistry* 8, 207–221.

Fontes, J.Ch., Gasse, F., Callot, Y., Plaziat, J.C., Carbonel, P., Dupeuple, P.A. & Kaczmarska, I. (1985) Freshwater to marine-like environments from Holocene lakes in northern Sahara. *Nature* 317, 608–610.

Freytet, P. & Plaziat, J.-C. (1982) Continental carbonate sedimentation and pedogenesis – Late Cretaceous and Early Tertiary of Southern France. In: Purser, B.H. (Ed.) *Contribution to Sedimentology*, Vol. 12. Stuttgart: Schweizerbart'sche Verlag.

Freytet, P. & Verrecchia, E.P. (1998) Freshwater organisms that build stromatolites: a synopsis of biocrystallization by prokaryotic and eukaryotic algae. *Sedimentology* 45, 535–563.

Freytet, P. & Verrecchia, E.P. (2002) Lacustrine and palustrine carbonate petrography: an overview. *Journal of Paleolimnology* 27, 221–237.

Gasse, F. (2000) Hydrological changes in the African tropics since the Last Glacial Maximum. *Quaternary Science Reviews* 19, 189–211.

Gasse, F., Fontes, J.Ch., Plaziat, J.C., Carbonel, P., Kaczmarska, I. , De Deckker, P., Soulié-Marsche, I., Callot, Y. & Dupeuple, P.A. (1987) Biological remains, geochemistry and stable isotopes for the reconstruction of environmental and hydrological changes in the Holocene lakes from North Sahara. *Palaeoecology, Palaeogeography, Palaeoclimatology* 60, 1–46.

Giralt, S., Julià, R. & Klerkx, J. (2001) Microbial biscuits of vaterite in Lake Issyl-Kul (Republic of Kyrgyzstan). *Journal of Sedimentary Research* 71, 430–435.

Håkanson, L. & Jansson, M. (1983) *Principles of Lake Sedimentology*. Berlin: Springer-Verlag.

Hardie, L.A., Smoot, J.P. & Eugster, H.P. (1978) Saline lakes and their deposits: a sedimentological approach. Special Publication 2, International Association of Sedimentologists. Oxford: Blackwell Scientific Publications, pp. 7–41.

Hutchinson, G.E. (1957) *A Treatise on Limnology – I Geography, Physics, and Chemistry*. New York: Wiley.

Kelts, K.R. & Hsü, K.J. (1978) Freshwater carbonate sedimentation. In: Lerman, A. (Ed.) *Lakes – Chemistry, Geology, Physics*. New York: Springer-Verlag, pp. 295–323.

Kling, G.W., Tuttle, M.L. & Evans, W.C. (1989) The evolution of thermal structure and water chemistry in Lake Nyos. *Journal of Volcanology and Geothermal Research* 39, 151–165.

Klappa, C.F. (1980) Rhizoliths in terrestrial carbonates: classification, recognition, genesis and significance. *Sedimentology* 27, 613–629.

Krumbein, W.E. (1975) Biogenic monohydrocalcite spherules in lake sediments of Lake Kivu (Africa) and the Solar Lake (Sinai). *Sedimentology* 22, 631–634.

Lambert, P. (1999) *La sédimentation dans le Lac de Neuchâtel (Suisse): processus actuels et reconstitution paléoenvironnementale de 1500 BP à nos jours.* PhD thesis, Université de Neuchâtel, Neuchâtel.

Leng, M.J. & Marshall, J.D. (2004) Palaeoclimate interpretation of stable isotope data from lake sediment archives. *Quaternary Science Reviews* 23, 811–831

Lockwood, J.P. & Rubin, M. (1989) Origin and age of the Lake Nyos maar, Cameroon. *Journal of Volcanology and Geothermal Research* 39, 117–124.

Magny, M. (1991) *Une approche paléoclimatique de l'Holocène: les fluctuations des lacs du Jura et des Alpes du Nord françaises.* PhD thesis, Université de Franche-Comté, Besançon.

Murphy, D.H. & Wilkinson, B.H. (1980) Carbonate deposition and facies distribution in a central Michigan marl lake. *Sedimentology* 27, 123–135.

Reeves Jr., C.C. (1978) Economic significance of playa lake deposits. Special Publication 2, International Association of Sedimentologists. Oxford: Blackwell Scientific Publications, pp. 279–290.

Risacher, F. (1992) Géochimie des lacs salés et croûtes de sel de l'Altiplano bolivien. *Sciences Géologiques Bulletin* 45, 135–198.

Rognon, P. (1985) Désert et désertification. *Total Information* 100, 4–10.

Schöne, B.R., Dunca, H., Mutvei, H. & Norlund, U. (2004) A 217-year record of summer air temperature reconstructed from freshwater pearl mussels (*M. margaritifera,* Sweden). *Quaternary Science Reviews* 23, 1803–1816.

Sebag, D., Verrecchia, E.P. & Durand, A. (1999) Biogeochemical cycle of silica in an apolyhaline interdunal Holocene lake (Chad, N'Guigmi region, Niger). *Naturwissenschaften* 86, 475–478.

Sebag, D., Verrecchia, E.P., Lee S.J. & Durand A. (2001) The natural hydrous sodium silicates from the northern bank of Lake Chad: occurrence, petrology and genesis. *Sedimentary Geology* 139, 15–31.

Stoffers, P. & Fischbeck, R. (1974) Monohydrocalcite in the sediments of Lake Kivu (East Africa). *Sedimentology* 21, 163–170.

Thompson, J.B., Ferris, F.G. & Smith, T.H.D. (1990) Geomicrobiology and sedimentology of the mixolimnion and chemocline in Fayetteville Green Lake, New York. *Palaios* 5, 52–75.

Tiercelin, J.-J. & Vincens, A. (Coordinators) (1987) The Baringo-Bogoria half graben, Gregory Rift, Kenya, 30 000 years of hydrological and sedimentary history. *Bull. Centres Rech. Explor.-Prod. Elf-Aquitaine* 11, 249–540.

Verrecchia, E.P. (2002) Géodynamique du carbonate de calcium à la surface des continents. In: Miskowsky, J.-C. (Ed.) *Géologie de la Préhistoire : méthodes, techniques, applications.* Paris: Editions Géopré, pp. 233–258.

Verrecchia, E.P. (2004) Multiresolution analysis of shell growth increments to detect natural cycles. In: Francus, P. (Ed.) *Image Analysis, Sediments and Paleoenvironments,* Vol. 7, *Developments in Paleoenvironmental Research.* Dordrecht: Kluwer Academic, pp. 273–293.

Wetzel, R.G. (2001) *Limnology, Lake and River Ecosystems,* 3rd edn, San Diego: Academic Press.

Chapter Ten

Terrestrial Evaporites

Allan R. Chivas

10.1 Introduction

Evaporites are rocks composed of minerals that precipitate at or near the Earth's surface from waters, upon their evaporation. The largest terrestrial occurrences are layered lacustrine deposits, commonly interbedded with variable amounts of clastic (silt- and mud-sized) sediment. Modern examples typically occur in hyperarid and arid environments in topographic depressions, commonly within internal drainage basins that lack an outlet to the oceans. Indeed, the lowest places in several continents, and which are below sea-level, are notable sites for evaporite accumulation: Dead Sea, Israel and Jordan (Asia); Qattara, Djibouti and Danakil Depressions (Africa; Aref et al., 2002); Death Valley, California (USA) (Figure 10.1A); Salina del Gualicho (eastern Argentina; Lombardi et al., 1994); and Lake Eyre (Australia; Magee et al., 1995). However, altitude alone is no criterion, as some closed basins at elevations of 2800 to 4000 m above sea level are major sites for evaporites, such as the Altiplano-Puna region of Bolivia, Chile and Argentina (Alonso et al., 1991) (Figure 10.1B, C), and parts of the Xizang-Qinghai (Tibet) Plateau (Chen and Bowler, 1986) (Figure 10.1D).

The most common evaporite minerals are halite (NaCl, common salt) and gypsum ($CaSO_4 \cdot 2H_2O$). Depending on the chemistry of the inflow waters, other relatively common terrestrial evaporite minerals include sodium sulphates, sodium carbonates and bicarbonates, and magnesium and potassium chlorides and sulphates. The largest of the terrestrial evaporite deposits may cover thousands of square kilometres, with a stratigraphic thickness of hundreds of metres (e.g. Salar de Uyuni, Bolivia; Risacher and Fritz, 2000). In the older rock record, such preserved deposits may be macroscopically and chemically difficult to discern from marine evaporites that formed from the evaporation of barred near-coastal or epicontinental seas, if they have similar mineralogy (e.g. Lowenstein et al., 1989).

Figure 10.1 (A) Playa system at Death Valley, California. The lowest point (Badwater) on the playa floor (83 m below sea level), is to the right of this view. Snowmelt reaches the playa floor through a series of large alluvial fans. (B) Salar de Cauchari, Jujuy Province, Puna, northern Argentina (Alonso, 1999). This playa is the site of El Porvenir bedded ulexite (boron) deposit. The playa floor is at an elevation of 3920 m. The small mesa-like features (*c.* 2 m high) in the middle distance are relics of older clastic lake sediments now highly eroded by salt etching and deflation. There is a halite crust on the far side of the playa. (C) Laguna Santa Rosa, part of the Salar de Maricunga, east of Copiapó, Chile (road in left foreground for near scale). The playa floor is at an altitude of 3780 m, with the main range of the Andes Mountains behind rising to over 6750 m above sea level. The location is at the southern end of the Atacama Desert, where snowmelt from the higher peaks is the principal source of water to the playa. (D) Halite crust, Dabuxan Lake, Qaidam Basin, China (2700 m above sea level). This crust is a dirty brown colour owing to the incorporation of loess-like airborne dust. Spade for scale.

Accordingly, the distinction between ancient marine and non-marine evaporites is itself a key topic, particularly with respect to hydrocarbon exploration (Warren, 1989; Hite and Anders, 1991; Bohacs et al., 2000).

There are also smaller-scale but widespread terrestrial evaporites, for example in the form of efflorescences and impregnations within soils (see Chapter 12), mineral precipitates around springs (e.g. carbonate tufa deposits; Chapter 6) and 'salt scalds' (halite precipitation). The latter occur in surface soils and are commonly associated with rising saline groundwater tables which, in some cases, are a function of land management practices such as removal of vegetation ('land clearing') or over-irrigation. Evaporative processes are also implicated in the precipitation of many pedogenic and non-pedogenic calcrete (Chapter 2) and silcrete (Chapter 4) duricrusts, and some lacustrine carbonates (Freytet and Verrecchia, 2002; and Chapter 9).

In this chapter emphasis is given to bedded evaporites that have precipitated from lake- or groundwater in ephemeral lacustrine or playa systems. Bedded terrestrial evaporites may be classified according to:

1   their depositional environment within the landscape (e.g. permanent lake, ephemeral lake, groundwater discharge, efflorescences);
2   the local environment of their precipitation (depositional subenvironments);
3   the broader hydrological position in the landscape (discharge, recharge or throughflow);
4   the chemistry of their input waters, which commonly determines their mineralogy;
5   by the origins of their solutes (e.g. bedrock weathering, airborne delivery as aerosols or dust, and marine incursions).

The first three schemes are amenable to field observation and decision, whereas the latter two require further chemical and isotopic measurements. This chapter reviews these classifications and concludes by offering some end-member classes of terrestrial evaporites that unify the field and chemical aspects. Indeed, there is a sound geological and geomorphological basis for the broad range of chemical and isotopic differences within evaporites.

10.2 Distribution, Field Occurrence and Geomorphological Relations

Modern terrestrial bedded evaporites occur on all continents, although sparingly in Europe (common only in Spain; Ordóñez et al., 1994; Gutiérrez-Elorza et al., 2002, 2005), in areas of closed drainage where the climate favours evaporation over precipitation (Figure 10.2). Accordingly, all the arid and semi-arid areas of the world, including hot deserts, cold deserts (e.g. central Asia) and seasonally cold dry areas (e.g. Great Plains of Canada; Last, 1989, 1992), host evaporites. There are limited deposits of mirabilite ($Na_2SO_4 \cdot 10H_2O$) and other sulphates in ice-free portions of Antarctica as precipitates within ponds and unconsolidated sediments, and as efflorescences caused by freeze concentration and evaporation (Bowser and Black, 1970; Keys and Williams, 1981).

10.2.1 General geomorphological setting

Evaporite precipitation spans a large variety of lacustrine environments, from permanent saline lakes to commonly dry flat basins, the latter the

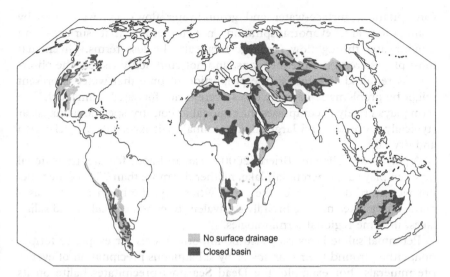

Figure 10.2 Distribution of areas without surface drainage and with interior basin (or endorheic) drainage. Modified by several authors (e.g. Cooke and Warren, 1973; Smoot and Lowenstein, 1991) after the original concept by de Martone and Aufrère (1928).

flattest of all landforms (Neal, 1975). A large array of names has been applied to the intermittent wet and dry basins: salina, salar, salt lake, saline lake, salt pan, dry pan, ephemeral lake, dry lake, playa lake, amongst many (Rosen, 1994; Briere, 2000). Shallow basins that are marginal to marine settings and which precipitate gypsum, dolomite and anhydrite are termed sabkhas (or the Arabic plural, sabkhat) and the type area is the Arabian Gulf (Curtis et al., 1963; Butler, 1969; Evans et al., 1969; Warren and Kendall, 1985). Although the mixing of continental and marine waters is implicated in their formation (Wood et al., 2002, 2005; Yechieli and Wood, 2002) this category of evaporite is not considered further here. However, it should be noted that some authors would wish to use the term sabkha also for some intra-continental evaporite basins (Handford, 1982; Warren and Kendall, 1985; Warren, 2006). Since Russell's (1885) work in the western USA, the most common usage has been in favour of playa or playa lake for the continental environment, and this is the usage adopted in this chapter. Playa is Spanish for beach or shore, although as a geological technical term in English it now largely means non-marine dry basin, beach or coastal flat.

A playa may be defined as an intra-continental arid-zone basin with negative water balance for one-half of each year, dry for over 75% of the time, with a capillary fringe close enough to the surface such that evaporation will cause water to discharge, usually resulting in evaporites (Briere, 2000; building upon criteria in Matts, 1965; Shaw and Thomas, 1989; and Rosen, 1994a). The negative water balance refers to the sum of inflows (precipitation, surface-water and groundwater flow) less water loss by evaporation and evapotranspiration, not forgetting that surface- and groundwater throughflow ('leakage') can also be loss terms. Given that most playas are in remote locations, and not commonly visited, the observations required to establish the proportion of time that water is present might be problematic in many cases, were it not for satellite imagery. The term playa largely encompasses the regional or country-specific salina, salar (typically more salt and larger than a salina), salt lake, salt pan, clay pan and dry lake.

A playa lake, following Briere (2000), is an arid-zone feature, transitional between playa and (perennial) lake, neither dry more than 75% of the time, nor wet more than 75% of the time. When dry, the basin qualifies as a playa. Playa lakes may be broadly equivalent to ephemeral lakes and saline lakes, in some regional terminologies.

Perennial saline lakes may also be important sites for evaporite formation, either around their margins or by subaqueous precipitation of evaporite minerals. For example, the Dead Sea now precipitates halite on its floor, beneath its lake water which has exceeded halite saturation, and about 400 years ago subaqueous carbonate laminites were also formed (Neev and Emery, 1967). The salinity of today's Dead Sea is approaching $350\,g\,L^{-1}$ TDS (total dissolved solids), tenfold that of modern seawater.

10.2.2 Depositional subenvironments and macromorphological characteristics

A full range of sedimentary environments equivalent to the more commonly described clastic deposits is also occupied by evaporite facies where climate and hydrology permit. Indeed, small changes in these parameters commonly lead to fresher water clastic and carbonate sediments interbedded with evaporites, such as in Great Salt Lake, Utah (Spencer et al., 1985). In other cases (e.g. Lake Eyre) irregular flood cycles are recorded as multiple silt–mud–gypsum triplets, because clastic bedload material is first deposited followed by evaporation of the lake water and precipitation of the dissolved load.

Most Australian playas have only a thin (centimetres to 1 m) surface halite crust that redissolves upon flooding and which reprecipitates by evaporation to dryness after flood-borne clastic deposition. By this mechanism the halite crust is preserved as the uppermost crust, with clastic sediment (up to 150 m thick) progressively accumulating 'beneath' the halite crust (Figure 10.3A). This is an example where the strict relative age of superposition is challenged, at least with respect to the age of the solutes and salts. However, evaporite sediments commonly behave as their clastic equivalents. Hardie et al. (1978) and Smoot and Lowenstein (1991),

Figure 10.3 (A) Surface halite crust, Lake Koorkoordine, Southern Cross, Western Australia. Small excavated pit with hammer hooked onto the original salt-crust surface. Immediately below the salt crust is a dark layer about 5 cm thick which is clastic sediment impregnated with ion monosulphides formed by sulphate-reducing bacteria (or algae) that have converted interstitial dissolved sulphate to sulphide. (B) Lake Eyre North, Australia (12 m below sea level), after partial flooding and evaporative retreat, showing facies zonation. Saline water body (W); halite salt crust (Sc); saline mudflat (Smf); dry mudflat (Dmf, also depicted in Figure 10.3F); shoreline sandflat (Sf); and gypseous sandy foreshore (S). Person for scale. (C) Western shoreline of Lake Frome, South Australia, showing halite-encrusted ephemeral stream. (D) Halite crust, Lake Frome, South Australia, showing polygonally disposed ridges formed by the pressure of salt crystallisation. (E) Halite crust with sinuous salt-crystallisation pressure ridges, Sickle Lake, Northern Territory, Australia. The trail of deep footprints behind the person are black, as the underlying iron monosulphide layer has been breached. (F) Pervasive mudcracked texture, Dry Mudflat facies, Lake Eyre North, Australia (April 1990). Gloves for scale. (G) Regressive strandlines, Lake Buchanan, Queensland, Australia. Also shown is a now-dry birdsfoot delta encroaching into the halite-encrusted playa floor. (H) Regressive shorelines, Lake Buchanan, Queensland, Australia. The playa is floored by halite with a surface brine pool visible in the top left corner. The lake in the foreground, Lake Constant, is so named because it rarely evaporates to dryness and is largely maintained by regional groundwater. (I) An island composed of Archaean bedrock draped by gypcrete in the halite-encrusted floor of Lake Lefroy, near Kambalda, Western Australia. The white outcrop in the foreground is uncemented very fine-grained gypsum, locally called *kopi*, that has been deflated from the playa floor, and probably recrystallised under the influence of percolating rainwater. (J) Carnallite ($MgCl_2 \cdot KCl \cdot 6H_2O$) crystals from the commercial evaporating ponds that use brines trapped within the Qarhan salt plain, Qaidam Basin, China.

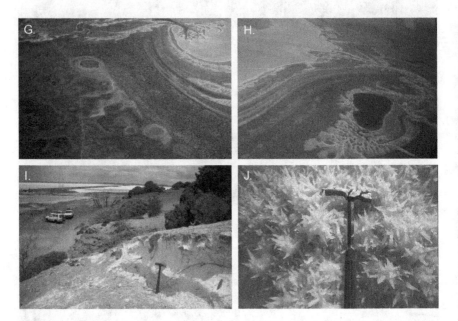

Figure 10.3 *Continued*

provide extensive descriptions of evaporite facies in the following sub-environments, which may broadly be grouped under the headings of lacustrine, fluvial and other environments.

1 Lacustrine deposits: (a) perennial saline lake, (b) salt crust or saline pan, (c) saline mudflat, (d) dry mudflat and (e) shoreline; commonly arranged in sequence outwards from the centre of a lake basin (Figure 10.3B and Figure 10.4).
2 Fluvial deposits (in isolation, and commonly as streams feeding the closed lake basins, Figure 10.3C): (f) alluvial fan-sandflat, (g) ephemeral stream floodplain and (h) perennial stream floodplain.
3 Other deposits (environments dominated by direct rainfall and/or groundwater discharge): (i) aeolian dunefield, (j) spring and (k) saline soils.

Perennial saline lakes may be very large (e.g. Lake Chad, Africa (Eugster and Maglione, 1974; Ghienne et al., 2002), the Dead Sea and Caspian Sea) and deep, although many are smaller and commonly shallow. Evaporite minerals (typically carbonates, halite and gypsum) occur as cumulus crystals that have precipitated at the brine–air interface, perhaps float for a time (as 'rafts'), before sinking to accumulate on the lake floor (Smoot and Lowenstein, 1991). Some minerals may redissolve during sinking or on the lake floor. Evaporite crusts are precipitated directly in the lake floor, commonly in shallow water, and include spectacular examples of vertically

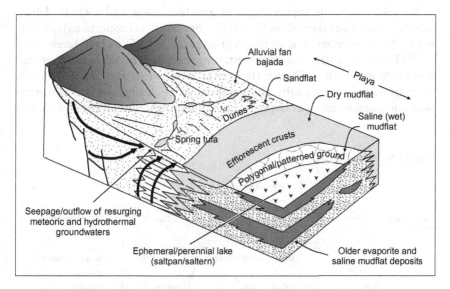

Figure 10.4 Playa depositional/evaporative facies arranged parallel to, and potentially concentrically in plan around the shorelines of an evaporating lake (from Eugster & Hardie, 1978; with additional information from Kendall, 1992; Warren, 2006). The figure shows the relationship with other geomorphological and hydrological features such as alluvial fans, dunes, spring tufa, and the sources and movement of water.

oriented gypsum crystals (up to 1 m high) (e.g. Yorke Peninsula, South Australia; Warren, 1982) and large chevron (i.e. V-shaped) crystals of halite forming beds up to 5 m thick (e.g. Lake McLeod, Western Australia; Logan, 1987). The mechanical subaqueous reworking of evaporate minerals leads to detrital evaporites, with abraded forms that have accumulated in ripple or cross-bedded forms.

The various playa environments include a central salt crust or saline pan, composed of dry salt, commonly halite (Figure 10.3D, E), but in some cases trona (Eugster, 1970), gypsum (Stoertz and Ericksen, 1974) or other sulphates such as mirabilite, epsomite or bloedite. The saline mudflat is typically moist clay to silt with surface salt efflorescences and intra-sediment (displacive) evaporite minerals. In some systems, these may be zoned on a broad scale, with more-soluble minerals towards the lowest central portion of the mudflat, caused by groundwater evaporation gradients (e.g. Saline Valley, California; Hardie, 1968).

Dry mudflats lie further shoreward where the groundwater table is deeper and evaporite minerals are limited to minor efflorescences and intra-sediment salts related to fluctuating water tables. Entire basins floored by dry mudflat facies, commonly called clay pans, occur if the water table is low. Repeated wetting and drying leads to a distinctive pervasive cracked surface (Smoot and Lowenstein, 1991; Figure 10.3F).

Shoreline deposits mimic those of any lake basin and include deltas, beaches, spits and raised shorelines (of higher lake-level conditions), including multiple 'regressive' strandlines (Figure 10.3G, H) as lake levels fall.

Within the lacustrine environment, the generalised flooding and evaporation cycle is shown in Figure 10.5 (after Lowenstein and Hardie, 1985), wherein earlier layered salts are preserved by enclosing clastic sedimentation; subaqueous salts (as hoppers and rafts) grow as evaporation proceeds;

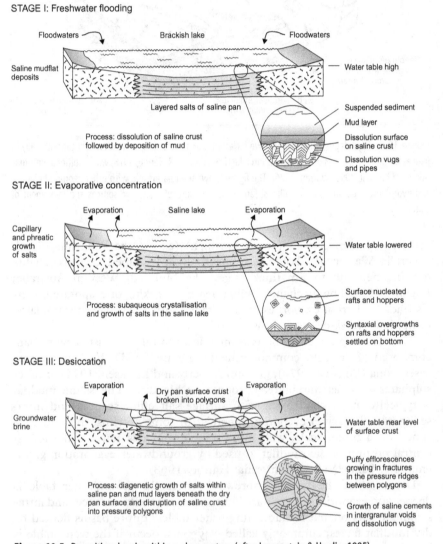

Figure 10.5 Depositional cycle within a playa system (after Lowenstein & Hardie, 1985).

displacive and diagenetic salts precipitate from saturated pore waters as brine infiltrates clastic sediments; and a polygonal surface halite crust forms, followed by further efflorescence in the pressure ridges of the polygons.

A fourth stage might be added to this sequence following prolonged aridity, wherein the groundwater table within the playa continues to fall, concentrating bitterns below the playa floor and leading to desiccation of its upper layers (Bowler, 1981, 1986). Aeolian activity may remove the surface layers, thereby deflating halite, gypsum (e.g. White Sands, New Mexico; Langford, 2003), clay and silt from the playa floor (Figure 10.3I). Crescentic, source-bordering downwind dunes called lunettes (Hills, 1940) are common adjacent to arid-zone playas (Bowler, 1973, 1977; Goudie and Thomas, 1986; Goudie and Wells, 1995). Many Australian lunettes are composed largely of gypsum, the deflated halite having being redissolved by meagre rainfall and returned in solution to the playa floor. In many cases worldwide, even greater amounts of clay and silt are deflated and transported large distances and beyond the playa's catchment. The loss of material by deflation from the stratigraphic record of playas is broadly understood, but difficult to recognise in cored sequences from within playa basins. Accordingly, the sedimentary history of playas can be expected to be incompletely preserved.

10.2.3 Hydrological setting

The quantity and frequency of water in playas and playa lakes may depend as much on groundwater regime as on surface delivery by inflowing streams and direct precipitation. The relative importance of groundwater processes commonly increases with settings that are typically more uniformly arid, of more subdued topography, and commonly of larger scale. Accordingly, groundwater fed and controlled playas are typical of large planated areas of Australia, the western and northern Sahara (e.g. Mali and Tunisia) and Namibia. Surface-water-fed playas are more typical of intermontane playa basins, such those of the Great Basin, USA.

Rosen (1994b) provides a useful classification of playa types (Figure 10.6) by hydrological setting based on examples from the western USA, and generally applicable elsewhere, except in the flattest of terrains which may have little surface flow. However, it is the interaction between groundwater and surface-water flows that is significant, and distinguished as throughflow playas, recharge playas and discharge playas (modified after Eakin et al., 1976). In this scheme, evaporites accumulate in the discharge playas and may form in the throughflow playas. Recharge playas may be subject to erosion by deflation. As indicated previously, if the climate were to become drier, the groundwater table would fall, and throughflow playas and later discharge playas could change their status to recharge playas and

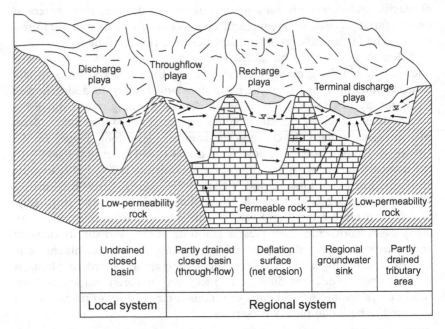

Figure 10.6 Hydrological classification of playa types and their topographic settings (after Rosen, 1994b, modified from Eakin et al., 1976). Note the position of the regional groundwater table with respect to each playa floor.

also be subject to deflation. In such a chain of hydrologically connected playas, the system as a whole may be hydrologically closed, but some individual playas and their local catchments may be hydrologically open, meaning that water and dissolved salts are partially transferred (commonly called 'leakage') towards a terminal basin.

In topographically subdued arid terrain where the groundwater table is invariably horizontal or nearly so, playas and their evaporites occur in depressions in the landscape if the groundwater table is close to the land surface. Such playas are effectively 'groundwater windows' or groundwater discharge zones (or complexes) and have been termed *boinkas* (Macumber, 1991; Jacobson et al., 1994) after a particular locality in eastern Australia. Such systems are hydrologically open, commonly have negligible or confused surface drainage, and may be enclosed or surrounded by regional aeolian dunes (not deflated from the playa floor) and downwind lunettes.

10.3 Micromorphological Characteristics

The textures and fabrics of evaporites at the hand-specimen to microscopic scale may be distinctive for several depositional facies. This section considers the morphology of one of the commonest evaporite minerals, gypsum; the facies identification of which may also be used to determine palaeoenvironments (see section 10.7). Once formed, gypsum is not particularly soluble but can be both chemically etched and physically abraded. Accordingly, its morphological identity may be traceable even following reworking, thereby providing a more complete record of playa processes than might the more soluble evaporite minerals.

Extensive research on the morphology of gypsum (e.g. Cody, 1976, 1979; Cody and Cody, 1988; Magee, 1991; Mees, 1999) and its experimental growth under controlled conditions show that variations in the presence of organic compounds, salinity, pH, NaCl content and Ca/SO_4 ratios, may affect the mineral's morphology (Simon and Bienfait, 1965; Edinger, 1973; Van der Voort and Hartmann, 1991). Such parameters are clearly too numerous to relate to specific crystallographic forms, although there is some consensus that several broad environments are represented.

1 Subaqueous lacustrine gypsum is commonly prismatic, typically a few millimetres in size but ranging from $100\,\mu m$ to several metres (Mees, 1999; Warren, 1982). The larger, vertically oriented crystals are commonly referred to as selenite.
2 Early diagenetic gypsum has a pyramidal (or hemi-bipyramidal) habit and grows displacively, commonly within siliciclastic sediment from interstitial porewater, and may be termed groundwater gypsum. Coalescing clusters or rosettes of this 'discoidal' form (up to a few centimetres for each blade) lead to 'desert roses', some of which are indeed pink if clay or iron oxides are incorporated.
3 Clastic gypsum has been reworked and abraded in a subaqueous environment. Some relict morphology may be identifiable allowing its initial origin as either a subaqueous or diagenetic precipitate to be recognised (Magee, 1991).
4 Aeolian gypsarenite forms by deflation of playa-floor gypsum, commonly of early diagenetic origin and accordingly is typically pyramidal in form with varying degrees of abrasion (Magee, 1991).
5 Pedogenic gypsum, commonly lenticular in habit, may form on playa floors (Chen, 1997; Chen et al., 1991) during periods of non-deposition during perhaps slightly more humid intervals. Pedogenesis also occurs on aeolian gypsarenite dunes forming fine lenticular or acicular crystals, commonly $50\,\mu m$ to $1\,mm$ long, called *kopi* in Australia (Jack, 1921; Figure 10.3I).

Further detail of the several morphological forms of gypsum and a useful key for their recognition is given by Magee (1991). Similarly, several distinct morphologies are displayed by halite (Gornitz and Schreiber, 1981; Schubel and Lowenstein, 1997; Schreiber and El Tabakh, 2000) although these fabrics will not necessarily be preserved.

10.4 Chemistry and Mineralogy

The initial chemical composition of an evaporating brine controls the sequence of minerals that precipitate from it by a process termed fractional crystallisation. As ocean water evaporates (Usiglio, 1849; Harvie et al., 1980) the sequence of crystallisation at 25°C and 1 atm pressure (Figure 10.7) is initially calcite (i.e. the least soluble mineral), followed sequentially by gypsum, halite, magnesium and potassium sulphates (e.g. epsomite, kainite) and chlorides (e.g. sylvite, carnallite; see Figure 10.3J), and finally borates. Some other economically important elements (e.g. Li, Br, I) are so soluble that they commonly form no separate mineral phases, but remain in residual solutions called bitterns which reach densities of $1.37\,g\,cm^{-3}$. Virtually all natural evaporite minerals are colourless (white or clear and translucent), although the final bitterns, as seen in commercial evaporating operations, may be brown owing to the presence of bromine and iodine. In some largely marine evaporite deposits, sylvite, halite and carnallite may be orange to red in colour owing to the presence of tiny crystals of haematite, or brown to black if clays are incorporated (e.g. Lowenstein and Spencer, 1990; Cendón et al., 1998).

Few non-marine evaporites follow the same crystallisation sequence as ocean water. Indeed, the compositions of non-marine surface waters are

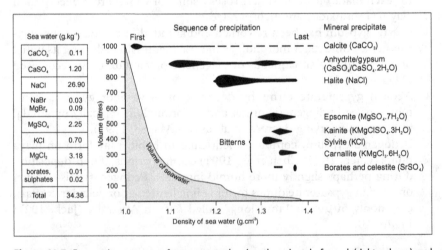

Figure 10.7 Evaporation sequence for seawater, showing the minerals formed (right column) and the brine density ($g\,cm^{-3}$) at which each mineral precipitates as 1000 L evaporates to dryness. Also shown are the weights of the major precipitated phases (left column) and which sum to the salinity of typical seawater, namely 35‰ (parts per thousand, or $g\,kg^{-1}$) (after Valyashkov, 1972).

highly variable and are governed by the initial rainwater composition (bicarbonate generated by dissolution of atmospheric CO_2 is important) and by leaching and exchange reactions with rocks and unconsolidated sediments through which the percolating waters pass prior to inflow into a playa basin. There may be further chemical reactions with fine-grained clastic sediments in the playa, and dissolution of earlier evaporites. It follows that the mineralogy of most terrestrial evaporites is governed by the chemistry of the rock types in their surface and groundwater catchment areas.

The chemical evolution of surface waters is well explained by the hydrological classification and brine evolution pathways of Eugster and Hardie (1978). Their scheme recognises five major continental water types, namely, (a) Ca–Mg–Na–(K)–Cl, (b) Na–(Ca)–SO_4–Cl, (c) Mg–Na–(Ca)–SO_4–Cl, (d) Na–CO_3–Cl and (e) Na–CO_3–SO_4–Cl (Figure 10.8, right side, which shows the minerals precipitated from these five types). The minerals that precipitate sequentially are governed by a series of 'chemical divides' (Hardie and Eugster, 1970) determined by three pathways, namely, (I) $HCO_3 \gg Ca + Mg$, (II) $HCO_3 \ll Ca + Mg$ and (III) $HCO_3 \geq Ca + Mg$. Thus, whether bicarbonate content (in molar terms) exceeds that of calcium + magnesium in playa inflow waters ultimately decides whether gypsum or sodium bicarbonate (e.g. nahcolite $NaHCO_3$ and trona $NaHCO_3 . Na_2CO_3$) precipitates.

Path II is most like that of seawater, and leads to halite, gypsum and/or epsomite and potash salts. Path I, typical of young volcanic terrains (e.g. Alkali Valley and Mono Lake, California; and Lakes Magadi and Natron, east African rift) produces halite, natron and trona. Recent evidence (Earman et al., 2005) suggests that the high bicarbonate levels required to form trona may be provided by magmatic CO_2 or from decay of organic matter. Path III, common for bedrock terrains dominated by carbonate and pre-existing evaporites, leads to terrestrial evaporites, with abundant carbonates, and depending on Mg:Ca ratio, precipitates high-Mg calcite, dolomite or magnesite, and ultimately sodium sulphates such as mirabilite.

The sequence of evaporite minerals progressively but predictably evolves as individual chemical components are consumed (in this case, precipitated as particular minerals) and the liquid residue continues to evaporate until the next, least soluble, phase precipitates. This process can be modelled chemically, and using the starting dilute-water chemical composition, the mineralogy of evaporite sequences can be predicted. A useful tool in this regard is the salt norm (SNORM) introduced by Jones and Bodine (1987), which calculates the equilibrium salt assemblage expected for any water as it evaporates to dryness at 25°C in the presence of an atmosphere with its normal CO_2 content. Such calculations also permit the tracing of the origin of dilute waters and brines, and whether they have formed by mixing of several source waters, or have already lost salts by precipitation in transit to the playa surface.

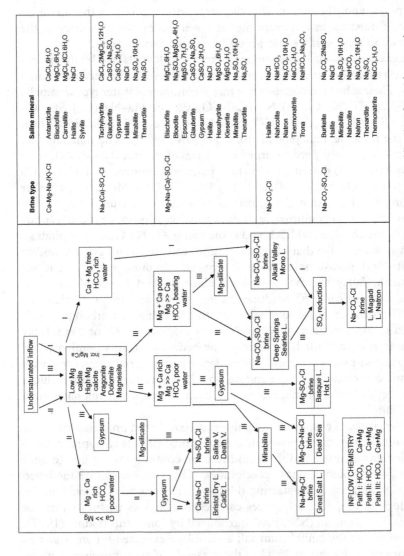

Figure 10.8 Brine evolution pathways (I, II and III) and a hydrological classification of progressively evaporating (direction of arrows) non-marine waters. For the several lakes and basins named as examples of each pathway: V, valley; L, lake. The right side of the diagram lists the major evaporite minerals associated with five principal brine types. (From Eugster & Hardie (1978), as presented by Warren (2006).)

10.5 Origin of Solutes

Considerable attention has been given to investigating the sources of the solutes from which evaporitic minerals precipitate, commonly represented as either marine or non-marine in origin (e.g. Hardie, 1984, 1990; Sonnenfeld, 1985). There are evaporites with mixed marine and terrestrial solute sources, with progressively more complicated mechanisms (e.g. recycling or partial dissolution of earlier evaporites, and marine aerosol delivery to terrestrial basins) being identified. The problem is particularly perplexing for older deposits (e.g. Taberner et al., 2000; Cendón et al., 2004). For Quaternary examples, the geomorphological setting is clearer, but not necessarily indicative, for as shown below, there are some terrestrial evaporites that receive the majority of their solutes from marine sources as airborne aerosols or marine-carbonate dust, a pathway not readily demonstrable by field observations alone.

A variety of chemical approaches to distinguish solute sources (e.g. Risacher et al., 2003; Jones and Deocampo, 2004) include general chemistry and salt norms (Jones and Bodine, 1987), Br/Cl ratios or Br content of halite (Valyashko, 1956; Dean, 1978; Holser, 1979; Hardie, 1984), Sr content of gypsum (Kushnir, 1980, 1981, 1982a,b; Rosell et al., 1998) and isotopic methods. Because there are modern examples of terrestrial evaporites with marine-like bulk chemistry, such compositional comparisons, although helpful, cannot exclusively distinguish sources. Isotopic methods, commonly the ratios between two isotopes of the same element, are more promising. However, some key and common elements within evaporites are not amenable to such study. Sodium (Na) and fluorine (F), for example, only have a single stable isotope each, and K, Ca and Mg show little natural isotopic variation. Chlorine isotope ratios ($^{37}Cl/^{35}Cl$) may provide information about closed-system progressive evaporation, or may identify episodes of brine mixing (Eggenkamp et al., 1995; Eastoe and Peryt, 1999) but do not independently readily distinguish specific solute origin.

10.5.1 Strontium isotopes

Strontium is chemically related to calcium and is used commonly to trace the source of calcium, by proxy, using the isotopic variations of strontium. The modern ocean contains about 8 ppm dissolved strontium, with a $^{87}Sr/^{86}Sr$ ratio of 0.7092 (Hildreth and Henderson, 1971; Elderfield, 1986). Young basaltic volcanic rocks have lower $^{87}Sr/^{86}Sr$ ratios, typically 0.704–0.706. Most rocks, particularly those from older continental areas, and groundwaters that flow through such rocks, have higher $^{87}Sr/^{86}Sr$ ratios, commonly with ratios of 0.73 or higher. These higher $^{87}Sr/^{86}Sr$ ratios occur because an isotope of rubidium, ^{87}Rb, decays to form ^{87}Sr, progressively increasing the $^{87}Sr/^{86}Sr$ ratio for older and more Rb- and therefore K-rich rocks. Accordingly, $^{87}Sr/^{86}Sr$ ratios in evaporite minerals (gypsum is suitable) may indicate whether Sr, and possibly Ca, are of marine or non-marine origin. Also, the $^{87}Sr/^{86}Sr$ ratio of ocean water is quite well determined for the past 600 Myr (Burke et al., 1982; DePaolo and Ingram, 1985; Veizer et al., 1999; McArthur et al., 2001) through analysis of well-preserved marine carbonate fossils, so that the marine and non-marine contributions may be estimated for older evaporites (although there may be difficulty in estimating a realistic mean $^{87}Sr/^{86}Sr$ ratio for continental material originally surrounding some ancient basins). There will also be instances where groundwaters draining continental bedrock have $^{87}Sr/^{86}Sr$ values similar to that of contemporaneous ocean water, making the distinction between marine and non-marine evaporites difficult, as is the case for some Turkish Cenozoic evaporites (Palmer et al., 2004).

In practice, most marine evaporites can be expected to receive varying amounts of continental run-off and solutes either as direct fluvial input or by submarine groundwater discharge. The unambiguous detection of small quantities of such terrestrial contributions is difficult, by either $^{87}Sr/^{86}Sr$, or by $\delta^{34}S$ and $\delta^{18}O$ of precipitated sulphates (Lu and Meyers, 2003). However, strontium isotopes usefully demonstrate the proportions of solutes delivered from sub-catchments, and their variations through time, within playa systems (e.g. Hart et al., 2004), particularly those developed as chains of linked playas with multiple sub-catchments draining a variety of rock types.

The extensive modern playa systems of western and southern Australia, despite their terrestrial setting, have $^{87}Sr/^{86}Sr$ ratios indicative of a marine solute origin (Chivas et al., 1987; McArthur et al., 1989) as do extensive regional calcrete deposits (Quade et al., 1995; Lintern et al., 2006), as discussed in following sections. However, some playa gypsum samples from the Yilgarn Craton of Western Australia, also demonstrate calcium isotope ratios ($^{40}Ca/^{42}Ca$) indicative of Precambrian calcium (a minor amount of the ^{40}Ca is radiogenic, a progeny of ^{40}K decay), thereby indicating a component of bedrock calcium from the Archaean basement (Nelson and McCulloch, 1989).

10.5.2 Sulphur isotopes

The sulphur isotopic ratio, $^{34}S/^{32}S$, expressed as $\delta^{34}S$ (in parts per thousand, ‰, relative to meteoritic sulphur) of ocean water has changed through time, and is reasonably well known for the past 900 Myr (Holser and Kaplan, 1966; Claypool et al., 1980; Strauss, 1997). Accordingly, older marine (and therefore, by default, also some terrestrial) evaporites may be identified from the geological record, although the global marine $\delta^{34}S$ curve has been constructed by assuming that various evaporites are derived from solely marine solutes, and there are likely to be some errors here. The measurement of $\delta^{34}S$ in marine barite (Paytan et al., 1999) and in structurally bound sulphate within marine carbonates (Kampschulte and Strauss, 2004) will lead to a better estimation of marine sulphate $\delta^{34}S$, and probable recognition of additional ancient terrestrial evaporites. Modern marine sulphate has a $\delta^{34}S$ value of about +21‰ (Rees et al., 1978), and gypsum precipitated therefrom has a $\delta^{34}S$ value of +22.65‰ (Thode and Monster, 1965). Sulphur isotopes are also fractionated by several in-lake (or marine) processes, such as degree of progressive crystallisation (Raab and Spiro, 1991; and therefore demonstrate a residual isotopic reservoir effect), bacterial sulphate-reduction (Pierre, 1985; Lyons et al., 1994), the proportion of sulphide burial in clastic sediments (Kampschulte and Strauss, 2004), and dissolution and recycling of sulphate minerals. Some systems such as the Dead Sea, which initially formed by ingress of Mediterranean sea water, exhibit all these processes during a complex evolution (Gavrieli et al., 2001; Torfstein et al., 2005).

Notwithstanding these limitations, the continental solute origin of most modern terrestrially located playas is readily demonstrated. For example, sulphates from Andean playas have $\delta^{34}S$ values of +3 to +9‰ (Carmona et al., 2000; Rech et al., 2003), typical of sulphur liberated by weathering of surrounding igneous rocks and by volcanic emissions. Paleocene to Miocene terrestrial evaporites from Spain have $\delta^{34}S$ values of +9 to +19‰, and are interpreted as deriving much of their sulphate by dissolution of earlier Triassic and perhaps Cretaceous evaporites (Birnbaum and Coleman, 1979; Utrilla et al., 1992). Similarly, the modern playa system at Chott el Djerid, Tunisia, nested above earlier evaporites, appears to have received its sulphate from progressive recycling of evaporites of Cretaceous, Eocene and Mio-Pliocene age (Drake et al., 2004).

By contrast, the playas, soils and shallow groundwaters of western and southern Australia have $\delta^{34}S$ values for gypsum, dissolved sulphate and alunite, that are zoned on a continental scale, with $\delta^{34}S$ values of +22‰ near the southern and western coasts, and values decreasing progressively inland, over distances of 1000 km, to $\delta^{34}S$ of +14‰ (Chivas et al., 1991). This pattern persists over all of southern Australia ($3 \times 10^6 km^2$), parallel to the southern coastline, and has been interpreted as marine-aerosol delivery of salts to the continent, dominated by modern oceanic inorganic sulphate nearer the coast, and by marine-derived dimethylsulphide (later also oxidised to sulphate) further inland. The sulphur isotope data also rule out a marine transgression as the source of sulphate, as this process would produce a $\delta^{34}S$ value of +22‰ for all gypsum samples irrespective of their distance from coastlines.

A similar marine-aerosol origin (cf. Calhoun and Bates, 1989; Calhoun et al., 1991) is offered for the surficial sulphate deposits of Namibia (Eckardt and Spiro, 1999), albeit over a smaller area. Furthermore, $\delta^{17}O$ data from both Namibian gypcretes (whose sulphate has deflated from playas; Eckardt et al., 2001) and some Atacama sulphates also clearly demonstrate an atmospheric component (Bao et al., 2000, 2001, 2004; Thiemens, 2006). Accordingly, these data strongly suggest that some playas, and on regional scales, obtain a large portion of their solutes from marine sources.

10.5.3 Boron isotopes

Boron has two stable isotopes, ^{10}B and ^{11}B, with an average relative abundance of 20% and 80%, respectively. The fractionation of $^{10}B/^{11}B$ (expressed as $\delta^{11}B$, in per mil) relates to boron's speciation as either $B(OH)_4^-$ or $B(OH)_3$, rather than any strong temperature or redox dependence. Most continental (e.g. granitic) rocks, and therefore rivers draining therefrom, have $\delta^{11}B$ values close to 0‰, whereas ocean water has a $\delta^{11}B$ value of +39‰, due to preferential adsorption of the $B(OH)_3$ species onto marine clays and carbonates.

Reconnaissance boron isotope measurements of borates from marine ($\delta^{11}B = +21$ to +29‰) and non-marine ($\delta^{11}B = -17$ to +3‰) evaporites (Swihart et al., 1986), show an expected 20‰ offset ($\Delta^{11}B_{mineral-H_2O} \sim -20$) from the parent waters, and clearly distinguish the evaporative environments. Further detailed $\delta^{11}B$ analyses of trace boron in halite, gypsum and co-existing brines (Vengosh et al., 1992) established that simple non-borate salts and their brines show similar distinctions.

The $\delta^{11}B$ values of brines from modern playa evaporites include data from Qaidam, western China ($\delta^{11}B$ of +1 to +15‰; Vengosh et al., 1995) and the central Andes ($\delta^{11}B$ of -18 to 0‰; Kasemann et al., 2004), and which are very different from the marine brine value ($\delta^{11}B$ of +40‰), which was once considered to have the highest $\delta^{11}B$ value in nature. However, the $\delta^{11}B$ values of the brines from Australian playas ($\delta^{11}B$ of +26 to +59‰; Vengosh et al., 1991a) and the Dead Sea ($\delta^{11}B$ of +56‰; Vengosh et al., 1991b) are even more positive than the seawater value. This reflects the original marine solute origin of these systems (despite their contemporary terrestrial settings) and a second cycle of boron adsorption on clay and carbonate minerals, and attendant isotopic fractionation, within these playa and lake systems.

Thus the boron-isotope distinction between modern marine and terrestrial solute origin is clear, even if in some cases the marine solutes now reside in terrestrial settings. Conversely, in the ancient rock record, $\delta^{11}B$ values indicative of a marine solute origin will, in a probable minority of cases, refer to deposits in terrestrial settings.

10.6 Mechanisms of Formation and Classification

The sequence of requirements for lacustrine evaporite formation is a source of solutes, their transport to an accumulation basin and their evaporation. The previous sections have outlined the several geomorphological environments where this occurs and the variety of chemical pathways and solute sources that produce a range of specific evaporite minerals. Accordingly, a classification is offered (Table 10.1) that groups evaporites by their depositional environment (marine or terrestrial) and by their solute origins.

Table 10.1 Classification of evaporites by solute sources and geological setting

Evaporite classification	Characteristics and sub-types
A. Fully marine	Marine setting and solutes. Desiccated ocean basin. No known modern example; Messinian evaporites from the Mediterranean. Many probable ancient examples exist, although detailed geochemical investigations commonly show at least a minor terrestrial solute contribution (i.e. B1 below)
B. Marine setting with a component of terrestrial solutes	B1. Deeper marine basins, e.g. Neogene evaporites, Spain (Playà et al., 2000) B2. Coastal sabkhat, e.g. Holocene Arabian Gulf
C. Terrestrial setting with largely marine solutes	C1. Marine incursions; either by transgression (e.g. the initial Dead Sea; Klein-Ben David et al., 2004) or by seepage (e.g. Lake Asal, Djibouti; Stieljes, 1973) C2. Marine aerosol delivery (e.g. western and southern Australian playas; Namibian playas; probably western Sahara playas (Mali)
D. Fully terrestrial. Terrestrial settings and solutes	D1. Single-cycle weathering origin (e.g. some young volcanic terrains; Lake Magadi, Kenya; Eugster, 1970) D2. Multiple-cycle weathering, including particularly dissolution of earlier marine or terrestrial evaporites (e.g. modern Qaidam Basin, China; and Chott el Djerid, Tunisia), both of which are nested above earlier terrestrial evaporite basins)

This table demonstrates the complexities of evaporative environments, and the difficulty in expecting that single (either geomorphological, mineralogical or geochemical) parameters might define differences between marine and non-marine evaporites, which in their simplest and originally proposed form, corresponds to categories A and D1, respectively. For example, evaporative environments may evolve through several categories. During Messinian times (late Pliocene; about 6 Ma), the Mediterranean Sea was tectonically blocked near the present Gibraltar Strait and the enclosed water body evaporated to near dryness (Butler et al., 1995; Krijgsman et al., 2002). Initially, the Balearic basin in the western Mediterranean precipitated a 2-km-thick sequence of halite and gypsum in <300,000 years (category A), over an area of 300,000 km^2, with brine surfaces up to 2000 m below sea level, and fed by inward brine leakage from the Atlantic Ocean. Later and marginal basins show a stronger geochemical imprint ($^{87}Sr/^{86}Sr$ and $\delta^{34}S$) of a significant terrestrial solute component (category B1; Playà et al., 2000). The Gulf of Carpentaria in northern Australia has oscillated between categories B and C during Quaternary sea-level changes (Playá et al., 2007).

It is clear that there is also a major twofold division of evaporites within contemporary terrestrial settings. The most commonly described (categories D1 and D2) are typical of mountainous areas (e.g. southwestern USA, Andes, Tibet) and largely derive their solutes from bedrock sources, which may include volcanic emanations. By contrast, in many deeply weathered planated terrains of the world, the evaporite solute/deposition cycle commonly occurs above the active rock-weathering zone, allowing that marine aerosols dominate the near-surface hydrochemistry (category C2). Such regions are typically west-coast (i.e. windward) areas of Australia and Africa (Namibia and western Sahara) and some coastal areas of northern Chile. The C2 category and solute-delivery mechanism is probably more common than has been recognised previously.

10.7 Palaeoenvironmental Significance

The presence of terrestrial evaporites in the older part of the geological record, for example in the Palaeozoic and beyond, has been taken to indicate the former presence of negative water-balance conditions, and commonly interpreted as palaeo-aridity. Of more interest, in the present context, is the detailed information that is now available on Quaternary continental water-balances from terrestrial sequences containing evaporites.

Many modern lake-basins display evidence for fluctuations in their past lake-water levels. Datable high lake-levels are commonly represented as carbonate-bearing terraces and, in chains of lakes, the water-balance and sequential sill heights will control downstream lake levels (Broecker and Orr, 1958; Benson, 1978; Spencer et al., 1985; Bacon et al., 2006 – for examples in western USA; Bookman et al., 2004; Haase-Schramm et al., 2004 – for the Dead Sea). The evaporative sections of such basins are commonly attainable by coring modern playa floors to recover sequences of interbedded evaporites and clastic sediments (e.g. Smith, 1979; Smith and Bischoff, 1997) and conventional facies analysis applied to discriminate palaeoenvironments. Such techniques use particle-size analysis of clastic materials, microfossils, and recognition of mineralogy, mineral textures and fabrics within evaporites, and chemistry of salts, pore water and fluid inclusions to elucidate lacustrine palaeochemistry (Spencer et al., 1985 – for Great Salt Lake, Utah; Chivas et al., 1986; Torgersen et al., 1986; Teller and Last, 1990; Chivas and De Deckker, 1991).

A knowledge of past lake-levels may be used to interpret variations in precipitation/evaporation – commonly called effective precipitation – and which is clearly dependent upon air temperature. Evaporite mineralogy is a key component of such studies. For example, in Death Valley, California, the interpreted dry periods are associated with glauberite, gypsum and calcite in palaeo-mudflat deposits, whereas abundant calcite, scarce $CaSO_4$ minerals and halite with mud layers, typify wet periods, including a perennial-lake phase (Li et al., 1997; Lowenstein et al., 1999). A crude 100 ka cyclicity is noted, possibly related to Milankovitch forcing.

Direct temperature estimates for evaporite sequences may be provided by amino acid palaeothermometry which produces effective diagenetic temperatures, commonly integrated over broad depositional phases. For the Bonneville Basin, Utah, Kaufman (2003) used the racemisation of aspartic and glutamic acids from ostracods to demonstrate substantial cooling (a reduction of ~ 10°C) during the Last Glacial Maximum compared with the Late Holocene, suggesting that reduced evaporation, rather than increased precipitation, may have led to the growth of glacial-age Lake Bonneville.

More specific palaeotemperature data for individual halite crystals may be gleaned from fluid inclusion homogenisation temperatures, a technique which has been substantially refined since the 1990s (Roberts and Spencer, 1995; Lowenstein and Brennan, 2001). From a core spanning 100 ka of saline deposition in Death Valley, Lowenstein et al. (1998) were able to reconstruct depositional temperatures for several intervals (with an expected lower value, at the Last Glacial Maximum), and to faithfully recover the known seasonal temperature range for a modern salt crust.

10.8 Relationship to other Terrestrial Geochemical Sediments

Many lake basins oscillate between an evaporative or dry status and being water filled, on a variety of timescales, from seasonal to glacial–interglacial. Therefore, evaporite minerals are commonly interbedded with lacustrine clastic sediments (Chapter 9) and in particular with authigenic carbonate minerals (aragonite, calcite, dolomite), which are commonly precipitated at moderate salinities during the early stages of evaporation.

Calcrete is a common component around the margins of playa basins, and has been widely documented in association with gypcrete in many palaeodrainage lines in the central Australian groundwater discharge zone (e.g. Jacobson et al., 1988; Arakel, 1991). Groundwater- or valley calcretes typically cement gravels in buried palaeochannels or the walls of active channels leading to playa shorelines. Pedogenic calcrete may develop within the deflation-related lunettes and dunes surrounding playas and assists in their preservation (Chapter 2).

10.9 Directions for Future Research

The general processes of terrestrial evaporite formation seem fairly well understood, although future research can be expected to deliver increasingly detailed and sophisticated studies, especially geochemical, on evaporite solute origins, and for palaeoenvironmental analysis. Barely mentioned in this chapter is the application of preserved biogenic remains (e.g. ostracods, molluscs, diatoms, charophytes) and their speciation and chemistry, to saline-lake palaeoenvironments, and this is an anticipated area of growth. Given likely future and current global and regional climate change, some evaporitic basins may well undergo reduced stream inflow, leading to their changed geomorphology and, for which, baseline studies of their recent past history will be required as background knowledge, to assist with conservation and management.

The continued application of geophysical methods, particularly remotely sensed imagery, to playa environmental changes and dynamics (e.g. Millington et al., 1989; Prata, 1990; Bryant et al., 1994; Bryant, 1999; White and Eckardt, 2006) is necessary. Advanced spectroradiometry and airborne γ-scintillometry can detect the mineral composition and chemical changes in playas. The author recalls participating in extensive flights over Australian playas and their palaeochannel margins in the early 1970s in the search for calcrete-hosted uranium ore bodies using multichannel γ-spectrometry (K, Th, U), and the inadvertent discovery of many alunite-bearing playas. Such work is common in the mineral exploration industry, but rarely proceeds to formal publication.

The discovery of terrestrial evaporites will continue, even in extraterrestrial settings. The recent discoveries and stunning images of Ca-sulphate and chloride evaporites on Mars and their reworked aeolian deposits (Squyres et al., 2004, 2006; Grotzinger et al., 2005; McLennan et al., 2005; Tosca et al., 2005) provide rich insights into possible new sedimentary and geomorphological environments, chemical pathways for evaporite formation, and additional analogues for their terrestrial counterparts (Chavdarian and Sumner, 2006).

References

Alonso, R.N. (1999) On the origin of La Puna borates. *Acta Geologica Hispanica* **34**, 141–166.

Alonso, R.N., Jordan, T.E., Tabbutt, K.T. & Vandervoort, D.S. (1991) Giant evaporite belts of the Neogene central Andes. *Geology* **19**, 401–404.

Arakel, A.V. (1991) Evolution of Quaternary duricrusts in Karinga Creek drainage system, central Australian groundwater discharge zone. *Australian Journal of Earth Sciences* **38**, 333–347.

Aref, M.A.M., El-Khoriby, E. & Hamdan, M.A. (2002) The role of salt weathering in the origin of the Qattara Depression, Western Desert, Egypt. *Geomorphology* **45**, 181–195.

Bacon, S.N., Burke, R.M., Pezzopane, S.K. & Jayko, A.S. (2006) Last glacial maximum and Holocene lake levels of Owens Lake, eastern California, USA. *Quaternary Science Reviews* **25**, 1264–1282.

Bao, H., Thiemens, M.H., Farquhar, J., Campbell, D.A., Lee, C. C-W., Heine, K. & Loope, D.B. (2000) Anomalous ^{17}O compositions in massive sulphate deposits on the Earth. *Nature* **406**, 176–178.

Bao, H., Thiemens, M.H. & Heine, K. (2001) Oxygen-17 excesses of the central Namib gypcretes: spatial distributions. *Earth and Planetary Science Letters* **192**, 125–135.

Bao, H., Jenkins, K.A., Khachaturyan, M. & Chong Díaz, G. (2004) Different sulfate sources and their post-depositional migration in Atacama soils. *Earth and Planetary Science Letters* **224**, 577–587.

Benson, L. V. (1978) Fluctuations in the level of pluvial Lake Lahontan for the past 40,000 years. *Quaternary Research* **9**, 300–318.

Birnbaum, S.J. & Coleman, M. (1979) Source of sulphur in the Ebro Basin (northern Spain); Tertiary nonmarine evaporite deposits as evidenced by sulphur isotopes. *Chemical Geology* **25**, 163–168.

Bohacs, K.M., Carroll, A.R., Neal, J.E. & Mankiewicz, P.J. (2000) Lake-basin type, source potential, and hydrocarbon character: an integrated sequence stratigraphic-geochemical framework. In: Gierlowski-Kordesch, E.H. & Kelts, K.R. (Eds). *Lake Basins Through Space and Time*. Studies in Geology 46. Tulsa, OK: American Association of Petroleum Geologists, pp. 3–34.

Bookman (Ken-Tor), R., Enzel, Y., Agnon, A. & Stein, M. (2004) Late Holocene lake levels of the Dead Sea. *Geological Society of America, Bulletin* **116**, 555–571.

Bowler, J.M. (1973) Clay dunes: their occurrence, formation and environmental significance. *Earth-Science Reviews* **9**, 315–338.

Bowler, J.M. (1977) Aridity in Australia: age, origins and expression in aeolian landforms and sediments. *Earth-Science Reviews* **12**, 279–310.

Bowler, J.M. (1981) Australian salt lakes – a palaeohydrologic approach. *Hydrobiologia* **82**, 431–444.

Bowler, J.M. (1986) Spatial variability and hydrologic evolution of Australian lake basins: Analogue for Pleistocene hydrologic change and evaporite formation. *Palaeogeography, Palaeoclimatology, Palaeoecology* **54**, 21–41.

Bowser, C.J. & Black, R.F. (1970) Geochemical evidence for the origin of mirabilite deposits near Hobbs Glacier, Victoria Land, Antarctica. *Mineralogical Society of America, Special Paper* **3**, 261–272.

Briere, P.R. (2000) Playa, playa lake, sabkha: Proposed definitions for old terms. *Journal of Arid Environments* **45**, 1–7.

Broecker, W.S. & Orr, P.C. (1958). Radiocarbon chronology of Lake Lahontan and Lake Bonneville. *Bulletin of the Geological Society of America* **69**, 1009–1032.

Bryant, R.G. (1999) Application of AVHRR to monitoring a climatically sensitive playa. Case study: Chott el Djerid, southern Tunisia. *Earth Surface Processes and Landforms* **24**, 283–302.

Bryant, R.G., Sellwood, B.W., Millington, A.C. & Drake, N.A. (1994) Marine-like potash evaporite formation on a continental playa: case study from Chott el Djerid, southern Tunisia. *Sedimentary Geology* **90**, 269–291.

Burke, W.H., Denison, R.E., Hetherington, E.A., Koepnick, R.B., Nelson, H.F. & Otto, J.B. (1982) Variation of seawater $^{87}Sr/^{86}Sr$ throughout Phanerozoic time. *Geology* **1**, 516–519.

Butler, G.P. (1969) Modern evaporite deposition and geochemistry of coexisting brines, the sabkha, Trucial Coast, Arabian Gulf. *Journal of Sedimentary Petrology* **39**, 70–89.

Butler, R.W.H., Lickorish, W.H., Grasso, M., Pedley, H.M. & Ramberti, L. (1995) Tectonics and sequence stratigraphy in Messinian basins, Sicily: constraints on the initiation and termination of the Mediterranean salinity crisis. *Geological Society of America, Bulletin* **107**, 425–439.

Calhoun, J.A. & Bates, T.S. (1989) Sulfur isotope ratios. Tracers of non-sea-salt sulfur in the remote atmosphere. In: Saltzman E.S. & Cooper W.J. (Eds) *Biogenic Sulphur in the Environment*. Washington, DC: American Chemical Society, pp. 369–379.

Calhoun, J.A., Bates, T.S. & Charlson, R.J. (1991) Sulfur isotope measurements of submicrometer sulfate aerosol particles over the Pacific Ocean. *Geophysical Research Letters* **18**, 1877–1880.

Carmona, V., Pueyo, J.J., Taberner, C, Chong, G. & Thirlwall, M. (2000) Solute imputs in the Salar de Atacama. *Journal of Geochemical Exploration* **69–70**, 449–452.

Cendón, D.I., Ayora, C., Pueyo J.J. (1998) The origin of barren bodies in the Subiza potash deposit, Navarra, Spain: Implications for sylvite formation. *Journal of Sedimentary Research* **68**, 43–52.

Cendón, D.I., Peryt, T.M., Ayora, C., Pueyo J.J. & Taberner, C. (2004) The importance of recycling processes in the Middle Miocene Badenian evaporite basin (Carpathian foredeep): palaeoenvironmental implications. *Palaeogeography, Palaeoclimatology, Palaeoecology* **212**, 141–158.

Chavdarian, G.V. & Sumner, D.Y. (2006) Cracks and fins in sulfate sand: Evidence for recent mineral-atmospheric water cycling in Meridiani Planum outcrops? *Geology* **34**, 229–232.

Chen, K. & Bowler, J.M. (1986) Late Pleistocene evolution of salt lakes in the Qaidam Basin, Qinghai Province, China. *Palaeogeography, Palaeoclimatology, Palaeoecology* **54**, 87–104.

Chen, X.Y. (1997) Pedogenic gypcrete formation in arid central Australia. *Geoderma* **77**, 39–61.

Chen, X.Y., Bowler, J.M. & Magee, J.W. (1991) Gypsum ground: a new occurrence of gypsum sediment in playas of central Australia. *Sedimentary Geology* **72**, 79–95.

Chivas, A.R. & De Deckker, P. (Eds) (1991) Palaeoenvironments of Salt Lakes. *Palaeogeography, Palaeoclimatology, Palaeoecology* **84**, 423 pp.

Chivas, A.R., De Deckker, P., Nind, M., Thiriet, D. & Watson, G. (1986a) The Pleistocene palaeoenvironmental record of Lake Buchanan: an atypical Australian playa. *Palaeogeography, Palaeoclimatology, Palaeoecology* **54**, 131–152.

Chivas, A.R., Torgersen, T. & Bowler, J.M. (Eds) (1986b) Palaeoenvironments of Salt Lakes. *Palaeogeography, Palaeoclimatology, Palaeoecology* **54**, 328 pp.

Chivas, A.R., McCulloch, M.T., Lyons, W.B., Donnelly, T.H. & Cowley, J.A. (1987) Isotopic tracers of the source of salts. In: *SLEADS (Salt Lakes, Evaporites and Aeolian Deposits) Workshop 1987*. Canberra: Australian National University, p. 10.

Chivas, A.R., Andrew, A.S., Lyons, W.B., Bird, M.I. & Donnelly, T.H. (1991) Isotopic constraints on the origin of salts in Australian playas. 1. Sulphur. *Palaeogeography, Palaeoclimatology, Palaeoecology* **84**, 309–332.

Claypool, G.E., Holser, W.T., Kaplan, I.R., Sakai, H. & Zak, I. (1980) The age curves of sulfur and oxygen isotopes in marine sulfate and their natural interpretation. *Chemical Geology* **28**, 199–260.

Cody, R.D. (1976) Growth and early diagenetic changes in artificial gypsum crystals grown within bentonite muds and gels. *Geological Society of America, Bulletin* **87**, 1163–1168.

Cody, R.D. (1979) Lenticular gypsum: occurrences in nature, and experimental determinations of effects of soluble green plant material on its formation. *Journal of Sedimentary Petrology* **49**, 1015–1028.

Cody, R.D. & Cody, A.M. (1988) Gypsum nucleation and crystal morphology in analog saline terrestrial environments. *Journal of Sedimentary Petrology* **58**, 247–255.

Cooke, R.V. & Warren, A. (1973) *Geomorphology in Deserts*. Los Angeles, CA: University of California Press, 374 pp.

Curtis, R., Evans, G., Kinsman, D.J.J. & Shearman, D.J. (1963) Association of dolomite and anhydrite in the Recent sediments of the Persian Gulf. *Nature* **197**, 679–680.

Dean, W.E. (1978) Trace and minor elements in evaporites, in Dean, W.E. & Schreiber, B.C. (Eds). *Marine Evaporites*. Short Course Notes 4. Tulsa, OK: Society of Economic Paleontologists and Mineralogists, pp. 86–104.

De Martone, E. & Aufrère, L. (1928) L'extension des régions privées d'écoulement vers l'océan. *Annales de Geographie* **38**, 1–24.

DePaolo, D.J. & Ingram, B.L. (1985) High resolution stratigraphy with strontium isotopes. *Science* **227**, 938–941.

Drake, N.A., Eckardt, F.D. & White, K. H. (2004) Sources of sulphur in gypsiferous sediments and crusts and pathways of gypsum redistribution in southern Tunisia. *Earth Surface Processes and Landforms* **29**, 1459–1471.

Eakin, T.E., Price, D. & Harrill, J.R. (1976) Summary appraisals of the nation's ground-water resources – Great Basin region. *U.S. Geological Survey Professional Paper* **813-G**.

Earman, S., Phillips, F.M. & McPherson, B.J.O.L. (2005) The role of 'excess' CO_2 in the formation of trona deposits. *Applied Geochemistry* **20**, 2217–2232.

Eastoe, C.J. & Peryt, T. (1991) Stable chlorine isotope evidence for non-marine chloride in Badenian evaporites, Carpathian mountain region. *Terra Nova* **11**, 118–123.

Eckardt, F.D. & Spiro, B. (1999) The origin of sulphur in gypsum and dissolved sulphate in the Central Namib Desert, Namibia. *Sedimentary Geology* **123**, 255–273.

Eckardt, F.D., Drake, N., Goudie, A.S., White, K. & Viles, H. (2001) The role of playas in pedogenic gypsum crust formation in the Central Namib Desert: a theoretical model. *Earth Surface Processes and Landforms* **26**, 1177–1193.

Edinger, S.E. (1973) The growth of gypsum. An investigation of the factors which affect the size and growth rates of the habit faces of gypsum. *Journal of Crystal Growth* **18**, 217–224.

Eggenkamp, H.G.M., Kreulen, R. & Van Gross, A.F.K. (1995) Chlorine stable isotope fractionation in evaporites. *Geochimica et Cosmochimica Acta* **59**, 5169–5175.

Elderfield, H. (1986) Strontium isotope stratigraphy. *Palaeogeography, Palaeoclimatology, Palaeoecology* **57**, 71–90.

Eugster, H.P. (1970) Chemistry and origin of the brines of Lake Magadi, Kenya. *Geological Society of London, Special Paper* **3**, 215–235.

Eugster, H.P. & Hardie, L.A. (1978) Saline lakes. In: Lerman, A. (Ed.) *Lakes: Chemistry, Geology, Physics*. New York. Springer-Verlag, pp. 237–293.

Eugster, H.P. & Maglione, G. (1979) Brines and evaporites of the Lake Chad basin, Africa. *Geochimica et Cosmochimica Acta* **43**, 973–981.

Evans, G., Schmidt, V., Bush, P. & Nelson, H. (1969) Stratigraphy and geologic history of the sabkha, Abu Dhabi, Persian Gulf. *Sedimentology* **12**, 145–159.

Freytet, P. & Verrecchia, E.P. (2002) Lacustrine and palustrine carbonate petrography: an overview. *Journal of Paleolimnology* **27**, 221–237.

Garrett, D.E. (2001) *Sodium Sulfate: Handbook of Deposits, Processing, Properties and Use*. San Diego: Academic Press, 365 pp.

Gavrieli, I. Yechieli, Y., Halicz, L., Spiro, B., Bein, A. & Efron, D. (2001) The sulfur system in anoxic subsurface brines and its implication in brine evolutionary pathways: the Ca-chloride brines in the Dead Sea area. *Earth and Planetary Science Letters* **186**, 199–213.

Ghienne, J.-F., Schuster, M., Bernard, A., Duringer, P. & Brunet, M. (2002) The Holocene giant Lake Chad revealed by digital elevation models. *Quaternary International* **87**, 81–85.

Gornitz, V.M. & Schreiber, B.C. (1981) Displacive halite hoppers from the Dead Sea: Some implications for ancient evaporite deposits. *Journal of Sedimentary Petrology* **51**, 787–794.

Goudie, A.S. & Thomas, D.S.G. (1986) Lunette dunes in southern Africa. *Journal of Arid Environments* **10**, 1–12.

Goudie, A.S. & Wells, G.L. (1995) The nature, distribution and formation of pans in arid zones. *Earth Science Reviews* **38**, 1–69.

Grotzinger, J.P., Arvidson, R.E., Bell, J.F., Calvin, W., Clark, B.C., Fike, D.A., Golombek, M., Greeley, R., Haldemann, A., Herkenhoff, K.E., Jolliff, B.L., Knoll, A.H., Malin, M., McLennan, S.M., Parker, T., Soderblom, L., Sohl-Dickstein, J.N., Squyres, S.W., Tosca, N.J. & Watters, W.A. (2005) Stratigraphy and sedimentology of a dry to wet eolian depositional system, Burns formation, Meridiani Planum, Mars. *Earth and Planetary Science Letters* **240**, 11–72.

Gutiérrez-Elorza, M., Desir, G. & Gutiérrez-Santolalla, F. (2002) Yardangs in the semiarid central sector of the Ebro Depresion (NE Spain). *Geomorphology* **44**, 155–170.

Gutiérrez-Elorza, M., Desir, G., Gutiérrez-Santolalla, F. & Marín, C. (2005) Origin and evolution of playas and blowouts in the semiarid zone of Tierra de Pinares (Duero Basin, Spain). *Geomorphology* **72**, 177–192.

Haase-Schramm, A., Goldstein, S.L. & Stein, M. (2004) U–Th dating of Lake Lisan (late Pleistocene Dead Sea) aragonite and implications for glacial East Mediterranean climate change. *Geochimica et Cosmochimica Acta* **68**, 985–1005.

Handford, C.R. (1982) Sedimentology and evaporite genesis in a Holocene continental sabkha playa basin – Bristol Dry Lake, California. *Sedimentology* **29**, 239–253.

Hardie, L.A. (1968) The origin of the Recent non-marine evaporite deposits of Saline Valley, Inyo County, California. *Geochimica et Cosmochimica Acta* **32**, 1279–1301.

Hardie, L.A. (1984) Evaporites: marine or non-marine? *American Journal of Science* **284**, 193–240.

Hardie, L.A. (1990) The roles of rifting and hydrothermal $CaCl_2$ brines in the origin of potash evaporites: An hypothesis. *American Journal of Science* **290**, 43–106.

Hardie, L.A. & Eugster, H. P. (1970) The evolution of closed-basin brines. *Mineralogical Society of America Special Publication* **3**, 273–290.

Hardie, L.A., Smoot, J.P. & Eugster, H.P. (1978) Saline lakes and their deposits: a sedimentological approach. In: Matter A. and Tucker M.E. (Eds) *Modern and Ancient Lake Sediments*. Special Publication 2, International Association of Sedimentologists. Oxford: Blackwell Scientific Publications, pp. 7–42.

Hart, W.S., Quade, J., Madsen, D.B., Kaufman, D.S. & Oviatt, C.G. (2004) The $^{87}Sr/^{86}Sr$ ratios of lacustrine carbonates and lake level history of the Bonneville paleolake system. *Geological Society of America, Bulletin* **116**, 1107–1119.

Harvie, C.E., Weare, J.H., Hardie, L.A. & Eugster, H.P. (1980) Evaporation of seawater: calculated mineral sequences. *Science* **208**, 498–500.

Hildreth, R.A. & Henderson, W.T. (1971) Comparison of Sr^{87}–Sr^{86} for sea-water strontium and the Eimer and Amend $SrCO_3$. *Geochimica et Cosmochimica Acta* **35**, 235–238.

Hills, E.S. (1940) The lunette: a new landform of aeolian origin. *The Australian Geographer* **3**, 1–7.

Hite, R.J. & Anders, D.E. (1991) Petroleum and evaporites. In: Melvin, J.L. (Ed.) *Evaporites, Petroleum and Mineral Resources*. Elsevier, Amsterdam, pp. 349–411.

Holser, W.T. (1979) Trace elements and isotopes in evaporites. In: Burns, R.G. (Ed.) *Marine Minerals. Mineralogical Society of America, Reviews in Mineralogy* **6**, 295–346.

Holser, W.T. & Kaplan, I.R. (1966) Isotope geochemistry of sedimentary sulfate. *Chemical Geology* **1**, 93–135.

Jack, R.L. (1921) The salt and gypsum resources of South Australia. *Bulletin of the Geological Survey of South Australia* **8**.

Jacobson, G., Arakel, A.V. & Chen, Y. (1988) The central Australian groundwater discharge zone: evolution of associated calcrete and gypcrete deposits. *Australian Journal of Earth Sciences* **35**, 549–565.

Jacobson, G., Ferguson, J. & Evans, W.R. (1994) Groundwater-discharge playas of the Mallee Region, Murray Basin, southeast Australia. In: Rosen, M.R. (Ed.) *Paleoclimate and Basin Evolution of Playa Systems*. Special Paper 289. Boulder, CO: Geological Society of America, pp. 81–96.

Jones, B.F. & Bodine, M.W. Jr (1987) Normative salt characterization of natural waters. In: Fritz, P. & Frape, S.K. (Eds) *Saline Water and Gases in Crystalline Rocks*. Special Paper 33. Geological Association of Canada, pp. 5–18.

Jones, B.F. & Deocampo, D.M. (2004) Geochemistry of saline lakes. In: Drever, J.I. (Ed.) *Surface and Ground Water Weathering and Soils*, Vol. 5, *Treatise on Geochemistry* (Eds Holland, H.D. and Turekian, K.K.). Oxford: Elsevier, Pergamon, pp. 393–424.

Kampschulte, A. & Strauss, H. (2004) The sulfur isotope evolution of Phanerozoic seawater based on the analysis of structurally substituted sulfate in carbonates. *Chemical Geology* **204**, 255–286.

Kasemann, S.A., Meixner, A., Erzinger, J., Viramonte, J.G., Alonso, R.N. & Franz, G. (2004) Boron isotope composition of geothermal fluids and borate minerals from salar deposits (central Andes/NW Argentina). *Journal of South American Earth Sciences* **16**, 685–697.

Kaufman, D.S. (2003) Amino acid paleothermometry of Quaternary ostracodes from the Bonneville Basin, Utah. *Quaternary Science Reviews* **22**, 899–914.

Kendall, A.C. (1992) Evaporites. In Walker, R.G. & James, N.P. (Eds). *Facies Models: Responses to Sea Level Change*. Geological Association of Canada, pp. 375–409.

Keys, J.R. & Williams, K. (1981) Origin of crystalline, cold desert salts in the McMurdo region, Antarctica. *Geochimica et Cosmochimica Acta* **45**, 2299–2309.

Klein-Ben David, O., Sass, E. & Katz, A. (2004) The evolution of marine evaporitic brines in inland basins: the Jordan–Dead Sea Rift valley. *Geochimica et Cosmochimica Acta* **68**, 1763–1775.

Krijgsman, W., Blanc-Valleron, M.-M., Flecker, R., Hilgen, F.J., Kouwenhoven, T.J., Merle, D., Orszag-Sperber, F. & Rouchy, J.-M. (2002) The onset of the Messinian salinity crisis in the Eastern Mediterranean (Pissouri Basin, Cyprus). *Earth and Planetary Science Letters* **194**, 299–310.

Kushnir, J. (1980) The coprecipitation of strontium, magnesium, sodium, potassium and chloride ions with gypsum. An experimental study. *Geochimica et Cosmochimica Acta* **44**, 1471–1482.

Kushnir, J. (1981) Formation and early diagenesis of varved evaporite sediments in a coastal hypersaline pool. *Journal of Sedimentary Petrology* **51**, 1193–1203.

Kushnir, J. (1982a) The partitioning of seawater cations during the transformation of gypsum to anhydrite. *Geochimica et Cosmochimica Acta* **46**, 433–446.

Kushnir, J. (1982b) The composition and origin of brines during the Messinian desiccation event in the Mediterranean basin as deduced from concentrations of ions coprecipitated with gypsum and anhydrite. *Chemical Geology* **35**, 333–350.

Langford, R.P. (2003) The Holocene history of the White Sands dune field and influences on eolian deflation and playa lakes. *Quaternary International* **104**, 31–39.

Last, W.M. (1989) Continental brines and evaporites of the northern Great Plains of Canada. *Sedimentary Geology* **64**, 207–221.

Last, W.M. (1992) Chemical composition of saline and subsaline lakes of the northern Great Plains, western Canada. *International Journal of Salt Lake Research* **1**, 47–76.

Li, J., Lowenstein, T.K. & Blackburn, I.R. (1997) Responses of evaporite mineralogy to inflow water sources and climate during the past 100 k.y. in Death Valley, California. *Geological Society of America, Bulletin* **109**, 1361–1371.

Lintern, M.J., Sheard, M.J. & Chivas, A.R. (2006) The source of pedogenic carbonate associated with gold-calcrete anomalies in the western Gawler Craton, South Australia. *Chemical Geology* **235**, 299–324.

Logan, B.W. (1987) The Macleod evaporite basin, Western Australia. *American Association of Petroleum Geologists, Memoir* **44**, 140 pp.

Lombardi, G., Brodtkorb, A., Romero, S., Aurisicchio, G., Schalamuk, A., Del Blanco, M., De Barrio, R., Marchioni, D. & Manili, M. (1994) The salt body of the Salina del Gualicho (Río Negro, Argentina). *Bolletino della Società Geologica Italiana* **112**, 1037–1057.

Lowenstein, T. & Brennan, S.T. (2001) Fluid inclusions in paleolimnological studies of chemical sediments. In: Last, W.M. & Smol, J.P. (Eds) *Tracking Environmental Change Using Lake Sediments*, Vol. 2, *Physical and Chemical Methods*. Dordrecht: Kluwer, pp. 189–216.

Lowenstein, T.K. & Hardie, L.A. (1985) Criteria for the recognition of salt-pan evaporites. *Sedimentology* **32**, 627–644.

Lowenstein, T.K. & Spencer, R.J. (1990) Syndepositional origin of potash evaporates: petrographic and fluid inclusion evidence. *American Journal of Science* **290**, 1–42.

Lowenstein, T.K., Spencer, R.J. & Zhang, P. (1989) Origin of ancient potash evaporites: Clues from the modern nonmarine Qaidam basin of western China. *Science* **245**, 1090–1092.

Lowenstein, T.K., Li, J., Brown, C.B. (1998) Paleotemperatures from fluid inclusions in halite: method verification and a 100,000 year paleotemperature record, Death Valley, CA. *Chemical Geology* **150**, 223–245.

Lowenstein, T.K., Li, J., Brown, C., Roberts, S.M., Ku, T.-L., Luo, S. & Yang, W. (1999) 200 k.y. paleoclimate record from Death Valley salt core. *Geology* **27**, 3–6.

Lu, F.H. & Meyers, W.J. (2003) Sr, S and O_{SO4} isotopes and the depositional environment of the upper Miocene evaporites, Spain. *Journal of Sedimentary Research* **73**, 444–450.

Lyons, W.B., Hines, M.E., Last, W.M. & Lent, R.M. (1994) Sulfate reduction rates in microbial mat sediments of differing chemistries: implications for organic carbon preservation in saline lakes. In: Renaut, R.W. & Last, W.M. (Eds) *Sedimentology and Geochemistry of Modern and Ancient Saline Lakes*. Special Publication 50. Tulsa, OK: Society of Economic Paleontologists and Mineralogists (Society for Sedimentary Geology), 13–20.

Macumber, P.G. (1991) *Interaction between Ground Water and Surface Systems in Northern Victoria*. Department of Conservation and Environment, Victoria, Australia, 345 pp.

Magee, J.W. (1991) Late Quaternary lacustrine, groundwater, aeolian and pedogenic gypsum in the Prungle Lakes, southeastern Australia. *Palaeogeography, Palaeoclimatology, Palaeoecology* **84**, 3–42.

Magee, J.W., Bowler, J.M., Miller, G.H. & Williams, D.L.G. (1995) Stratigraphy, sedimentology, chronology and palaeohydrology of Quaternary lacustrine deposits at Madigan Gulf, Lake Eyre, South Australia. *Palaeogeography, Palaeoclimatology, Palaeoecology* **113**, 3–42.

McArthur, J.M., Turner, J., Lyons, W.B. & Thirlwall, M.F. (1989) Salt sources and water-rock interaction on the Yilgarn Block, Australia: isotopic and major element tracers. *Applied Geochemistry* **4**, 79–92.

McArthur, J.M., Howarth, R.J. & Bailey, T.R. (2001) Strontium isotope stratigraphy: LOWESS Version 3: best fit to the marine Sr-isotope curve for 0–509 Ma and accompanying look-up table for deriving numerical age. *Journal of Geology* **109**, 155–170.

McLennan, S.M., Bell III, J.F., Calvin, W.M., Christensen, P.R., Clark, B.C., de Souza, P.A., Farmer, G.J., Farrand, W.H., Fike, D.A., Gellert, R., Ghosh, A., Glotch, T.D., Grotzinger, J.P., Hahn, B., Herkenhoff, K.E., Hurowitz, J.A., Johnson, J.R., Johnson, S.S., Jolliff, B., Klingelhöfer, G., Knoll, A.H., Learner, Z., Malin, M.C., McSween, Jr., H.Y., Pocock, J., Ruff, S.W., Soderblom, L.A., Squyres, S.W., Tosca, N.J., Watters, W.A., Wyatt, M.B. and Yen, A. (2005) Provenance and diagenesis of the evaporite-bearing Burns Formation, Meridiani Planum, Mars. *Earth and Planetary Science Letters* **240**, 95–121.

Mees, F. (1999) Distribution patterns of gypsum and kalinstrontite in a dry lake basin of the southwestern Kalahari (Omingwa pan, Namibia). *Earth Surface Processes and Landforms* **24**, 731–744.

Millington, A.C., Drake, N.A., Townshend, J.R.G., Quarmby, N.A., Settle, J.J. & Reading, A.J. (1989) Monitoring salt playa dynamics using Thematic Mapper data. *IEEE Transactions on Geoscience and Remote Sensing* **27**, 754–761.

Motts, W.S. (1965) Hydrologic types of playas and closed valleys and some relations of hydrology to playa geology. In: J.T. Neal (Ed.) *Geology, Mineralogy and Hydrology of U.S. Playas*. Environmental Research Papers 96. Bedford, MA: Airforce Research Laboratory, pp. 73–105.

Neal, J.T. (1975) Introduction. In: Neal J.T. (Ed.) *Playas and Dried Lakes*. Benchmark Papers in Geology. Stroudsberg, PA: Dowden, Hutchinson and Ross, pp. 1–5.

Neev, D. & Emery, K.O. (1967) The Dead Sea: depositional processes and environments of evaporites. *Israel Geological Survey Bulletin* **41**, 1–147.

Nelson, D.R. & McCulloch, M.T. (1989) Petrogenetic applications of the $^{40}K-^{40}Ca$ radiogenic decay scheme – a reconnaissance study. *Chemical Geology* **79**, 275–293.

Ordoñez, S., Sánchez Moral, S. García del Cura, M.A., Rodríguez Badiola, E. (1994) Precipitation of salts from Mg^{2+}-(Na^+)-SO_4^{2-}-Cl^- playa-lake brines: the endorheic saline ponds of La Mancha, central Spain. In: Renaut, R.W. & Last, W.M. (Eds) *Sedimentology and Geochemistry of Modern and Ancient Saline Lakes.* Special Publication 50. Tulsa, OK: Society of Economic Paleontologists and Mineralogists (Society for Sedimentary Geology), pp. 61–71.

Palmer, M.R., Helvaci, C. & Fallick, A.E. (2004) Sulphur, sulphate oxygen and strontium isotope composition of Cenozoic Turkish evaporites. *Chemical Geology* **209**, 341–356.

Paytan, A., Kastner, M., Campbell, D. & Thiemens, M.H. (1998) Sulfur isotopic composition of Cenozoic seawater sulfate. *Science* **282**, 1259–1262.

Pierre, C. (1985) Isotopic evidence for the dynamic redox cycle of dissolved sulphur compounds between free and interstitial solutions in marine salt pans. *Chemical Geology* **53**, 191–196.

Playà, E., Ortí, F. & Rosell, L. (2000) Marine to non-marine sedimentation in the upper Miocene evaporites of the Eastern Betics, SE Spain: sedimentological and geochemical evidence. *Sedimentary Geology* **133**, 135–166.

Playá, E., Cendón, D.I., Travé, A., Chivas, A.R. & García, A. (In press) Non-marine evaporites with both inherited marine and continental signatures: the Gulf of Carpentaria, Australia, at ~70 ka. *Sedimentary Geology.*

Prata, A.J. (1990) Satellite-derived evaporation from Lake Eyre, South Australia. *International Journal of Remote Sensing* **11**, 2051–2068.

Quade, J., Chivas. A.R. & McCulloch, M.T. (1995) Strontium and carbon isotope tracers and the origins of soil carbonate in South Australia and Victoria. *Palaeogeography, Palaeoclimatology, Palaeoecology* **113**, 103–117.

Raab, M. & Spiro, B. (1991) Sulfur isotopic variations during seawater evaporation with fractional crystallization. *Chemical Geology* **86**, 323–333.

Rech, J.A., Quade, J. & Hart, W.S. (2003) Isotopic evidence for the source of Ca and S in soil gypsum, anhydrite and calcite in the Atacama Desert, Chile. *Geochimica et Cosmochimica Acta* **67**, 576–586.

Rees, C.E., Jenkins, W.J. & Monster, J. (1978) The sulphur isotopic composition of ocean water sulphate. *Geochimica et Cosmochimica Acta* **42**, 377–381.

Risacher, F. & Fritz, B. (2000) Bromine geochemistry of Salar de Uyuni and deeper salt crusts, central Altiplano, Bolivia. *Chemical Geology* **167**, 373–392.

Risacher, F., Alonso, H. & Salazar, C. (2003) The origin of brines and salts in Chilean salars: a hydrochemical review. *Earth-Science Reviews* **63**, 249–293.

Roberts, S.M. & Spencer, R.J. (1995) Paleotemperatures preserved in fluid inclusions in halite. *Geochimica et Cosmochimica Acta* **59**, 3929–3942.

Rosell, L., Ortí, F., Kasprzyk, A., Playà, E. & Peryt, T.M. (1998) Strontium geochemistry of Miocene primary gypsum: Messinian of southeastern Spain and Sicily and Badenian of Poland. *Journal of Sedimentary Research* **68**, 63–79.

Rosen, M.R. (Ed.) (1994) *Paleoclimate and Basin Evolution of Playa Systems.* Special Paper 289. Boulder, CO: Geological Society of America, 117 pp.

Rosen, M.R. (1994) The importance of groundwater in playas: a review of playa classifications and the sedimentology and hydrology of playas. In: Rosen, M.R. (Ed.) *Paleoclimate and Basin Evolution of Playa Systems.* Special Paper 289. Boulder, CO: Geological Society of America, pp. 1–18.

Russell, I.C. (1885) Playa-lakes and playas. *U.S. Geological Survey, Monograph* **11**, 81–86.

Schreiber, B.C. & El Tabakh, M. (2000) Deposition and early alteration of evaporites. *Sedimentology* 47(supplement 1), 215–238.

Schubel, K.A. & Lowenstein, T.K. (1997) Criteria for the recognition of shallow-perennial-saline-lake halites based on recent sediments from Qaidam Basin, western China. *Journal of Sedimentary Research* **67**, 74–87.

Shaw, P.A. & Thomas, D.S.G. (1997) Pans, playas and salt lakes. In: Thomas D.S.G. (Ed.) *Arid Zone Geomorphology: Process, Form and Change in Drylands*, 2nd edn. Chichester: Wiley, pp. 293–317.

Simon, B. & Bienfait, M. (1965) Structure et mechanisme de croissance du gypse. *Acta Crystallographica* **19**, 750–756.

Smith, G.I. (1979) Subsurface stratigraphy and geochemistry of Late Quaternary evaporites, Searles Lake, California. *U.S. Geological Survey Professional Paper* **1043**, 130 pp.

Smith, G.I. & Bischoff, J.L. (1997) An 800,000-year paleoclimate record from core OL-92, Owens Lake, southeast California. *Geological Society of America, Special Paper* 317.

Smoot, J.P. & Lowenstein, T.K. (1991) Depositional environments of non-marine evaporites. In: Melvin J.L. (Ed.) *Evaporites, Petroleum and Mineral Resources*. Elsevier, Amsterdam, pp. 189–347.

Sonnenfeld, P. (1985) Comment on 'Evaporites: Marine or non-marine?' *American Journal of Science* **285**, 661–672.

Spencer, R.J., Eugster, H.P. & Jones, B.F. (1985) Geochemistry of Great Salt Lake, Utah II: Pleistocene-Holocene evolution. *Geochimica et Cosmochimica Acta* **49**, 739–747.

Squyres, S.W., Grotzinger, J.P., Arvidson, R.E., Bell, J.F., Calvin, W., Christensen, P.R., Clark, B.C., Crisp, J.A., Farrand, W.H., Herkenhoff, K.E., Johnson, J.R., Klingelhofer, G., Knoll, A.H., McLennan, S.M., McSween, H.Y., Morris, R.V., Rice, J.W., Rieder, R., Soderblom, L.A. (2004) *In-situ* evidence for an ancient aqueous environment at Meridiani Planum, Mars. *Science* **306**, 1709–1714.

Squyres, S.W., Knoll, A.H., Arvidson, R.E., Clark, B.C., Grotzinger, J.P., Jolliff, B.L., McLennan, S.M., Tosca, N., Bell, J.F., Calvin, W.M., Farrand, W.H., Glotch, T.D., Golombek, M.P., Herkenhoff, K.E., Johnson, J.R., Klingelhoefer, G., McSween, H.Y., Yen, A.S. (2006) Two years at Meridiani Planum: results from the Opportunity Rover. *Science* **313**, 1404–1407.

Stieljes, L. (1973) Evolution tectonique récente du rift d'Asal. *Revue de Géographie Physicale et de Géologie Dynamique* **15**, 425–436.

Stoertz, G.E. & Ericksen G.E. (1974) Geology of salars in northern Chile. *U.S. Geological Survey Professional Paper* **811**, 65 pp.

Strauss, H. (1997) The isotopic composition of sedimentary sulphur through time. *Palaeogeography, Palaeoclimatology, Palaeoecology* **132**, 97–118.

Swihart, G.H., Moore, P.B. & Callis, E.L. (1986) Boron isotopic composition of marine and nonmarine evaporite borates. *Geochimica et Cosmochimica Acta* **50**, 1297–1301.

Taberner, C., Cendón, D.I., Pueyo, J.J. & Ayora, C. (2000) The use of environmental markers to distinguish marine vs. continental deposition and to quantify the significance of recycling in evaporite basins. *Sedimentary Geology* **137**, 213–240.

Teller, J.T. & Last, W.M. (1990) Palaeohydrological indicators in playas and salt lakes, with examples fom Canada, Australia, and Africa. *Palaeogeography, Palaeoclimatology, Palaeoecology* **76**, 215–240.

Thiemens, M.H. (2006) History and applications of mass-independent isotope effects. *Annual Review of Planetary Sciences* **34**, 217–262.

Thode, H.G. & Monster, J. (1965) Sulfur isotope geochemistry of petroleum, evaporites and ancient seas. *American Association of Petroleum Geologists, Memoir* **4**, 367–377.

Torfstein, A., Gavrieli, I. & Stein, M. (2005) The sources and evolution of sulfur in the hypersaline Lake Lisan (paleo-Dead Sea). *Earth and Planetary Science Letters* **236**, 61–77.

Torgersen, T, De Deckker, P., Chivas, A.R. & Bowler, J.M. (1986) Salt lakes: a discussion of processes influencing palaeoenvironmental interpretation and recommendations for future study. *Palaeogeography, Palaeoclimatology, Palaeoecology* **54**, 7–19.

Tosca, N.J, McLennan, S.M., Clark, B.C., Grotzinger, J.P., Hurowitz, J.A., Knoll, A.H., Schröder, C. & Squyres, S.W. (2005) Geochemical modeling of evaporation processes on Mars: Insights from the sedimentary record at Meridiani Planum. *Earth and Planetary Science Letters* **240**, 122–148.

Usiglio, J. (1849) Analyse de l'eau de la Mediterranée sur les côtes de France. *Annales de Chemie et de Physique* **27**, 92–107.

Utrilla, R., Pierre, C., Ortí, F. & Pueyo, J.J. (1992) Oxygen and sulphur isotope compositions as indicators of the origin of Mesozoic and Cenozoic evaporites from Spain. *Chemical Geology* **102**, 229–244.

Valyashko, M.G. (1956) Geochemistry of bromine in the processes of salt deposition and the use of bromine content as a genetic and prospecting criterion. *Geokhimiya* **6**, 570–589.

Valyashko, M.G. (1972) Playa lakes – a necessary stage in the development of a salt-bearing basin. In: Richter-Bernberg, G. (Ed.) *Geology of Saline Deposits.* Paris: UNESCO, pp. 41–51.

Van der Voort, E. & Hartmann, P. (1991) The habit of gypsum and solvent interaction. *Journal of Crystal Growth* **112**, 137–149.

Veizer, J., Ala, D., Azmy, K., Bruckschen, P., Buhl, D., Bruhn, F., Carden, G. A.F., Diener, A., Ebneth, S., Godderis, Y., Jasper, T., Korte, G., Pawellek, F., Podlaha, O.G. & Strauss, H. (1999) $^{87}Sr/^{86}Sr$, $\delta^{13}C$ and $\delta^{18}O$ evolution of Phanerozoic seawater. *Chemical Geology* **161**, 59–88.

Vengosh, A., Chivas, A.R., McCulloch, M.T., Starinsky, A. & Kolodny, Y. (1991a) Boron isotope geochemistry of Australian salt lakes. *Geochimica et Cosmochimica Acta* **55**, 2591–2606.

Vengosh, A., Starinsky, A., Kolodny, Y. & Chivas, A.R. (1991b) Boron isotope geochemistry as a tracer for the evolution of brines and associated hot springs from the Dead Sea, Israel. *Geochimica et Cosmochimica Acta* **55**, 1689–1695.

Vengosh, A., Starinsky, A., Kolodny, Y., Chivas, A.R. & Raab, M. (1992) Boron isotope variations during fractional evaporation of sea water: New constraints on the marine vs. nonmarine debate. *Geology* **20**, 799–802.

Vengosh, A., Chivas, A.R., Starinsky, A., Kolodny, Y., Zhang, B. & Zhang, P. (1995) Chemical and boron isotope compositions of non-marine brines from the Qaidam Basin, Qinghai, China. *Chemical Geology* **120**, 135–154.

Warren, J.K. (1982) Hydrologic setting, occurrence, and significance of gypsum in late Quaternary salt lakes, South Australia. *Sedimentology* **29**, 609–637.

Warren, J.K. (1989) *Evaporite Sedimentology: Importance in Hydrocarbon Accumulation.* Englewood Cliffs, NJ: Prentice Hall, 285 pp.

Warren, J.K. (2006) *Evaporites: Sediments, Resources and Hydrocarbons.* Berlin: Springer-Verlag, 1035 pp.

Warren, J.K. & Kendall, C.G.S.C. (1985) Comparison of sequences formed in marine sabhka (subaerial) and salina (subaqueous) settings; modern and ancient. *Bulletin of the American Association of Petroleum Geologists* **69**, 1013–1023.

White, K. & Eckardt, F. (2006) Geochemical mapping of carbonate sediments in the Makgadikgadi Basin, Botswana, using moderate resolution remote sensing data. *Earth Surfaces Processes and Landforms* **31**, 665–681.

Wood, W.W., Sanford, W.E. & Al Habshi, A.R.S. (2002) Source of solutes to the coastal sabkha of Abu Dhabi. *Geological Society of America, Bulletin* **114**, 259–268.

Wood, W.W., Sanford, W.E. & Frape, S.K. (2005) Chemical openness and potential for misinterpretation of the solute environment of coastal sabkhat. *Chemical Geology* **215**, 361–372.

Yechieli, Y. & Wood, W.W. (2002) Hydrogeological processes in playas, sabkhas, and saline lakes. *Earth-Science Reviews* **58**, 343–365.

Chapter Eleven

Beachrock and Intertidal Precipitates

Eberhard Gischler

11.1 Introduction: Nature and General Characteristics

Beachrock is a friable to well-cemented sedimentary rock that results from rapid lithification of sand and/or gravel by calcium carbonate cement precipitation in the intertidal zone. It occurs predominantly on tropical ocean coasts, but is also found in temperate realms that extend up to 60° latitude. In contrast to the implications of the name, beachrock precipitation phenomena are not restricted to beaches but also occur on reef ridges, tidal flats and in tidal channels. Intertidal beachrock may be confused with other sediments lithified in the intertidal and subtidal zones, such as hardened crusts or certain reef limestones.

The first scientific mention of cemented beach deposits would appear to be by Chamisso (1821, pp. 107–108) in his description of the Radak and Ralik Islands in the Pacific. Subsequently, Lyell (1832, p. 259) noted the occurrence of human skeletons cemented into an 'indurated beach' on the island of Guadeloupe, West Indies. Other famous 19th and early 20th century scholars, such as Moresby (1835, p. 400), Darwin (1842, p. 16), Dana (1875, pp. 122–123), Gardiner (1898, pp. 443–444) and Daly (1920) discussed beachrock occurrence and speculated on its formation. The first modern studies of beachrock included detailed petrographical and mineralogical investigations (Ginsburg, 1953a; Stoddart and Cann, 1965), and since then it has become clear that petrographic examination of thin-sections in conjunction with mineralogical and geochemical analyses, and radiometric dating of cements, are essential techniques in deciphering beachrock formation. To date, the origin of beachrock is still not fully understood as witnessed by the fact that it has been variously attributed to physico-chemical precipitation of carbonate cement, biologically induced cementation, or a combination of both.

11.2 Occurrence and Distribution

Beachrock exposures are quite conspicuous where they form more or less linear and elongated features in the intertidal zone (Figure 11.1). These exposures parallel the coast, and bedded layers usually dip gently (<10°) towards the sea. Even so, outcrops are usually not continuous over distances >100 m. Typically occurrences along the beach appear irregular and patchy and exposed beachrock is often fragmented into rectangular blocks (Figure 11.2A,B). Beachrock thickness is usually less than 5 m with individual layers of up to a few decimetres thickness. In cases of beach movement, beachrock outcrops may become stranded to form isolated ridges. A prominent isolated Holocene example under several metres of water near Bimini, Bahamas, has been repeatedly interpreted as being a paved road of the sunken continent of Atlantis, even though this interpretation has been conclusively disproven by geologists, who have consigned it to the world of myths (see, e.g. Shinn, 2004).

Another typical feature of beachrock exposures is their continuous destruction by fracturing and mechanical erosion through grinding, scraping and wearing away by friction and impact of sediment, i.e. by abrasion or corrasion (e.g. McLean, 1967). During beach erosion and accretion, broken-off fragments of older beachrock may become incorporated into younger beachrock (Strasser and Davaud, 1986). Beachrock may also be eroded by chemical dissolution via low pH fluids such as rainwater, and by marine waters when CO_2 content is increased due to organic matter oxidation (Revelle and Emery, 1957). Bioerosion (Ginsburg, 1953b; Neumann, 1966) may also play a role in the destruction of beachrock occurrences through boring by microorganisms such as algae, bacteria and fungi, boring by sponges, bivalves, sipunculids and polychaetes, and grazing by invertebrates such as sea urchins, gastropods and chitons. To date, there are no systematic data on beachrock destruction rates.

Beachrock occurs predominantly on tropical to subtropical ocean coasts and islands (e.g. Russell, 1962; Krumbein, 1979; Scoffin and Stoddart, 1983), but it is also found in the Mediterranean (Alexandersson, 1972; El-Sayed, 1988; Strasser et al., 1989), the Black and Caspian Seas (Zenkovitch, 1967, pp. 183–186), South Africa (Siesser, 1974; Cooper, 1991) and as far north as 57° latitude (Knox, 1973; Kneale and Viles, 2000). Beachrock is also reported to form on the coasts of freshwater lakes (Binkley et al., 1980; Jones et al., 1997). Beachrock occurrences from polar regions, i.e. north and south of 60° latitude have not been reported. Collectively, this distributional evidence identifies warm climates with pore waters rich in calcium carbonate as essential cement-precipitation prerequisites for beachrock formation.

Figure 11.1   Photographs of beachrock outcrops. (A) Seaward dipping layers of beachrock on the north-west coast of Basse Terre, Guadeloupe, West Indies. This is the area that Charles Lyell (1832) mentioned in his '*Principles of Geology*'. Beachrock consists largely of quartz and volcanic sand cemented by calcium carbonate. (B) Extensive outcrop of beachrock, which also exhibits significant erosion. Windward side of Halfmoon Cay, Lighthouse Reef, Belize. (C) Beachrock ridge on southern shore of Kuramathi island, Rasdu Atoll, Maldives. Note flat and karstified top of outcrop, which resulted from meteoric dissolution.

Figure 11.2 Photographs of beachrock outcrops. (A) Beachrock outcrop with slight inclination of layers towards the sea. Kubbar Island, Kuwait, Arabian–Persian Gulf. (B) Close-up of same location, which shows rectangular fracturing of beachrock. (C) Cemented low beach ridge, which is partly eroded. Tidal flats on western coast of Andros Island, Bahamas. Note tidal channel in background.

11.3 Macro- and Micromorphological Characteristics

11.3.1 General observations

The composition of beachrock constituent particles is more or less similar to that of the adjacent shore. Constituent particles on tropical coasts and islands are often marine-derived carbonate grains (Figure 11.3); however, grains from siliciclastic, magmatic and metamorphic rocks are also known to form beachrock (e.g. Knox, 1973; Siesser, 1974; Figure 11.1A). Grain-size ranges from sand to gravel of moderately to very well sorted sediment with sorting usually better than that of the adjacent subtidal sediments. A common feature is the stronger cementation of fine-grained as opposed to coarse-grained layers (Ginsburg, 1953a,b; Gischler and Lomando, 1997) (Figure 11.3C), reflecting the relative ease of filling smaller pore spaces with cement.

Cementation of beachrock usually occurs in the marine diagenetic environment. However, beachrock may also be subject to meteoric diagenesis (Figure 11.4A). In both marine and meteoric environments, cements may grow under phreatic conditions, when pore space is completely filled with precipitating fluids, or, under vadose conditions, when precipitating fluids only occupy the pore space partially. Cement mineralogy and morphology is indicative of the diagenetic environment (e.g. Longman, 1980; Harris et al., 1985; Scoffin, 1987, pp. 89–131; Tucker and Wright, 1990, pp. 314–364).

11.3.2 Marine cements in beachrock

Beachrock cements are usually dominated by aragonite and high-magnesium calcite (10–20 mol% $CaCO_3$), formed in the marine-phreatic diagenetic environment (e.g. Land, 1970; Bricker, 1971; Scoffin and Stoddart, 1983). The aragonite cement includes isopachous fringes of needles that are up to 100 μm long (e.g. Ginsburg, 1953a; Moore, 1973; Beier, 1985) and often overlie a dark layer at their base that consists of aragonite platelets rich in iron and sulphur (Strasser et al., 1989) (Figures 11.5A and 11.6A, B). The dark layer may also be part of micrite envelopes around constituent carbonate grains (Bathurst, 1966; Purdy, 1968), which form due to centripetal boring of carbonate grains by microalgae, fungi or bacteria, the boreholes of which are filled by fine-grained carbonate cement (Reid and Macintyre, 2000). Small aragonite (<10 μm long) needles may form 'micritic' cement in beachrock (e.g. Webb et al., 1999). High-magnesium calcite cements often include microcrystalline ('micritic') cement with crystals less than 5 μm in diameter (e.g. Meyers, 1987) (Figure 11.5C).

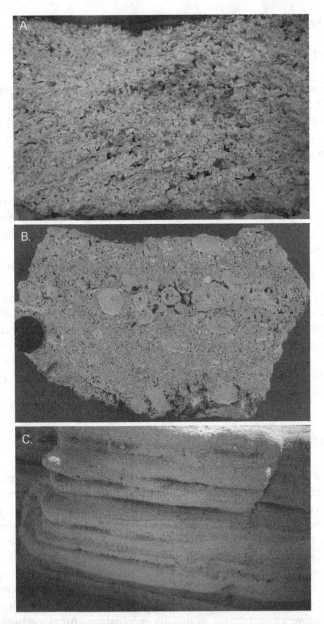

Figure 11.3 Outcrop and hand specimens of beachrock. (A) Freshly broken piece of beachrock with mostly sand-sized constituent particles. Kuramathi island, Rasdu Atoll, Maldives. Height of picture is 5 cm. (B) Beachrock with mostly cobble-sized fragments of coral. Long Cay, Glovers Reef, Belize. (C) Close-up of fossil beachrock outcrop from Ras Al-Julayah, southern Kuwait. Note how fine-grained layers are better cemented than coarse-grained ones. (For overview of outcrop see Figure 11.9A.)

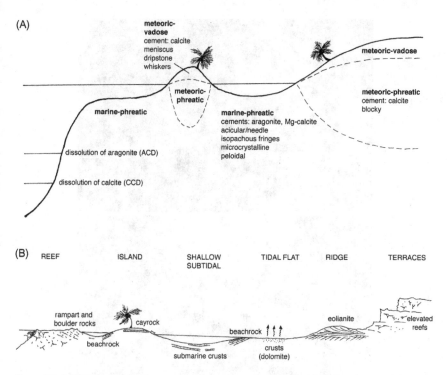

Figure 11.4 (A) Diagenetic environments and typical cements (modified from Flügel, 1978). (B) Formation of beachrock in relation to other cemented coastal deposits (schematic).

Less common are cement fringes of high-magnesium calcite blades or scalenohedral crystals (up to 70 μm long; Figure 11.7A) (e.g. Moore, 1973). Peloidal cements with approximately 40 μm diameter peloids (Figures 11.5B and 11.6C), and equant crystal crusts of high-magnesium calcite also occasionally occur (e.g. Pigott and Trumbly, 1985; Gischler and Lomando, 1997). Cements indicative of the marine-vadose diagenetic environment are not as common and may reflect marine cementation in the wave spray zone. In the vadose environment, typical meniscus or gravitational ('dripstone') cement fabrics are common (e.g. Land, 1970; Bricker, 1971). Highly acicular 'whisker' type cement fabrics are also observed (Bricker, 1971; Gischler and Lomando, 1997).

11.3.3 Meteoric cements in beachrock

Cements from meteoric waters may develop in beachrock during precipitation from percolating calcium-carbonate-saturated rain water during low

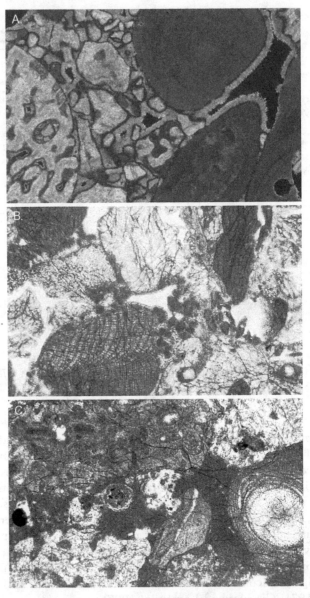

Figure 11.5 Photomicrographs of marine beachrock cements. (A) Isopachous crust of aragonite needle cement. Light particles are coral fragments; dark grains are coralline algae. Kubbar Island, Kuwait. Diameter of picture 140 µm. (B) Peloidal high-magnesium calcite cement. Constituent particles are coral and coralline algal fragments, as in (A). Long Cay, Glovers Reef, Belize. Diameter of picture is 1 mm. (C) Microcrystalline high-magnesium calcite cement among fragments of coral and coralline algae. Long Cay, Glovers Reef, Belize. Diameter of picture is 1 mm.

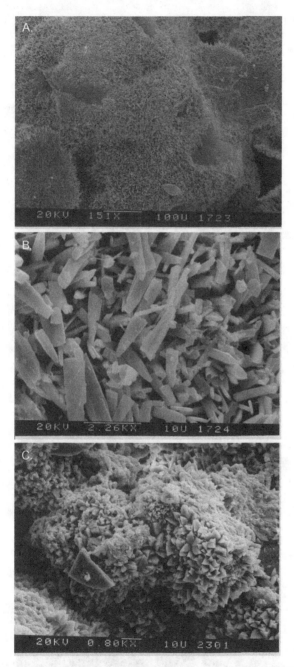

Figure 11.6 Scanning electron microscopy images of marine beachrock cements. (A) Overview of coatings of aragonite needle cement on constituent grains in beachrock. Note that parts of grain surfaces are not as yet covered by cement. Halfmoon Cay, Lighthouse Reef, Belize. (B) Close-up of aragonite needle cement. Aragonite crystals show hexagonal cross-sections and pointed ends. Same sample. (C) Peloidal magnesian calcite cement in beachrock. Cement peloids are composed of rhombohedral crystals on the outside. Hunting Cay, southern Belize Barrier Reef.

Figure 11.7 Scanning electron microscopy images of marine beachrock and meteoric cayrock cements. (A) Blades of high-magnesium calcite on constituent particles in beachrock. Northeast Sapodilla Cay, southern Belize Barrier Reef. (B) Dissolution cavity is lined with blocky and bladed low-magnesium calcite in cayrock. Cay Bokel, Turneffe Islands, Belize. (C) Low-magnesium calcite needle fibre (whisker) cement in cayrock. Harry Jones Point, Turneffe Islands, Belize.

tide exposure, or during longer times of subaerial exposure, for example in the case of Pleistocene lowstands of global sea-level. Cements are typically low-magnesium calcite with <5 mol% $MgCO_3$ (e.g. Tucker and Wright, 1990). These cements exhibit the above-mentioned dripstone and whisker fabrics when precipitated in the meteoric-vadose diagenetic environment (Figures 11.7C and 11.8A). Blocky cement fabrics are found in the meteoric-phreatic environment (e.g. Land, 1970; Vollbrecht and Meischner, 1993) (Figures 11.7B and 11.8B, C).

Beachrock exceptionally occurs on freshwater lake shores. For example, beachrock with highly acicular low-magnesium calcite cements is found on Michigan Marl Lake in North America indicating consolidation in the freshwater meteoric-vadose diagenetic environment (Binkley et al., 1980). Additionally, Jones et al. (1997) reported silica-bound beachrock occurrences on the shore of Lake Taupo, New Zealand. This is a special case, however, as hot spring water rich in amorphous silica flows into the lake, moves along the shore, and gets washed onshore to percolate through the beach to form beachrock cement.

11.3.4 Techniques of investigation

Various techniques are applied in order to decipher beachrock formation. These include polarisation microscopy of thin-sections to identify cements and constituent grains. Quantification of grains, cement and pore-space in thin-section by point-counting is desirable. However, very few studies exist as point-counting of tiny cement rims is not only time-consuming but also may be highly inaccurate (Halley, 1978). In any event, a comparison of thin-section grain-sizes with those of unconsolidated sediment sieving results requires conversion factors (Harrell and Eriksson, 1979). Staining of thin-sections with solutions such as alizarine or Feigl's solution are used in order to identify aragonite, low-magnesium calcite, and high-magnesium calcite cements (Friedman, 1959). Relative percentages of carbonate mineral abundance in ground samples may be estimated quantitatively by using X-ray diffractometry (Milliman, 1974, pp. 21–27). Moore (1973), Meyers (1987) and Strasser et al. (1989) have performed geochemical analyses of trace element abundances of calcium, magnesium and strontium in beachrock cements. Apart from calcium and magnesium, strontium is particularly useful, as marine cements have relatively high Sr contents (1000–2000 ppm in high-magnesium calcite and 7000–13,000 ppm in aragonite) as compared with meteoric cements (<500 ppm in low-magnesium calcite) (Kinsman, 1969). Stable isotopes of carbon and oxygen ($\delta^{13/12}C$ and $\delta^{18/16}O$) are also used in order to distinguish between marine-derived and meteoric-derived beachrock cements (e.g. Beier, 1985; Holail and Rashed, 1992; Vollbrecht and Meischner, 1996). Carbon and oxygen

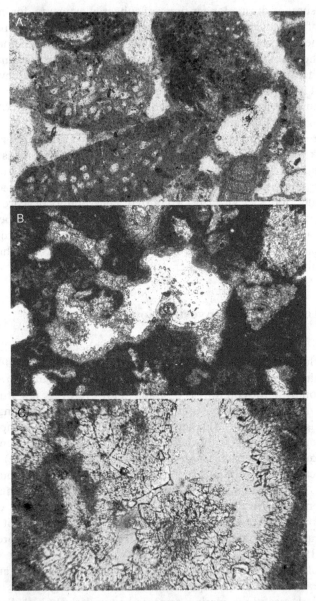

Figure 11.8 Photomicrographs of meteoric cements. (A) Low-magnesium calcite cement with meniscus fabrics in cayrock. Constituent grains are *Halimeda* fragments. Harry Jones Point, Turneffe Islands, Belize. Diameter of picture is 1 mm. (B) Blocky and bladed low-magnesium calcite line a dissolution cavity in cayrock. Cay Bokel, Turneffe Islands, Belize. Diameter of picture is 1.2 mm. (C) Close-up of same sample. Diameter of picture is 750 μm.

isotopes in marine-derived cements exhibit $\delta^{13}C$ values of about 0 to +5‰ PDB (i.e. relative to the PeeDee belemnite standard) and $\delta^{18}O$ values of about −1 to +1‰ PDB, whereas meteoric cements are significantly lighter by as much as 10‰ in carbon and by up to 5‰ in oxygen. Ideally, both trace element and isotope contents of cements and cement-precipitating pore fluids sampled in wells are measured and compared as, for example, in the beachrock studies of Moore (1973), Hanor (1978), Pigott and Trumbly (1985) and Meyers (1987). Radiometric age dating of beachrock may bring about important results with regard to time of formation (e.g. Siesser, 1974; McLean et al., 1978; Strasser et al., 1989; Kindler and Bain, 1993; Gischler and Lomando, 1997). However, these ages should only be used as a rough approximation because they reflect a mixture of time-averaged grains and different cement generations. With the development of mass spectrometric age dating, the very small amounts of sample required lends itself to the age dating of beachrock cements.

11.4 Chemical Considerations

In order to understand the formation of beachrock, some fundamentals of the marine carbonate system (CO_2–H_2O–$CaCO_3$) and calcium carbonate precipitation and dissolution have to be considered. Only limited aspects of the subject can be discussed here as the topic is complex and treated at length in various textbooks (e.g. Lippmann, 1973; Berner, 1980; Morse and Mackenzie, 1990).

Atmospheric carbon dioxide is dissolved in seawater where it forms carbonic acid, which dissociates into bicarbonate and carbonate ions, respectively (reaction 1). Dissociation is strongly dependent upon pH.

$$CO_2 + H_2O \leftrightarrow H_2CO_3 \leftrightarrow HCO_3^- + H^+ \leftrightarrow 2H^+ + CO_3^{2-} \qquad (1)$$

The alkalinity of seawater is defined by the concentrations of bicarbonate and carbonate ions as well as boric acid (HBO_3^{2-}). Carbonate alkalinity depends on the presence of HCO_3^- and CO_3^{2-}. Under normal marine conditions and a pH of about 8.2, HCO_3^- is the most abundant carbonate ion in seawater. Its concentration in normal marine seawater is $142\,mg\,L^{-1}$. Only the sulphate ($2712\,mg\,L^{-1}$) and chlorine ($19,353\,mg\,L^{-1}$) anions reach higher concentrations. Calcium has an average concentration of $413\,mg\,L^{-1}$ and is the third most common cation in seawater after sodium ($10,760\,mg\,L^{-1}$) and magnesium ($1294\,mg\,L^{-1}$) (Milliman, 1974, p. 6). Calcium may react with bicarbonate to form calcium carbonate (reaction 2). Quantitatively, aragonite and high-magnesium calcite or magnesian calcite (10–$20\,mol\%\ MgCO_3$) are important as marine precipitates. A magnesian calcite with $12\,mol\%$ $MgCO_3$ is thermodynamically equivalent to aragonite.

$$Ca^{2+} + 2HCO_3^- \leftrightarrow CaCO_3 + H_2O + CO_2 \qquad (2)$$

In the above equation the removal of CO_2, for example during photosynthesis, will enhance precipitation of $CaCO_3$. Likewise, the reverse reaction of respiration or decay of organic matter (for simplicity expressed as CH_2O) will result in dissolution of calcium carbonate (reaction 3).

$$CO_2 + H_2O \leftrightarrow CH_2O + O_2 \qquad (3)$$

Additionally, increases in temperature, salinity and pH will lead to calcium carbonate precipitation, whereas an increase in pressure will result in dissolution of $CaCO_3$. The latter is responsible for calcium carbonate dissolution in the deep sea.

As a consequence of the latitudinal temperature gradient, saturation of seawater with regard to $CaCO_3$ increases from polar regions towards the Equator. In tropical and subtropical latitudes, seawater is several times supersaturated with respect to $CaCO_3$ (Kleypas et al., 1999). Several processes prevent the spontaneous precipitation of calcium carbonate, including the overcoming of nucleation energy, i.e. the presence of abundant nuclei, and inhibitory effects of other ions such as sulphate, phosphate and magnesium, as well as organic matter. Phosphate adheres to crystal growth sites and prevents nucleation. Similarly, organic matter may coat growing crystals and further prevent nucleation (Suess, 1970). With regard to precipitation of calcite, the relatively small magnesium ion is surrounded by a large hydration sphere, which has to be dehydrated before Mg^{2+} can fit into the crystal lattice (Lippmann, 1973). As a consequence, a number of prerequisites apart from calcium carbonate supersaturation are necessary to allow for precipitation, even in low latitudes. These include stable substrates or, more specifically, stable pore space, rather low sedimentation rates, high rates of flushing by pore waters, and time. With regard to mineralogy it is far from clear as to what controls whether aragonite or magnesian calcite is precipitated in the marine environment (Tucker and Wright, 1990, p. 325). In general, however, aragonite precipitation is favoured as Mg^{2+} inhibits calcite growth. Given and Wilkinson (1985) argued that in addition to Mg/Ca ratios of pore waters, supply rates of CO_3^{2-} determine the mineralogy of marine precipitates. Supply rates of carbonate ions are high in intertidal beach and reef settings where CO_2 degasses due to high water agitation. As a consequence, HCO_3^- dissociates to CO_3^{2-}, which favours aragonite over magnesian calcite formation. Only when porosity becomes increasingly filled by aragonite growth and the rate of flushing of pore waters decreases will magnesian calcite start to precipitate (Given and Wilkinson, 1985; Tucker and Wright, 1990, pp. 326–327).

In meteoric waters, ion concentrations amount to $15 \, mg \, L^{-1}$ calcium and $58 \, mg \, L^{-1}$ bicarbonate as identified by Milliman (1974, p. 6) for average river water, i.e. ion concentrations are significantly lower as compared with seawater. Consequently, diagenesis in the meteoric realm is considered to be much slower than in the marine environment. However, several studies have shown that the degree of meteoric diagenesis is strongly dependent on the availability of cement-precipitating meteoric fluids and hence on rainfall rates and climate (e.g. Halley and Harris, 1979; Pierson and Shinn, 1985). The dominant carbonate mineral in the meteoric diagenetic environment is low-magnesium calcite as Mg/Ca ratios are usually very low in meteoric waters. Because meteoric fluids are commonly undersaturated with regard to calcium carbonate, dissolution is a common phenomenon. Aragonite and magnesian calcite are more soluble than calcite and are preferentially dissolved. Continued dissolution may lead to supersaturation of pore waters with respect to $CaCO_3$ resulting in the precipitation of low-magnesium calcite. During dissolution of aragonite, grains and cement may be entirely dissolved, with moulds filled in subsequently by calcite cement. Traces of the initial grain or cement architecture may be preserved when dissolution and reprecipitation act on a very fine scale. During the transformation from magnesian calcite to calcite, Mg^{2+} is leached from the crystal and the initial grain or cement architecture remains unaffected (Scoffin, 1987, pp. 108–111).

The significantly different cement morphologies observed, including acicular and microcrystalline cements, in the marine diagenetic realm versus equant crystals in the meteoric environment is commonly explained as a function of rates of cement growth and nucleation. Again, availability of CO_3^{2-} appears to play a major role, in that high supply rates of carbonate ions favour rapid growth in direction of the c-axis of the crystal-forming acicular cements, whereas low supply rates result in low growth-rates and equant crystal morphologies. In microcrystalline cements, growth rates are high, but due to the presence of many nuclei, crystals are not able to grow to larger sizes (Given and Wilkinson, 1985; Tucker and Wright, 1990, pp. 326–327).

11.5 Mechanisms of Formation

Beachrock forms due to cementation of beach sediments under a thin cover of sediment in the intertidal zone. The sediment cover is important because in order to precipitate cement, a stable substrate (i.e. stable pore space) is required. Erosion of the cover of unlithified sediment leads to beachrock exposure. The rapid formation of beachrock in the intertidal zone was well-known to 19th century writers (e.g. Moresby, 1835; Gardiner, 1898). For example, inhabitants of Indo-Pacific islands were known to harvest beachrock for building stone where new occurrences formed on the same beach within a few years. Further reports confirming rapid formation include the observation of 1919 hurricane beach deposits being cemented within a time period of one year (Daly, 1920), as well as the findings of World War II wreckage (e.g. Emery et al., 1954, p. 44; Frankel, 1968) and unaltered coconut husks (Easton, 1974) cemented into beachrock. High-resolution radiometric dating of cements in coral reef slopes has shown that marine aragonite cements, comparable to those observed in beachrock, reach growth rates of 80–100 $\mu m\,yr^{-1}$ (Grammer et al., 1993).

The mechanisms used to explain beachrock cementation include both physico-chemically and biologically induced precipitation of calcium carbonate. Physico-chemical models explain precipitation of carbonate cement by soaking of beaches during high tides and evaporation of sea water during low tides (e.g. Ginsburg, 1953a; Stoddart and Cann, 1965; Hanor, 1978). The strictly intertidal position of beachrock lends strong evidence to this model. Degassing of CO_2 during the process is of importance, and degassing has been suggested to be particularly important during the higher water agitation accompanying both falling (Meyers, 1987) and higher (Pigott and Trumbly, 1985) tides. Water agitation might indeed be an important factor in beachrock formation in addition to CO_2 pore water saturation states. McKee (1958) noted that appreciable amounts of beachrock are only found on seaward island sides of Kapingamarangi Atoll, Caroline Islands. A large-scale study on Belize shelf and atoll island beachrock also showed that the vast majority of beachrock exposures occurred on windward beaches of reef islands, suggesting that beachrock cementation only occurred where beaches experienced intensive and per-sistent flushing by sea water (Gischler and Lomando, 1997). Variations in pore water pressure during pumping of sea water through the beach may also lead to calcium carbonate precipitation. Beachrock formation in Grand Cayman island is interpreted by Moore (1973) to be largely a product of cementation under mixed meteoric-marine conditions. In contrast, Hanor (1978) used thermodynamic calculations to indicate that precipitation of beachrock cement cannot be induced by mixing of marine and meteoric water. Instead, these calculations favour CO_2 degassing as a means of supersaturating pore-water with calcium carbonate to the point of inducing cement precipitation.

There is also evidence suggesting the importance of biological processes in beachrock formation. Algal coatings may form zones of surface stability on the beach, stabilising grain movements below it so that they can be preferentially cemented (Davies and Kinsey, 1973). Withdrawal of CO_2 by photosynthesis may furthermore induce calcium carbonate precipitation. Krumbein (1979) noted the occurrence of high concentrations of organic matter during initial stages of Gulf of Aqaba beachrock formation and sug-gested that anaerobic and later aerobic decay processes were responsible for cement precipitation. Decay processes may include ammonification leading to higher pH, hydrolysis of urea (reaction 4) forming ammonium carbonate and eventually calcium carbonate (Macintyre and Marshall, 1985), or sulphate reduction under anoxic conditions, which elevates alkalinity and eventually may result in calcium carbonate precipitation (reaction 5).

$$CO_2 + 2NH_3 \leftrightarrow H_2O + CO(NH_2)_2$$
$$CO(NH_2)_2 + 2H_2O \rightarrow (NH_4)_2CO_3$$
$$(NH_4)_2CO_3 + Ca^{2+} + 2Cl^- \rightarrow CaCO_3 + 2NH_4Cl \qquad (4)$$

$$2CH_2O + SO_4^{2-} \rightarrow 2HCO_3^- + H_2S$$
$$2HCO_3^- + Ca^{2+} \leftrightarrow CaCO_3 + H_2O + CO_2 \qquad (5)$$

Further suggestions of a biological influence on beachrock formation are provided by Webb et al. (1999). These workers found microbial filaments in both beachrock cement and microbialites within beachrock cavities on Heron Island, Great Barrier Reef. They stress the importance of acid organic macromolecules with Ca^{2+} binding carboxyl groups in biofilms in forming nucleation zones for cements. Indeed, the common dark zones at the base of isopachous fringes of acicular aragonite beachrock cement may reflect the occurrence of organic mucus (Davies and Kinsey, 1973). Nuclei of marine peloids, which may also form beachrock cement, are composed of bacterial clumps according to the scanning electron microscopy (SEM) study of Chafetz (1986). Recently, Neumeier (1999) observed beachrock cementation by microcrystalline $CaCO_3$ in experiments inoculated by microorganisms, whereas sterile experiments exhibited only minor cementation.

11.6 Palaeoenvironmental Significance

As beachrock predominantly occurs in tropical and subtropical latitudes, one could be tempted to use pre-Quaternary occurrences as rough palaeogeographical indicators. Indeed, there are several reports of Tertiary (e.g. Tanner, 1956), Mesozoic (e.g. Moore et al., 1972, Cretaceous), Palaeozoic (e.g. Tanner, 1956, Permian) and even Precambrian beachrock occurrences (Donaldson and Ricketts, 1979). Even so, beachrock may also form in higher, temperate latitudes, and as far as 57°N, so its potential as a palaeogeographical indicator is rather limited.

The use of intertidal beachrock as a sea-level indicator has much higher palaeoenvironmental significance (Hopley, 1986). McLean et al. (1978), Kindler and Bain (1993) and Dickinson (2001), for example, reconstructed Holocene sea-level stands based on fossil beachrock occurrences in the Pacific and western Atlantic, respectively. Vollbrecht and Meischner (1993, 1996) reconstructed several Pleistocene highstands of sea-level on the island of Bermuda, based on detailed petrographical, mineralogical and geochemical investigations of stacked beaches, which were subject to frequently changing diagenetic environments as evidenced in multiple cement generations. Hopley (1986) pointed out, however, that there are also problems with beachrock as a sea-level indicator. First, beachrock occurrence is not restricted rigorously to the intertidal zone because it may still form in the supratidal marine spray zone and in the shallow subtidal realm. Second, the sea-level datum obtained from beachrock in macrotidal regions may not be very precise. Third, beachrock may be confused with other deposits, which are being cemented near or close to the coast, as discussed below (Figure 11.4B). Beachrock may also act as a means of preserving landforms. Bain (1988) and Cooper (1991) discussed how intertidal cementation may be a significant factor in protecting beaches from erosion. However, there are no systematic or quantitative studies on the long-term effect of coastal protection by beachrock.

11.7 Relationship to other Modes of Lithification

Sewell (1932, pp. 454–457) and Kuenen (1933, pp. 86–88; 1950, pp. 434–435) noted that intertidal 'beach sandstone' may be easily confused with other coastal cemented deposits and coined the term 'cay sandstone' for supratidal cemented deposits on reef islands (from the Spanish *cayo*, meaning island). Cay sandstone or cayrock (Figure 11.9B,C) may be distinguished from beachrock based on its horizontal bedding, very good sorting, and, most importantly, by the predominance of meteoric calcite cements (Hattin and Dodd, 1978; Marshall and Jacobsen, 1985; Gischler and Lomando, 1997). On reef islands, cayrock is often reported to be rich in phosphate due to downward percolation of porewaters through guano (Hopley, 1986; Baker et al., 1998). Another supratidal cemented deposit that is not uncommon close to beaches is aeolianite (e.g. Ward, 1973, 1975; Kindler and Hearty, 1995; see Chapter 5). As in cayrock, meteoric calcite cementation predominates in this geochemical sediment type. Sedimentary structures resulting from wind transport of constituent grains (Figure 11.10A), such as cross-stratification, are also important identifying features, and are not observed in cayrock. Hardened crusts and layers are found, for example, in the supratidal flats of the Bahamas (Figure 11.10C). These contain calcite, aragonite, and dolomite, and form due to evaporation and precipitation during extensive periods of subaerial exposure (Shinn et al., 1965, 1969). The formation of dolomite $[CaMg(CO_3)_2]$ on tidal flats is facilitated because precipitation of gypsum $(CaSO_4 \cdot 2H_2O)$ increases the Mg/Ca ratio and decreases the sulphate concentration of fluids. All the above-mentioned supratidal occurrences usually occur close to beachrock. However, they may be readily distinguished from beachrock by either their occurrence, texture, mineralogy, diagenesis or a combinations of these.

Figure 11.9 Photographs of outcrops of beachrock and other cemented beach deposits. (A) Middle Holocene outcrop of fossil beachrock, in part eroded from undercutting and which probably documents a higher Holocene sea-level. Near Ras Al-Julayah, southern Kuwait. (B) Outcrop of cayrock, in which exposure resulted from coastal erosion during a hurricane. Cay Bokel, Turneffe Islands, Belize. (C) Outcrop of cayrock on the western shore of the island of Hurasdhoo, lagoon of Ari Atoll, Maldives.

Figure 11.10 Photographs of outcrops of other cemented beach deposits. (A) Pleistocene coastal deposit on the coast of Cat Cay, western margin of Great Bahama Bank. Note inclined bedding and root structures. (B) Elevated Holocene reef terraces on the southwestern coast of Barbados, West Indies. (C) Indurated dolomite crust on tidal flats. Near western coast of Andros Island, Bahamas.

Certain intertidal and subtidal cementation phenomena are not as easy to distinguish from beachrock formation. Rampart and boulder rocks in Pacific reefs, mainly consisting of cemented coral branches and fragments larger than 10 cm, sometimes exhibit bedding inclined towards the coast (McLean et al., 1978; Scoffin and McLean, 1978), and may be confused with beachrock. Rampart and boulder rocks usually occur intertidally, but the upper flat surface of the deposit reaches into the supratidal zone. Similarly, elevated reefs and reef terraces have been confused with intertidal beachrock (Hopley, 1986) (Figure 11.10B). Finally, cemented layers of sediment that form on tidal flats (as reported by Taylor and Illing, 1969), and on the shallow seafloor (as observed by Shinn (1969) in the Arabian-Persian Gulf) can be confused with intertidal beachrock. A general identification problem with these intertidal and subtidal occurrences is not only their close proximity, but sometimes even transitional nature to beachrock, and, most important, the identical diagenetic marine-phreatic characteristics of cements.

11.8 Directions for Future Research

The number of studies on beachrock is enormous. For example, the geosciences publication search engine GeoRef® that covers the time span from 1785 until today currently lists close to 500 papers on beachrock. A great number of these have identified controlling factors in beachrock formation including physico-chemical and biological processes and combinations of both. However, unanswered questions still remain. For example, if physico-chemical (inorganic) processes are generally sufficient to induce calcium carbonate cementation, why are beachrock outcrops so patchy in distribution, even on the windward beaches of reef islands? The same patchy distribution argues against organic processes as a prerequisite for precipitation of $CaCO_3$ cement, unless and until it can be shown that specific organic processes are peculiar to beachrock formation and therefore are not ubiquitous in occurrence as is the presence of microbes. It would seem that aspects of both inorganic and organic processes have to be taken into account to explain the origin of beachrock. To date, however, exclusively inorganic or organic models do not suffice to explain the distribution of intertidal marine cementation. Further elucidation of the controlling factors will need to combine field observations including sampling of pore waters from wells, detailed petrographic analyses of constituent grains and cements, mineralogy and geochemistry of cements and pore-water, age dating of cements and microbiological investigations. Further directions of research should also include topics such as rates of beachrock destruction and the effect of coastal protection through beachrock, because systematic studies on these society-relevant aspects of beachrock are largely lacking.

References

Alexandersson, T. (1972) Intergranular growth of marine aragonite and Mg-calcite: evidence of precipitation from supersaturated seawater. *Journal of Sedimentary Petrology* **42**, 441–460.

Bain, R.J. (1988) Exposed beachrock: its influence on beach processes and criteria for recognition. *Proceedings of the 4th Symposium on Geology of the Bahamas*, San Salvador, pp. 33–44.

Baker, J.C., Jell, J.S., Hacker, J.L.F. & Baublys, K.A. (1998) Origin of recent insular phosphate rock on a coral cay – Raine Island, northern Great Barrier Reef, Australia. *Journal of Sedimentary Research* **68**, 1001–1008.

Bathurst, R.G.C. (1966) Boring algae, micrite envelopes and lithification of molluscan biosparites. *Journal of Geology* **5**, 15–32.

Beier, J.A. (1985) Diagenesis of Quaternary Bahamian beachrock: petrographic and isotopic evidence. *Journal of Sedimentary Petrology* **55**, 755–761.

Berner, R.A. (1980) *Early Diagenesis, a Theoretical Approach*. Princeton, NJ: Princeton University Press.

Binkley, K.L., Wilkinson, B.H. & Owen, R.M. (1980) Vadose beachrock cementation along a southeastern Michigan marl lake. *Journal of Sedimentary Petrology* **50**, 953–962.

Bricker, O.P. (Ed.) (1971) *Carbonate Cements. Part I. Beachrock and Intertidal Cement*. Baltimore: Johns Hopkins Press, pp. 1–43.

Chafetz, H.S. (1986) Marine peloids: a product of bacterially induced precipitation of calcite. *Journal of Sedimentary Petrology* **56**, 812–817.

Chamisso, A.V. (1821) Bemerkungen und Ansichten von dem Naturforscher der Expedition. In: Kotzebue, O.V. (Ed.) *Entdeckungs-Reise in die Süd-See und nach der Berings-Strasse zur Erforschung einer nordöstlichen Durchfahrt*, Vol. 3. Weimar: Hofmann.

Cooper, J.A.G. (1991) Beachrock formation in low latitudes: implications for coastal evolutionary models. *Marine Geology* **98**, 145–154.

Daly, R.A. (1920) Origin of beachrock. *Carnegie Institution Washington Yearbook* **18**, 192.

Dana, J.D. (1875) *Corals and Coral Islands*. London: Sampson Low, Marston, Low and Searle.

Darwin, C.R. (1842) *Structure and Distribution of Coral Reefs*. London: Smith Elder.

Davies, P.J. & Kinsey, D.W. (1973) Organic and inorganic factors in recent beachrock formation, Heron Island, Great Barrier Reef. *Journal of Sedimentary Petrology* **43**, 59–81.

Dickinson, W.R. (2001) Paleoshoreline record of relative Holocene sea levels on Pacific islands. *Earth-Science Reviews* **55**, 191–234.

Donaldson, J.A. & Ricketts, B.D. (1979) Beachrock in Proterozoic dolostone of the Belcher Islands, Northwest Territories, Canada. *Journal of Sedimentary Petrology* **49**, 1287–1294.

Easton, W.H. (1974) An unusual inclusion in beachrock. *Journal of Sedimentary Petrology* **44**, 693–694.

El-Sayed, M.K. (1988) Beachrock cementation in Alexandria, Egypt. *Marine Geology* **80**, 29–35.

Emery, K.O., Tracey, J.I. & Ladd, H.S. (1954) Geology of Bikini and nearby atolls. *U.S. Geological Survey Professional Paper*, **260-A**.

Flügel, E. (1978) *Microfacies Analysis of Limestones*. Berlin: Springer.

Frankel, E. (1968) Rate of formation of beachrock. *Earth and Planetary Science Letters* **4**, 439–440.

Friedman, G.M. (1959) Identification of carbonate minerals by staining methods. *Journal of Sedimentary Petrology* **28**, 87–97.

Gardiner, J.S. (1898) The coral reefs of Funafuti, Rotuma and Fiji together with some notes on the structure and formation of coral reefs. *Proceedings of the Cambridge Philosophical Society* **9**, 417–503.

Ginsburg, R.N. (1953a) Beachrock in south Florida. *Journal of Sedimentary Petrology* **23**, 85–92.

Ginsburg, R.N. (1953b) Intertidal erosion on the Florida Keys. *Bulletin of Marine Science Gulf and Caribbean* **3**, 55–69.

Gischler, E. & Lomando, A.J. (1997) Holocene cemented beach deposits in Belize. *Sedimentary Geology* **110**, 277–297.

Given, R.K. & Wilkinson, B.H. (1985) Kinetic control of morphology, composition, and mineralogy of abiotic sedimentary carbonates. *Journal of Sedimentary Petrology* **55**, 109–119.

Grammer, G.M., Ginsburg, R.N., Swart, P.K., McNeill, D.F., Jull, A.J. & Prezbindowsky, D.R. (1993) Rapid growth rates of syndepositional marine aragonite cements in steep marginal slope deposits, Bahamas and Belize. *Journal of Sedimentary Petrology* **63**, 983–989.

Halley, R.B. (1978) Estimating pore and cement volumes in thin section. *Journal of Sedimentary Petrology* **48**, 642–650.

Halley, R.B. & Harris, P.M. (1979) Freshwater cementation of a 1,000 year old oolite. *Journal of Sedimentary Petrology* **49**, 969–988.

Hanor, J.S. (1978) Precipitation of beachrock cements: mixing of marine and meteoric waters *vs.* CO_2-degassing. *Journal of Sedimentary Petrology* **48**, 489–501.

Harrell, J.A. & Eriksson, K.A. (1979) Empirical conversion equations for thin-section and sieve derived distribution parameters. *Journal of Sedimentary Petrology* **49**, 273–280.

Harris, P.M., Kendall, G.S.C. & Lerche, I. (1985) Carbonate cementation – a brief review. In: Schneidermann, N. & Harris, P.M. (Eds) *Carbonate Cements*. Special Publication 36. Tulsa, OK: Society of Economic Paleontologists and Mineralogists, pp. 79–95.

Hattin, D.E. & Dodd, J.R. (1978) Holocene cementation of carbonate sediments in the Florida Keys. *Journal of Sedimentary Petrology* **48**, 307–312.

Holail, H. & Rashed, M. (1992) Stable isotopic composition of carbonate-cemented recent beachrock along the Mediterranean and the Red Sea coasts of Egypt. *Marine Geology* **106**, 141–148.

Hopley, D. (1986) Beachrock as sea-level indicator. In: van de Plassche, O. (Ed.) *Sea-level Research*. Great Yarmouth: Galliard Printers, pp. 157–173.

Jones, B., Rosen, M.R. & Renaut, R.W. (1997) Silica-cemented beachrock from Lake Taupo, north island, New Zealand. *Journal of Sedimentary Research* **67**, 805–814.

Kindler, P. & Bain, R.J. (1993) Submerged upper Holocene beachrock from San Salvador Island, Bahamas: implications for recent sea-level history. *Geologische Rundschau* **82**, 242–247.

Kindler, P. & Hearty, P.J. (1995) Pre-Sangamonian eolianites in the Bahamas? New evidence from Eleuthera Island. *Marine Geology* **127**, 73–86.

Kinsman, D.J.J. (1969) Interpretation of Sr^{+2} concentrations in carbonate minerals and rocks. *Journal of Sedimentary Petrology* **39**, 486–508.

Kleypas, J.A., Buddemeier, R.W., Archer, D., Gattuso, J.-P., Langdon, C. & Opdyke, B.N. (1999) Geochemical consequences of increased atmospheric carbon dioxide on coral reefs. *Science* **284**, 118–120.

Kneale, D. & Viles, H.A. (2000) Beach cement: incipient $CaCO_3$-cemented beachrock development in the upper intertidal zone, north Uist, Scotland. *Sedimentary Geology* **132**, 165–170.

Knox, G.J. (1973) An aragonite-cemented volcanic beach rock near Bilbao, Spain. *Geologie en Mijnbouw* **53**, 9–12.

Krumbein, W.E. (1979) Photolithotropic and chemoorganotrophic activity of bacteria and algae as related to beachrock formation and degradation (Gulf of Aqaba, Sinai). *Journal of Geomicrobiology* **1**, 139–203.

Kuenen, P.H. (1933) *Geology of coral reefs. Snellius Expedition V. Geological Results, Part 2.* Utrecht: Kemink en Zoon.

Kuenen, P.H. (1950) *Marine Geology.* New York: Wiley.

Land, L.S. (1970) Phreatic versus vadose meteoric diagenesis of limestones: evidence from a fossil water table. *Sedimentology* **14**, 175–185.

Lippmann, F. (1973) *Sedimentary Carbonate Minerals.* Berlin: Springer-Verlag.

Longman, M.W. (1980) Carbonate diagenetic textures from nearsurface diagenetic environments. *Bulletin of the American Association of Petroleum Geologists* **64**, 461–487.

Lyell, C. (1832) *Principles of Geology, being an Attempt to Explain the Former Changes of the Earth's Surface, by Reference to Causes now in Operation,* Vol. 2, London: John Murray.

Macintyre, I.G. & Marshall, J.F. (1985) Submarine lithification in coral reefs: some facts and misconceptions. *Proceedings 6th International Coral Reef Symposium* **1**, 263–272.

Marshall, J.F. & Jacobsen, G. (1985) Holocene growth of a mid-Pacific atoll: Tarawa, Kiribati. *Coral Reefs* **4**, 11–17.

McKee, E.D. (1958) Geology of Kapingamarangi Atoll, Caroline Islands. *Bulletin of the Geological Society of America* **69**, 241–278.

McLean, R.F. (1967) Origin and development of ridge-furrow system in beachrock in Barbados, West Indies. *Marine Geology* **5**, 181–193.

McLean, R.F., Stoddart, D.R., Hopley, D. & Polach, H. (1978) Sea level change in the Holocene on the northern Great Barrier Reef. *Philosophical Transactions of the Royal Society of London* **A291**, 167–186.

Meyers, J.H. (1987) Marine vadose beachrock cementation by cryptocrystalline magnesian calcite – Maui, Hawaii. *Journal of Sedimentary Petrology* **57**, 755–761.

Milliman, J.D. (1974) *Recent Sedimentary Carbonates. Part 1. Marine Carbonates.* Berlin: Springer.

Moore, C.H.Jr. (1973) Intertidal carbonate cementation on Grand Cayman, West Indies. *Journal of Sedimentary Petrology* **43**, 591–602.

Moore, C.H.Jr., Smitherman, J.M. & Allen, S.H. (1972) Pore system evolution in a Cretaceous carbonate beach sequence. *International Geological Congress* **24**(6), 124–136.

Moresby, R. (1835) Extracts from Commander Moresby's report on the northern atolls of the Maldives. *Journal of the Royal Geographical Society* **5**, 398–404.

Morse, J.W. & Mackenzie, F.T. (1990) *Geochemistry of Sedimentary Carbonates.* Developments in Sedimentology 48. Amsterdam: Elsevier.

Neumann, A.C. (1966) Observations on coastal erosion in Bermuda and measurements of the boring rate of the sponge *Cliona lampa*. *Limnology Oceanography* **11**, 92–108.

Neumeier, U. (1999) Experimental modelling of beachrock cementation under microbial influence. In: Camoin, G.F. (Ed.) *Microbial Mediation in Carbonate Diagenesis. Sedimentary Geology* **126**, 35–46.

Pierson, B.J. & Shinn, E.A. (1985) Cement distribution and carbonate mineral stabilization in Pleistocene limestones of Hogsty Reef, Bahamas. In: Schneidermann, N. & Harris, P.M. (Eds) *Carbonate Cements*. Special Publication 36. Tulsa, OK: Society of Economic Paleontologists and Mineralogists, pp. 153–168.

Pigott, J.D. & Trumbly, N.I. (1985) Distribution and origin of beachrock cements, Discovery Bay (Jamaica). *Proceedings of the 5th International Coral Reef Symposium*, Tahiti, Vol. 3, pp. 241–247.

Purdy, E.G. (1968) Carbonate diagenesis: an environmental survey. *Geologica Romana* **7**, 183–227.

Reid, R.P. & Macintyre, I.G. (2000) Microboring versus recrystallization: further insight into the micritization process. *Journal of Sedimentary Research* **70**, 24–28.

Revelle, R. & Emery, K.O. (1957) Chemical erosion of beach rock and exposed reef rock. *U.S. Geological Survey Professional Paper* **260-T**, 699–709.

Russell, R.J. (1962) Origin of beach rock. *Zeitschrift für Geomorphologie* **6**, 1–16.

Scoffin, T.P. (1987) *An Introduction to Carbonate Sediments and Rocks*. London: Blackie.

Scoffin, T.P. & McLean, R.F. (1978) Exposed limestones of the northern province of the Great Barrier Reef. *Philosophical Transactions of the Royal Society of London* **A291**, 119–138.

Scoffin, T.P. & Stoddart, D.R. (1983) Beachrock and intertidal cements. In: Goudie, A.S. & Pye, K. (Eds) *Chemical Sediments and Geomorphology*. London: Academic Press, pp. 401–425.

Sewell, R.B.S. (1932) The coral coasts of India. *Geographical Journal* **79**, 449–465.

Shinn, E.A. (1969) Submarine lithification of Holocene carbonate sediments in the Persian Gulf. *Sedimentology* **12**, 109–144.

Shinn, E.A. (2004) A geologist's adventures with Bimini beachrock and Atlantis true believers. *Skeptical Inquirer* **January/February**, 38–44.

Shinn, E.A., Ginsburg, R.N. & Lloyd, R.M. (1965) Recent supratidal dolomite from Andros Island, Bahamas. In: Pray, L.C. & Murray, R.C. (Eds) *Dolomitization and Limestone diagenesis*. Special Publication 13. Tulsa, OK: Society of Economic Paleontologists and Mineralogists, pp. 112–123.

Shinn, E.A., Lloyd, R.M. & Ginsburg, R.N. (1969) Anatomy of a modern carbonate tidal-flat, Andros Island, Bahamas. *Journal of Sedimentary Petrology* **39**, 1202–1228.

Siesser, W.G. (1974) Relict and recent beachrock from southern Africa. *Geological Society of America Bulletin* **85**, 1849–1854.

Stoddart, D.R. & Cann, J.R. (1965) Nature and origin of beachrock. *Journal of Sedimentary Petrology* **35**, 243–273.

Strasser, A. & Davaud, E. (1986) Formation of Holocene limestone sequences by progradation, cementation, and erosion: two examples from the Bahamas. *Journal of Sedimentary Petrology* **56**, 422–428.

Strasser, A., Davaud, E. & Jedoui, Y. (1989) Carbonate cements in Holocene beachrock: example from Baihret el Biban, southeastern Tunisia. *Sedimentary Geology* **62**, 89–100.

Suess, E. (1970) Interaction of organic compounds with calcium carbonate – I. Association phenomena and geochemical implications. *Geochimica et Cosmochimica Acta* **34**, 157–168.

Tanner, W.F. (1956) Examples of probably lithified beachrock. *Journal of Sedimentary Petrology* **26**, 307–312.

Taylor, J.C.M. & Illing, L.V. (1969) Holocene intertidal calcium carbonate cementation, Qatar, Persian Gulf. *Sedimentology* **12**, 69–107.

Tucker, M.E. & Wright, V.P. (1990) *Carbonate Sedimentology*. Oxford: Blackwell.

Vollbrecht, R. & Meischner, D. (1993) Sea level and diagenesis: a case study on Pleistocene beaches, Whalebone Bay, Bermuda. *Geologische Rundschau* **82**, 248–262.

Vollbrecht, R. & Meischner, D. (1996) Diagenesis in coastal carbonates related to Pleistocene sea level, Bermuda platform. *Journal of Sedimentary Research* **66**, 243–258.

Ward, W.C. (1973) Influence of climate on the early diagenesis of carbonate eolianites. *Geology* **1**, 171–174.

Ward, W.C. (1975) Petrology and diagenesis of carbonate eolianites of northeastern Yucatán peninsula. *American Association Petroleum Geologists Studies in Geology* **2**, 500–571.

Webb, G.E., Jell, J.S. & Baker, J.C. (1999) Cryptic intertidal microbialites in beachrock, Heron Island, Great Barrier Reef: implications for the origin of microcrystalline beachrock cement. In: Camoin, G.F. (Ed.) *Microbial Mediation in Carbonate Diagenesis. Sedimentary Geology* **126**, 317–334.

Zenkovitch, V.P. (1967) *Processes of Coastal Development*. Edinburgh: Oliver & Boyd.

Chapter Twelve

Sodium Nitrate Deposits and Efflorescences

Andrew S. Goudie and Elaine Heslop

12.1 Introduction

Sodium nitrate deposits are less widespread than the other major types of chemical crust or sediment that are found in different parts of the world. Indeed, the only deposits of any great spatial extent and thickness are those of portions of the hyper-arid Atacama Desert in South America. These materials are, in the words of Ericksen (1983, p. 366), 'so extraordinary that, were it not for their existence, geologists could easily conclude that such deposits could not form in nature'. Ericksen notes a series of features of the deposits that defy rational explanation. These include their restricted distribution in a very salty area, their occurrence in a wide variety of topographic settings, the abundance of nitrate minerals, and the presence of a series of other minerals, such as perchlorate, that do not occur in any other saline complexes and the origin of which is obscure.

Sodium nitrate is also known from dryland situations other than the Atacama, although nowhere does it attain the same significance. It is, for example, known from numerous caves in the southwestern USA (Hill, 1977, 1981; Hill and Eller, 1977) and Argentina (Forti and Buzio, 1990), and on open sites from the Mojave Desert of California (Ericksen et al., 1988; Böhlke et al., 1997), the Basin and Range Province of the USA (Kirchner, 1996), and numerous locations in Antarctica (Johannesson and Gibson, 1962; Claridge and Campbell, 1968; Keys and Williams, 1981; Matsuoka, 1995). It has also been recorded from various building surfaces in Switzerland (Arnold and Zehnder, 1988), Bukhara in Uzbekistan (Cooke, 1994) and from Cologne in Germany (Laue et al., 1996). Most deposits are not by any means pure sodium nitrate, and Ericksen et al. (1988), for example, report that in the caliche deposits of the Atacama, the principal minerals in addition to nitratite (sodium nitrate) are halite, darapskite, glauberite, gypsum, anhydrite, borax, tincalconite and trona. Because of the predominant development of nitrate deposits in the Atacama, this chapter will concentrate on them, but more general aspects of the nature of sodium nitrate deposits and their role in weathering will also be considered.

12.2 The Distribution, Field Occurrence and Geomorphological Relations of the Atacama Nitrate Deposits

The sodium nitrate deposits of the Atacama are called caliche. This leads to some confusion given that, in North America, calcium carbonate crusts are also given this name (Goudie, 1972). Carbonate crusts are now more generally and sensibly called calcrete (Chapter 2). Caliche occurs primarily in the provinces of Tarapacá and Antofagasta in northern Chile (Figure 12.1), although it also extends northwards into Peru. The high grade salitre deposits have been mined commercially since the nineteenth century and processed at many officinas (Figure 12.2) linked by an extensive network of railways. Production started in the 1820s, reached a peak in the 1880s, when the area accounted for c. 90% of total world production (during which in some years 3 million metric tonnes of sodium nitrate were exported), and then fell markedly after World War 1 when synthetic fertilizers could be manufactured by newly available technologies. Total nitrate production from 1830 to 1970 amounted to 135 million tonnes of $NaNO_3$. By 1980, the Chilean nitrates held only 0.14% of the world market for fixed nitrogen. Some production still goes on, however, and involves the extraction of secondary materials, such as iodates, lithium and potassium nitrate.

The nitrates occur as a band, up to 30 km wide, along the eastern (inland) side of the Coastal Range (Ericksen, 1981). They extend from c. 19°30'S to 26°S, a distance of about 700 km. The lower grade deposits are more extensive than this, and much of the Coastal Range is encrusted with nitrate-bearing saline-cemented regolith (Figure 12.3). The deposits extend over a considerable altitudinal range (some up to 4000 m above sea level, but most below 2000 m). They occur in a wide range of topographic situations from tops of hills and ridges to the centres of broad valleys, many of which are occupied by closed salt lakes (salars). Some of the salars contain nitrate crusts, including the Pampa Blanca and Pampa Lina in the Basquedano district and the Salars del Carmen and de Lagunas. However, nitrates are especially well developed on the lower slopes of hills and piedmont plains. The deposits also occur in and on all types of rock and superficial sediment in the area, and there appears to be little lithological control of their different types and mineral assemblages. High-grade nitrate is found on such diverse rock types as granite, andesite, rhyolite, limestone, sandstone and shale, but is most extensive in unconsolidated surface materials.

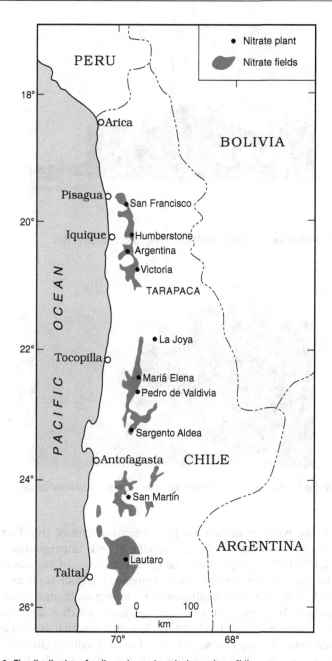

Figure 12.1 The distribution of sodium nitrate deposits in northern Chile.

Figure 12.2 An abandoned *officina* in the Atacama near Iquique, northern Chile.

Figure 12.3 The hyperarid, salt mantled landscape of the Atacama inland from Iquique.

Two major types of nitrate ore have been recognised (Ericksen, 1981, 1983): regolith cement ('alluvial caliche') and impregnated bedrock ('bedrock caliche'). A typical alluvial caliche consists of various zones. At the top is *chuca*, a powdery to poorly cemented surface layer around 10–30 cm thick. This consists of silt, sand, rock fragments and some gypsum and anhydrite. It consists of a surface soil from which a large portion of the soluble saline materials has been leached and to which aeolian additions have taken place. It may contain a layer of soluble salts, including thenardite, bloedite and humberstonite. Beneath this, typically, is the 0.5–2.0 m thick transitional *costra* zone, which is moderately to firmly cemented and has polygonal structures. It in turn is underlain by the 1–3 m thick 'caliche' zone, which is firmly cemented and contains layers of high purity white

Figure 12.4 Cemented regolith on a raised beach south of Iquique. Beer can for scale. Note the disintegrated beach pebbles.

nitrate-rich material called 'caliche blanco'. The caliche grades downwards into *conjelo* (saline-cemented regolith containing little sodium nitrate) or into *coba* (loose, uncemented regolith).

Bedrock caliche consists of saline impregnations and layers (from a few centimetres to several tens of centimetres in thickness) that have gradually forced open cracks and fissures in the rock by crystallisation processes (Figure 12.4). In some cases the fractured bedrock may be so transformed that it is converted into a mosaic of isolated rock fragments in a matrix of 'caliche blanco'. The micromorphology of nitrates from the Antofagasta region is described and illustrated by Pueyo et al. (1998), but overall the amount of information available on this aspect of caliche deposits is modest. The main features that can be noted, however, are:

1 the frequent complex mix of crystals of different composition, with, for example, layers of relatively pure sulphate minerals (humberstonite, bloedite, etc.);
2 evidence for multiple episodes of precipitation and dissolution (e.g. solution cavities), and formation of desiccation polygons;
3 evidence for repeated fracturing and displacive growth, including the incorporation of microscopic fragments of saline minerals into larger crystals or grains of soda niter and halite;
4 incorporation of silt and other detrital material, together with remains of birds (feathers, guano, egg fragments, etc.) (Ericksen, 1981, p. 18);
5 the presence of veins of purer nitrate (Ericksen and Mrose, 1972);
6 evidence of extensive fracturing in the vicinity of faults;

7   the formation of hard, spheroidal cakes of anhydrite, called *losa*, pro-
    duced at the surface by prolonged slow leaching (Ericksen, 1981);
8   the presence of small, chalcedony nodules.

12.3 The Chemistry and Mineralogy of the Nitrate

Some of the Atacama nitrate deposits are relatively pure sodium nitrate
(nitratite or soda niter). This is the case with some near-surface veins in
bedrock (Ericksen and Mrose, 1972) and with some *caliche blanco*. However,
most of the deposits are impure and contain substantial amounts of other
salts, including a range of sulphates, chlorides, nitrates, borates, iodates,
perchlorate and chromates. Ericksen (1981), Pueyo et al. (1998) and Searl
and Rankin (1993), provide a detailed list of such minerals (Table 12.1),
although some of Searl and Rankin's identifications have been challenged
by Ericksen (1994). One of the nitrate minerals, humberstonite, derives its
name from Humberstone, one of the nitrate towns in the area (Mrose et al.,
1970). Of the impurities, sodium chloride (halite) appears to be the most
important and may often exceed the percentage of nitrate (see e.g., Penrose,
1910, p. 14). A detailed chemical analysis of the Maria Elena nitrate deposits is provided by Collao et al. (2002). They report (p. 181):

> The ore is largely composed of nitratite and halite. Nitrate contents vary
> between 7.5 to more than 20 wt% and NaCl between 8.6 to 40 wt%. Locally
> abundant are the sulphates thenardite, anhydrite, bloedite, polyhalite, hexa-
> hidrite and glauberite, which reflect part of 7.5 to 19.7 wt% of Na_2SO_4.

Sodium nitrate is highly soluble in comparison with many other salts.
At 35°C its solubility in water is 49.6%, whereas that of sodium sulphate

Table 12.1 Salt minerals present in caliche deposits of the Atacama (after Ericksen, 1981)

Halides	Nitrates	Iodates and chromates	Borates	Sulphates
Halite	Soda niter, Niter, Darapskite, Humberstonite	Lautarite, Bruggenite, Dietzite, Tarapacacite, Lopezite	Ulexite, Probertite, Ginorite, Hydroboracite, Kaliborite	Thenardite, Glauberite, Bloedite, Kieserite, Epsomite, Gypsum, Anhydrite, Bassanite

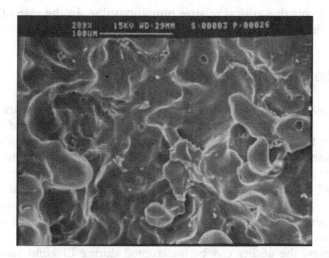

Figure 12.5 The hygroscopic and deliquescent nature of sodium nitrate crystals observed under the scanning electron microscope in the laboratory. The scale bar is 100 μm.

is 33.4%, of sodium chloride 26.6% and of calcium sulphate just 0.21%. This has two implications. One of these is that it can only persist in extremely arid environments, and the other is that large volumes of the salt can be precipitated within rock pores, where it can cause severe rock disintegration and brecciation (q.v.). Sodium nitrate is also hygroscopic (it can absorb moisture from the air) and deliquescent (it can dissolve in that moisture). When the relative humidity exceeds a critical level, water condenses on the surface of the mineral and dissolves it in that water. Then, if the critical relative humidity decreases below that critical level, the nitrate will effloresce or crystallize again. Thus cycling across this critical relative humidity is a major control of the frequency of salt crystallisation cycles in the salt weathering process. Figure 12.5 provides an illustration of the hygroscopic and deliquescent nature of sodium nitrate, showing amorphous sodium nitrate gel which has begun to deliquesce from crystals that have been grown on calcite and have absorbed moisture from the air under ambient conditions of relative humidity and temperature found in the laboratory in Oxford.

The hygroscopic and deliquescent properties of sodium nitrate also mean that nitrate speleothems, of which there are some examples, are transient features that come and go with the seasons (Hill and Forti, 1997). The threshold temperature for the transition from deliquescence to crystallisation takes place at 42°C (Pantony, 1961). The critical relative humidity (RH) at which the liquid phase starts to appear on crystals of sodium nitrate is 73.8% at 25°C for pure sodium nitrate, but impurities such as sodium chloride lower this critical relative humidity value. In the case of sodium chloride the value is 62.2% RH (Tereschenko and Malyutin, 1985).

12.4 The Aridity and Age of the Atacama

Because of the ready solubility of sodium nitrate, it has been argued that it exists to such an extent in the Atacama as it is drier than any of the other deserts in the world. Sub-tropical subsiding air, the upwelling of cold off-shore waters, and the rain-shadow effect of the Andes, create this aridity. The most intense aridity occurs in northern Chile, which receives less than 10 mm of rainfall per annum. Indeed, some stations, such as Calama, receive on average less than 2 mm. The climate station at Quillagua (mean annual rainfall 0.05 mm) can lay claim to be the driest place on Earth (Middleton, 2001).

However, the aridity can be interrupted during El Niño years, when large rainfall events occur (Bendix et al., 2000), causing extensive flooding (Magilligan and Goldstein, 2001). These will cause dissolution and remobilisation of soluble materials (Ericksen, 1994). There is also evidence from the Atacama and from the neighbouring Altiplano that there have been some relatively moist phases during the Late Pleistocene and early Holocene (see e.g., Betancourt et al., 2000; Holmgren et al., 2001). These too may have played a role in the solution, remobilisation and translocation of saline material.

Another of the reasons that has been given for the development of nitrates in the Atacama is that the desert is one of the oldest of the world's deserts (Clarke, 2006) and that thus there has been an extended period for large amounts of nitrate to build up even from modest rates of inputs. There is evidence that the Atacama's aridity may have started in the Eocene and became profound in the middle to late Miocene (Alpers and Brimhall, 1988). There were perhaps two crucial factors responsible for the initiation of the desert: the uplift of the Andes Cordillera during the Oligocene and early Miocene, and the development around 15–13 Ma of the cold offshore Peruvian current as a result of ice build-up in Antarctica. The former produced a rain-shadow effect and helped to stabilize the southeastern Pacific anticyclone, whereas the latter provided the cold waters that are necessary for hyperaridity to develop. One line of evidence for this early initiation of the Atacama, in comparison with many of the world's deserts, is the existence of gypsum crusts preserved beneath an ignimbrite deposit that has been dated to c. 9.5 Ma (Hartley and May, 1998). The ready solubility of gypsum implies the existence of aridity ever since that time. Lake-basin studies also indicate drying in the Late Miocene (Saez et al., 1991; Alonso et al., 1999; Diaz et al., 1999; Gaupp et al., 1999; May et al., 1999). However, there is some controversy relating to this issue, on sedimentological grounds.

Hartley and Chong (2002) have argued that the development of hyperaridity was a late Pliocene phenomenon, associated, as in other deserts, with global climate cooling. They recognised, however, that a semi-arid climate persisted from 8 to 3 Ma, punctuated by a phase of increased aridity at around 6 Ma. It is clear, however, that the major nitrate deposits of the Atacama are indicators of profound and long-continued aridity.

12.5 Mechanisms of Formation and Accumulation

There are many theories that have been developed over the past 150 years to account for the development of the Chilean nitrate deposits, and the issue is still far from resolved. This longstanding debate is the product of three factors identified by Ericksen (1981). First, a lack of accurate geological descriptions of the deposits and second, a limited understanding about processes that might form or supply large quantities of nitrate in a desert environment. Finally, these problems are compounded by a 'tendency on the part of some authors to ignore geological data that would make their theories untenable' (Ericksen, 1981, p. 1). Eriksen (1981, p. 21) provides a review of some of the early ideas about the derivation of the nitrate.

12.5.1 The seaweed and guano theories

In the late 19th century, one view (see e.g., Forbes, 1861) was that the nitrates were the remnants of ancient seaweed deposits left stranded by falling sea levels. There are indeed raised beaches in northern Chile, which indicate that sea levels have been higher in the past, but the main argument behind this theory was that this would account for the iodine present in the caliche. Darwin (1890 edition, p. 347) also argued:

> ... from the manner in which the gently inclined, compact bed follows for so many miles the sinuous margin of the plain, there can be no doubt that it was deposited from a sheet of water: from the fragments of embedded shells, from the abundant iodic salts, from the superficial saliferous crust occurring at a higher level and being probably of marine origin ... there can be little doubt that this sheet of water was, at least originally, connected with the sea.

However, the seaweed theory was dismissed by Newton (1869) on the grounds that it would not explain the presence of certain other salts in the caliche and that there were no marine deposits at the altitudes at which the caliche deposits occurred.

A related theory was that the nitrates were remnants of ancient guano deposits (Gautier, 1894). Once again, there is much guano along the current coastline, but one objection to the theory is the heights at which the caliche occurs. Another is the limited amount of phosphate which they contain (Newton, 1896). However, this theory was championed in the 20th century by Penrose (1910).

12.5.2 The capillary concentration theory

Mueller (1968) argued that one of the unique features of the Atacama was what he described as its 'climatic asymmetry'. Basically, he indicated that in the Andes to the east there was relatively high precipitation, whereas at lower altitudes in the west there was hyper-aridity. Weathering in the high rainfall zone produced solutes that accumulated in the closed basins at lower altitudes and formed salt deposits. Waterlogged sumps were zones of chloride and sulphate precipitation. Higher zones around the sumps were fed by capillary concentration, and it is this that led to the formation of zones of nitrate accumulation.

12.5.3 The role of atmospheric deposition

The idea that the nitrates and related salts are the result of atmospheric deposition is now one that receives relatively widespread support. Claridge and Campbell (1968, p. 429) proposed what they called 'a new general theory for the origin of principal nitrate deposits':

> The atmosphere contains small quantities of nitrogenous material as well as the compounds of iodine, sulphur and chlorine derived from the ocean surface . . . and these compounds, in varying proportions and states of oxidation form part of the precipitation from the atmosphere throughout the world. In most parts of the world, the compounds are intercepted by organisms or pass by leaching through the soil and are returned to the sea. It is only the almost complete absence of biological activity and leaching in such places as the Antarctic Continent and the deserts of northern Chile that nitrate soils can accumulate on a large scale.

Ericksen (1981) also argued that some of the nitrates could have an atmospheric origin, with their source originating from the Pacific. However, he recognised that volcanic emissions from the Andes could have played a role, as could biological activity and bedrock weathering. He recognised that atmospherically derived materials would be leached, redistributed and enriched in their highly soluble components and would accumulate on old land surfaces that have had little or no modification since the Miocene, on lower hillsides and at breaks in slopes as the result of leaching and redistribution by rainwater, and in saltpans. In 1983, however, Ericksen admitted that his ideas had changed, and argued that most of the nitrate was formed 'by fixation of atmospheric nitrogen by microorganisms in playa lakes and associated moist soils' (p. 372). Subsequent stable isotopic studies, using N, O and S isotopes, by Ericksen and co-workers (see Böhlke et al., 1997, p. 135) 'support the hypothesis that some high-grade caliche-type nitrate-rich deposits in some of the Earth's hyperarid deserts represent long-term accumulations of atmospheric deposition . . . in the relative absence of soil leaching or biologic recycling.' This was confirmed by Michalski et al. (2002). Böhlke and Michalski (2002) investigated the oxygen isotopic composition of nitrates from both the Atacama and from the Mojave and argued (p. 1) that:

> the magnitude of the non-mass-dependent isotope effect in the Atacama desert nitrate deposits is consistent with the bulk of the nitrate having been derived from atmospheric nitrate deposition, with relatively little having formed by oxidation of reduced N compounds on the earth's surface. The proportion of the microbial end member is larger in the Mojave deposits, possibly because of slightly higher rainfall and more biological activity.

Similarly, Michalski et al. (2001) argued that the oxygen isotope composition of Antarctic soil nitrate indicates that those deposits are entirely due to atmospheric deposition. Arias (2003) also supported an atmospheric origin for the Atacama deposits, and suggested that the nitrates resulted from the decay of marine algae concentrations and inland transport of aerosols in sea-spray and fog.

Searl and Rankin (1993) recognised that there were many possible sources for the salts, and that there may have been complex and multiple phases of dissolution, reprecipitation and recrystallisation. However, they believed (pp. 331–2) that the mineralogical assemblage of the deposits 'is consistent with salts being sourced from highly evolved groundwater of Andean origin with smaller components of marine aerosol and volcanigenic atmospheric fallout.' In a response, Ericksen (1994) was sceptical of some of their views, and suggested that because of their great diversity in topographic position it 'is unlikely that flooding of the nitrate deposits by ephemeral lakes fed by Andean waters could have been a factor in their formation' (p. 849). He also remarked that 'chemical evolution of Andean waters during transport to or near the sites of the nitrate deposits, could not have been a significant factor in concentrating the saline constituents of the nitrate deposits'. In turn, Searl (1994, p. 851) responded 'The Andes are an active volcanic terrain and as such would seem an obvious source of saline ingredients. Unpublished sulphur isotope data for sulphates associated with the nitrate ore tend to support a largely Andean source for the sulphur with additional localised sulphur sources from remobilised older evaporites and sulphide ores.' Volcanic and geothermal processes of nitrate delivery were also proposed by Chong (1994).

12.5.4 The fog-source model

Ericksen (1981, p. 9) argued that the coastal fogs of northern Chile (*caman-chaca*) are saline and so may have been important sources of constituents of the nitrate deposits. He provided an analysis of fog condensate from near Antofagasta and showed that it contained some $162\,mgL^{-1}$ of dissolved salts, including $19\,mg/lL^{-1}$ of NO_3. Ericksen recognised that some pollution of the sample could have occurred, and that it was dangerous to infer too much from one sample. Fortunately, some systematic and careful analyses have been made of Chilean fog chemistry in recent years (Schemenauer and Cereceda, 1992). These tend to show rather lower dissolved salt contents than those given by Ericksen, and a mean NO_3 concentration of just $1.6\,mg/lL^{-1}$ was estimated (Eckardt and Schemenauer, 1998). However, some ridge sites in close proximity to the ocean have a high frequency of fogs, with as many as 189 days in the year (Cereceda and Schemenauer, 1991). Nevertheless, the quantity of fog water that is deposited falls off sharply as one moves inland, with a flux of about $8.5\,Lm^{-2}\ day^{-1}$at the coast, and $1.1\,Lm^{-2}\ day^{-1}$12 km inland (Cereceda et al., 2002). The purity of fog water and the limited amounts that are deposited away from the coast, suggest that it can only be a minor contributor to the nature of the nitrate deposits. Indeed, S and Sr isotopic studies of Atacama aerosols and sediments confirms this view, and Rech et al. (2003) suggest that the spatial distribution of high-grade nitrate deposits corresponds to areas that receive the lowest fluxes of salts derived from the sea or from salars. They argue that high inputs of marine or local salar salts will tend to dilute atmospheric nitrate fall-out, which they regard as the main source of not only the nitrates, but also of the perchlorate and iodate that the caliche contains.

12.6 Sodium Nitrate in Weathering

Caliche mantles the slopes in the regions in which it occurs in the Atacama, and contributes to the generally subdued and rounded nature of many of the slopes. It acts as a carapace over extensive areas, although large tracts are now pitted and mounded because of the exploitation of the deposits for fertilizers and other purposes. The caliche is also often associated with the development of extensive tracts of polygonal patterned ground (Ericksen, 1981). However, one of the main geomorphological roles of the sodium nitrate is to cause bedrock weathering and volume displacement (Searl and Rankin, 1993).

In the Atacama, various observers have postulated that salt weathering is an active process that leads to rock brecciation (Searl and Rankin, 1993), slope planation (Abele, 1983) and tafoni development (Tricart and

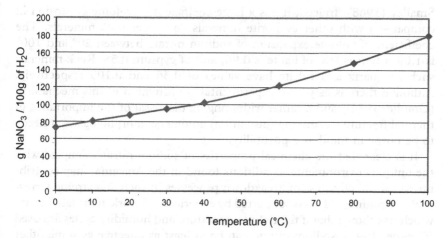

Figure 12.6 The solubility of sodium nitrate in water (drawn from data in Dean and Lange, 1992).

Cailleux, 1969; Heslop, 2003). Three properties of sodium nitrate favour it as an effective salt-weathering agent. First, it is highly soluble in water; second, it is both hygroscopic and deliquescent; finally, it has a high coefficient of thermal expansion on heating. These properties mean that sodium nitrate has the potential to be an effective salt-weathering agent through the three widely cited mechanisms of salt weathering: crystallisation, hydration and thermal expansion.

As we have noted, sodium nitrate is highly soluble. As Figure 12.6 shows, the solubility of sodium nitrate, in common with most salts, increases with temperature. This is significant for the Atacama as it favours the process of salt weathering under hot desert conditions: from a theoretical perspective, the increase in solubility with temperature means that at higher temperatures, more solute will be able to dissolve in the solvent. This, coupled with the high solubility of sodium nitrate, means that when crystallisation takes place, a large volume of salt is available to crystallise out.

The fact that sodium nitrate is both a hygroscopic and deliquescent salt is highly significant for salt-weathering processes operating in the Atacama. The absorption of water from the air permits the salt to undergo cyclic crystallisation by hygroscopic reaction under variations in environmental conditions (Zehnder and Arnold, 1989). The aridity of the Atacama is tempered by the humid coastal fog, which provides appreciable amounts of water as condensate. It is possible that the presence of this fog allows the delicate balance between the need for moisture to enable the cyclic crystallisation/ hydration necessary for salt weathering to take place, without an excess of water which would result in the leaching and removal of the nitrates.

It is also possible that sodium nitrate could contribute to rock disintegration by its thermal expansion on heating, as proposed by Cooke and

Smalley (1968). Indeed, it has a large coefficient of volume expansion in comparison with other evaporite minerals and other rock minerals. The coefficient of volume expansion of sodium nitrate between 20° and 100° is 1.08, whereas that of halite is 0.96, and of gypsum 0.58. Rock minerals such as quartz and calcite have values of 0.36 and 0.105 respectively. Although there is as yet no experimental confirmation of this mechanism (Goudie, 1974), and a rather wide-ranging dismissal of the importance of thermal fracture weathering (insolation) among geomorphologists, it needs to be borne in mind as a possibility.

It is clear that the chemical properties of sodium nitrate, coupled with the unique environmental conditions found in the Atacama, may contribute to the effectiveness of breakdown processes in such an extreme terrestrial environment. This is supported by experimental work in the laboratory, which has shown that if the right temperature and humidity cycles are used (Goudie, 1993), sodium nitrate can be at least as effective as some other common salts at causing rock breakdown and decay of concrete (Malone et al., 1997). Indeed, Goudie et al. (2002) used a temperature cycle derived from rock surface field monitoring in the Atacama, together with simulated fog application, which showed conclusively that under simulated Atacama conditions, rapid weathering can occur.

12.7 Directions for Future Research

The relatively sparse geographical spread of nitrates in deserts is intriguing, and although the special environmental conditions in the Atacama clearly explain their particularly fine development in that area, it is possible that further detailed mineralogical work may lead to the identification of nitrate deposits in other hyperarid areas (such as the Namib, the Libyan Desert or the Rub 'Al Khali). It is also evident that stable isotope studies in recent years have been immensely productive in determining the origin of deposits, and such work could usefully be extended to areas other than the Atacama. Within the Atacama itself, there is still a relative paucity of detailed micromorphological work, and further studies need to be conducted on the relationships between the nitrate deposits and underlying bedrock types, and the variability that exists between deposits in differing geomorphological situations.

12.8 Conclusion

Nitrate deposits, although found in Antarctica and some arid areas as surface efflorescences, are developed to a unique extent in the Atacama Desert of Chile. In reality the deposits have a complex mineralogy, and contain large amounts of sulphates, chlorides and other nitrates. Diverse theories have been advanced to explain their formation. Considerable controversy and debate still exist, but it is likely that they are the result of complex processes of deposition, remobilisation and translocation. It is probable that atmospheric inputs have played a major role, although the role of fog may be less than previously thought. Whatever their precise mode of formation may be, and it is likely that more than one theory may be valid, the deposits must owe their unique degree of development in the Atacama to the relative longevity of aridity in the area and to the extreme degree which that aridity attains. The Atacama is the world's driest desert and one of its oldest. Further research is required on the micromorphology of nitrates and it is evident that isotopic studies have considerable potential to help identify the sources of nitrates in drylands.

References

Abele, G. (1983) Flacenhafte Hanggestaltung und Hangzerschneidung im chilenish-peruanischen trockengebeit. *Zeitschrift für Geomorphologie, Supplementband* **48**, 197–201.

Alpers, G.N. & Brimhall, G.H. (1988) Middle Miocene climatic change in the Atacama Desert, northern Chile: evidence from supergene mineralisation at La Escondida. *Bulletin Geological Society of America* **100**, 1640–1646.

Alonso, R.N., Jordan, T.E., Tubbutt, K.T. & Vandorvoort, D.S. (1999) Giant evaporite belts of the Neogene Central Andes. *Geology* **19**, 401–404.

Arias, J. (2003) On the origin of saltpetre, northern Chile coast. *Abstract, XVI INQUA Congress*, Reno, Nevada, paper 6–9, 24 July.

Arnold, A. & Zehnder, K. (1988) Decay of stone materials in salts in humid atmosphere. *Proceedings of the 6th International Conference on the Deterioration and Conservation of Stone*, Nicolas Copernicus University, Torun, pp. 138–1468.

Bendix, J. Bendix, A. & Richter, M. (2000) El Niño 1997/1998 in Nordperu: Anzeichen eines Ökosystem-Wandels? *Petermanns Geographische Mitteilungen* **144**, 20–31.

Betancourt, J.L. Latorre, C., Recho, J.A., Quade, J. & Rylander, K.A. (2000) A 22,000-year record of monsoonal precipitation from northern Chile's Atacama Desert. *Science* **289**, 1542–1546.

Böhlke, J. & Michalski, G. (2002) Atmospheric and microbial components of desert nitrate deposits indicated by variations in $\delta^{18}O$ $\delta^{17}O$ $\delta^{16}O$ isotope ratios. Fall Meeting, Supplement of Abstracts. *Eos (Transactions of the American Geophysical Union)* **83**(47), 1 pp.

Böhlke, J.K., Ericksen, G.E. & Revesz, K. (1997) Stable isotope evidence for an atmospheric origin of desert nitrate deposits in northern Chile and Southern California, USA. *Chemical Geology* **136**, 135–152.

Cereceda, P. & Schemenauer, R.S. (1991) The occurrence of fog in Chile. *Journal of Applied Meteorology* **30**, 1097–1105.

Cereceda, P., Osses, P., Larrain, H., Farías, M. Lagos, M., Pinto, R. & Schemenauer, R.S. (2002) Advective, orographic and radiation fog in the Tarapacá region, Chile. *Atmospheric Research* **64**, 261–271.

Chong, G.D. (1994) The nitrate deposits of Chile. In: Reutter, K.-J., Scheuber, E. & Wigger, P.J. (Eds) *Tectonics of the southern Central Andes*. Berlin: Springer, pp. 303–316.

Claridge, G.G.C. & Campbell, I.B. (1968) Origin of nitrate deposits. *Nature* **217**, 428–430.

Clarke, J.D.A. (2006) Antiquity of aridity in the Chilean Atacama Desert. *Geomorphology* **73**, 101–114.

Colleo, S., Arce, E. & Andia, A. (2002) Mineralogy, chemistry and fluid inclusions in the nitrate ore deposits of Maria Elena, II Region, Chile. *Boletin de la Sociedad Chilena de Quimica* **47**, 181–190.

Cooke, R.U. (1994) Salt weathering and the urban water table in deserts. In: Robinson, D.A. & Williams, R.B.G. (Eds) *Rock Weathering and Landform Evolution*. Chichester: Wiley, pp. 193–205.

Cooke, R.U. & Smalley, I.J. (1968) Salt weathering in deserts. *Nature* **220**, 1226–1227.

Darwin, C. (1890) *Coral Reefs, Volcanic Islands, South American Geology*. London: Ward Lock.

Dean, J.A. & Lange, N.A. (1992) *Lange's Handbook of Chemistry*, 14th edn. London: McGraw-Hill.

Diaz, G.G., Mendoza, M., Garcia-Veigac, J., Pueyo, J.J. & Turner, P. (1999) Evolution and geochemical signatures in a Neogene forearc evaporitic basin: the Salar Grande (Central Andes of Chile). *Palaeogeography, Palaeoclimatology, Palaeoecology* **151**, 39–54.

Eckardt, F.D. & Schemenauer, R.S. (1998) Fog water chemistry in the Namib Desert, Namibia. *Atmospheric Environment* **32**, 2595–2599.

Ericksen, G.E. (1981) Geology and origin of the Chilean nitrate deposits. *U.S. Geological Survey Professional Paper* **1188**, 1–37.

Ericksen, G.E. (1983) The Chilean nitrate deposits. *American Scientist* **71**, 366–374.

Ericksen, G.E. (1994) Discussion of a petrographic study of the Chilean nitrates. *Geological Magazine* **131**, 849–852.

Ericksen, G.E. & Mrose, M.E. (1972) High purity veins of soda-niter, $NaNO_3$, and associated saline minerals in the Chilean nitrate deposits. *U.S. Geological Survey Professional Paper* **800-B**, B43–B49.

Ericksen, G.E., Hosterman, J.W. & St. Amand, P. (1988) Chemistry, mineralogy, and origin of the clay-hill nitrate deposits, Amagosa River valley, Death valley region, California, U.S.A.. *Chemical Geology* **67**, 85–102.

Forbes, D. (1981) On the geology of Bolivia and southern Peru. *Quarterly Journal of the Geological Society of London* **17**, 7–62.

Forti, P. & Buzio, A. (1990) La nitratite della Grotta de las Manos (Argentina). *Revista Italiana di Mineralogia e Palaeontologia* **63**, 3–9.

Gaupp, R., Kòltt, A. & Worner, G. (1999) Palaeoclimatic implications of Mio-Pliocene sedimentation in the high-altitude intra-arc Lauca Basin of northern Chile. *Palaeogeography, Palaeoclimatology, Palaeoecology* **151**, 79–100.

Gautier, A. (1894) Sur un gisement de phosphates de chaux et d'alumine contenant des espèces rares ou nouvelles et sur la genèse des phosphates et nitres naturels. *Annales de Mines Serie 9* **5**, 5–53.

Goudie, A.S. (1972) On the definition of calcrete deposits. *Zeitschrift für Geomorphologie* **16**, 464–468.

Goudie, A.S. (1974) Further experimental investigation of rock weathering by salt and other mechanical processes. *Zeitschrift für Geomorphologie, Supplementband* 21, 1–12.

Goudie, A.S. (1993) Salt weathering simulation using a single-immersion technique. *Earth Surface Processes and Landforms* **18**, 369–376.

Goudie, A.S., Wright, E. & Viles, H.A. (2002) The roles of salt (sodium nitrate) and fog in weathering: a laboratory simulation of conditions in the northern Atacama Desert, Chile. *Catena* **48**, 255–266.

Hartley, A.J. & Chong, G. (2002) Late Pliocene age for the Atacama Desert: Implications for the Desertification of western South America. *Geology* **31**, 43–46.

Hartley, A.J. & May, G. (1998) Miocene gypcretes from the Calama Basin, northern Chile. *Sedimentology* **45**, 351–364.

Heslop, E.E.M. (2003) *Clast breakdown in the Atacama Desert, northern Chile: An integrated field and laboratory approach.* Unpublished PhD thesis, University of Oxford.

Hill, C.A. (1977) Niter and soda-niter in a lava tube, Socorro County, New Mexico. *Cave Research Fund Annual Report* **19**, 15.

Hill, C.A. (1981) Minerology of cave nitrates. *National Speleological Society Bulletin* **43**, 127–132.

Hill, C.A. & Eller, P.G. (1977) Soda niter in earth cracks of Wupatki National Monument. *National Speleological Society Bulletin* **39**, 113–116.

Hill, C. & Forti, P. (1997) *Cave minerals of the World*, 2nd edn. Huntsville, AL: National Speleological Society.

Holmgren, C.A. Betancourt, J.L., Rylander, K.A., Roque, J., Tovar, O., Zeballos, H., Linares, E. & Quade, J. (2001) Holocene vegetation history from fossil rodent middens near Arequipa, Peru. *Quaternary Research* **56**, 242–251.

Johannesson, J.K. & Gibson, G.W. (1962) Nitrate and iodate in Antarctica salt deposits. *Nature* **194**, 567–568.

Keys, J.R. & Williams, K. (1981) Origin of crystalline, cold desert salts in the McMurdo Region, Antarctica. *Geochimica et Cosmochimica Acta* **45**, 2299–2309.

Kirchner, G. (1996) Cavernous weathering in the Basin and Range area, Southwestern USA and northern Mexico. *Zeitschrift für Geomorphologie, Supplementband* **106**, 73–97.

Laue, S., Bläuer Böhm, C. & Jeannette, D. (1996) Salt weathering and porosity. *Proceedings 8th International Conference on Deterioration and Conservation of Stone.* Berlin: Ernst und Sohn, pp. 513–533.

Magilligan, F.J. & Goldstein, P.S. (2001) El Niño floods and culture change: a late Holocene flood history for the Rio Moquegua, southern Peru. *Geology* **29**, 431–434.

Malone, P.G., Poole, T.S., Wakeley, L.D. & Burkes, J. P. (1997) Salt related expansion reactions in Portland-cement-based wasteforms. *Journal of Hazardous Materials* **52**, 237–246.

Matsuoka, N. (1995) Rock weathering processes and landform development in the Sor Rondane Mountains, Antarctica. *Geomorphology* 12, 323–339.

May, G., Hartley, A., Stuart, F.M. & Chong, G. (1999) Tectonic signatures in arid continental basins: an example from the Upper Miocene-Pleistocene Calama Basin, Andean Forearc, northern Chile. *Palaeogeography, Palaeoclimatology, Palaeoecology* 151, 55–77.

Michalski, G.M., Holve, M., Feldmeier, H., Bao, H., Bockheim, J.G., Reheis, M. & Thiemens, M.H. (2001) Tracing the atmospheric source of desert nitrates using $\Delta^{17}O$. *Abstract, 11th Annual V.M. Goldschmidt Conference*, Hot Springs, VA, 20–24 May.

Michalski G., Savarino, J., Böhlke J.K. & Thiemens, M. (2002) Determination of the Total Oxygen Isotopic composition of nitrate and the calibration of a $\Delta^{17}O$ nitrate reference material. *Analytical Chemistry* 74, 4989–4993.

Middleton, N.J. (2001) *Going to Extremes*. London: Channel 4 Books.

Mrose, M.E., Fahey, J.J. & Ericksen, G.E. (1970) Mineralogical studies of the nitrate deposits of Chile III. Humberstonite, $K_3N_7Mg_2$ $(SO_4)_6$ $(NO_3)_2{\cdot}6H_2O$, a new saline mineral. *American Mineralogist* 55, 1518–1533.

Mueller, G. (1968) Genetic histories of nitrate deposits from Antarctica and Chile. *Nature* 219, 1131–1134.

Newton, W. (1869) The origin of nitrate in Chili. *Geological Magazine* NS, Decade 4(3), 339–342.

Pantony, D.A., (1961) Sodium nitrate. In: Mellor, J.W. (Ed.) *Comprehensive Treatise on Inorganic and Theoretical Chemistry*, Vol. ii, Supplement ii, *The Alkali Metals*, Part I. London: Longman, pp. 1206–1258.

Penrose, R.A.F. (1910) The nitrate deposits of Chile. *Journal of Geology* 18, 1–32.

Pueyo, J.J., Chong, G. & Vega, M. (1998) Mineralogia y evolucion de las salmueras madres en el yacimento de nitratos Pedro de Valdivia, Antofagasta, Chile. *Revista Geologica de Chile* 25, 3–15.

Rech, J.A., Quade, J. & Hart, W.S. (2003) Isotopic evidence for the source of Ca and S in soil gypsum, anhydrite and calcite in the Atacama Desert, Chile. *Geochimica et Cosmochimica Acta* 67, 576–586.

Saez, A., Cabrera, L., Jensen, A. & Chong, G. (1991) Late Neogene lacustrine record and palaeogeography in the Quillagua-Llamara basin, Central Andean fore-arc (northern Chile). *Palaeogeography, Palaeoclimatology, Palaeoecology* 151, 5–37.

Schemenauer, R.S. & Cereceda, P. (1992) The quality of fog water collected for domestic and agricultural use in Chile. *Journal of Applied Meteorology* 31, 275–290.

Searl, A. (1994) Discussion of a petrographic study of the Chilean nitrates. *Geological Magazine* 131, 849–852.

Searl, A. & Rankin, S. (1993) A preliminary petrographic study of the Chilean nitrates. *Geological Magazine* 130, 319–333.

Tereschenko, O. & Malyutin S. (1985) Hygroscopicity of sodium nitrate (Chile saltpetre). *Journal of Applied Chemistry (USSR)* 85, 810–813.

Tricart, J. & Cailleux, A. (1969) *Le Modélé des régions sèches*. Paris: Sedes.

Zehnder, K. & Arnold, A. (1989) Crystal Growth in Salt Efflorescence. *Journal of Crystal Growth* 97, 513–521.

Chapter Thirteen

Analytical Techniques for Investigating Terrestrial Geochemical Sediments

John J. McAlister and Bernie J. Smith

13.1 Introduction

Geochemical sediments and weathering products are evident over vast areas of the tropical and sub-tropical regions of the world. Different lithologies respond to similar environmental conditions in various ways to produce heterogeneous, complex geochemical sediments and generally less stable weathering products. In many parts of the world, especially the tropics and sub-tropics, these are important palaeoclimatic and palaeoenvironmental indicators. For example, erosion-resistant indurated duricrusts may form positive relief features, and deep sequences of these deposits are often associated with long periods of landscape stability. From an economic point of view, terrestrial sediments such as calcretes (Chapter 2) and silcretes (Chapter 3) are used extensively for road and building foundations in many developing countries (Goudie and Pye, 1983). Precise and consistent identification, characterisation and evaluation of these products are therefore important and have been accomplished using a wide range of analytical techniques. However, despite the availability of numerous instrumental procedures, reliance is generally placed on only a few. This is possibly due to a lack of resources, but in some cases may also be due to ignorance of their existence. The choice of analytical technique is not, however, the only decision to be taken. It is also important to remember that, no matter how much the precision and accuracy of analytical instruments continues to improve, the quality of information obtained is highly dependent on sample preparation prior to analysis.

This review attempts to bridge the gap that can exist between geomorphologists and analytical chemists by raising awareness, not only of a range of analytical techniques, but also various preparation procedures chosen specifically to improve the partitioning of geochemical sediments into their component parts and to enhance the sensitivity of analyses. In doing so, it sets out to address such questions as which techniques may be applied to different geochemical sediments and weathering products, how are samples best prepared prior to analysis, how techniques work and what information can be obtained from the results? The chapter is divided into three substantive sections that deal with chemical, mineralogical and isotopic techniques. Within each section, special attention is given to identifying appropriate sample preparation methods that allow us not only to characterise samples quantitatively but also to examine where the components are held. This relates especially to the various size fractions and geochemical phases present within samples, and how the latter can allow us to examine element mobility under differing environmental conditions. The emphasis throughout is on linking preparation and analysis to desired outputs. This systematic, output-led approach is essential for all analytical projects and especially when analysing small irretrievable samples collected in remote locations and often unique exposures.

13.2 Elemental Analysis

13.2.1 Spectroscopic techniques for analysis of elemental composition

In most research studies, the first step is to identify objectives and decide the nature of the information (data) required for them to be met. This, in turn, dictates the analytical instrumentation that is capable of delivering these data in an appropriate form (e.g. quantitative or qualitative). The most widely used techniques for the analysis of geochemical sediments and weathering products are those that rely on spectroscopy. These are based on the interaction of the prepared sample with some form of electromagnetic radiation. This leads to the term spectroscopy since the energy of this radiation depends on its frequency. Analysis is carried out in the ultraviolet, visible, infrared, X-ray, microwave and radio-frequency regions of the electromagnetic spectrum, where an X-ray photon is 10,000 times more energetic than an ultraviolet or a visible one.

These techniques fall into two categories: those considered as routine (e.g. atomic absorption and emission spectroscopy, X-ray fluorescence) and a growing number of microanalytical surface techniques (e.g. laser microprobe mass analysis [LAMMA] and sensitive high-resolution ion microprobe [SHRIMP]). Each analytical technique requires specific sample preparation prior to analysis, as summarised in Table 13.1.

Established techniques

There are a number of established techniques that are used on a routine basis in many laboratories.

Table 13.1 Characterisation of a selected range of analytical techniques

Technique†	Analysis		Pretreatment			Output‡			
	Mineralogical	Chemical	Grind	Polish	Digest	Quantitative	Morphogical	Surface	Depth
XRF		****	Y			***			*
AAS		****	Y		Y	****			*
ICP-AES		****	Y		Y	****			*
ICP-MS		****	Y		Y	****			*
IC		****	Y		Y	****			
EPMA	***	***		Y		**	****	**	*
ESEM	***	**		Y		*	****	**	*
LAMMA	**	**					*	*	*
XPS		**				***		***	****
SHRIMP		****				***	***	***	***
PIXE		****	Y	Y		****	***	**	*
RLMP		*				**		**	**
XRD	****	**	Y	Y		*	****	**	
SEM	***	**		Y		*	****	***	
HRTEM	****	**		Y		*	**	**	***
FTIR	****	**	Y			**	**	**	
Thin-section	****			Y			****		

**** = Excellent *** = V. Good ** = Good * = Good ✝ Y = Poor Y = Yes.

†XRD, X-ray diffraction; XRF, X-ray fluorescence; AAS, atomic absorption spectrometry; ICP–AES, inductively coupled plasma–atomic emission spectrometry; ICP–MS, Inductively coupled plasma/mass spectrometry; IC, ion chromatography; EPMA, electron probe microanalysis; SEM, scanning electron microscope; ESEM, environmental scanning electron microscope; HRTEM, high-resolution transmission electron microscopy; LAMMA, laser microprobe mass analysis; XPS, X-ray photo-electron spectroscopy; RLMP, Raman laser microprobe analysis; SHRIMP, sensitive high resolution ion microprobe. PIXE, proton-induced X-ray emission; FTIR, Fourier transform infrared.

‡Quantitative: ability to carry out quantitative analysis; morphological, ability to examine morphology of the surface.

Atomic emission (AES) and absorption spectroscopy (AAS)

Atomic spectroscopy is a quantitative technique for the determination of elements in sample solutions. Analysis of atoms or elementary ions is only possible in a gaseous medium where individual atoms or ions are well separated. Therefore the first and most important step is atomisation, whereby a sample solution is volatilised into an atomic vapour using an appropriate heat source. When an atom absorbs energy from this source, an outer electron is promoted from the 'ground state' to a less stable 'excited state'. On return to the ground state, a photon of light energy is emitted with an intensity related to the concentration of the element of interest. The efficiency and reproducibility of atomisation determines the sensitivity, precision and accuracy of the technique. Sensitivities lie in the parts per million (ppm) and parts per billion (ppb) range, with additional advantages being speed, high sensitivity, convenience and moderate instrument costs.

Atomic emission uses flame and three types of plasma source (inductively coupled, direct current and microwave induced) to atomise and excite atoms (Lajunen, 1992; Vandecasteele and Block, 1993). As the number of atoms in the excited state increases, AES sensitivity also increases. In atomic absorption, the only function of the heat source is to convert the sample aerosol into an atomic vapour. Ground state atoms absorb energy of a specific wavelength from a light source (hollow cathode lamp) that emits the spectrum of the element of interest as it enters the excited state. Sensitivity of this technique is proportional to the number of atoms in the ground state. In AAS, atomisation can be achieved by burning the sample solution in a flame or by injecting it into an electro-thermally heated graphite furnace. In the latter, individual atoms have a longer residence time in the optical path and much smaller sample volumes can be analysed at a very high temperature (Skoog, 1985; Lajunen, 1992; Vandecasteele and Block, 1993). Conversion of samples into a homogeneous slurry using ultrasound prior to its introduction into a graphite furnace is gaining popularity. High silica matrices can be homogenised into a polytetrafluoroethylene (PTFE) slurry using nitric acid and the highly volatile silicon fluoride formed is vaporised prior to analysis (Cave et al., 2000). Current instruments for flame AES and AAS are time-saving, robust, reliable and accurate. Inductively coupled plasma (ICP) provides low background signals, more precise sample introduction, no chemical interference, good detection limits and allows for a more comprehensive analysis. Both these techniques provide the quality of data required for the majority of routine research studies on geochemical sediments and weathering products.

Inductively coupled plasma–mass spectroscopy (ICP–MS)

This technique combines the analytical capabilities of mass spectroscopy and the efficient sample atomisation capabilities of a plasma torch. Sample solutions are pneumatically nebulised in a stream of argon into the plasma, and ions are extracted into a vacuum system that is compatible with that used for the operation of a mass spectrometer. This instrumental combination is used mainly for simultaneous multi-element and isotopic ratio determinations. Analytical limitations may be introduced by matrix-induced changes in ion signal intensity and these effects are more problematic than when ICP–AES is used alone. Dissolved solids can cause matrix interference and the extent of matrix effects depends on the element to be analysed, matrix operating conditions and the type of instrument used. Several techniques have been introduced to analyse solids without dissolution, and these include laser ablation, slurry nebulisation and electrothermal vaporisation (Lajunen, 1992; Vandecasteele and Block, 1993; Cave et al., 2000). Rare earth elements (REE) are analysed routinely using ICP–MS, but care must be exercised when choosing the sample decomposition method (Liu et al., 1998). Laser energy is used for *in situ* solid microsampling of materials (LA–ICP–MS), analysis of dissolution-resistant geochemical sediments and to study spatial distribution of trace elements and isotopic composition in a microscale area of sample surface (Guo and Lichte, 1995). However LA–ICP–MS is less superior than, for example, sensitive high-resolution ion microprobe (SHRIMP) for the determination of isotopic ratios across the mass range and for spatial resolution. Applications include zircon geochronology (Feng et al., 1993), heavy isotope analysis such as U/Pb and Pb/Pb dating on detrital and metamorphic materials (provided Pb concentration is >3 ppm) and REE distribution in fossil materials (Feng, 1994; Meisel, 2000). This technique is very sensitive and detection limits for more than 60 elements are between 0.03 and $0.3 \mu g L^{-1}$ (ppb) and for the halogens, phosphate and sulphur, between $0.001–1.0 mg L^{-1}$ (ppm) (Date and Gray, 1989). More sensitive detectors for these instruments are now available, but column separation of the major elements is necessary before isotopic ratios can be determined. This technique provides a means of carrying out less routine and more complex research studies (Taylor, 2001; Thomas, 2003).

X-ray fluorescence (XRF) and electron probe microanalysis (EPMA)

These methods are based on the existence of discrete energy levels known as K,L,M,N shells for the inner electrons surrounding the central nucleus of an atom. When sufficient energy is applied to the atoms, electrons are ejected from an inner shell and are replaced by electrons from an outer one. During each step, a photon of electromagnetic radiation is emitted with a wavelength in the X-ray region of the spectrum. The spectrum produced shows the relationship between wavelength and atomic number of the element. Element concentrations are determined by comparing the number of X-ray counts emitted from a prepared sample to that of a certified standard. The source is an X-ray tube, where, on bombardment of an anode with fast electrons, the X-ray spectrum is emitted. X-rays are both absorbed and scattered by the sample and the emitted spectrum is collimated and directed towards a rotating analysing crystal which disperses X-rays according to the Bragg Equation, $n\lambda = 2d\sin\theta$ (Norrish and Chappel, 1977; Potts, 1987; Bain et al., 1994; Jenkins, 1999) – see section 13.3.1 for a full explanation of terms. X-ray fluorescence is a rapid and often non-destructive method for qualitative analysis. However, for quantitative analysis it is destructive and considerable effort may be required for sample preparation. This technique is useful for classification analysis of a large number of samples and analysis of all elements from sodium to uranium. Hand-held high performance XRF analysers are now available at relatively low cost and are easy and safer to use since they contain an X-ray source and high-resolution detectors as compared to equipment that uses radioisotopes (Thomsen and Schatzlein, 2002). These instruments can analyse geochemical sediments, weathering products and fine coatings and are particularly useful for *in situ* analysis of, for example, rock varnish.

In electron probe microanalysis, a beam of electrons from an electron gun is directed at a specimen and the X-rays generated are analysed by a crystal spectrometer (Reed, 1993; Laue and Dalton, 1994; Scott, 1995). There are various ways in which qualitative information is obtained, including spot and overall analyses of the whole or part of a sample. Quantitative analysis involves measuring the number of X-ray counts from a sample and comparing this to a known standard over a fixed time interval. Back-scatter electron images (BSE) reflect the composition of the sample and this BSE effect is stronger for heavy minerals (ores, garnet, spinel) than for light ones (quartz, silicates). This technique can be used, for example, to analyse the cores and rims of rock-forming minerals with respect to growth history and to map element distribution at various scales.

Ion chromatography (IC)

Ion chromatography resulted from of the merging of two significant areas of development, chromatography and ion exchange (Small, 1989). This technique comes under the classification of liquid–solid methods, in which a liquid (eluent) is passed through a stationary solid phase to a flow-through detector. The stationary phase is composed of small diameter (5 μm) uniform particles packed in a cylindrical column 5–30 cm long. A high-pressure pump forces the eluent through the column at a rate of 1–2 mL min^{-1}. The sample to be separated is injected into the flowing eluent prior to the column and carried to a detector. Sample components move through the column at different rates and therefore enter the detector at different times where they are sensed and a chromatogram produced. Ion chromatography with conductometric detection has radically changed the analysis of common water-soluble anions and cations (F, Cl, Br, NO_2 NO_3 SO_4 PO_4 Ca, Mg, Na and K). Marked improvements such as more efficient column resins and suppressed conductometric detection have led to improved resolving power, faster separations, higher sensitivity and a wider spectrum of ion affinities all in one single chromatographic run. All these factors have made IC a powerful and relatively cheap analytical technique. With the appropriate detector, analysis of transition and rare earth elements may be carried out (Smith and Chang, 1983; Wilson and Gent, 1983; Haddad and Jackson, 1990). Ion chromatography is an excellent technique for the determination of water-soluble salt concentrations and is particularly suited to the analysis of evaporites.

Selected microanalytical techniques

There is an increasing interest in microanalytical procedures that provide structural and chemical information at micrometre and nanometre surface depths. *In situ* analyses have become very important following the realisation that a vast amount of information can be obtained with respect to origin and evolution of terrestrial geochemical sediments and weathering products by measuring elemental variations at this scale. A comprehensive review is given by Becker and Dietze (1998). These techniques are not used on a routine basis due to high costs, the absence of sufficiently small depth resolution, and difficulties in obtaining a vacuum environment that maintains surfaces in good condition. In all analytical techniques in surface microanalytical procedures, sensitivity and specificity must be taken into account. Specificity is achieved when the signal from the bulk sample is small compared with that from the surface and such techniques are then considered to be surface sensitive. A very important question is: what information is required? The answer to this will dictate the level of preparation needed (Benedict et al., 1989; Mitra, 2003; Mukhopadhyay, 2003).

Environmental scanning electron microscope (ESEM)

This instrument is a powerful analytical tool and, like most SEM instruments, it can image surfaces at very high spatial resolution (Danilatos,

1988; Li et al., 1995; Schalek and Drzal, 2000). As primary electrons are emitted from the gun system, secondary electrons on the specimen surface are accelerated towards a detector. Collisions between the electrons and the gas molecules used to surround the sample (water vapour, air, N_2, Ar, O_2, etc.) liberate more electrons and hence more signals are produced. Positive ions formed in the gas neutralise the excess electron charge that builds up on the specimen. This instrument, unlike other conventional SEMs, does not require the sample chamber to be kept under high vacuum conditions, the specimen does not have to be electrically conductive and wet samples can be analysed. When water vapour is used as the gas for the specimen chamber, wet samples are held in their hydrated state. By keeping the specimen at a constant temperature and increasing or decreasing the partial pressure of the water vapour, condensation or dehydration occurs respectively. A wide range of geological samples can be examined without specimen preparation or coating (Watt et al., 2000).

X-ray photoelectron spectroscopy (XPS)

Chemical characteristics and the oxidation state of elements in the near-surface layer (~5 nm) of a sample are recorded by photoelectrons that are produced by an X-ray beam. When this technique is combined with intermittent ion sputtering, data on depth distribution can be obtained. X-ray photoelectron spectroscopy goes beyond elemental analysis to provide chemical information such as distinguishing Si–Si from Si–O bonds. Elements from Li to U may be analysed with detection levels at 0.5% under high vacuum conditions. Raster scanning techniques produce images with a spatial resolution of 26 μm and depth profiles of 1 μm thick are possible (Mossotti et al., 1987; Wilson and Burns, 1987).

Proton induced X-ray emission (PIXE)

Proton induced X-ray emission is a powerful technique used to identify and quantify elements ranging from Na to U in a single spectrum. Like other spectroscopic techniques, it is based on the physics of an atom and not its chemistry. The exciting beam in particle-induced techniques consists of protons that are energetically light ions, making this a non-destructive technique. Interaction of these ions with atoms of materials results in the emission of characteristic X-rays and/or Auger electrons for each element. Deceleration of protons in the sample is slow, smooth and predictable, with little scattering and deflection. This produces low levels of continuum background, making detection limits about two orders of magnitude better than techniques using electron beams. Detection limits are orders of magnitude lower than those for EPMA and, in some cases, analysis is possible outside the vacuum (Moschini and Valkovic, 1996; Cohen et al., 2001). Sensitivities range from 1 to 100 ppm depending on the element, energy and total charge of the incident particles, X-ray filters and the quality of detector used. Most samples can be analysed in their original state. However, as only the top 10–50 μm are being probed, the area irradiated by the beam (usually 1–10 mm) should be representative of the whole sample. It is therefore very important that the samples are made homogeneous by grinding to a very fine powder (1–2 μm). These powders should be thoroughly mixed with analytical grade carbon powder and pressed into pellets. Polished uncovered sections of samples used for polarising microscopy can also be analysed (Tesmer and Nastasi, 1995).

Laser microprobe mass analysis (LAMMA)

In this technique, a small volume of sample (0.5 μm diameter) is evaporated by a powerful laser system. Ions are generated essentially at the same instant, since laser ionisation pulse time is short, and analysed using mass spectroscopy. Analysis is qualitative or semi-quantitative in the parts per million range for all elements and isotopes, and samples must be particles or thin films. Laser light is generally emitted at 1064 nm in the infrared range since this wavelength couples easily with samples containing significant amounts of transition elements (Longerich et al., 1993). Controlled ablation of materials with low transition element concentrations (e.g. calcite and feldspar) can be achieved at wavelengths between 532 and 266 nm (Jenner et al., 1994). Carbonate materials are transparent to laser light, but they may be ablated at 1064 nm if the laser pulse has sufficient energy (Feng, 1994). Careful matrix matching with an internal standard is critical for quantitative analysis of small samples, both chemically and especially mineralogically (Williams and Jarvis, 1993).

Sensitive high-resolution ion microprobe (SHRIMP)

This instrument is designed for *in situ* surface analysis of elements and isotopic compositions at the micron scale. Mineral grains may be separated, mounted in an epoxy resin or thin sectioned and gold coated prior to analysis. Sensitive high-resolution ion microprobe is different from other ion probes as it has very high mass resolution, i.e. it can accurately measure (a few parts per million) the isotopic composition of trace elements and has detection limits of parts per billion for most elements. This instrument focuses high-energy oxygen ions onto an area 5–30 μm diameter on the surface of the sample to produce 'sputtering'. The secondary ions are gathered using electrostatic lenses and ejected through the slit of a double focusing mass spectrometer where they are separated according to their relative abundances. This technique allows elemental and/or isotopic analysis with spectral resolution of 5–30 μm and the isotopic composition of Mg, Si, S, Ca, Ti, Cr, Fe, Sr, Hf, Pb and U and most elements in the Periodic Table to be analysed. Trace elements in sedimentary rocks can be monitored from their formation to their destruction including their change in isotopic composition. Detection limits for elements are down to 5 ppb and isotopic ratios can be measured to within one part per thousand (Compston et al., 1982; Eldridge et al., 1987; Sato et al., 1998; Fletcher et al., 2000).

13.2.2 Sample preparation for elemental analysis

General considerations

For indurated/cemented samples, mechanical grinding is usually the first preparation step, even though it can introduce contamination from the mill. For example, hard steels contain chromium, and tungsten carbide releases both tungsten and cobalt. Agate mills produce minimal contamination (Bain and Smith, 1994), but there are difficulties in grinding materials that are extremely hard, such as silcrete. Generally, grinding time should be kept to a minimum and particle size reduced sufficient only to ensure homogeneity and ready attack by chemical reagents. Grinding should not be confused with disaggregation and segregation. In non- or weakly-indurated samples, light grinding with a rubber tipped pestle is often sufficient to reduce them to their constituent particles. Ultrasonic treatment using an analar ammonium hydroxide solution as a dispersing agent and a combination of sieve and centrifugal techniques allows sand, silt and clay fractions to be separated for a more detailed analysis. Chemical and physical changes can also occur on heating (Hesse, 1971) and samples for total element analysis should be dried at 105°C and those for water soluble and exchangeable ions at 30–35°C. Contamination may arise during any stage and therefore sample preparation should be kept as simple as possible. Pure water and high analytical grade reagents must be used at all times, glass and polyware should be rinsed and soaked in 10–20% (v/v) and 1–5% (v/v) nitric acid respectively.

Sample extraction and digestion using open and closed systems

Most readily mobilised elements such as calcium, phosphorus and potassium must be extracted in acid after pretreatment, often at high temperature (Kamp and Krist, 1988). Total element analysis involves complete dissolution using concentrated mineral acids or fusion in a platinum crucible with a suitable flux such as carbonate, peroxide or borate (Ingamells, 1966; Shapiro, 1967; Verbeek et al., 1982).

Commonly used acid digestion procedures liberate ions into homogeneous solution at the molecular level. It is essential to understand the basic chemistry involved, since control of sample environment, rate, type and reproducibility of reaction can be influenced by many factors, especially temperature. Element solubility is particularly critical. For example, if chlorides become soluble during digestion they may precipitate the element of interest, and calcium precipitation can cause several other elements to co-precipitate. Element stability is also crucial since metal ions can complex or precipitate from solution. Choice of reagent is therefore important since complexing properties vary. Perchloric acid and hydrogen peroxide are, for example, poor complexing agents, whereas hydrofluoric and hydrochloric acids stabilise numerous elements in solution. Because of the explosion potential of dry perchlorate salts and its reaction with organic matter, perchloric acid is regarded as a safety hazard, especially in microwave radio frequency (RF) digestion systems. Hydrogen peroxide has a similar explosive potential in the presence of organic matter and, like perchloric acid, it should be added after predigesting samples with nitric acid. Organic matter should never be removed from small samples such as rock chips by heating in hydrogen peroxide. If the peroxide is allowed to dry on these hot surfaces, an explosion will result. Certain combinations must also be avoided. For example, a mixture of perchloric and sulphuric acids can be explosive in the presence of organic matter since sulphuric acid becomes dehydrated, whereas a perchloric/nitric mix works well in this case. Nitric is the primary acid used in most digestion procedures since most nitrates are soluble and it is an excellent oxidising agent at high temperatures (Kingston and Jassie, 1988). Hydrochloric acid complexes a number of ions, stabilises them in solution and ensures that they do not adsorb onto particulate matter or vessel walls. Hydrofluoric acid can be included in an acid mix to dissolve silicates. It is a non-oxidising acid and depends on the complexing properties of the fluoride ion, which dramatically changes the redox potential of many elements. A minimum volume of hydrofluoric acid should be used since fluorides are very insoluble and must be complexed with boric acid before analysis (Matthes et al., 1983). Hydrofluoric acid may cause problems with the ICP–AES technique if a quartz plasma torch is used. In this case the fluoride should be complexed or a hydrofluoric acid resistant ICP torch should be used. Care must be taken not to over dilute nitric acid as this could result in incomplete digestion of organic matter.

Wet digestion procedures can be carried out using open or closed systems. Systematic errors can occur in open systems by contamination from digestion vessels, loss of elements by volatilisation and adsorption onto vessel surfaces. These problems led to the development of pressurised closed systems using, for example, 'PTFE bombs' comprising a stainless steel or nickel body with a PTFE inner vessel (Bernas, 1968). Such techniques remove volatilisation and contamination from external sources, improve decomposition, shorten reaction times, reduce reagent volumes, and lower blank values in comparison to open systems. However, above 240°C, PTFE loses its mechanical strength and cannot be used in excess of 200°C, but high-pressure, computer controlled ashing systems with quartz inner vessels are now available to overcome these problems.

Microwave RF heating procedures have also become established as standard methods for elemental analysis (Papp and Fischer, 1987; Berman, 1988; Kemp and Brown, 1990; Kingston and Haswell, 1995; Walter and Kingston, 1995; Zlotorzynski, 1995; Kingston and Walter, 1997). Microwaves heat samples from the inside, which is in contrast to normal heating. The latter can cause a 'skin effect' where elements from the outer part of the sample are removed faster and prevent release of those from the inside (Zlotorzynski, 1995). The main advantages of microwave digestion are speed (generally less than 30 min for most acid extractions), less acid consumption, lower contamination and no volatilisation loss (Cave et al., 2000). Total dissolution in hydrofluoric acid using microwave heating has been compared to *aqua regia* and HNO_3 digestion using reflux techniques for environmental samples (Sastre et al., 2002). Modern microwave digestion systems use ceramic supporting vessels with TFM (tetrafluoroethylene modified) liners and are capable of reaching 1000 W power and 75-bar pressure (Perkin Elmer Corporation). They also monitor pressure and temperature, and the programme will automatically close down if safe levels are exceeded. However, in any closed-system acid digestion procedure, analysts should be wary of vigorous reactions. For example, high carbonate calcretes or other samples containing easily oxidisable organic matter should be predigested. Finally, sample duplicates, a certified reference material (CRM) and a blank containing identical reagents and no sample should be processed with each sample batch.

In contrast to acid digestion, water soluble ions in samples containing salts can be removed either by shaking or by ultrasonic dispersion in an appropriate extractant, filtered to <0.2 μm using a cellulose acetate membrane and examined by techniques such as ion chromatography (IC). Extraction with deionised water is preferred for IC since acid digestion introduces a large excess of anions, lowers solution pH and disrupts multiple equilibria between eluent species and the column. Ion chromatography is principally used for anion analysis, but cations may be analysed after filtration, selective removal of the analyte or removal of interfering matrix components (Haddad and Jackson, 1990). Fusion techniques generally offer a better alternative to acid digestion for anion determination since some of the flux materials are identical to the eluent components, for example carbonate and borate (Wilson and Gent, 1983). Sample ashing in air or total combustion in oxygen can also be used to convert some non-metallic elements such as F, Cl, S and P into volatile gaseous compounds that can be collected in suitable absorber solutions and injected directly into the IC for analysis (Evans and Moore, 1980). Other clean-up procedures use cation or anion exchange resins (Haddad and Jackson, 1990), dialysis (Cox and Twardowski, 1980; Cox et al., 1988), electrodialysis (Petterson et al., 1988; Haddad and Jackson, 1990) and commercially available disposable cartridge columns (Haddad and Jackson, 1990). Ultra-trace-analyses require high sensitivity ultraviolet detection systems and complex preconcentration techniques (Haddad and Jackson, 1990).

Selective extraction

Most research on geochemical sediments and weathering products has used total element analysis that provides little information on element-bearing phases and hence limited understanding of their geochemistry. This applies both to 'target' elements that form the main components in a given type of mineral deposit and are of economic interest, and especially to 'path-finder' or indicator elements that may act as a guide to mineralisation. Included in the latter are trace elements in geochemical sediments and weathering products with distributions that are determined by the physico-chemistry of the source medium, and by crystal-chemical factors such as their ionic radii, valances and electron configuration. Identification of elements present in different phases can indicate how they are held and what controls their mobility and fixation. These phases include water soluble, exchangeable, carbonate, amorphous Fe and Mn, crystalline Fe and Mn, organic, sulphide and residual. Identification of phase compositions can be accomplished by selective extraction procedures developed initially by environmental and soil chemists primarily to extract various forms of Fe, Al, Mn and trace elements.

Trace elements become associated with solid sample phases by exchange reactions, occlusion into iron and manganese oxides and association with background silica matrices. This often occurs during chemical weathering when secondary concentrations of various elements are produced that may occur as residual, absolute or supergene accumulations of primary mineralisation. A major sink for element adsorption is the 'oxide' phase (metal hydroxides, oxy-hydroxides and hydrous oxides) of any geochemical sediment or weathering product, and this can provide an insight into, for example, pedogenic conditions during profile formation. Due to the complexity of these interactions, approximation techniques have to be applied.

The reactions responsible for element partitioning, particularly transition elements (e.g. Fe, Mn, Cu) that exist in various oxidation states are strongly controlled by redox potential (Eh) and hydrogen ion activity (pH) (Rose, 1975). Redox reactions determine the mobility of many inorganic compounds and pH controls mineral dissolution, precipitation and complexation reactions (Dzombak and Morel, 1990). Other factors influencing element partitioning and sample composition are the nature of the original substrate and the degree of weathering. Selective extraction exposes these phases within a sample to a sequence of solutions of increasing concentrations via a stepwise procedure under strict specific conditions (Tessier et al., 1979; Ure et al., 1993; Quevauviller et al., 1994; Hall et al., 1996; McAlister and Smith, 1999; McAlister et al., 2003).

Interest in these techniques has increased in recent years, especially for geochemical exploration in areas such as the intertropical belt where lateritic profiles are potential sources of rich metal deposits (Antropova et al., 1992; Hall et al., 1996). When studying lateritic materials, especially those containing magnetite and haematite, a final extraction using an oxalate/ascorbic acid solution is essential for complete dissolution and release of elements from these minerals (McAlister and Smith, 1999; Figure 13.1). Selective extraction can also provide information on transport, mobilisation and trapping mechanisms, but there are some basic and operational problems related to the selection of elements from a specific phase or binding form (Pickering, 1981; Van Valin and Morse, 1982; Martin et al., 1987; Nirel and Morel, 1990). Selective extraction techniques using radioactive tracers have examined the specificity of different extractants to each phase and concluded that the latter should be operationally defined and no general agreement exists as to which solutions should be used to extract elements from the various phases (Tessier et al., 1979). Ultimately, choice of extractants must depend on study aim, type of sample and elements of interest. Further complications may arise through contamination from extractants that form insoluble compounds and cause burner and plasma clogging during analysis. Despite these limitations, chemical extraction remains effectively the most useful tool to examine element partitioning in solid phases, since it constitutes a differential approach, even if only between operationally defined solid phases. This is illustrated by a detailed analysis of rock varnish removed from an Entrada Sandstone outcrop, collected in Moab, Utah (Table 13.2).

Sample preparation for surface analytical techniques

Surface analytical techniques such as X-ray fluorescence (XRF) generally use pressed powder or fused borosilicate glass disc samples, although new methods of sample preparation continue to be developed (Buhrke et al., 1998). Samples are chipped and ground to a powder using, for example, a Tema mill that contains tungsten carbide rings. Time of grinding depends

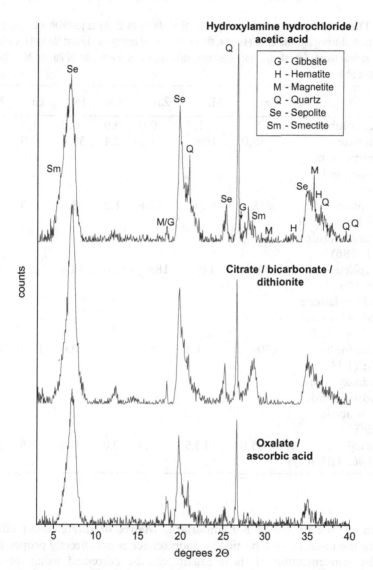

Figure 13.1 Diffractograms showing the importance of correct choice of extractant for the selective dissolution of crystalline Fe from a laterite. Note, for example, the importance of oxalate/ascorbic acid extraction for isolating crystalline Fe minerals. (Modified from McAlister and Smith, 1999.)

very much on the type and degree of induration of the geochemical sediment to be analysed. For example, a typical silcrete would require a much longer grinding time than a laterite. Pressed powder samples are widely used to analyse heavy metals where maximum sensitivity is required and

Table 13.2 Selective extraction of a rock varnish highlighting its ability to partition a sample into its constituent phases (ppm). Note, for example, the prominence of amorphous Fe and Mn and the affinities of the various phases for selected trace elements, including the concentration of Pb and Ni within the carbonate phase

	Fe	Mn	Zn	Cu	Pb	Cr	Ni
Water soluble	0	1.0	0.0	0.6	0.0	1.5	0.0
Carbonate (ammonium acetate pH 5.0)	39.0	168	7.2	2.4	30.6	0.0	96.0
Amorphous Mn (hydroxylamine hydrochloride pH 4.86)	258	1620	13.8	1.2	18.0	0.0	19.8
Amorphous Fe/ Mn (25 M hydroxylamine hydrochloride/ 0.25 M HCl)	1800	348	186	6.0	38.4	8.4	30.6
Crystalline Fe/ Mn (1 M hydroxylamine hydrochloride/ 25% acetic acid)	1500	14.4	6.6	3.0	0.0	18.0	11.4
Residual (HNO₃/HF/HCl)	85.0	12.5	7.5	0.0	0.0	7.5	10.0

depth of penetration is not important. Matrix or interelement effects, where the intensity of a particular fluorescence is not directly proportional to the concentration of the element, can be corrected using different mathematical models. Matrix effects may be suppressed when preparing glass discs by the addition of a heavy absorber (Potts, 1987; Alvarez, 1990).

Sample preparation for electron probe microanalysis (EPMA) requires sample chips or friable material to be impregnated in an epoxy resin. Sample surfaces are polished using progressively finer grades of diamond paste, with samples mounted onto aluminium stubs and coated with carbon prior to analysis. Samples should be clean, flat polished (2.5 cm diameter,

1 μm smoothness) and on solid mounts or uncovered petrographic thin-sections.

13.3 Mineralogical Analysis

13.3.1 Analytical methods for mineralogical analysis

X-ray diffraction (XRD)

X-ray diffraction is one of the most useful techniques for the identification of clay and associated minerals and has been discussed extensively (Klug and Alexander, 1974; Brindley and Brown, 1980; Wilson, 1987; Cullity and Rock, 2001). The basic principles are therefore discussed in brief. When a monochromatic X-ray beam of wavelength λ is projected onto a crystal lattice comprised of a competitive order of atoms arranged in parallel planes, separated by a distance *d*, at an angle θ, diffraction occurs. Diffraction occurs only when the distance travelled by the rays reflected from successive planes differs by a complete number of wavelengths *n*. The conditions of Braggs Law are satisfied by different *d*-spacings in polycrystalline materials when the angle θ is varied. This constructive interference is known as diffraction and its direction depends on the size and shape of the unit cell of a crystal and its intensity on the nature of the crystal structure. Instrumentation consists of an X-ray generator, a goniometer for sample rotation and measurement of diffraction angles, and an X-ray counter to detect, amplify and measure diffracted radiation. Copper *k* α radiation is the most popular due to its high intensity, but in instruments that do not have background correction facilities and where samples with high iron concentrations are analysed, cobalt *k* α radiation should be used (Brindley and Brown, 1980; Wilson, 1987; Cullity and Rock, 2001).

Scanning electron microscopy (SEM)

This technique is used to study the surface or near-surface characteristics of specimens, and is one of the most versatile and widely used instruments in science. In this technique, a beam of electrons from a thermionic emission type tungsten filament is accelerated to 20–40 KeV, demagnified and reduced in diameter to 2–10 nm on point of contact with a sample. The fine beam is scanned across the sample and a detector counts the number of low-energy secondary electrons or the radiation given off from each point on the surface (McHardy and Birnie, 1987; Newbury et al., 1987; Goodhew and Humphreys, 1988). Electron emission from a sample is divided into two energy regions, secondary and back-scatter. Secondary electrons have low energy and only those generated within a few nanometers of the surface escape and participate in the signal. Back-scattered electrons have energies close to the incident electron beam, escape from greater depths and their yield depends on the atomic number of the element. Topographical images can resolve features on the order of 2 nm and compositional analysis is provided using energy-dispersive X-ray spectrometry that shows spatial distribution of specific elements on a submicron scale.

High-resolution transmission electron microscopy (HRTEM)

In transmission electron microscopy, samples are bombarded with a highly focused beam of single-energy electrons that are transmitted through the sample. This electron signal is magnified by a series of magnetic lenses and observed either by electron diffraction or direct electron imaging. The former is used to determine the crystallographic structure and the latter provides information about the microstructure of the material. Transmission electron microscopy has undergone major developments to allow surface imaging to be carried out at the atomic level. Resolution in the Ångstrom range can be achieved by instruments which use coherent waves of a short wavelength and imaging lenses with small aberrations (Williams and Carter, 1996). High resolution with 300 kV acceleration potential can produce images of atomic lattices, and energy dispersive spectrometry performs qualitative and quantitative elemental analysis from fluorine to uranium.

Fourier transform infrared (FTIR) and Raman laser microprobe (RLMP) microscopy

Infrared and Raman spectroscopy, coupled with optical microscopy, provide vibrational data that allow us to chemically characterise geochemical sediments and weathered samples with lateral resolutions of 10–20 μm and 1–2 μm respectively. Fourier transform infrared spectroscopy involves the absorption of IR radiation, where the intensity of the beam is measured before and after it enters the sample as a function of the light frequency. Fourier transform infrared is very sensitive, fast and provides good resolution, very small samples can be analysed and information on molecular structure can be obtained. Weak signals can be measured with high precision from, for example, samples that are poor reflectors or transmitters or have low concentrations of active species, which is often the case for geochemical sediments and weathered materials. Samples of unknown composition can be identified using a database search and compositional and structural inhomogeneity at the micron level can be examined. Raman spectroscopy uses laser light scattering to measure vibrational frequencies of molecules, and both techniques provide a 'fingerprint' that reveals the oxidation state, molecular structure and speciation of the molecules that make up the material. Raman spectroscopy provides further information on crystalline phase, degree of order, strain and grain size. This technique can distinguish carbonate from elemental carbon and between molecular and crystalline structures, for example calcite versus aragonite (Woodward, 1967; Baker and Von Endt, 1988; Fadini and Schnepel, 1989; Smith, 1995; Pelletier, 1999; Wang, 1999; Kuebler et al., 2001; Wang et al., 2003). This technique combines Raman and Fourier transform spectroscopy and uses a Nd-YAG laser that emits radiation with a wavelength of 1064 nm to eliminate fluorescence signals. Non-destructive analysis with spatial mapping is possible and vacuum conditions are not necessary. Lasers provide a more intense frequency source, better resolution and easier focusing plus collimation of the radiation. Solid, powder and liquid samples may be analysed, however, limits of detection and sensitivity are poor and catalogues of Raman spectra are not as complete as those for infrared.

13.3.2 Thin and polished sections

Thin-section analysis of sedimentary materials gives a detailed description of their framework, porosity and grain size, and provides a means of interpreting the fine-grained minerals that make up their structure. However, these details are not easily obtained unless a good quality section is prepared and this depends on obtaining a perfect first surface preparation. This can be achieved by ensuring that the grinding lap on both mechanical and hand operated equipment is perfectly flat (Hutchison, 1974; Williams et al., 1982). Preparation procedures depend on sample type. For example, some geochemical sediments may be prepared without prior treatment, whereas weathering products are usually friable and porous and require vacuum impregnation with an epoxy resin. Water-soluble samples such as evaporites that dissolve, and dried mud and clays that collapse, should be impregnated in Araldite MY753 and HY956 or Santolite (Allman and Lawrence, 1972; Camuti and McGuirel, 1999). Pore spaces may be highlighted by adding Keyplast Blue A dye during the preparation. Surfaces should be prepared with abrasives using liquid paraffin as a lubricant, cleaned with kerosene or methylated spirit and mounted onto glass slides using molten Santolite. Samples consisting of cemented sands (e.g. many silcretes, calcretes and aeolianites), should be hand ground on a zinc lap using, for example, alumina abrasive and impregnated three or four times to ensure a suitable surface finish (Hutchison, 1974). Coherent samples are cut and ground to remove saw marks and surface impregnated on a hot plate at 120°C. Fine grinding may be carried out mechanically using different grades of silicon carbide or calcined alumina as abrasive. A variety of equipment is available for grinding and polishing thin-sections that range from multispecimen automated machines through variable speed to popular low-cost automated versions.

Samples should be mounted on thoroughly cleaned and ground glass slides with a uniform layer of the same epoxy resin that was used for impregnation. Araldite AY 105 with 935F hardener is very suitable since its refractive index is 1.55 and this value is best for the distinction of rock-forming minerals such as quartz, alkali feldspar and plagioclase. If sections are to be further analysed using, for example, EPMA, then slides of a correct size should be used that fit the mounting stage of that particular instrument. Thickness is reduced to approximately 200 μm using a diamond lap attachment (impregnated diamond wheel). Further reduction in thickness to approximately 40 μm is achieved by mechanical grinding using, for example, 600 grade silicon carbide or calcined alumina abrasive. A final thickness of 30–35 μm is obtained by hand grinding on a glass surface using 1500 grade calcined alumina as abrasive; hand grinding at this stage practically eliminates plucking. Sections can also be mechanically finished using, for example, a Logitech grinder and 600-grade carborundum abrasive powder. If specimens are to be examined using normal petrographic techniques, conventional covering methods using Canada Balsam can be used. (Hutchinson, 1974). However, if opaque minerals under reflected light or EPMA studies are required, then a polished thin-section is required and an epoxy resin replaces Canada Balsam. Polished sections allow the study of transparent and opaque minerals, reflectivity, microhardness and micro-surface chemical analysis. A high quality surface preparation is critical if high quality polished sections are to be obtained. Polishing can be achieved using successively finer grades of diamond paste, for example, 6 to 3 to 1 μm and the final finish is obtained using 0.25 μm diamond paste (Allman and Lawrence, 1972).

13.3.3 Sample preparation

Size fractionation is a very important preparatory step in mineralogical analysis and it is essential to know the particle size characteristics of material being analysed if results are to be reproducible. Preparation for most samples therefore begins with disaggregation and/or grinding. However, it must be emphasised that clay minerals can be easily damaged by mechanical treatment and it should be kept to a minimum. Samples may be mixed with a volatile inert organic liquid (e.g. isopropyl alcohol) prior to grinding to minimise overheating and damage to mineral lattices. Sample dispersion is necessary before fractionation, and care must be exercised when choosing a dispersing agent. Hydroxides of alkali metals cause spurious peaks when used prior to X-ray diffraction analysis, and ammonium hydroxide solution has been found to be more suitable (McKenzie and Mitchell, 1972; McAlister et al., 1988). Organic matter should be removed prior to analysis and oxidising agents such as hydrogen peroxide may form complex oxalato-aluminates or ferrates that must be removed by thoroughly washing the sample residue in deionised water (Farmer and Mitchell, 1963).

Ion exchange reactions are of fundamental importance for mineralogical analysis and a ground sample leached with, for example, ammonium acetate pH 4.5 can remove the cation of this solution and in turn liberate an equivalent amount of other exchangeable cations (H^+, Na^+, K^+, Ca^{2+}, Mg^{2+}) present in the sample. Care should be exercised when pretreating high carbonate calcretes due to their effervescence in acid solutions. Clay minerals should be homoionic and the exchangeable cation used to saturate their exchange complex has an important bearing on hydroscopic moisture content and on their identification. For example, magnesium ions allow relatively uniform interlayer adsorption of water by swelling-layer silicates, whereas potassium ions restrict adsorption by occupying sites on the basal oxygen sheets (Walker, 1961). As a result of this, cation saturation can be used to distinguish expanding from non-expanding clay minerals. Supplementary information for the identification of swelling clay minerals can be obtained by 'solvation' whereby organic complexes are formed by reacting clay minerals with liquids such as ethylene glycol or glycerol. There are different methods of applying these organic liquids, but the most reproducible technique is achieved by mixing the solvating agent directly with the clay suspension when it is in its maximum hydration state (Novich and Martin, 1983). Response of clay minerals to heat treatments and relative humidity changes should ideally be carried out on an XRD instrument that has a controlled heating stage or specimen chamber. However, these facilities may not be available and furnace heating followed by rapid scanning between 3–15° 2θ where the major basal spacings occur is a possible alternative. Humidity control can be achieved by equilibrating mounted samples over saturated salt solutions. However, low humidity control is difficult to achieve.

Clay minerals are frequently platy and a high degree of preferred orientation is required when preparing sample mounts for XRD analysis. This is related to their layer lattice characteristics and atomic sequence normal to the surface of the clay plate. There are numerous ways of preparing orientated mounts, and these include precipitation onto glass slides (Brown, 1953), suction onto unglazed ceramic tiles (Gibbs, 1965; Rich, 1975; Rhoton et al., 1993) and suction onto membrane filters (McAlister and

Smith, 1995a,b). The latter, more sensitive, technique was developed especially for samples with low clay concentrations (see Figure 13.2), where clay suspensions are precipitated onto 25 mm, 0.1 μm membrane filters. Preconcentration amongst fine-grained minerals is reduced using this technique, which can present a problem when slide mounts are used (Gibbs, 1965). Other advantages include speed, simplicity, low cost and the fact

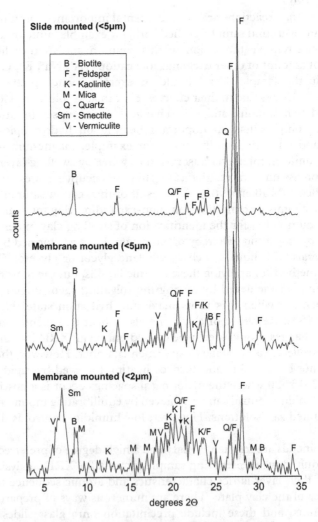

Figure 13.2 Diffractograms comparing XRD analysis from glass slide and membrane mounted samples of the same weathered granite from Meniet (central Algeria). Results show the greater sensitivity of the membrane technique and the effect of size fraction prior to analysis.

that these mounts can also be subsequently examined by SEM and other surface microanalytical techniques.

True random orientation mounts are difficult to prepare, and numerous techniques for filling powder sample holders have been described (Brindley and Brown, 1980). For example, powder samples have been embedded in plastic (Brindley and Kurtossy, 1961), polyester foam (Thompson et al., 1972), mixed with powdered cork (Wilson, 1987), pasted in acetone prior to smearing onto glass slides (Paterson et al., 1986) and spray dried (Hillier, 2002).

Sample preparation for SEM analysis includes grinding, polishing, etching (ionic, chemical or electrolytic) plus various mechanical and frozen fracture methods. Sample surfaces can be ultrasonically cleaned (where possible), ground and polished using artifical or natural abrasive powders. Chemical, electrochemical and etching methods have also been used (Brady, 1971; Goodhew, 1973). Prepared samples are mounted onto sonically cleaned aluminium stubs using adhesives such as silver paint, silver pastes, conductive epoxies or double-sided tape. Emphasis is placed on the importance of close contact between sample and stub in order to minimise charge build-up, image distortion and loss of sample. Drying procedures may cause some clays to shrink and methods such as freeze drying (Green-Kelly, 1973) and critical point drying (Boyd and Tamarin, 1984) have been used. Surfaces must finally be coated with a fine layer of gold to make them electrically conductive in order to remove surplus electrons and hence reduce a build-up of negative charge (Postek et al., 1980).

Sample preparation for transmission electron microscopy is similar to SEM, except that much smaller samples are required. This may employ the use of equipment such as a low-speed diamond saw for precision cutting, grinding apparatus with speeds between 50 and 500 rpm and equipment to make sample discs of 3 mm diameter. A Multiprep system for parallel, precise angle and site-specific polishing techniques and a dimpler lapping instrument may be used (e.g. VCR Group Model D500i) prior to ion milling. Ion milling eliminates the shadowing effect of the sample holder, radiation damage and specimen heating. Very thin sample specimens are cleaned and placed on a grid or glued to a special sample holder (Buseck et al., 1989; Mukhopadhyay, 2003; Spence, 2003). Sample thickness depends on how the sample affects electron scattering – the higher the atomic number of the elements in the sample, the thinner the final sample specimen needs to be (Goodhew, 1973; Mukhopadhyay, 2003). Sample storage and handling are important since it is the top layer (down to 20 nm) that is analysed. Non-reactive surface samples can be analysed without special coatings on introduction into the vacuum chamber (Mukhopadhyay, 2003).

Sample preparation for FTIR and RLMP have significant differences. Since the former is an absorption process, geochemical sediments (2.5 μm) must be ground thoroughly with KBr at 1–3% by weight and pressed into a fine transparent pellet. This process is necessary as samples may absorb some bands and light will not penetrate the sample. These complications are avoided with Raman measurements, since this technique is based on a scattering phenomenon and samples may be placed on a microscope slide for viewing (Adar, 2001; Sommer, 2002).

13.4 Isotopic Analysis

13.4.1 Introduction

An isotope is one of two or more species of an atom that have the same atomic number but differ in mass number. Therefore, isotopes of the same element differ from each other by the number of neutrons present in their nuclei and have slightly different chemical and physical properties due to their mass differences. When an isotope (the parent) loses particles to form an isotope of a new element (the daughter), a spontaneous process known as radioactive decay occurs. The rate of decay is expressed as an isotope's half-life and most unstable (radioactive) isotopes have short half-lives. However, some isotopes decay slowly and several of these are used as geological clocks. The majority of elements in the natural state have two or more isotopes, with the exceptions of Be, Al, P and Na. Elements with atomic number >83 and a few of the lighter ones (e.g. ^{40}K) are radioactive and are referred to as unstable isotopes since they decay over time and form other isotopes. Stable isotopes do not decay to other isotopes at the geological timescale, but may themselves be produced by the decay of radioactive isotopes. Isotopes of light elements, for example, H, O, C, N and S, have large enough mass differences between them to fractionate. Isotopic fractionation is caused by equilibrium and kinetic effects. In the former, forward and backward reactions are identical and the heavier isotopes generally become enriched in the species with higher energy state, for example sulphate is enriched with ^{34}S relative to sulphide. In contrast, kinetic isotopic fractionation occurs when forward and backward reactions are not identical and reaction rates depend on the masses of the isotopes and their vibrational energies. Bonds between lighter isotopes are more easily broken than in the heavier ones and reactions occur more readily and light isotopes become concentrated in the products.

13.4.2 Analysis of stable isotopes

Stable isotopes of the elements oxygen, hydrogen, sulphur, nitrogen and carbon can provide useful insights into weathering mechanisms, ecosystem dynamics, hydrothermal ore-forming systems, climate change and atmospheric processes (Rundel et al., 1989; Fricke and O'Niell, 1999; Alexandre et al., 2004). Materials such as silicate ($^{18}O/^{16}O$, $^{2}H/^{1}H$), carbonate ($^{13}C/^{12}C$, $^{18}O/^{16}O$), sulphide ($^{34}S/^{32}S$) and sulphate ($^{34}S/^{32}S$, $^{18}O/^{16}O$) minerals, plus organic matter ($^{13}C/^{12}C$, $^{15}N/^{14}N$) and waters ($^{2}H/^{1}H$, $^{18}O/^{16}O$, $^{34}S/^{32}S$), can be automatically measured using gas isotope-ratio mass spectroscopy. Sample preparation involves quantitative conversion or production of pure gas from the sample. The gas is cryogenically or chromatographically purified and introduced to the mass spectrometer where it is ionised to produce positively charged species. The different masses are dispersed in a magnetic field and the ratios of the isotopes present in the ionised gas are measured. Geochemical sediments are oven-dried (80°C, 24 h) or, if they contain nitrogen in the ammonicial form, they must be freeze-dried before analysis. Sample size and total element concentration have a major effect on isotopic analysis. Ideally samples should contain, for example, 200 µg of total N and 800 µg of total C. However, these may be analysed at 25 and 200 µg concentrations respectively. Dried samples are ground to < 250 µm particle size using an efficient ball mill to ensure homogeneity. This is especially important in the case of small samples.

Isotopic compositions are recorded as 'delta' (δ) values in parts per thousand (‰) enrichments or depletions relative to a standard of known composition. Analytical precisions for oxygen, carbon, nitrogen and sulphur are in the range of 0.05–0.2‰ and hydrogen is in the range of 0.2–1.0‰ (Fritz and Fontes, 1986; Clauer and Chaudhuri, 1992; Coplen, 1996; Werner and Brand, 2001).

13.4.3 Analysis of unstable isotopes

Isotopes of the same element are extremely difficult to separate and impossible by one-step chemical techniques since they possess the same chemical properties. Mass differences in the isotopes cause extremely small differences in their physical properties and this is the basis of physical separation techniques. Separation is carried out via a large number of stages known as a 'cascade', with each stage having a higher level of purity than the previous one. An enriched fraction from any one stage can become the raw material for the next (Faure, 1986; Vegors and Nieschmidt, 2001).

After the separation and purification stages, isotopic abundances are measured in most cases by mass spectrometry. Exceptions include short-lived radioactive isotopes, where decay rates are measured by fission track dating that measures the abundance of ^{238}U by induced fission. Ionisation is carried out using a thermal method for a solid source or electron bombardment for a gas source. Efficiency of ionisation determines the amount of sample required for analysis, and efficiency may range from almost 100%

down to 0.1% for a number of elements. Electron bombardment of a gas stream causes electron–molecule collisions, one of the electrons is knocked out of its orbit and this ionises the molecule or atom. Other techniques use ion sputtering where solid samples are bombarded with positive ions. The resulting ions are focused and accelerated by electrostatic lenses, passed through a magnetic field where they are deflected according to their charge/mass ratio in the mass analyser (Al-Aasm et al., 1990; Dikin, 1995).

Various geochronological methods are available for dating geochemical sediments and these include U-series dating for Quaternary sedimentary carbonate and silica materials. To date some geochemical sediments it may be possible to isolate unusual minerals such as glauconite that contain K and allow a K–Ar dating technique to be used. In order to relate materials that are less easily dated (e.g. silcrete) to the radiometric timescale, they may need to be bracketed within time zones determined by dating appropriately selected igneous rocks (Ludwig and Paces, 2002). Strontium-isotope measurements can be used to date marine fossils and processes such as dolomitisation. Dating between 10,000 and billions of years is possible using $^{40}Ar/^{39}Ar$ with a precision of 1–0.5%. Tephrachronology may be used to date volcanic ash and tuff within samples up to 15 million years old, ^{14}C dates material containing organic carbon up to a maximum of 100,000 years, and ^{210}Pb has a short half-life that permits dating over a period of 150–200 years (Geyh and Schleicher, 1990; Nelson, 1994; Mojzsis and Harrison, 1999; Ludwig, 2000; Link et al., 2003).

Advances have been made in geochemistry and geochronology, especially in the use of cosmogenic nuclides by the introduction of accelerator mass spectrometry (AMS). This is a leading technique for detection of long-lived radionuclides such as ^{10}Be, ^{14}C, ^{36}Al and ^{129}I. This instrument uses an ion accelerator and its beam transport system to give an ultrasensitive technique that can provide several stages of mass and charge analysis and to ultimately count individual atoms. Accelerator mass spectrometry is unaffected by half-life, since atoms and not radiation are the result of this decay, and it is unaffected by almost all background interference that limits conventional mass spectrometry. Sample size is reduced to a few milligrams after processing and a blank must be included when analysing unknown samples (Tuniz et al., 1998; Fifield, 1999).

13.5 Conclusions

This chapter has presented detailed information on the more common analytical instrumentation and preparation procedures used in chemical and mineralogical analysis, and an overview of selected microanalytical techniques that still require development in this field. It must be emphasised that the quality of results and the information that can be obtained from any instrumental technique is highly dependent on the analyst choosing the best method for sample preparation, and carrying this out with accuracy and precision. Factors such as contamination during grinding, digestion procedures and carefully chosen extraction protocols for selective extraction studies are of vital importance. A wide range of analytical techniques is available. However, a number of these are not yet fully developed for the analysis of terrestrial geochemical sediments or economically viable for a number of academic institutions. Nevertheless, methodological development and availability will no doubt improve and it is important that as researchers we are fully aware of the wide range of techniques available along with their strengths and limitations.

References

Adar, F. (2001) Evolution and revolution of Raman instrumentation – applications of available technologies to spectroscopy and microscopy. In: Lewis, I.R. & Edwards, H.G.M. (Eds) *Handbook of Raman Spectroscopy*. New York: Marcel Dekker, pp. 11–40.

Al-Aasm, I.S., Taylor, B.E. & South, B. (1990) Stable isotope analysis of multiple carbonate samples using selective acid extraction. *Chemical Geology* **80**, 119–125.

Alexandre, A., Meunier, J.D., Llorens, E., Hill, S.M. & Savin, S.M. (2004) Methodological improvements for investigation silcrete formation: petrography, FT–IR and oxygen ratio of silcrete quartz cement, Lake Eyre Basin (Australia). *Chemical Geology* **211**, 261–274.

Allman, M. & Lawrence, D.F. (1972) *Geological Laboratory Techniques*. London: Blandford Press.

Alvarez, M. (1990) Glass disc fusion method for the X-ray fluorescence analysis of rocks and silicates. *X-Ray Spectrometry* **19**, 203–206.

Antropova, L.V., Goldberg, L.S., Voroshilov, N.A. & Ryss, Ju. S. (1992) New methods of regional exploration for blind mineralization: application in the USSR. *Journal of Geochemical Exploration* **43**, 157–166.

Bain, D.C. & Smith, B.E.L. (1994) Chemical analysis. In: Wilson, M.J. (Ed.) *Clay Mineralogy: Spectroscopic and Chemical Determinative Methods*. London: Chapman and Hall, pp. 300–332.

Bain, D.C., McHardy, W.J. & Lachowski, E.E. (1994) X-ray spectroscopy and microanalysis. In: Wilson M.J (Ed.) *Clay Mineralogy: Spectroscopic and Chemical Determinative Methods*. London: Chapman and Hall, pp. 260–299.

Baker, M.T. & Von Endt, D.W. (1988) Use of FTIR microspectrometry in examinations of artistic and historic works. *Proceedings of the Materials Research Symposium* **123**, 71–76.

Becker, J.S. & Dietze, H.J. (1998) Inorganic trace analysis by mass spectrometry. *Spectrochimica Acta* **53B**, 1475–1506.

Benedict, J.P., Klepeis, S.J., Vandygrift, W.G. & Anderson, R. (1989) A method of precision specimen preparation for both SEM and TEM analysis. *ESMA Bulletin* **19**, G.W. Bailey (Ed.) San Francisco Press, p. 712.

Berman, S.S. (1988) Acid digestion of marine samples for trace element analysis using microwave heating. *Analyst* **113**, 159–163.

Bernas, B.J. (1968) A new method for decomposition and comprehensive analysis of silicates by atomic spectrometry. *Analytical Chemistry* **40**, 1682–1686.

Boyd, A. & Tarmarin, A. (1984) Improvement to critical point drying technique for SEM. *Scanning* **6**, 30–35.

Brady, G.S. (1971) *Materials Handbook*, 10th edn. New York: McGraw-Hill.

Brindley, G.W. & Brown, G (1980) *Crystal Structures of Clay Minerals and their X-ray Identification*. London: Mineralogical Society.

Brindley, G.W. & Kurtossy, S.S. (1961) Quantitative determination of kaolinite by X-ray diffraction. *American Mineralogist* **46**, 1205–1215.

Brown, G. (1953) A semi-micro method for preparation of clays for X-ray study. *Journal of Soil Science* 4, 229–232.

Buhrke, V.E., Creasy, L.E., Croke, J.F., Feret, F., Jenkins, R., Kanare, H.M. & Kocman, V. (1998) Specimen preparation in X-ray fluorescence. In: *Practical Guide for Preparation of Specimens for X-ray Fluorescence and X-ray Diffraction Analysis*. New York: Wiley-VCH, pp. 59–122.

Buseck, P., Cowley, J. & Eyring, L. (Eds) (1989) *High Resolution Transmission Electron Microscopy*. Oxford: Oxford University Press.

Camuti, K.S. & McGuirel, P.T. (1999) Preparation of polished thin sections from poorly consolidated regolith. *Sedimentary Geology* **128**, 171–178.

Cave, M.R., Butler, O., Cook, J.M., Cresser, M.S., Garden, L.M. & Miles, D.L. (2000) Environmental analysis. *Journal of Analytical Atomic Spectrometry* **15**, 181–235.

Clauer, N. & Chaudhuri, S. (1992) *Isotopic Signatures and Sedimentary Records*. Lecture Notes in Earth Sciences, No. 43, New York: Springer-Verlag.

Cohen, D.D., Siegele, R., Orlic, I. & Stelcer, E. (2001) Long term accuracy and precision of PIXE and PIGE measurements for thin and thick sample analysis. *Proceedings of the 9th International PIXE Conference*, Guelph, 8–12 June.

Compston, W., Williams, I.S. & Black, L.P. (1982) Use of the ion microprobe in geological dating. In: *BMR 82, Yearbook of Bureau of Mineral Resources, Geology and Geophysics*. Canberra: Australian Government Publishing Services, pp. 39–42.

Coplen, T.B. (1996) New guidelines for reporting stable hydrogen, carbon and oxygen isotope-ratio data. *Geochimica et Cosmochimica Acta* **60**, 3359–3360.

Cox, J.A. & Twardowski, Z. (1980) Donnan dialysis, matrix normalisation for the volumetric determination of metal ions. *Analytica Chimica Acta* **119**, 39–45.

Cox, J.A., Dabex-Zlotorzynska, E., Saari, R. & Tanaka, N. (1988) Ion exchange treatment of complicated samples prior to ion chromatographic analysis. *Analyst* **113**, 109–250.

Cullity, B.D. & Rock, B.R. (2001) *Elements of X-Ray Diffraction*, 3rd edn. London: Prentice Hall.

Danilatos, G.D. (1988) Foundations of environmental scanning electron microscopy. *Advances in Electronic and Electron Physics* **71**, 109–250.

Date, A.R. & Gray, A.L. (1989) *Applications of Inductive Plasma Mass Spectrometry*. Glasgow: Blackie.

Dikin, A. (1995) *Radiogenic Isotope Geochemistry.* Cambridge University Press, Cambridge.

Dzombak, D.A. & Morel, F.M.M. (1990) *Surface Complexation Modelling: Hydrous Ferric iron Oxide.* New York: John Wiley & Sons.

Eldridge, C.S., Compston, W., Williams, I.S., Walshe, J.L. & Both, R.A. (1987) In situ microanalysis of $^{34}S/^{32}S$ ratios using an ion microprobe SHRIMP. *International Journal of Mass Spectroscopy and Ion Processes* 76, 65–83.

Evans, K.C. & Moore, B.B. (1980) Combustion ion chromatographic determination of chlorine in silicate rocks. *Analytical Chemistry* 58, 1908–1912.

Fadini, A. & Schnepel, F.M. (1989) *Vibrational Spectroscopy.* Ellis Horwood, Chichester.

Farmer, V.C.& Mitchell, B.D. (1963) Occurrence of oxalates in soil clays following hydrogen peroxide treatment. *Soil Science* 96, 221–229.

Faure, G. (1986) *Principles of Isotopic Geology,* 2nd edn. New York: John Wiley & Sons.

Feng, R. (1994) *In situ* trace element determination of carbonates by laser probe inductively coupled plasma mass spectrometry using non-matrix matched standardization. *Geochimica et Cosmochimica Acta* 58, 1615–1623.

Feng, R., Machado, N. & Ludden, J. (1993) Lead chronology of zircon by laser probe-inductively coupled mass spectrometry (LP–ICP–MS). *Geochimica et Cosmochimica Acta* 57, 3479–3486.

Fifield, L.R. (1999) Accelerator mass spectrometry and its applications. *Reports on Progress in Physics* 62, 1223–1274.

Fletcher, I.R., Rasmussen, B. & McNaughton, N.J. (2000) SHRIMP U–Pb geochronology of authigenic xenotime and its potential for dating sedimentary basins. *Australian Journal of Earth Sciences* 47, 845–859.

Fricke, H.C. & O'Niell, J.R. (1999) The correlation between $^{18}O/^{16}O$ ratios of meteoric water and surface temperature: its use in investigating climate change over geological time. *Earth and Planetary Science Letters* 170, 181–196.

Fritz, P. & Fontes, J.Ch. (1986) *Handbook of Environmental Isotope Geochemistry,* Vol. 2, New York: Elsevier.

Geyh, M.A., Schleicher, H. (1990) *Absolute Age Determination: Physical and Chemical Methods and Their Application.* Dordrecht: Springer Verlag.

Gibbs, R.J. (1965) Error due to segregation in quantitative clay mineral X-ray diffraction mounting techniques. *American Mineralogist* 50, 741–751.

Goodhew, P.J. (1973) Specimen preparation in materials science. In: Glauent, A.M. (Ed.) *Practical Methods in Electron Microscopy.* New York: N. Holland Publishing.

Goodhew, P.J. & Humphreys, F.J. (1988) *Electron Microscopy and Analysis.* New York: Taylor and Francis.

Goudie, A.S. & Pye, K. (Eds) (1983) *Chemical Sediments and Geomorphology.* London: Academic Press.

Green-Kelly, R. (1973) The preparation of clay soils for the determination of structure 2. *Journal of Soil Science* 24, 277–283.

Guo, X. & Lichte, F.E. (1995) Analysis of rocks, soils and sediments for the chalcophile elements by laser ablation inductively coupled plasma-mass spectrometry. *Analyst* 120, 2707–2711.

Haddad, P.R. & Jackson, P.E. (1990) *Ion Chromatography, Principals and Applications*. Amsterdam: Elsevier.

Hall, G.E.M., Vaive, J.E., Beer, R.Y. & Hoashi, M. (1996) Selected leaches revisited, with emphasis on the amorphous Fe oxyhydroxide phase extraction. *Journal of Geochemical Exploration* **56**, 59–78.

Hamilton, J.A., Royko, B., Burtt, M. & Nichol, I. (1992) *Geochemical Exploration Applied to Base Metal and Gold Exploration in Ontario*. Kingston, Ontario: Department of Geological Sciences Report, Queen's University.

Hesse, P.R. (1971) *A Textbook of Soil Chemical Analysis*. London: John Murray.

Hillier, S. (2002) Spray drying for X-ray powder diffraction specimen preparation. *IUCR Commission on Powder Diffraction, Newsletter* **27**, 7–9.

Hutchison, C.S. (1974) *Laboratory Handbook of Petrographic Techniques*. New York: John Wiley & Sons.

Ingamells, C.O. (1966) Absorptiometric methods in rapid silicate analysis. *Analytical Chemistry* **38**, 1228–1234.

Jenkins, R. (1999) *X-Ray Fluorescence Spectrometry*, 2nd edn. Chichester: John Wiley & Sons.

Jenner, G.A., Foley, S., Jackson, S.E., Green, B.J., Fryer, B.J. & Longerich, P. (1994) Determination of partition coefficients for trace elements in high pressure-temperature experimental run products by laser ablation microprobe-inductively coupled plasma-mas spectrometry. *Geochemica et Cosmochimica Acta* **58**, 5099–5103.

Kamp, H.H. & Krist, H. (Eds) (1988) *Laboratory Manual for the Examination of Water, Wastewater and Soil*. Weinheim: Verlagsgesellschaft mbH.

Kemp, A.J. & Brown, C.J. (1990) Microwave digestion of carbonate rock samples for chemical analysis. *Analyst* **115**, 1197–1199.

Kingston, H.M. & Haswell, S. (1995) *Microwave Enhanced Chemistry*. Washington DC: American Chemical Society.

Kingston, H.M.& Jassie, L.B. (1988) *Introduction to Microwave Sample Preparation*. Washington, DC: American Chemical Society.

Kingston, H.M. & Walter, P.J. (1997) The art and science of microwave sample preparation for trace and ultratrace analysis. In: Montaser, A. (Ed.) *Inductively Coupled Plasma Mass Spectrometry*. New York: Wiley-VCH, pp. 33–53.

Klug, H.P. & Alexander, L.E. (1974) *X-Ray Diffraction Procedures*. New York: John Wiley & Sons.

Kuebler, K.E., Wang, A., Abbott, K., Haskin, L.A. (2001) Can we detect carbonate and sulfate minerals on the surface of Mars by Raman spectroscopy? *32nd Lunar and Planetary Science Conference*, Houston, TX, Abstract 1889.

Lajunen, L.H.T. (1992) *Spectrochemical Analysis by Atomic Absorption and Emission*. London: Royal Society of Chemistry.

Laue, S.J. & Dalton, J.A. (1994) Electron microprobe analysis of geological carbonates. *American Mineralogist* **79**, 745–749.

Li, M.J., Rodgers, K. & Rust, C.A. (1995) Environmental scanning microscopes. *Advanced Materials and Processes* **7**, 24–25.

Link, P.K., Pocatells, I.D. & Mahoney, I.B. (2003) *Isotopic Determination of Sediment Provenance: Techniques and Applications*. GSA Abstracts, Vol. 35, No. 6. Boulder, CO: GSA, Geological Society of America, p. 466.

Liu, H., Liu, Y. & Zhang, Z. (1998) Determination of ultra-trace rare earth elements in chondritic meteorites by inductively coupled plasma mass spectrometry. *Spectrochimica Acta* **53B**, 1399–1404.

Longerich, H.P., Jackson, S.E., Fryer, B.J. & Strong, D.F. (1993) The laser ablation microprobe-inductively coupled plasma-mass spectrometer. *Geoscience Canada* **20**, 21–27.

Ludwig, K.R. (2000) Processes and challenges in the numerical evaluation of geochronological data, Abstracts and proceedings of the Beyond 2000 New Frontiers. In: Woodhead, J.D., Hergt, J.M., Noble, W.P. (Eds) *Isotope Geoscience Conference, Lorne Australia*. Victoria, Australia: Eastern Press, pp. 177–199.

Ludwig, K.R. & Paces, J.B. (2002) Uranium series dating of pedogenic silica and carbonate, Crater Flat, Nevada. *Geochimica et Cosmochimica Acta* **66**, 487–506.

Martin, J.M., Nirel, P.M.V. & Tomas, A. (1987) Sequential extraction techniques: promises and problems. *Marine Chemistry* **22**, 313–341.

Matthes, S.A., Farrell, R.F. & Mackie, A.J. (1983) *Technical Process Report 120*. Washington, DC: Bureau of Mines.

McAlister, J.J. & Smith, B.J. (1995a) A pressure membrane technique to prepare clay samples for X-ray diffraction analysis. *Journal of Sedimentary Research* **A65**, 569–571.

McAlister, J.J. & Smith, B.J. (1995b) A rapid preparation technique for X-ray diffraction analysis of clay minerals in weathered rock material. *Microchemical Journal* **52**, 53–61.

McAlister, J.J. & Smith, B.J. (1999) Selectivity of ammonium acetate, hydroxylamine hydrochloride and oxalate/ascorbic acid solutions for the speciation of Fe, Mn, Zn, Cu, Ni and Al in early Tertiary paleosols. *Microchemical Journal* **63**, 415–426.

McAlister, J.J., Svehla, G. & Whalley, W.B. (1988) A comparison of various pre-treatment and instrumental techniques for the mineral characterisation of chemically weathered basalt. *Microchemical Journal* **38**, 211–231.

McAlister, J.J., Smith, B.J. & Curran, J.A. (2003) The use of sequential extraction to examine iron and trace metal mobilisation and the case hardening of building sandstone: a preliminary investigation. *Microchemical Journal* **74**, 5–18.

McHardy, W.J. & Birnie, A.C. (1987) Scanning electron microscopy. In: Wilson, M.J. (Ed.) *A Handbook of Determinative Methods in Clay Mineralogy*. New York: Chapman and Hall, pp. 174–208.

McKenzie, R.C. & Mitchell, B.D. (1972) Soils. In: McKenzie, R.C. (Ed.) *Differential Thermal Analysis*, Vol. 2. London: Academic Press, pp. 276–297.

Meisel, T. (2000) Determination of rare earth elements (REE) in geological samples by sodium peroxide sintering and ICP-MS detection Agilent Technologies, *ICP–MS Journal* **8**, 6–10.

Mitra, S. (Ed.) (2003) *Sample Preparation Techniques in Analytical Chemistry*. Chichester: John Wiley & Sons.

Mojzsis, S.J. & Harrison, T.M. (1999) Geochronological studies of the oldest known marine sediments. *9th Annual V.M. Goldschmidt Conference, Abstract 7602*, Houston, Lunar and Planetary Institute.

Moschini, G. & Valkovic, V. (1996) Particle induced x-ray emission and its analytical applications. *Nuclear Instrumental Methods* **B109**, 1–705.

Mossotti, V.G., Lindsey, J.R. & Hochella, M.F. Jr. (1987) Effect of acid rain environment on limestone surfaces. *Materials Performance* **26**, 47–52.

Mukhopadhyay, S.M. (2003) Sample preparation for microscopic and spectroscopic characterization of solid surfaces and films. In: Mirta, S. (Ed.) *Sample Preparation Techniques in Analytical Chemistry*. Chichester: John Wiley & Sons, pp. 377–410.

Nelson, D.R. (1994) *Compilation of SHIRMP U–Pb Zircon Geochronology Data*. Perth: Geological Survey Western Australia.

Newbury, D.E., David, C., Echlin, J.P., Fiori, C.E. & Goldstien, J.I. (1987) *Advanced Scanning Election Microscopy and X-Ray Microanalysis*. New York: Plenum Press.

Nirel, P.M.V. & Morel, F.M.M. (1990) Pitfalls of sequential Extractions. *Water Research* **24**, 1055–1056.

Norrish, K. & Chappell, B.W. (1977) X-ray fluorescence spectrometry. In: Zussmann J. (Ed.) *Physical Methods in Determinative Minerology*, 2nd edn. London: Academic Press, pp. 201–272.

Novich, B.E. & Martin, R.T. (1983) Solvation methods for expandable layer clays. *Clays and Clay Mineralogy* **31**, 235–238.

Papp, C.S.E. & Fischer, L.B. (1987) Application of microwave digestion to the analysis of peat. *Analyst* **112**, 337–338.

Paterson, E., Bunch, J.L. & Duthie, D.M.L. (1986) Preparation of randomly orientated samples for X-ray diffractometry. *Clay Mineralogy* **21**, 101–106.

Pelletier, M.J. (1999) *Analytical Applications of Raman Spectroscopy*. Oxford: Blackwell Science.

Petterson, J.M., Johnston, H.G. & Lund, W. (1988) The determination of nitrate and sulphate in 50% sodium hydroxide solution by ion chromatography. *Talanta* **35**, 245–247.

Pickering, W.F. (1981) Selective chemical extraction of soil components and bound metal species. *Critical Reviews in Analytical Chemistry* **2**, 233–266.

Postek, M.T., Howard, K.S., Johnson, A.H. & McMichael, K.L. (1980) *Scanning Electron Microscopy, A Students Handbook*. Williston, Vermont: Ladd Research Industries.

Potts, P.J. (1987) Principals and practices of wavelength dispersive spectrometry. In: *A Handbook of Silicate Rock Analysis, X-ray Fluorescence Analysis*. Glasgow: Blackie, pp. 226–285.

Quevauviller, P., Rauret, G. & Muntau, H., Ure, A.M., Rubio, R., Lopez-Sanchez, J.F., Fiedler, H.D. & Griepink, B. (1994) Evaluation of sequential extraction procedures for the determination of extractable trace elements in sediments. *Fresenius Journal of Analytical Chemistry* **349**, 808–814.

Reed, S.J.B. (1993) *Electron Probe Microanalysis*, 2nd edn. Cambridge: Cambridge University Press.

Rhoton, F.E., Grissinger, E.H. & Bingham, J.M. (1993) An improved suction apparatus for placing clay suspensions onto ceramic tiles. *Journal of Sedimentary Research* **63**, 763–765.

Rich, C.I. (1975) Determination of the amount of clay needed for X-ray diffraction analysis. *Proceedings of the Soil Science Society of America* **39**, 161–162.

Rose, A.W. (1975) The mode of occurrence of trace elements in soils and stream sediments applied to geochemical exploration. *Proceedings of the 5th International. Geochemistry Exploration Symposium*, Vancouver, BC., pp. 691–705.

Rundel, P.W., Ehlermyer, J.R. & Nayg, A. (Eds) (1989) *Stable Isotopes in Ecological Research.* New York: Springer-Verlag.

Sastre, J., Sahuquillo, A., Vidal, M. & Rauret, G. (2002) Determination of Cd, Cu, Pb, and Zn in environmental samples: microwave-assisted total digestion versus *aqua regia* and nitric acid extraction. *Analytica Chimica Acta* **462**, 59–72.

Sato, T., Yanase, N., Williams, I.S., Compston, W., Zaw, M., Payne, T.E. & Airey, P.L. (1998) Uranium micro-isotopic analysis of weathered rock by sensitive high resolution ion microprobe (SHRIMP). *Radiochimica Acta* **82**, 335–340.

Schalek, R.L. & Drzal, L.T. (2000) Characterisation of advanced materials using an environmental SEM. *Journal of Advanced Materials* **32**, 32–38.

Scott, V.D. (1995) *Quantitative Electron Probe Microanalysis.* Princeton, NJ: Prentice Hall.

Shapiro, L. (1967) Rapid analysis of rocks and minerals by a single solution method. *U.S. Geological Survey Professional Paper* **575B**, B187–B191.

Skoog, D.A. (1985) *Principals of Instrumental Analysis.* New York: Sanders College Publishing.

Small, H. (1989) *Ion Chromatography.* New York: Plenum Press.

Smith, B.C. (1995) *Fundamentals of Fourier Transform Spectroscopy.* Boca Raton, FL: CRC Press.

Smith, F.C. & Chang, R.C. (1983) *The Practice of Ion Chromatography.* New York: John Wiley & Sons.

Spence, J.C.H. (2003) *High Resolution Electron Microscopy.* Monographs on the Physics and Chemistry of Materials, No. 60, London: Clarendon Press.

Sommer, A.J. (2002) Mid-infrared transmission microscopy, In: Griffiths, P. and Chalmers, J. (Eds), *Handbook of Vibrational Spectroscopy*, Vol. 2. New York: John Wiley & Sons.

Taylor, H.E. (2001) *Inductively Coupled Plasma–Mass Spectrometry. Practices and Techniques.* New York: Academic Press.

Tesmer, J.R. & Nastasi, M. (1995) *Handbook of Modern Ion Beam Materials Analysis.* Pittsburg: Materials Research Society.

Tessier, A., Campbell, P.G.C. & Bisson, M. (1979) Sequential procedure for the speciation of particulate trace metals. *Analytical Chemistry* **51**, 844–851.

Thomas, R. (2003) *Practical Guide to ICP-MS.* New York: Marcel Dekker.

Thompson, A.P., Duthie, D.M.L. & Wilson, M.J. (1972) Random orientated powders for quantitative determination of clay minerals. *Clay Mineralogy* **9**, 345–348.

Thomsen, V. & Schatzlein, D. (2002) Advances in field-portable XRF. *Spectroscopy* **17**, 14–21.

Tuniz, C., Baird, J.R., Fink, D. & Herzog, G.F. (1998) *Accelerator Mass Spectrometry, Ultra Sensitive Analysis for Global Science.* Boca Raton, FL: CRC Press.

Ure, A.M., Quevauviller, Ph., Mantau, H. & Griepink, B. (1993) Speciation of heavy metals in solids and sediments. An account of the improvement and harmonisation of extraction techniques undertaken under the auspices of the BCR of the Commission of the European Communities. *International Journal of Environmental and Analytical Chemistry* **51**, 135–151.

Van Valin, R. & Morse, J.W. (1982) An investigation of methods normally used for selective removal and characterisation of trace metals in sediments. *Marine Chemistry* **11**, 535–564.

Vandecasteele, C. & Block, C.B. (1993) *Modern Methods for Trace Element Determination*. John Wiley & Sons, New York.

Vegors, S.H. & Nieschmidt, E.B. (2001) *Laser Separation of Isotopes, 43rd Meeting Idaho Acadamy of Science March 29–31*. Albertson College of Idaho.

Verbeek, A.A., Mitchell, M.C. & Ure, A.M. (1982) The analysis of small samples of rocks and soils by atomic absorption and emission spectroscopy after lithium metaborate fusion/nitric acid dissolution procedure. *Analytica Chimica Acta* 135, 215–228.

Walker, G.G. (1961) Vermiculite minerals. In: Brown, G. (Ed.) *The X-ray Identification and Crystal Structures of Clay Minerals*. London: Mineralogical Society, pp. 297–324.

Walter, P.J. & Kingston, H.M. (1995) *Total Microwave Processing Using Microwave Technologies*. Cincinnati, OH: Federation of Analytical Chemistry and Spectroscopy Societies.

Wang, A. (1999) Some grain size effects on Raman scattering for in situ measurements on rocks and soils-experimental tests and modelling. *30th Lunar and Planetary Science Conference*, Houston, TX.

Wang, A., Kuebler, K.E., Jolliff, B.L. & Haskin, L.A. (2003) Fe–Ti–Cr oxides in Martian meteorite EETA 740001 studied by point counting procedure using Raman spectroscopy. *Abstract 1742, 34th Lunar and Planetary Science Conference*, League City, TX.

Watt, G.R., Griffin, B.J. & Kinny, P.D. (2000) Charge contrast imaging of geological materials in the environmental scanning electron microscope. *American Mineralogist* 85, 1784–1794.

Werner, R.A. & Brand, W.A. (2001) Referencing strategies and techniques in stable isotope ratio analysis. *Rapid Communications in Mass Spectrometry* 15, 501–519.

Williams, B.B. & Carter, C.B. (1996) *Transmission Electron Microscopy*. New York: Plenum Press.

Williams, H., Turner, F.J. & Gilbert, C.M. (1982) *Petrography: An Introduction to the Study of Rocks in Thin Sections*, 2nd edn. San Francisco: W.H. Freeman and Company.

Williams, J.G. & Jarvis, K.E. (1993) Preliminary assessment of laser ablation inductively coupled plasma mass spectrometry for quantitative multielement determination of silicates. *Journal of Analytical and Atomic Spectrometry* 8, 25–34.

Wilson, M.J. (1987) X-ray powder diffraction methods. In: Wilson, M.J. (Ed.) *A Handbook of Determinative Methods in Clay Mineralogy*. New York: Chapman and Hall, pp. 26–98.

Wilson, Y.M.K. & Burns, G. (1987) X-ray photoelectron spectroscopy of ancient murals and tombs at Beni Hasan, Egypt. *Canadian Journal of Chemistry* 65, 1058–1064.

Wilson, S.A. & Gent, C.A. (1983) Determination of chloride in geological samples by ion chromatography. *Analytica Chimica Acta* 148, 299–303.

Woodward, L.A. (1967) General introduction. In: Woodward, L.A. (Ed.) *Raman Spectroscopy Theory and Practice*. New York: Plenum Press, pp. 1–43.

Zlotorzynski, A. (1995) The application of microwave radiation to analytical and environmental chemistry. *Critical Reviews in Analytical Chemistry* 25, 43–76.

Chapter Fourteen

Geochemical Sediments and Landscapes: General Summary

Sue J. McLaren and David J. Nash

The overall aim of this final chapter is to draw together some of the salient characteristics, similarities and differences between the various types of terrestrial geochemical sediment covered in this volume. For further details and references to the appropriate literature please see the relevant chapters.

As the preceding chapters have demonstrated, there is a wide variety of geochemical sediments that exist in surface or near-surface terrestrial environments, ranging from surficial deposits such as rock varnish to silcrete lenses forming up to 100 m below ground level. In general, these can be grouped into those that actually alter a host material through processes of weathering, lithification, replacement and/or displacement of the parent rock, soil or saprolite fabric (e.g. laterites, Widdowson, Chapter 3) through to those that form coatings (e.g. on rocks, Dorn, Chapter 8), crusts (e.g. nitrate deposits, Goudie and Heslop, Chapter 12), accumulations (e.g. lacustrine deposits, Verrecchia, Chapter 9), or are direct precipitates (e.g. speleothems, Fairchild et al., Chapter 7). However, for any given type of geochemical sediment, variability in terms of nature, form and scale appears to be the overarching characteristic. Such variability is due to the large number of processes (operating at different spatial and temporal scales) affecting the types and rates of alteration and accumulation of the various precipitates.

Terrestrial geochemical sediments form largely as a result of the mobilisation, enrichment and precipitation of various minerals in the presence of water. These minerals are precipitated into or onto sediments, soil profiles, boulders and/or bedrock, and most commonly comprise carbonates (e.g. calcretes, aeolianites, tufas, travertine and beachrock; see Wright, Chapter 2; McLaren, Chapter 5; Viles and Pentecost, Chapter 6; and Gischler, Chapter 11, respectively); silica (see Nash & Ullyott, Chapter 4) or evaporites (see Chivas, Chapter 10). As noted in Chapter 1, purely physico-chemical processes dominate the formation of many of these sediments (e.g. speleothems). However, it is becoming ever more evident how important biogenic processes can be in the formation of many geochemical sediments, most notably tufas, travertines, pedogenic calcretes and rock coatings. Furthermore, for deposits such as calcrete, aeolianite and laterite, it may be the case that a combination of both sets of processes are in operation, whereby the precipitation of cements and minerals involves mechanisms that can include weathering, CO_2 degassing, evaporation, evapotranspiration as well as other biological mechanisms. As has been proposed for calcretes (see Wright, Chapter 2), it may, in future, be more appropriate to think of many geochemical sediments as forming along a process spectrum from dominantly biogenic to dominantly physico-chemical in origin. The macro- and micromorphology of terrestrial geochemical sediments are equally complex, and are functions of: (a) their environment and mode of formation (with diagenetic and other processes occurring under vadose, phreatic, pedogenic, lacustrine or marine environments); (b) the processes that control precipitation (biogenic and/or physicochemical); (c) the amount of water present; (d) the chemistry and rate of flow of interstitial fluids through the sediment bodies; as well as (e) the nature of the host materials.

The preceding chapters have demonstrated that geochemical sediments form in a wide range of locations across the globe and in different positions within landscapes. Some of these materials (e.g. beachrock, evaporites and lacustrine sediments) are still forming at the present day, whereas others (e.g. silcretes) mainly reflect deposits that accumulated at various time periods throughout the Quaternary and further back in geological history. As noted in Chapter 1, many deposits are restricted to specific climates, with warm tropical and sub-tropical environments being the most conducive for the formation of geochemical sediments such as calcretes, evaporites and beachrock. There are very few examples of terrestrial geochemical sediments that have formed under cold climatic regimes, with notable exceptions being speleothems, aeolianites and certain non-pedogenic calcretes.

Some geochemical sediments are restricted to specific locations within a landscape, as it is the material that is being cemented that is as much part of the deposit as the geochemical processes themselves. For example, beachrock can form only in the intertidal zone along tropical beaches or around some lakes. Aeolianite is the name given specifically to cemented dune sands whether they are in desert or coastal locations. Many deposits will form in/on a wide range of host deposits and in different geomorphological locations (such as calcretes, silcretes and laterites). Others are more likely to be found in locations where specific processes operate or are found in association with certain types of bedrock. Examples of this type of geochemical sediment are tufas and travertines, that preferentially form in locations where there are carbonate-saturated springs, rivers or lake waters and most commonly where the bedrock is limestone. Other types of deposit, such as calcrete and silcrete, vary in nature, being partly dependent upon the specific hydrological settings, such as lacustrine, pan or groundwater, in which they are forming.

In terms of geomorphological context, the majority of terrestrial geochemical sediments tend to form on stable exposed surfaces (e.g. speleothems). This is particularly the case where formation is linked to relatively slow soil-forming processes (e.g. in the development of pedogenic calcretes and silcretes). Over time, protection of the materials beneath the chemical crust may lead to the development of inverted relief, whereby crusts that originally developed in topographic depressions or valleys, for example, may end up forming resistant hill tops after phases of erosion. Alternatively, rather than requiring a pre-existing stable surface, diagenetic changes such as the precipitation of various minerals and lithification processes may actively enhance the stabilisation of the landscape. In the case of aeolianite, for example, the cementation of dune sands results in the preservation of considerably more stable relict deposits. Where water is abundant, geochemical sediments may form in rivers, lakes (ephemeral or long-standing), or underground in cave environments.

The persistence of any geochemical sediment in the natural environment is, to a great extent, determined by the degree of induration or cementation of the precursor host material. However, the nature, and in particular the solubility, of the cementing or binding agent is also critical. As a result, geochemical sediments range from those that are predominantly ephemeral features, such as nitrate deposits or evaporites, through to some highly indurated duricrusts such as laterites or silcretes. The more soluble minerals (evaporites) can exist only in hyperarid through to arid environments, which can make them of use as palaeoclimatic indicators. In contrast, the more resistant and less soluble duricrusts can persist, and sometimes be reworked, in landscapes for significant periods of geological time.

The usefulness of terrestrial geochemical sediments in a palaeoenvironmental context is highly variable, and often complicated by the fact that many deposits did not form over a short discrete period of time under stable environmental conditions but, instead, represent the sum of many stages of formation. To a greater or lesser extent, the deposits discussed in this book have the ability (either at present or in the near future) to provide information about:

1 past climates in terms of palaeotemperatures and the amount of water around at the time that the deposit formed (e.g. Chapters 2–12);
2 the provenance of deposits and minerals through the study of isotopes in the precipitates as well as the host sediment's clasts (e.g. Chapters 2, 4–7, 9, 10 and 12);

3   the role that C3/C4 vegetation may play in biogenic processes (e.g. Chapters 2, 5, 6 and 9);
4   the palaeosalinity of formational waters in lakes and playas (e.g. Chapters 9 and 10);
5   palaeorelief studies when trying to reconstruct the palaeogeography of areas (e.g. Chapters 2, 3 and 4);
6   evidence of human activity preserved within the deposits such as evidence of middens and lithic artefacts in cemented dune sands (e.g. Chapter 5);
7   reconstructing past sea levels (e.g. Chapters 5 and 11);
8   more specifically, the use of speleothems to test Croll–Milankovitch theory and to calibrate the timing of palaeoclimatic fluctuations through precise U-series dating (e.g. Chapter 7).

The problem, however, with most geochemical sediments (with the exception of speleothems and, more recently, calcrete) is the difficulty in being able to date the deposits accurately (and ideally using a multiproxy dating approach). Being able to obtain reliable geochronologies will allow a much better understanding of the rates at which various processes operate, and help to establish when particular deposits formed and how their development links to palaeoclimates and climate change. In addition, advancements in microanalytical techniques may help in the identification of the degree of complexity in the stages of precipitate development and to see if any secondary alteration or overprinting has occurred. In terms of rock coatings, studies of microlaminations may also open up a new level of information about the processes operating and environmental change. Understanding these factors will allow us then to develop better and more sophisticated models of how and when these deposits formed.

Thus, many of the outstanding questions surrounding terrestrial geochemical sediments are partly a result of limitations in the techniques available to date and analyse materials, which should improve with advancements in technology and further research. As McAllister and Smith (Chapter 13) suggest, a better understanding and awareness of the techniques that are now and will soon be available will enhance our studies of geochemical sediments and how they relate to the landscape in which they have formed.

Index

Page numbers in *italic* indicate figures; page numbers in **bold** indicate tables.